ENVIRONMENTAL SCIENCE

ENVIRONMENTAL SCIENCE

A STUDY OF INTERRELATIONSHIPS
seventh edition

Eldon D. Enger

Delta College

Bradley F. Smith

Western Washington University

Boston   Burr Ridge, IL   Dubuque, IA   Madison, WI   New York   San Franciso   St. Louis
Bangkok   Bogotá   Caracas   Lisbon   London   Madrid
Mexico City   Milan   New Delhi   Seoul   Singapore   Sidney   Taipei   Toronto

McGraw-Hill Higher Education

*A Division of The **McGraw-Hill** Companies*

ENVIRONMENTAL SCIENCE: A STUDY OF INTERRELATIONSHIPS,
SEVENTH EDITION

 This book is printed on recycled, acid-free paper containing 10% postconsumer waste.

1 2 3 4 5 6 7 8 9 0 VNH/VNH 0 9 8 7 6 5 4 3 2 1 0

ISBN 0–07–230495–2 (hardcover)
    0–07–290908–0 (softcover)

Vice president and editorial director: *Kevin T. Kane*
Publisher: *Michael D. Lange*
Senior sponsoring editor: *Margaret J. Kemp*
Senior developmental editor: *Kathleen R. Loewenberg*
Marketing manager: *Michelle Watnick*
Senior project manager: *Kay J. Brimeyer*
Production supervisor: *Sandy Ludovissy*
Design manager: *Stuart D. Paterson*
Senior photo research coordinator: *Lori Hancock*
Supplement coordinator: *Brenda A. Ernzen*
Compositor: *GAC—Indianapolis*
Typeface: *10/12 Times Roman*
Printer: *Von Hoffmann Press, Inc.*

Cover/interior design: *Jamie O'Neal*
Front cover image: *Tony Stone Images*
Back cover image: *Animals/Animals*
Photo research: *Shirley Lanners*
All part opener photos: © *Barry Barker*

The credits section for this book begins on page 423 and is considered
an extension of the copyright page.

Library of Congress Cataloging-in-Publication Data

Enger, Eldon D.
    Environmental science : a study of interrelationships / Eldon D.
Enger, Bradley F. Smith. — 7th ed.
        p.   cm.
        ISBN 0–07–230495–2 (hardcover).   — ISBN 0–07–290908–0 (softcover)
    1. Environmental sciences.   I. Smith, Bradley Fraser.   II. Title.
GEI05.E54   2000
628—dc21   99-21073
                                                                CIP

INTERNATIONAL EDITION   ISBN 0–07–116996–2
Copyright © 2000. Exclusive rights by The McGraw-Hill Companies, Inc. for manufacture
and export. This book cannot be re-exported from the country to which it is consigned by
McGraw-Hill. The International Edition is not available in North America.

www.mhhe.com

To my sons Kyle and Brook Enger with love and respect.

To my children, Morgan and Ian Smith, who have made me very proud.

BRIEF CONTENTS

PART ONE
Interrelatedness 2

PART TWO
Ecological Principles and Their Application   34

PART THREE
Energy 122

PART FOUR
Human Influences on Ecosystems 184

PART FIVE
Pollution and Policy   306

CONTENTS

PART FIVE
Pollution and Policy   306

Chapter 16
Risk and Cost: Elements of Decision Making   308

Chapter 17
Air Pollution   331

Chapter 18
Solid Waste Management and Disposal   357

Chapter 19
Regulating Hazardous Materials   374

Essential Study Partner CD-ROM

A study partner that engages, investigates, and reinforces what you are learning from your textbook. You'll find the **Essential Study Partner** for Enger/Smith's *Environmental Science: A Study of Interrelationships* to be a complete, interactive student study tool featuring animations and learning activities. From quizzes to interactive diagrams, you'll find that there has never been a better study partner to ensure the mastery of core concepts in environmental science.

The unit pop-up menu is accessible at anytime within the program. Clicking on the current unit will bring up a menu of other units available in the program.

The topic menu contains an interactive list of the available topics. Clicking on any of the listings within this menu will open your selection and will show the specific concepts presented within this topic. Clicking any of the concepts will move you to your selection. You can use the UP and DOWN arrow keys to move through the topics.

To the right of the arrows is a row of icons that represent the number of screens in a concept. There are three different icons, each representing different functions that a screen in that section will serve. The screen that is currently displayed will highlight yellow and visited ones will be checked.

Along the bottom of the screen you will find various navigational aids. At the left are arrows that allow you to page forward and backward through text screens or interactive exercise screens. You can also use the LEFT and RIGHT arrows on your keyboard to perform the same function.

The film icon represents an animation screen.

The activity icon represents an interactive learning activity.

The page icon represents a page of informational text.

PREFACE

Environmental science is an interdisciplinary field. Because environmental disharmonies occur as a result of the interaction between humans and the natural world, we must include both when seeking solutions to environmental problems. It is important to have a historical perspective, appreciate economic and political realities, recognize the role of different social experiences and ethical backgrounds, and integrate these with the science that describes the natural world and how we affect it. *Environmental Science: A Study of Interrelationships* incorporates all of these sources of information when discussing any environmental issue. Furthermore, the authors have endeavored to present a balanced view of issues, diligently avoiding personal biases and fashionable philosophies.

Environmental Science: A Study of Interrelationships is intended as a text for a one-semester, introductory course for students with a wide variety of career goals. They will find it interesting and informative. The central theme is interrelatedness. No text of this nature can cover all issues in depth. What we have done is to identify major issues and give appropriate examples that illustrate the complex interactions that are characteristic of all environmental problems. There are many facts presented in charts, graphs, and figures that help to illustrate the scope of environmental issues. However, this is not the core of the text, since the facts will change.

ORGANIZATION AND CONTENT

This book is divided into five parts and twenty chapters. It is organized to provide an even, logical flow of concepts and to provide clear illustrations of the major environmental issues of today.

Part 1 establishes the theme of the book in chapter 1 by looking at the kinds of environmental issues typical of different regions of North America. In each region, the specific issues selected involve scientific, social, political, and economic components typical of environmental problems. Chapter 2 focuses on the philosophical base needed to examine environmental issues by discussing various ethical and moral stands that shape how people approach environmental issues.

Part 2 provides an understanding of the ecological principles that are basic to organism interactions and the flow of matter and energy in ecosystems. The nature of food chains and how they affect the flow of matter and energy are discussed. Other topics included are: the efficiency of energy flow through ecosystems, the intricacies of organism-to-organism interaction, and the creative role of natural selection in shaping ecological relationships. Principles of population structure and organization are also developed in this section, with particular attention to the implications of these principles to growth and impact of human populations.

Part 3 focuses on energy. A major emphasis is on the historically important, nonrenewable fossil fuels that have stimulated economic success of the developed economies of the world. Renewable sources of energy are discussed, but with the recognition that they currently are a small part of the world energy picture. Weapons production and nuclear power plants use enormous amounts of energy that can be released from the nucleus of the atom. Both of these uses have caused fear among the public related to the dangers of radiation and the adequacy of waste disposal. These issues are discussed in this section.

Part 4 emphasizes the impact of human activity on natural ecosystems. As human populations grow, and technology changes, the magnitude of human actions becomes more apparent. The natural ecosystems on land and water are modified to meet human needs. The heavy use of pesticides in agriculture is discussed in this section.

Part 5 deals with the major types of pollution. Pollution affects the health and welfare of humans and other organisms. Air pollution, solid waste, and hazardous and toxic substances are discussed in this section. The cost of pollution cannot always be measured in financial terms, but may be reflected in the mental and physical health of the populace. Government often uses risk assessment and analysis to help set policy regarding pollution and other environmental issues. Separate chapters deal with the ways in which government sets policy and evaluates risk.

NEW TO THIS EDITION

1. The text has been edited throughout and rewritten where needed to include the most recent data and ways of thinking about environmental issues.

2. Many new illustrations were developed and many others were modified to improve their ability to convey information.

3. Several chapters have been substantially reorganized. Chapter 11, "Human Impact on Resources and Ecosystems," has been modified to add separate sections on agricultural ecosystems and the discussion on managing aquatic ecosystems was divided into freshwater and marine sections. Chapter 12, "Land-Use Planning," was completely reorganized and updated. Chapter 15, "Water Management," was significantly reorganized and many new and revised illustrations were included. Chapter 20 was completely rewritten with emphasis on environmental policy, personal responsibility for the environment, regulations, and the impact of multinational organizations.

4. Many new topics or boxed readings have been introduced into chapters that did not require major changes in chapter organization.

- Chapter 1 has new material on salmon concerns in the Northwest.
- Chapter 2 has additional material on environmental ethics and three new boxed readings on environmental philosophy, the Kyoto Protocol, and a code of environmental ethics and conduct.
- Chapter 4 has several new boxed readings on habitat conservation plans, ecological roles of humans, human impact on nutrient cycles, and the reintroduction of wolves into Yellowstone.
- Chapter 5 has two new readings on destruction of rainforests and rainforest products.
- Chapter 7 has new boxed readings on the impact of AIDS on populations, and Malthus' essay on population.
- Chapter 8 was updated throughout and has three new boxed readings on daily gasoline price fluctuations, and China's energy policy.
- Chapter 9 has a new boxed reading on the Three Gorges Dam in China and added material on hydroelectric power.
- Chapter 12 features a new boxed reading on the urbanization of the world population, including a chart of cities with the most rapidly growing populations.
- Chapter 13 has a new section on geologic processes—plate tectonics and a new boxed reading on world soil degradation. A new illustration denotes the various components in soil.
- Chapter 14 has increased coverage of sustainable and organic farming practices, world use of fertilizer and pesticides, and new subtitles in sections to make it easier for students to follow the text. A section on protecting soil and water was also added.
- Chapter 16 has expanded coverage of market-based incentives and extended product responsibility. Three new readings on corporate recycling of wood in urban environments, sustainable development, and valuing ecosystem services have been added.
- Chapter 17 has expanded coverage of secondhand smoke and two new boxed readings on global warming and indoor air pollution.
- Chapter 18 has 14 new or significantly modified illustrations.
- Chapter 19 has a new boxed reading on hazardous waste management in China.
- Chapter 20 features new readings on the Wise Use Movement, environment policy in the European Union, and the International Whaling Commission.

SPECIAL FEATURES AND LEARNING AIDS

1. Each of the five parts of the text begins with an **introduction** that places the upcoming chapters in context for the reader by recalling previously discussed material and by describing the organization of the chapters to come. A world map with political boundaries can be found immediately following this preface. We believe that this will help the reader to more fully understand and appreciate global environmental issues.

2. Each chapter begins with a set of learning **objectives,** an **outline,** and a **conceptual diagram**—all of which give the student a broad overview of the interrelated forces that are involved in the material to be discussed. The student is encouraged to refer to these resources while reading and reviewing the chapter.

3. Chapters conclude with an **Issues and Analysis** case study, **summary,** a list of **key terms, review questions, critical thinking questions,** and a list of topics that correspond to specific Internet links on the accompanying web site. The case studies have been specifically selected to allow the reader to apply the chapter concepts to actual situations. Review questions are related to the chapter objectives, and thus serve to reinforce understanding of basic concepts and principles.

4. To dramatize and clarify text material, each chapter includes a number of **tables, charts, graphs, maps, drawings,** or **photographs.** Each illustration has been carefully chosen to provide a pictorial image or an organized format for showing detailed information, which helps the reader comprehend the chapter material.

5. Each chapter also includes **boxed readings.** These provide an in-depth consideration of a specific situation that is relevant to the content, an alternative viewpoint, or a wider world view of the issues discussed in the chapter.

6. The text concludes with several appendixes that deal with the following topics: critical thinking, the metric system, the periodic table of the elements, some thoughts on what you can do to make the world a better place in which to live, and how to write to public officials. In addition there is a complete glossary and index.

USEFUL ANCILLARIES

1. An **Instructor's Manual** accompanies the text. It includes chapter outlines, objectives, and key terms; a range of test and discussion questions; suggestions for demonstrations; and suggestions for audiovisual materials and other teaching aids. The Instructor's Manual also provides **additional case studies** for instructors who wish to use additional concrete examples of how the concepts in the chapter can be applied to the real world. It is available on disk only, in either Mac or IBM format.

2. A set of **one hundred transparencies** is also available to users of the text. The transparencies duplicate text figures that clarify essential ecological, political, economic, social, and historical concepts.

3. **Computerized Testing Software** rounds out the supplementary materials. Available for either Windows® or Mac-

Intosh®, this software allows for easy test generation using the questions found in the printed testbank.

4. *Environmental Science Visual Resource Library* (VRL) (0-07-290683-9) © 1999. This dual platform CD-ROM allows the user to search with key words or terms and access hundreds of images to illustrate classroom lectures, with just the click of a mouse. Contains images from three McGraw-Hill textbooks and over 400 additional photographs.

5. *Comprehensive Web Site.* Visit our web site at http://www.mhhe.com/environmentalscience/ and discover a variety of valuable resources for both instructor and student. Examples include: chapter-by-chapter Internet links (updated regularly) that correspond to the end-of-chapter topics listed in the text, sample syllabi, laboratory exercises, classroom activities, current global environmental events in the news, practice quizzes, career information, and more.

RELATED TITLES OF INTEREST/ PACKAGING OPPORTUNITIES

Annual Editions: Environment 99/00 (0-07-228498-6) © 1999. Editor: John L. Allen, University of Connecticut, Storrs—31 articles that address the current state of Earth and the changes it faces; world population and hunger; present and future energy needs and problems; endangered species; natural resources; and pollution.

Taking Sides: Clashing Views on Controversial Environmental Issues, 8th edition (0-07-303184-4) © 1999. Editor: Theodore D. Goldfarb, State University of New York, Stony Brook—18 issues debating general philosophical and political issues; the environment and technology; disposing of wastes; and the environment and the future.

STUDENT ANCILLARIES

1. *Environmental Science Essential Study Partner* (ESP) (0-07-303867-9) © 2000. This tutorial CD-ROM contains high-quality 3-D animations, interactive study activities, illustrated overviews of key topics in environmental science, and supplementary quizzes and exams that students will find extremely valuable.

2. Life Science Living Lexicon (CD-ROM or paperback print version). (CD = 0-697-37993-0; Print = 0-697-12133-X) © 1996. Author: William Marchuk. A powerful, interactive CD-ROM that contains a complete lexicon of life science terminology. Conveniently assembled on an easy-to-use CD-ROM are components such as a glossary of common biological roots, prefixes, and suffixes; a categorized glossary of common biological terms; and a section describing the classification system.

3. *Field and Laboratory Activities Manual* (0-07-290913-7) © 2000. Authors: Enger and Smith. Newly revised with additional contributors, this laboratory manual features a variety of exercises on environmental topics. Supporting resources, directions, and interesting activities to accompany the manual may be found on the textbook's web site.

4. *Comprehensive Web Site.* Visit our web site at http://www.mhhe.com/environmentalscience/ and discover a variety of valuable resources for both instructor and student. Examples include: chapter-by-chapter Internet links (updated regularly) that correspond to the end-of-chapter topics listed in the text, sample syllabi, laboratory exercises, classroom activities, current global environmental events in the news, practice quizzes, career information, and more.

Sources: Notable Selections in Environmental Studies, 2nd edition (0-07-303186-0) © 2000. Editor: Theodore D. Goldfarb, State University of New York, Stony Brook. Overviews of environmental issues focus specifically on energy, environmental degradation, population issues and the environment, human health and the environment, and environment and society.

The Dushkin Student Atlas of Environmental Issues, 1st edition (0-697-36520-4) © 1997. Editor: John Allen, University of Connecticut, Storrs—covers recent agricultural, industrial, and demographic changes in every world region; 48 full-color maps illustrate global patterns in the physical and human environments; and presents human impact on the air, fresh water and the oceans, the biosphere, and land.

You Can Make a Difference: Be Environmentally Responsible, 2nd edition (0-07-292416-0) © 1999. Author: Judy Getis. This book is organized around the three parts of the biosphere: land, water, and air. Each section contains descriptions of the environmental problems associated with that part of the biosphere. In the section on the atmosphere, for example, such problems as acid rain and the greenhouse effect are discussed. Immediately following each discussion or "challenge" are suggested ways that individuals can help solve or alleviate them—i.e., the "response."

Environmental Ethics: Divergence and Convergence, 2nd edition (0-07-006180-7) © 1998. Authors: Botzler and Armstrong. This anthology surveys diverse approaches to environmental ethics by leading writers from a variety of disciplines, and provides a historical survey of thought on our responsibility to the environment. The perspectives are represented by articulate spokespersons and are accompanied by appraisals of their respective strengths and weaknesses. New topics include biodiversity, ecological restoration, environmental justice, and genetic engineering. A new section in the appendix on conflict resolution was requested by students.

The Age of Environmentalism (0-07-060841-5) © 1997. Author: J. E. de Steiguer. A scholarly treatment of environmental thought, this work provides information on the changing social and economic factors which gave rise to the modern environmental era, and examines a broad array of academic disciplines—biology, history, theology, philosophy, computer science and economics—which have contributed to modern environmentalism.

Environmental Problem Solving: A Case Study Approach (0-07-027686-2) © 1997. Author: Isobel W. Heathcote. This book focuses on real environmental case studies that illustrate an interaction of several disciplines and require analytical techniques to help students gain the "hands-on" experience needed to handle today's environmental issues.

ACKNOWLEDGEMENTS

We would like to thank our many colleagues who have reviewed all or part of *Environmental Science: A Study of Interrelationships*. Their valuable input has contributed significantly to the quality of this text.

Donald D. Adams, SUNY Plattsburgh
Matthew R. Auer, Indiana University
Robert P. Benard, American International College
Robert B. Buerger, University of North Carolina, Wilmington
Sai Chidambaram, Canisius College
William F. Christman, U. S. Corps of Engineers
David T. Corey, Midlands Technical College
William C. Culver, St. Petersburg Junior College
Lisa Danko, Mercyhurst College
David G. Fisher, Maharishi University of Management
Chris Fox, Catonsville Community College
Deborah Freile, University of St. Thomas
Thaddeus Gish, St. Mary's College
Gian Gupta, University of Maryland, Eastern Shore

A. Scott Helgeson, Des Moines Area Community College
Theresa Hoffman-Till, Northern Virginia Community College, Annandale
Huey-Min Hwang, Jackson State University
Jason W. Kelsey, Marymount University
Ned J. Knight, Linfield College
R. Steven Konkel, Eastern Kentucky University
Michael J. Lemke, Pace University
Roberte M.D. Makowski, Wilmington College
Terry R. Martin, Kishwaukee College
Muhammad Adel Miah, University of Arkansas at Pine Bluff
Muthena Naseri, Moorpark College
Kenneth E. Nuss, University of Northern Iowa
Sharon O'Reilly, Whatcom Community College
Chuks Ogbonnaya, Mountain Empire College
John H. Peck, St. Cloud State University
Pete Peterson, Central Community College
Jon K. Piper, Bethel College
Fred H. Schindler, Indian Hills Community College
Harold F. Sears, University of South Carolina at Union
John G. Shiber, University of Kentucky, Prestonsburg Community College
Chuck Smead, Kankakee Community College
Donald M. Spano, University of Southern Colorado
Margaret Summerfield, United World College
Liette Vasseur, Saint Mary's University
Larry Williams, Benedict College
Michael Windelspecht, Floyd College

Eldon D. Enger

ELDON D. ENGER

Eldon D. Enger is a professor of biology at Delta College, a community college near Saginaw, Michigan. He received his B.A. and M.S. degrees from the University of Michigan. Professor Enger has over 30 years of teaching experience, during which he has taught biology, zoology, environmental science, and several other courses. He has been very active in curriculum and course development. Recent activities include the development of a learning community course in stream ecology, which involves students in two weekend activities including canoeing and camping, and a plant identification course that incorporates weekend field activities with backpacking and camping. In addition, he was involved in the development of an environmental regulations course and an environmental technician curriculum.

Professor Enger is an advocate for variety in teaching methodology. He feels that if students are provided with varied experiences, they are more likely to learn. In addition to the standard textbook assignments, lectures, and laboratory activities, his classes are likely to include writing assignments, student presentation of lecture material, debates by students on controversial issues, field experiences, individual student projects, and discussions of local examples and relevant current events. Textbooks are very valuable for presenting content, especially if they contain accurate, informative drawings and visual examples. Lectures are best used to help students see themes and make connections, and laboratory activities provide important hands-on activities.

Professor Enger has been a Fulbright Exchange Teacher to Australia and Scotland, received the Bergstein Award for Teaching Excellence and the Scholarly Achievement Award from Delta College, and participated as a volunteer in an Earthwatch Research Program in Costa Rica, studying the behavior of a bird known as the long-tailed manakin. He has also visited Australia, New Zealand, New Guinea, Fiji, Puerto Rico, Mexico, Canada, Morocco, many areas in Europe, and much of the United States. During these travels he has spent considerable time visiting coral reefs, ocean coasts, mangrove swamps, alpine tundra, prairies, tropical rainforests, cloud forests, deserts, temperate rainforests, coniferous forests, deciduous forests, and many other special ecosystems. This extensive experience provides the background to look at environmental issues from a broad perspective.

Professor Enger is married, has two college-aged sons, and enjoys a variety of outdoor pursuits such as cross-country skiing, hiking, hunting, fishing, camping and gardening. Other interests include reading a wide variety of periodicals, beekeeping, singing in a church choir, and preserving garden produce.

Bradley F. Smith

BRADLEY F. SMITH

Bradley F. Smith is the Dean of Huxley College of Environmental Studies at Western Washington University in Bellingham, Washington. Prior to assuming the position as Dean in 1994, he served as the first Director of the Office of Environmental Education for the U.S. Environmental Protection Agency in Washington, D.C. from 1991 to 1994. Dean Smith also served as the Acting President of the National Environmental Education and Training Foundation in Washington, D.C. and as a Special Assistant to the EPA Administrator.

Before moving to Washington, D.C., Dean Smith was a professor of political science and environmental studies for fifteen years, and the executive director of an environmental education center and nature refuge for five years.

Dean Smith has considerable international experience. He was a Fulbright Exchange Teacher to England and worked as a research associate for Environment Canada in New Brunswick, Canada. He is a frequent speaker on environmental issues worldwide and serves on the International Scholars Program for the U.S. Information Agency. He also served as a U.S. representative on the Tri-Lateral Commission on environmental education with Canada and Mexico. In 1995, he was awarded a NATO Fellowship to study the environmental problems associated with the closure of former Soviet military bases in Eastern Europe. Dean Smith is an Adjunct Professor at Far Eastern State University in Vladivostok, Russia and is a member of the Russian Academy of Transport. He also serves as a commissioner for the International Union for the Conservation of Nature (IUCN).

Nationally, Dean Smith serves as a member/advisor for many environmental organizations' boards of directors, advisory councils, and executive committees, including President Clinton's Council for Sustainable Development (Education Task Force).

Dean Smith holds B.A. and M.A. degrees in Political Science and Public Administration and a Ph.D. from the School of Natural Resources and Environment at the University of Michigan.

Dean Smith lives with his wife Daria, daughter Morgan, son Ian, and English setter Skye, along Puget Sound south of Bellingham. He is an avid outdoor enthusiast.

Scale 1 : 90 000 000

Robinson Projection

West of Greenwich

PART ONE

Interrelatedness

Environmental science is an interdisciplinary study that describes problems caused by human use of the natural world. It also seeks remedies for these problems. To learn about this complex field of study, it helps to understand three things: First, it is important to understand the natural processes (both physical and biological) that operate in the world. Second, it is important to appreciate the role that technology plays in our society and its capacity to alter natural processes as well as solve problems caused by human impact. Third, it helps to understand the complex social processes that characterize human populations. When we integrate that understanding with a knowledge of technology and natural processes, we can fully appreciate our role in the natural world.

Chapter 1 introduces the central theme of interrelatedness by analyzing some environmental issues in North America region by region. **Chapter 2** discusses the differences that can exist among individuals in a society and the different behaviors people exhibit, depending on whether they are acting as individuals, as part of a corporation, or as part of government.

chapter 1

Environmental Interrelationships

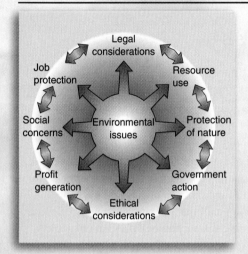

OBJECTIVES

After reading this chapter, you should be able to:

- Understand why environmental problems are complex and interrelated.
- Realize that environmental problems involve social, political, and economic issues, not just scientific issues.
- Understand that acceptable solutions to environmental problems are not often easy to achieve.
- Understand that all organisms have an impact on their surroundings.
- Understand what is meant by an ecosystem approach to environmental problem solving.
- Recognize that different geographic regions have somewhat different environmental problems, but the process for resolving them is the same and involves compromise.

CHAPTER OUTLINE

The Field of Environmental Science

The Interrelated Nature of Environmental Problems

Environmental Close-Up: *Science Versus Policy*

Global Perspective: *Fish, Seals, and Jobs*

An Ecosystem Approach

Regional Environmental Concerns

Environmental Close-Up: *The Greater Yellowstone Ecosystem*

Environmental Close-Up: *Ecosystem Size*

    The Wilderness North
    The Agricultural Middle
    The Dry West
    The Forested West
    The Great Lakes and Industrial Northeast
    The Diverse South

THE FIELD OF ENVIRONMENTAL SCIENCE

Environmental science is an interdisciplinary area of study that includes both applied and theoretical aspects of human impact on the world. Since humans are generally organized into groups, environmental science must deal with politics, social organization, economics, ethics, and philosophy. Thus, environmental science is a mixture of traditional science, societal values, and political awareness. (See figure 1.1.)

Environmental science as a field of study is evolving, but it is rooted in the early history of civilization. Many ancient cultures expressed a reverence for the plants, animals, and geographic features that provided them with food, water, and transportation. These features are still appreciated by modern people. The following quote from Henry David Thoreau (1817–1862) is over a century old but is consistent with current environmental philosophy:

> I wish to speak a word for Nature, for absolute freedom and wildness, as contrasted with a freedom and culture merely civil . . . to regard man as an inhabitant, or a part and parcel of Nature, rather than a member of society.

The current interest in the state of the environment began with philosophers like Thoreau and received an additional push by the organization of the first Earth Day on April 22, 1970. Subsequent Earth Days reaffirmed this commitment, as have similar Earth Days since then. As a result of this continuing interest in the state of the world and how people affect it, environmental science is now a standard course on many college campuses and is also a part of most high school course offerings. Most of the concepts covered by environmental science courses had previously been taught in ecology, conservation, or geography courses. Environmental science incorporates the scientific aspects of these courses with input from the social sciences, such as economics, sociology, and political science, creating a new interdisciplinary field.

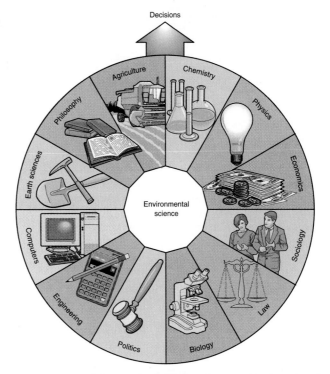

FIGURE 1.1 Environmental Science The field of environmental science involves an understanding of scientific principles, economic influences, and political action. Environmental decisions often involve compromise. A decision that may be supportable from a scientific or economic point of view may not be supportable from a political point of view without modification. Often political decisions relating to the environment may not be supported by economic analysis.

THE INTERRELATED NATURE OF ENVIRONMENTAL PROBLEMS

Environmental science is by nature an interdisciplinary field. The world *environmental* is usually understood to mean the surrounding conditions that affect people and other organisms. In a broader definition, **environment** is everything that affects an organism during its lifetime. From a human perspective, environmental issues involve concerns about science, nature, health, employment, profits, politics, ethics, and economics.

Most social and political decisions are made with respect to political jurisdictions, but environmental problems do not necessarily coincide with these artificial, political boundaries. For example, air pollution may involve several local units of government, several states or provinces, and even different nations. The forest fires that raged in Mexico in 1998 had a severe impact on air quality in Texas. On a more local level, the air pollution problems in Juarez, Mexico, are also problems in El Paso, Texas. But the issue is more than air quality and human health. Lower wage rates and less strict environmental laws have influenced some industries to move to Mexico for economic reasons. Mexico and many other developing nations are struggling to improve their environmental image and need the money generated by foreign investment to improve the conditions and the environment in which their people live.

Air pollutants produced in the major industrial regions of the United States drift across the border into Canada, where acid rain damages lakes and forests. A longstanding dispute exists between the United States and Canada over this issue. Canada claims that the United States should be doing more to reduce emissions that cause

ENVIRONMENTAL CLOSE-UP

Science Versus Policy

Scientific knowledge and government policy do not always agree. The scientific community can advise governments but cannot insist that certain policies be adopted. Governments may halt some scientific research because they control funding sources, or they may introduce regulations that make continuing the research difficult. For example, much federal money was spent on alternative energy research during the Carter presidency, but many of these projects were not in favor during the Reagan and Bush presidencies. Funding was reduced and much of the research into alternative fuels stopped. Conversely, the passage of the 1990 Clean Air Act during the Bush presidency mandated that alternative-fuel automobiles be used in some cities with severe air-pollution problems.

Government policy may be contrary to prevailing scientific opinion for economic or political reasons. For many years during the Rea-gan administration, most scientists in the United States and Canada agreed that the burning of high-sulfur coal and other acid-producing fuels was responsible for acid rain, which was leading to the deaths of lakes in parts of Canada and the northeastern United States. The administration continued to insist that the information was not conclusive and that the problem should be studied in greater detail.

The Clinton administration faced a similar dilemma regarding the debate over global warming. While the scientific community strongly supported the position that global warming is for real and that it is due in large part to human causes, Clinton was cautious in moving too fast to reduce emissions in the United States. This is in part due to the potential economic consequences involved. It is difficult to separate scientific knowledge, governmental policy, and economic policy. Can you identify similar examples of this debate in your community?

GLOBAL PERSPECTIVE

Fish, Seals, and Jobs

In 1995, the Canadian government announced a moratorium on cod fishing along the east coast of Canada. The cod industry contributes $700 million a year and 31,000 jobs to the Canadian economy, primarily in Newfoundland. At the same time, the government announced that it would begin a program to encourage the harvesting of harp seals by helping develop markets for seal products. Environmental groups opposed the harvesting of the seals.

How do all these pieces fit together? It is thought that the low numbers of cod in the North Atlantic are partially the result of an increasing population of harp seals. While overfishing, larger nets, and other factors have also contributed to the decline of the cod, it is true that harp seals feed on fish that could have been harvested. The current harp seal population along the Atlantic Coast has more than doubled since the 1970s to 4.8 million in 1997 and is projected to reach 6 million by 2000 if there are no hunts. The increase in harp seals is at least partly the result of actions during the 1970s by environmental groups that sought to stop the killing of seals because they considered the harvesting method inhumane. The traditional method involves clubbing the young seals to death. In 1996, the Canadian government increased the seal hunt quota from 186,000 to 287,000. In 1997 scientists from seven countries met in St. John's, Newfoundland to consider the inter-actions between harp seals and fisheries in the Northwest Atlantic. The conclusions of the meeting supported the earlier findings of the dramatic population growth amongst the seals and also noted that the animals were now growing more slowly and the pregnancy rate is lower than in the 1980s. These are the effects you would likely see when food becomes more difficult to find.

The St. John's meeting reinforced the fact that the diet of harp seals in the nearshore waters of the Labrador-Newfoundland shelf is dominated by Arctic cod. It was not, however, determined whether or not harp seals were affecting commercial fishing stocks—and Atlantic cod, in particular—on the Labrador-Newfoundland shelf. This was because there is a need for and estimate of the amount of juvenile cod in both inshore and offshore areas, and for an assessment of the amounts of cod which are being taken by the other important predators such as Greenland halibut, whales, and seabirds.

It appears that what might be seen initially as a number of isolated and unrelated factors are really interrelated in a way that affects the economy of an entire region.

acid rain, and the United States claims it is doing as much as it can. In another example, farmers who use water from the Colorado River for irrigation reduce the quality and quantity of water entering Mexico. This causes political friction between Mexico and the United States.

The issue of declining salmon stocks in the Pacific Northwest of the United States and British Columbia, Canada is another example of political friction over a shared natural resource. It has been calculated that on the U.S. side of the salmon issue alone there are five federal Cabinet level departments, two federal agencies, five federal laws in question as well as numerous Tribal treaties, commissions and court decisions. All of this is in addition to many state level departments, commissions, and rulings. If all of this was not sufficient you still have the international bodies such as the United Nations and the international treaties that impact the fate of the salmon. Considering all this attention, maybe it is not that surprising that the plight of the salmon is in such a dangerous state. (See figure 1.2.)

Because of all these political, economic, ethical, and scientific links, solving environmental problems is complex. The problems seldom have simple solutions. However, international organizations such as the International Joint Commission have had major bearing on the quality of the environment over broad regions of the world.

The International Joint Commission was established in 1909 when the Boundary Waters Treaty was signed between the United States and Canada. The treaty was established in part to provide that the "boundary waters and waters flowing across the boundary shall not be polluted on either side to the injury of health or property of the other." The Commission has been instrumental in identifying areas of concern and encouraging the cleanup of polluted sites that affect the quality of the Great Lakes and other boundary waters. In general, the two governments have listened to the Commission's advice and have responded by initiating cleanup activities.

The first worldwide meeting of heads of state directed to concern for the environment took place at the Earth Summit, formally known as the United Nations Conference on Environment and Development (UNCED) in Rio de Janeiro in 1992. Most countries have also signed agreements on sustainable development and biodiversity. In 1997, representatives from 125 nations met in Kyoto, Japan for the Third Conference of the United Nations Framework Convention on Climate Change. This conference, commonly referred to as the Kyoto Conference on Climate Change, resulted in commitments from the participating nations to reduce their overall emissions of six greenhouse gases (linked to global warming) by at least 5 percent

Harvest
Overfishing has contributed to the decline of many fish populations. Often this excessive exploitation is caused by fishery managers trying to access harvestable hatchery salmon or other abundant fish in areas that contain depleted wild salmon populations.

Hatcheries
Hatchery fish that augment harvest levels can interbreed with wild fish, resulting in the loss of genetic diversity. Hatchery fish can also spread disease and compete with wild fish for food and habitat.

Habitat
Rural and Urban: Salmon face multiple, complex threats in the developed lower regions of watersheds. Problems include low water flows, pollution, degraded physical habitats, and migration barriers such as culverts.

Forests: Improper forest practices and road construction and maintenance are the biggest threat to salmon in the upper watershed. Department of Natural Resources in Washington receives 12,000 applications for forest practices annually.

Hydropower
Dams can block fish migration to and from the ocean, kill fish passing through turbines, delay migration, and increase predation. Dams can also cause inadequate flow downstream.
There are 1,018 dams on Washington rivers. The Columbia River hosts 150 hydroelectric projects and 250 reservoirs — more than half the length of the river is blocked to salmon and steelhead.

FIGURE 1.2 The Four H's: Human Activities that Affect Wild Salmon Survival The interrelated nature of environmental problems is evident with regards to concern about diminishing numbers of wild salmon in the Pacific Northwest of the United States and in British Columbia, Canada. This diagram portrays the plight of wild salmon in the State of Washington but the issue is region wide.

Source: Washington Department of Natural Resources Newsletter, Winter 1998, Department of Natural Resources, Washington, D.C.

below 1990 levels between 2008 and 2012. (See figure 1.3.) The Kyoto Protocol, as the agreement was called, was viewed by many as one of the most important steps to date in environmental protection and international diplomacy. It may be years before we will know if all countries that signed these agreements will meet their commitments to environmental improvement, but they have at least stated their intention to do so.

The United Nations, through the United Nations Educational, Scientific, and Cultural Organization (UNESCO) and the United Nations Environment Programme (UNEP), has supported many environmental programs. A recent undertaking is the International Environmental

Education Programme (IEEP). This program recognizes the need for both formal environmental education in schools and the informal education that occurs through the media and groups of interested citizens. Conferences on environmental education were first held during the 1970s and continue to the present.

AN ECOSYSTEM APPROACH

The natural world is organized into interrelated units called ecosystems. An **ecosystem** is a region in which the organisms and the physical environment form

United Nations Framework Convention on Climate Change

Third Session, Conference of the Parties

Kyoto, 1 - 10 December 1997

FIGURE 1.3 The Kyoto Protocol to the United Nations Framework Convention on Climate Change was the first time in history that such a protocol contained legally binding reduction targets for all major greenhouse gases. The protocol represented a major step forward in international efforts to avert the threat of climate change.

an interacting unit. Weather affects plants, plants use minerals in the soil, and affect animals, animals spread plant seeds, plants secure the soil, and plants evaporate water, which affects weather.

Ecosystems sometimes have fairly discrete boundaries, as is the case with a lake or island. Sometimes the boundaries are indistinct, as in the transition from grassland to desert. Grassland gradually becomes desert depending on the historical pattern of rainfall in an area.

An ecosystem approach requires a look at the way the natural world is organized. Where do the rivers flow? What are the prevailing wind patterns? What are the typical plants and animals in the area? How does human activity affect nature? The task of an environmental scientist is to recognize and understand the natural interactions that take place and to integrate these with the uses humans must make of the natural world.

To illustrate the interrelated nature of environmental issues, we will look at several regions of North America and highlight some of the key features and issues of each.

REGIONAL ENVIRONMENTAL CONCERNS

No region is free of environmental concerns. Most regions tend to focus on specific, local environmental issues that apply directly to them. For example, protecting endangered species is a concern in many parts of the world. It has become a matter of particular concern in the Pacific Northwest where an endangered species known as the northern spotted owl appears to required undisturbed, mature forests for its survival. In most metropolitan areas the

ENVIRONMENTAL CLOSE-UP

The Greater Yellowstone Ecosystem

In 1872, the U.S. government established Yellowstone National Park as the world's first national park. It was a huge area that protected unique natural features such as geysers, hot springs, rivers, lakes, and mountains. It was also a preserve for many kinds of wildlife such as grizzly bears, elk, moose, and bison. At the time it was established, the park was thought to be of adequate size to protect the scenic resources and the wildlife. Since that time, the lands surrounding the park have been converted to a variety of uses, including cattle grazing, timber production, hunting, and mining.

Fortunately, most of the lands surrounding Yellowstone National Park and the adjacent Grant Teton National Park are still under government control as national forests, national wildlife refuges, and other state, local, or federal bodies. Some of the park wildlife, particularly the grizzly bear and bison, often wander across the park boundaries. The grizzly in particular needs large regions of wilderness to survive as a species.

Many people assert that it is essential that these lands be integrated into a Greater Yellowstone Ecosystem management plan en-

Indigo bunting

compassing about 7.3 million hectares. The plan is based on more natural boundaries than the original boundaries established in 1872. This would require changes in the way much of the land surrounding Yellowstone is currently being used. The trade-offs are significant. Logging, mining, hunting, and grazing would be stopped or significantly reduced. This would result in a loss of jobs in those industries. Proponents argue that additional jobs would be created in the tourist and related service industries. The advantages, they argue, would be equal to or greater than the economic losses caused by stopping current uses.

ENVIRONMENTAL CLOSE-UP

Ecosystem Size

Many people interested in songbirds have documented a significant decrease in the numbers of certain species of songbirds. Species particularly affected are those that migrate from North America to South America and require relatively large areas of undisturbed forest. Many of these birds are being negatively affected by human activities that fragment large patches of forest into many smaller patches. This creates more edges between forested and nonforested habitants. Bird and animal species that use edge habitats, which have a mixture of closed forest and open areas, supplant the songbirds, which require large patches of undisturbed forest. Species are affected most severely when both their northern and southern habitats are disturbed.

A survey of migrating sharp-shinned hawks indicates greatly reduced numbers in recent years. It is thought that since sharp-shinned hawks use small songbirds as their primary source of food, the reduction in their numbers is directly related to the reduction in migratory songbirds.

Sharp-shinned hawk

---

problem of endangered species is purely intellectual, since the construction of cities has essentially destroyed the previously existing ecosystem. Here we present a number of regional vignettes to illustrate the complexity and interrelatedness of environmental issues. (See figure 1.4.)

The Wilderness North

Much of Alaska and Northern Canada can be characterized as **wilderness**—areas with minimal human influence. Much of this land is owned by governments, not by individuals, so government policies have a large effect on what happens in these regions. These areas have important economic values in their trees, animals, scenery, and other natural resources. Exploitation of the region's natural resources involves significant trade-offs. Usually, a portion of the natural world is altered permanently, but the area altered is so small that many people consider it insignificant. Because of the severe climate, northern wilderness areas tend to be very sensitive to insults and take a long time to repair damage done by unwise exploitation. Mining, oil exploration, development of hydroelectric projects, and harvesting of timber all require roads and other human artifacts, involve the insertion of new technologies into native culture, and generate economic benefits.

In the past, many short-term political and economic decisions failed to look at long-term environmental implications. Today, however, people are concerned about these remaining wilderness areas. Politicians are more willing to look at the scientific and recreational values of wilderness as well as the economic value of exploitation.

Native people, who consider much of this region to be their land, have become increasingly sophisticated in negotiating with state, provincial, and federal governments to protect rights they feel they were granted in treaties. They are sensitive to changes in land use or government policy that would force changes in their traditional way of life.

Native people, other concerned citizens, business interests, and environmental activists have become increasingly sophisticated in influencing decisions made by government. The process of compromise is often difficult and does not always assure wise decisions, but most governments now realize they must listen to the concerns of their citizens and balance economic benefits with social and cultural benefits. (See figure 1.5.)

The Agricultural Middle

The middle of the North American continent is dominated by intensive agriculture. This means that the original, natural ecosystems have been replaced by managed, agricultural enterprise. It is important to understand that this area was at one time wilderness. Today, you would need to search very hard to find regions of true wilderness in Iowa, Missouri, or southern Manitoba. Some special areas have been set aside to preserve fragments of the original natural plant and animal associations, but most of the land has been converted to agriculture wherever practical.

The economic value generated by this use of a rich soil resource is tremendous, and consequently most of the land is privately owned. Governments cannot easily control what happens on these privately held lands. But governments indirectly encourage certain activities through departments of agriculture that encourage agricultural research, grant special subsidies to farmers in the form of guaranteed prices for their products and other special payments, and develop markets for products. Yet because the economic risks involved in farming are great, the number of farmers constantly declines. There are a number of reasons for farm failures, including drought, disease, and lack of markets.

One of the major, nonpoint pollution sources (pollution that does not have an easily identified point of origin) is agriculture. Air pollution in the form of dust is an inevitable result of tilling the land. Soil erosion occurs when soil is exposed to wind and moving water and leads to siltation of rivers, impoundments, and lakes. Fertilizers and other agricultural chemicals blow or are washed from the areas where they are applied. Nutrients washed from

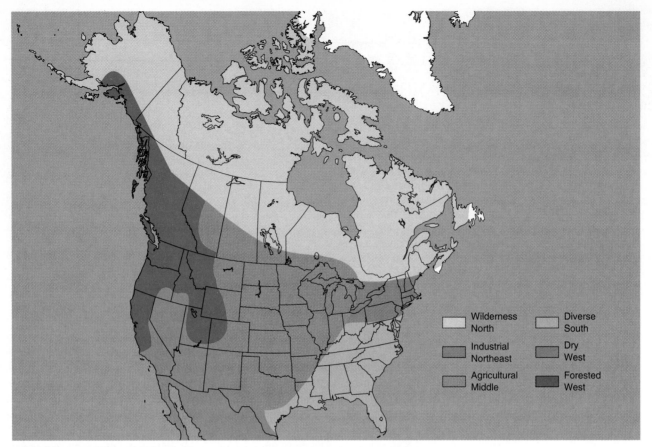

Wilderness North

Diverse South

Industrial Northeast

Dry West

Agricultural Middle

Forested West

FIGURE 1.4   Regions of North America Because of natural features of the land and the uses people make of the land, different regions of North America face different kinds of environmental issues. Certainly within each region people face a large number of specific issues, but certain kinds of issues are more important in some regions than in others.

Walrus harvesting

A clear-cut forest

Grizzly bear fishing for salmon

FIGURE 1.5   The Wilderness North Protection of wilderness is a major issue of this region. The major points of conflict involve the government role in managing these lands, the protection of the rights and beliefs of native people, and the desire of many to exploit the mineral and other resources of the region.

the land enter rivers and lakes where they encourage the growth of algae, lowering water quality. The use of pesticides causes concern about human exposure, effects on wild animals that are accidentally exposed, and residues in farm products.

Since many communities in this region rely on groundwater for drinking water, the use of fertilizers and pesticides, and their potential for entering the groundwater as a result of unwise or irresponsible use, is a concern. In addition, many farmers use

groundwater for irrigation, which lowers the water table and leaves less groundwater for other purposes.

In an effort to stay in business and preserve their way of life, farmers must use modern technology. Careful use of

A well-kept farm

Agricultural chemicals

Ship loaded with grain

FIGURE 1.6 **The Agricultural Middle** The rich soil resource of this region has been converted to managed agricultural activity. The use of pesticides and fertilizer and exposure of the land to erosion cause concern about pollution of surface and groundwater. Most farmers still maintain that these practices are essential in modern agriculture and that they can be used safely and with minimal pollution.

these tools can reduce their impact; irresponsible use causes increased erosion, water pollution, and risk to humans. (See figure 1.6.)

The Dry West

Where rainfall is inadequate to support agriculture, ranching and raising livestock are possible. This is true in much of the drier portions of western North America. Because much of the land is of low economic value, most is still the property of government, which encourages its use by providing water for livestock and irrigation at minimal cost, offering low rates for grazing rights, and encouraging mining and other development.

Many people believe that government agencies have seriously mismanaged these lands. They assert that the agencies are controlled by special interest groups and powerful politicians sensitive to the needs of ranchers, that they subsidize ranchers by charging too little for grazing rights, and that they allow destructive overgrazing because of the economic needs of ranchers. Ranchers argue that they need access to government-owned land, cannot afford significantly increased grazing fees, and that changing government policies would destroy a way of life that is important to the regional economy.

Water is an extremely valuable resource in this region. It is needed for municipal use and for agriculture. Many areas, particularly the river valleys, have fertile soils that can be used for intensive agriculture. Cash crops such as cotton, fruits, and vegetables can be grown if

water is available for irrigation. Because water tends to evaporate from the soil rapidly, long-term use of irrigated lands often results in the buildup of salts in the soil, thus reducing fertility. Irrigation water flowing from fields is polluted by agricultural chemicals that make it unsuitable for other uses such as drinking. As cities in the region grow, an increasing conflict arises between urban dwellers who need water for drinking and other purposes, and ranchers and farmers who need the water for livestock and agriculture. Increased demand for water will result in shortages, and decisions will have to be made about who will ultimately get the water. If the urban areas get the water they want, some farmers and ranchers will go out of business. If the agricultural interests get the water, urban growth and development will have to be limited and expensive changes will have to be made to conserve domestic water use.

Because population density is low in most of this region, much of the land has a wilderness character. Increasingly, a conflict has developed between the economic management of the land for livestock production and the desire on the part of many to preserve the "wilderness." Designating an area as wilderness means that certain uses are no longer permitted. This offends individuals and groups who have traditionally used the area for grazing, hunting, and other pursuits. A long history of use and abuse of this land by overgrazing, modification to encourage plants valuable for livestock, and the introduction of grasses for livestock has significantly altered the region so that it cannot truly be

called wilderness. The low population density does, however, provide a remoteness and wild character that many seek to preserve. (See figure 1.7.)

The Forested West

The coastal areas and mountain ranges of the western United States and Canada receive sufficient rainfall that coniferous forests are the dominant vegetation. Since most of these areas are not suitable for farmland, they have been maintained as forests with some grazing activity in the more open forests. Governments and large, commercial timber companies own large sections of these lands. Government forest managers (U.S. Forest Service, Bureau of Land Management, Environment Canada, and various state and provincial departments) historically have sold timber-cutting rights at a loss and are thought by many to be too interested in the production of forest products at the expense of other, less tangible values. In 1993, the U.S. Forest Service was directed to stop below-cost timber sales.

This policy change has become a major issue in the old-growth forests of the Pacific Northwest where timber interests maintain that they must have access to government-owned forests in order to remain in business. Many of these areas have significant wilderness, scenic, and recreational value. Environmental interests point out that it makes no sense to complain about the destruction of tropical rainforests in South America while North America makes plans to cut large areas of previously uncut, temperate rainforest.

Overgrazed land

Bryce Canyon

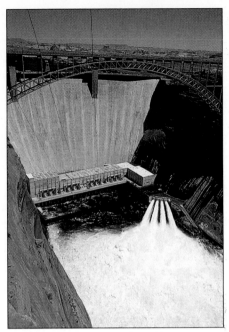
Irrigation water provided from Glen Canyon Dam.

Wilderness area

FIGURE 1.7 **The Dry West** Water is a key issue in this region. Both city dwellers and rural ranchers and farmers need water, and conflict results when there is not enough water to satisfy the desires of all. In addition, much of the land in this region is owned by government. This raises concerns about how government manages the land and how government policy affects the people of the region.

Are the intangible values of preserving an ancient forest ecosystem as important as the economic values provided by timber and jobs?

Environmental interests are concerned about the consequences logging would have on organisms that require mature, old-growth forests for their survival. Grizzly bear habitat in Alaska and British Columbia could be altered significantly by logging; the northern spotted owl has become a symbol of the conflict between logging and preservation in Oregon and Washington; and preservation of coastal redwood forests has become an issue in northern California. (See figure 1.8.)

The Great Lakes and Industrial Northeast

While much of the West and Central region of North America is characterized by low population densities and small towns, major portions of the Great Lakes and Northeast are dominated by large metropolitan complexes that generate social and resource needs that are difficult to satisfy. Many of these older cities formed around industrial centers that have declined, leaving behind poverty, environmental problems in abandoned industrial sites, and difficulties with solid waste disposal, air quality, and land-use priorities. Interspersed among the major metropolitan areas are small towns, farmland, and forests.

One of the major resources of the region is its access to water transport. The Great Lakes and eastern seacoast are extremely important to commerce, since ships can travel throughout the area by way of the St. Lawrence Seaway and the Great Lakes through a series of locks and canals that bypass natural barriers. Because of the importance of shipping in this region, harbors have been constructed and waterways have been deepened by dredging. The waterways are maintained at considerable government expense.

One of the greatest problems associated with the industrial uses of the Great Lakes and East Coast is contamination of the water with toxic materials. In some cases, unthinking or unethical individuals have dumped toxins directly into the water. In other cases, small, accidental spills or leaks over long periods of time have contaminated the sediments in harbors and bays.

A major concern about these pollutants is that they bioaccumulate (see chapter 14) in the food chain. The concentrations of some chemicals in the fat tissue of top predators, such as lake trout and fish-eating birds, can be a million times higher than the concentration in the water. Because of this, government agencies have issued consumption advisories for some fish and shellfish in contaminated areas. Since many kinds of fish can swim great distances, advisories for the Great Lakes warn against eating certain fish taken anywhere within the lakes, not just from the site of contamination. Similarly, Chesapeake Bay has been subjected to years of thoughtless pollution, resulting in reduced fish and shellfish populations and advisories against consuming some organisms taken from the bay.

Water always generates considerable recreational value. Consequently, conflicts arise between those who want to use the water for industrial and shipping purposes and those who wish to use it for recreation.

Cut logs being hauled

U.S. Forest Service ownership

Native elk

FIGURE 1.8 **The Forested West** The cutting of forested areas for timber production destroys the previous ecosystem. Some see the trees as a valuable resource that provides jobs and building materials. Others see the forest ecosystem as a natural resource that should be preserved. In addition, government ownership of much of this land has generated considerable political debate about what the appropriate use of the land should be.

Since so much of the North American population is concentrated in this region, the economic value of recreational use is extremely high. Pressure is great to clean up contaminated sites and prevent the pollution of new ones. Contaminated areas do not enhance tourism.

Most of these older, large cities had no plan to shape their growth. As a result, open space for people is limited and urban dwellers have few opportunities to interact with the natural world. Children who grow up in these cities often do not know that milk comes from a cow—they have never seen, smelled, or touched a cow. Consequently, urban people have difficulty understanding the feeling rural people have for the land. These urban dwellers may never have an opportunity to experience wilderness. Their major environmental priorities are cleaning up contaminated sites, providing more parks and recreation facilities, reducing air and water pollution, and improving transportation. (See figure 1.9.)

The Diverse South

In many ways, the South is a microcosm of all the regions previously discussed. The petrochemical industry dominates the economies of Texas and Louisiana, and forestry and agriculture are significant elements of the economy in other parts of the region. Major metropolitan areas thrive, and much of the area is linked to the coast either directly or by the Mississippi River and its tributaries. The environmental issues faced in the South are as diverse as those in all the other regions.

Some areas of the South (particularly Florida) have had extremely rapid population growth, which has led to groundwater problems, transportation problems, and concerns about regulating the rate of growth. Growth means money to developers and investors, but it requires municipal services, which are the responsibility of local governments. Too many people and too much development also threaten remaining natural ecosystems.

Poverty has been a problem in many areas of the South. This creates a climate

Inner-city decay in Chicago

Harbor in Superior, Wisconsin

Central Park in New York City

FIGURE 1.9   The Great Lakes and the Industrial Northeast Industry, waterways, and population centers are the defining elements of this region. The historically extensive use of the Great Lakes and coastal areas of the Northeast for industry, because of the ease of providing water transportation, has resulted in many older cities with poor land-use practices. Rebuilding cities, providing recreational opportunities for urban dwellers, and repairing previous environmental damage are important issues. The water resources of the region provide transportation, recreation, and industrial opportunities.

Miami metropolitan area

Everglades

Chemical plant on lower Mississippi

FIGURE 1.10   The Diverse South Poverty has been a historically important problem in the region. Often the creation of jobs was considered more important than the environmental consequences of those jobs. The use of coastal areas for industry has resulted in pollution of coastal waters. The heavy use of the Mississippi River for transportation and industry has caused pollution problems. In addition, the desirable climate in the South has resulted in intense pressure to develop new housing for those who want to move to the region. Unwise development of housing on fragile coastal sites has resulted in damage to buildings by storms and the actions of the oceans. This causes intense debate on land use.

that encourages state and local governments to accept industrial development at the expense of other values. Often, jobs are more important than the environmental consequences of the jobs; low-paying jobs are better than no jobs.

The use of the coastline is of major concern in many parts of the South. The coast is a desirable place to live, which encourages unwise development on barrier islands and in areas that are subject to flooding during severe weather. In addi-

tion, industrial activity along the coast has resulted in the loss of wetlands. (See figure 1.10.)

Summary

Artificial political boundaries create difficulties in managing environmental problems because most environmental units, ecosystems, do not coincide with political boundaries. Therefore, a regional approach to solving environmental problems, one that incorporates natural geographic units, is ideal. Each region of the world has certain environmental issues that are of primary concern because of the mix of population, resource use patterns, and culture.

Environmental problems become issues when someone finds a situation offensive. This inevitably leads to a confrontation between groups that have different views on what constitutes an environmental problem. Many social, economic, ethical, and scientific inputs shape a person's opinions. The process of environmental decision making must take all of these inputs into account and arrive at an acceptable compromise.

Environmental problems are people problems. They occur because the uses of natural resources, which some people feel are justified, result in a diminished environment for others in the region. Environmental problems are defined by the person who perceives the problem. When perceptions differ, conflict occurs. Environmental decisions inevitably involve economic considerations because someone is receiving value from the resources being used or someone perceives an economic loss because a use has been withdrawn.

- Some argue that economic considerations should not be important when making environmental decisions; others argue that economic considerations can resolve all environmental issues.
- Some argue that regulation is necessary to protect resources; others argue that regulation hinders valuable use of resources.
- Some consider nonhuman organisms as important as humans; others feel that humans have a special place in nature.
- Some are against change; others recognize that change must occur.

With all these differing opinions, compromise is the only way to resolve the conflicts. The social institution of government must play a role. Economic evaluation is important. Recognition of the validity of opposing points of view is essential. The field of environmental science seeks to find that middle ground.

Review Questions

1. Describe why finding solutions to environmental problems is so difficult. Do you think it has always been as complicated?
2. Describe what is meant by an ecosystem approach to environmental problem solving. Is this the right approach?
3. List two key environmental issues for each of the following regions: the wilderness North, the agricultural middle, the forested West, the dry West, the Great Lakes and industrial Northeast, and the South. How are the issues changing?
4. Define environment and ecosystem and provide examples of these terms from your region.
5. Describe how environmental conflicts are resolved.
6. Select a local environmental issue and write a short essay presenting all sides of the question. Is there a solution to this problem?

Critical Thinking Questions

1. Imagine you are a United States Congressional representative from a western state and a new wilderness area is being proposed for your district. Who might contact you to influence your decision? What course of action would you take? Why?
2. How do you weigh in on the issue of jobs or the environment? What limits do you set on economic growth? Environmental protection?
3. Imagine you are an environmentalist in your area that is interested in local environmental issues. What kinds of issues might these be?
4. Imagine that you lived in the urban east and that you were an advocate of wilderness preservation. What disagreements might you have with residents of the wilderness North or the arid West. How would you justify your interest in wilderness preservation to these residents?
5. You are the Superintendent of Yellowstone National Park and want to move to an ecosystem approach to managing the park. How might an ecosystem approach change the current park? How would you present your ideas to surrounding landowners?
6. Look at the issue of global warming from several different disciplinary perspectives—economics, climatology, sociology, political science, agronomy. What might be some questions that each discipline could contribute to our understanding of global warming?

Key Terms

ecosystem 7   environmental science 5
environment 5   wilderness 9

Related Internet Sites

Visit our web site at http://www.mhhe.com/environmentalscience/ for specific information on each of the topics below.

Search engine for environmental organizations

CEPA and the globalization of environmental protection

Coastal ecosystems programs in the U.S.

Seven environmental regions of the U.S.

Facts for the Pacific region of the U.S.

U.S. Fish and Wildlife Service clickable map

Environment and technology

Searchable index of environmental issues

International Global Change Institute information on climate

Saving the baby harp seal

Marine mammal conservation

chapter 2

Environmental Ethics

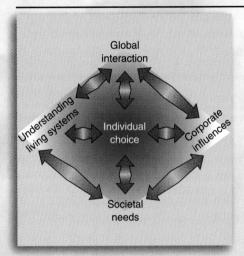

OBJECTIVES

After reading this chapter, you should be able to:

- Differentiate between ethics and morals.
- Define personal ethics.
- Explain the connection between material wealth and resource exploitation.
- Describe how industry must exploit resources and consume energy to produce goods.
- Explain how corporate behavior is determined.
- Describe the influential power that corporations wield because of their size.
- Explain why governmental action was necessary to force all companies to meet environmental standards.
- Describe the factors associated with environmental justice.
- Describe what has been the general attitude of society toward the environment.
- Explain the relationship between economic growth and environmental degradation.
- List three conflicting attitudes toward nature.

CHAPTER OUTLINE

Views of Nature
    Environmental Ethics

Environmental Close-Up: *What Is Ethical?*

Environmental Close-Up: *Code of Environmental Ethics and Conduct*

Environmental Attitudes

Societal Environmental Ethics

Envirnomental Close-Up: *Naturalist Philosophers*

Environmental Close-Up: *Environmental Philosophy*

Corporate Environmental Ethics

Environmental Close-Up: *A Corporate Perspective*

Global Perspective: *Chico Mendes and Extractive Reserves*

Environmental Justice

Individual Environmental Ethics

Global Environmental Ethics

Global Perspective: *International Trade in Endangered Species*

Issues and Analysis: *Antarctica— Resource or Refuge?*

Global Perspective: *Earth Summit*

Global Perspective: *The Kyoto Protocol*

VIEWS OF NATURE

The most beautiful object I have ever seen in a photograph in all my life, is the planet Earth seen from the distance of the moon, hanging in space, obviously alive. Although it seems at first glance to be made up of innumerable separate species of living things, on closer examination every one of its things, working parts, including us, is interdependently connected to all the other working parts. It is, to put it one way, the only truly closed ecosystem any of us know about.

–Lewis Thomas

One of the marvels of recent technology is that we can see the Earth from the perspective of space, a blue sphere unique among all the planets in our solar system. (See figure 2.1.) Looking at ourselves from space, it becomes obvious, says ecologist William Clark of Harvard University, that only as a global species, "pooling our knowledge, coordinating our actions, and sharing what the planet has to offer—do we have any prospect for managing the planet's transformation along pathways of sustainable development."

Many people see little value in an undeveloped river and feel it is unreasonable to leave it flowing in a natural state. Many rivers throughout the world have been "controlled" to provide power, irrigation, and navigation at the expense of the natural world. Many people think that to not use these resources would be wasteful.

In the U.S. Pacific Northwest, there is a conflict over the value of old-growth forests. Economic interests want to use the forests for timber production and feel that to not do so would cause economic hardship. They argue that the trees are going to die anyway and they might as well be used for the betterment of the community. Others feel that all the living things that make up the forest have a value we do not yet appreciate. Removing the trees would destroy something that took hundreds of years to develop and may never be replaced.

Interactions between people and their environment are as old as human civilization. The problem of managing those

FIGURE 2.1   The Earth as Seen from Space Political, geographical, and nationalistic differences among humans do not seem so important from this perspective. In reality, we all share the same "home."

interactions, however, has been transformed today by unprecedented increases in the rate, scale, and complexity of the interactions. At one time, pollution was a local, temporary event. Today, pollution may involve several countries—as with the concern over acid deposition in Europe and in North America—and affect multiple generations. The debates over chemical and radioactive waste disposal are examples of the increasingly international nature of pollution. For example, many European countries are concerned about the transportation of radioactive and toxic wastes across their borders. What were once straightforward confrontations between ecological preservation and economic growth now involve multiple linkages that blur the distinction between right and wrong. For example, the greenhouse effect is thought to result from energy consumption, agricultural practices, and climatic change.

Many people believe that we have entered an era characterized by global change that stems from the interdependence between human development and the environment. They argue that self-conscious, intelligent management of the earth is one of the greatest challenges facing humanity as we approach the twenty-first century. To meet this challenge, they believe, a new environmental ethic must evolve.

Environmental Ethics

Ethical issues dealing with the environment are different from other kinds of ethical problems. Depending on your perspective, an environmental ethic could encompass differing principles and beliefs.

Environmental ethics is a topic of applied ethics which examines the moral basis of environmental responsibility. In these environmentally conscious times, most people agree that we need to be environmentally responsible. Toxic waste contaminates groundwater, oil spills destroy shore lines, and fossil fuels produce carbon dioxide thus adding to the greenhouse effect. The goal of environmental ethics, then, is not to convince us that we should be concerned about the environment— most of us already are. Instead, environmental ethics focuses on the moral foundation of environmental responsibility, and how far this responsibility extends. There are three primary theories of moral responsibility to the environment. Although each supports environmental responsibility, their approaches are different.

The first of these theories is **anthropocentric,** or human-centered. Environmental anthropocentrism is the view that all environmental responsibility is derived from human interests alone. The assumption here is that only human beings are morally significant persons and have a direct moral standing. Since the environment is crucial to human well-being and human survival, then we have an indirect duty towards the environment, that is, a duty which is derived from human interests. This involves the duty to assure that the earth remains environmentally hospitable for supporting human life, and that its beauty and resources are preserved so human life on earth continues to be pleasant. Some have argued that our indirect environmental duties are derived both from the immediate benefit which living people receive from the environment, and the benefit that future generations of people will receive. But, critics have maintained that since future generations of people do not yet exist, then, strictly speaking, they cannot have rights any more than a dead person can have rights. Nevertheless, both parties to this dispute

What Is Ethical?

Ethics is one branch of philosophy. Ethics seeks to define fundamentally what is right and what is wrong, regardless of cultural differences. For example, most cultures have a reverence for life and hold that all individuals have a right to live. It is considered unethical to deprive an individual of life.

Morals differ somewhat from ethics because morals reflect the predominant feelings of a culture about ethical issues. For example, in almost all cultures, it is certainly unethical to kill someone; however, when a country declares war, most of its people accept the necessity of killing the enemy. Therefore, it is a moral thing to do even though ethics says that killing is wrong. No nation has ever declared an immoral war.

Environmental issues require a consideration of ethics and morals. For example, because there is currently enough food in the world to feed everyone adequately, it is unethical to allow some people to starve while others have more than enough. However, the predominant mood of those in the developed world is one of indifference. They don't feel morally bound to share what they have with others. In reality, this indifference says that it is permissible to allow people to starve. This moral stand is not consistent with a purely ethical one.

As we can see, ethics and morals are not always the same; thus, it is often difficult to clearly define what is right and what is wrong.

Some individuals view the world's energy situation as serious and have reduced their consumption. Others do not believe there is a problem and so have not modified their energy use. Still others do not care what the situation is. They will use energy as long as it is available. Other issues are population and pollution. Is it ethical to have more than two children when the world faces overpopulation? Should an industry persuade the public to vote "no" on a particular legislative bill because it might reduce profits, even though its passage would improve the environment? The stand we take on such issues often depends on our position. An industrial leader, for example, would probably not look upon pollution as negatively as someone who is active outdoors. In fact, any business leaders view the behavior of active preservationists as immoral because it restricts growth and, in some cases, causes unemployment.

Most ethical questions are very complex. Ethical issues dealing with the environment are no different. It is important to explore environmental issues from several points of view before taking a stand. One point to consider is the difference between the short-term and long-term effects of a course of action.

When we take an ethical stand, we become open to attack from those who disagree with our stand. Often, individuals are portrayed as villains for pursuing a course of action they consider righteous.

---

acknowledge that environmental concern derives solely from human interests.

A second theory of moral responsibility to the environment is **biocentric.** According to the broadest form of the life-centered theory, all forms of life have an inherent right to exist. Some biocentric thinkers give species a hierarchy of values. Some, for example, believe we have greater responsibility to protect animal species than plant species. Others determine the rights of various species depending on the harm they do to humans. For example, they see nothing wrong in killing pest species such as rats or mosquitoes. Some go further and believe that each individual organism, not just each species, has a basic right to survive. Individuals who support the animal rights movement tend to place more value on the individuals of animal species than on plant species. Trying to decide what types of species or individuals should be protected from early extinction or death resulting from human

activities is an ethical dilemma. It is hard to know where to draw the line and be ethically consistent.

The third approach to environmental responsibility, called **ecocentrism,** maintains that the environment deserves direct moral consideration, and not one which is merely derived from human (and animal) interests. In ecocentrism it is suggested that the environment has direct rights, that it qualifies for moral personhood, that it is deserving of a direct duty, and that it has inherent worth. The environment, by itself, is considered to be on a moral par with humans.

The position of ecocentrism is the view advocated by the ecologist and writer Aldo Leopold in his essay "The Land Ethic" from *A Sand County Almanac and Sketches Here and There* (1949):

"All ethics so far evolved rest upon a single premise: that the individual is a member of a community of

interdependent parts. The land ethic simply enlarges the boundaries of the community to include soils, waters, plants, and animals, or collectively the land . . . a land ethic changes the role of *Homo sapiens* from conqueror of the land-community to plain member and citizen of it . . . It implies respect for his fellow-members, and also respect for the community as such."

What Leopold put forth in "The Land Ethic" was viewed by many as a radical shift in how humans perceive themselves in relation to the environment. Originally we saw ourselves as conquerors of the land. Now according to Leopold we need to see ourselves as members of a community which also includes the land.

Leopold also wrote that "a thing is right when it tends to preserve the integrity, stability, and beauty of the biotic community. It is wrong when it tends otherwise. . . . We abuse land because we

Code of Environmental Ethics and Conduct
(draft discussion document—March 1995)

Produced by a multi-stakeholder group of Canadian national organizations, associations, and agencies and sponsored by the Canadian International Development Agency (CIDA)

Vision Statement:

We recognize that every human being is a part of the community of life on Earth, and that humans are subject to the same immutable ecological laws as all other species. This community links all human societies, present and future generations, and all other parts of Nature. It embraces both cultural and natural diversity.

Because all life depends on the uninterrupted functioning of natural systems, ecological responsibility among all people is necessary for the survival, security, equity, and dignity of the world's communities. Human culture must be built upon a profound respect for Nature, a sense of being at one with Nature, and recognition that human affairs must proceed in harmony and balance with Nature.

Basic Premises Governing the Human-Nature Relationship:

1. Humankind is a part of Nature and the diversity of living forms.

2. The sustainability of our life support systems is incompatible with current consumptive growth patterns and the rate of human population growth.

3. Humans have altered the natural systems of the Earth to such an extent that many extinctions have already taken place, and the existence of countless species, including our own is in jeopardy.

In accepting these basic ecological premises we agree to the following ethical principles:

1. We should have a profound respect for Nature in all its diverse manifestations.

2. All species are unique, and have value regardless of any perceived value that they may have for humans.

3. All persons should take responsibility for their impact on Nature.

The following are basic principles of appropriate conduct:

1. Development can only be sustainable in the long term if it ensures the integrity of the planet's life support systems, while simultaneously meeting basic human needs. Real sustainable development must involve an effort to integrate development into the natural environment, rather than the reverse.

2. For future generations and the survival of the planet's living systems we must plan for the long term.

3. We should pursue and adopt values that address current consumption patterns and the growth of human populations.

4. The human use of natural ecosystems and human activities must be managed rather than the systems themselves.

5. We must avoid actions that contribute to the extinction of any species, or that seriously degrade habitats.

6. Ecological expertise should be fully incorporated into the development process.

7. Upstream activities have downstream effects which must be addressed in the design and planning stage of any development project.

8. Restoration of degraded ecosystems combined with protection of remaining natural areas are important aspects of sustainable development initiatives.

For further information or additional copies contact:
Jean Arnold
FALLS BROOK CENTRE
RR#1, Hartland, New Brunswick
Canada, E0J 1N0
Tel: (506) 375-8143
Fax: (506) 375-4221

---

regard it as a commodity belonging to us. When we see land as a community to which we belong, we may begin to use it with love and respect."

As traditional political and nationalistic boundaries begin to fade or shift globally, there are also new variations of environmental thought and ethics evolving. Some of the new thoughts on environmental ethics are founded on an awareness that humanity is part of nature and that nature's many parts are interdependent. In any natural community, the well-being of the individual and of each species is tied to the well-being of the whole. In a world increasingly without environmental borders,

nations, like individuals, should have a fundamental ethical responsibility to respect nature and to care for the Earth, protecting its life-support systems, biodiversity, and beauty and caring for the needs of other countries and future generations. (See Environmental Close-Up: Code of Environmental Ethics and Conduct.)

Environmental ethicists argue that to consider environmental protection as a "right" of the planet is a natural extension of the concept of human rights. Many also argue that an environmental ethic considers one's actions toward the environment as a matter of right and wrong, rather than one of self-interest.

ENVIRONMENTAL ATTITUDES

There are many different attitudes about the environment, most of which fall under one of three headings: (*a*) the development ethic, (*b*) the preservation ethic, and (*c*) the conservation ethic. Each of these ethical positions has its own code of conduct against which ecological mortality may be measured. (See figure 2.2.)

The **development ethic** is based on individualism or egocentrism. It assumes that the human race is and should be the master of nature and that the earth and its

Preservation

Development

Conservation

Recreation

FIGURE 2.2 **The Views of Nature** Individuals envision the same resources used differently.

resources exist for our benefit and pleasure. This view is reinforced by the work ethic, which dictates that humans should be busy creating continual change and that things that are bigger, better, and faster represent "progress," which itself is good. This philosophy is strengthened by the idea that, "if it can be done, it should be done," or that our actions and energies are best harnessed in creative work.

Examples of the development ethic abound. The notion that bigger is better is certainly not new to us, nor is the belief that if something can be done or built it should be. The dream of upward mobility is embodied in this ethic. In some circles, questioning growth is considered almost unpatriotic. In the development ethic, nature has only instrumental value; that is, the environment has value only insofar as human beings economically utilize it. Only in the past fifty to one hundred years have the by-products and waste associated with development been considered.

The **preservation ethic** considers nature special in itself. Nature, it is argued, has intrinsic value or inherent worth apart from human appropriation. Preservationists

have different reasons for wanting to preserve nature. Some have a religious belief regarding nature. They hold a reverence for life and respect the right of all creatures to live, no matter what the social and economic costs. Some preservationists' interest in nature is primarily aesthetic or recreational. They believe that nature is beautiful and refreshing and should be available for picnics, hiking, camping, fishing, or just peace and quiet.

In addition to the religious and recreational preservationists, there are also preservationists whose reasons are essentially scientific. They argue that the human species depends on and has much to learn from nature. Rare and endangered species and ecosystems, as well as the more common ones, must be preserved because of their known or assumed long-range, practical utility. In this view, natural diversity, variety, complexity, and wilderness is thought to be superior to humanized uniformity, simplicity, and domesticity. Scientific preservationists want to lock up not all the land but only what they consider important to future generations.

The third environmental ethic is referred to as the **conservation ethic.** It is related to the scientific preservationist view but extends the rational consideration to the entire earth and for all time. It recognizes the desirability of decent living standards, but it works toward a balance of resource use and resource availability. The conservation ethic stresses a balance between total development and absolute preservation. It stresses that rapid and uncontrolled growth in population and economics is self-defeating in the long run. The goal of the conservation ethic is one people living together in one world, indefinitely.

SOCIETAL ENVIRONMENTAL ETHICS

Society is composed of a great variety of people with differing viewpoints. This variety can be distilled into a set of ideas that reflect the prevailing attitudes of society. The collective attitudes can be analyzed from an ethical point of view. Western, developed societies have long acted as if the earth has unlimited reserves of natural resources, an unlimited ability to assimilate wastes, and a limitless ability to accommodate unchecked growth.

The economic direction and rationale of developed nations have been that of continual growth. Unfortunately, this growth has not always been carefully planned or even desired. This "growthmania" has resulted in the use of our nonrenewable resources for comfortable homes, well-equipped hospitals, convenient transportation, fast-food outlets, VCRs, home computers, and battery-operated toys, among many other things. In economic statistics, such "growth" measures out as "productivity," and life looks rosy for all. But the question arises, "What is enough?" Many poor societies have too little, but rich societies never say, "Halt! We have enough." As the Indian philosopher and statesman Mahatma Ghandi said, "The earth provides enough to satisfy every person's need, but not every person's greed."

Growth, expansion, and domination remain the central sociocultural objectives

ENVIRONMENTAL CLOSE-UP

Naturalist Philosophers

The philosophy behind the environmental movement had its roots in the last century. Among many notable conservationist philosophers, several stand out: Ralph Waldo Emerson, Henry David Thoreau, John Muir, Aldo Leopold, and Rachel Carson.

In Emerson's first essay, *Nature,* published in 1836, he claimed that "behind nature, throughout nature, spirit is present." Emerson was an early critic of rampant economic development, and he sought to correct what he considered to be the social and spiritual errors of his time. In his *Journals,* published in 1840, Emerson stated that "a question which well deserves examination now is the Dangers of Commerce. This invasion of Nature by Trade with its Money, its Credit, its Steam, its Railroads, threatens to upset the balance of Man and Nature."

A naturalist who held beliefs similar to Emerson's was Henry David Thoreau. Thoreau's bias fell on the side of "truth in nature and wilderness over the deceits of urban civilization." The countryside around Concord, Massachusetts, fascinated and exhilarated him as much as the commercialism of the city depressed him. It was near Concord that Thoreau wrote his classic, *Walden,* which describes a year in which he lived in the country to have direct contact with nature's "essential facts of Life." In his later writings and journals, Thoreau summarized his feelings toward nature with prophetic vision:

> But most men, it seems to me, do not care for Nature and would sell their share in all her beauty, as long as they may live, for a stated sum—many for a glass of rum. Thank God, man cannot as yet fly, and lay waste the sky as well as the earth! We are safe on that side for the present. It is for the very reason that some do not care for these things that we need to continue to protect all from the vandalism of a few. (1861).

John Muir combined the intellectual ponderings of a philosopher with the hard-core, pragmatic characteristics of a leader. Muir believed that "wilderness mirrors divinity, nourishes humanity, and vivifies the spirit." Muir tried to convince people to leave the cities for a while to enjoy the wilderness. However, he felt that the wilderness was threatened. In the 1876 article entitled, "God's First Temples: How Shall We Preserve Our Forests," published in the Sacramento *Record Union,* Muir argued that only government control could save California's finest sequoia groves from the "ravages of fools." In the early 1890s, Muir organized the Sierra Club to "explore, enjoy, and render accessible the mountain regions of the Pacific Coast" and to enlist the support of the government in preserving these areas. His actions in the West convinced the federal government to restrict development in the Yosemite Valley, which preserved its beauty for generations to come.

Another thinker as well as a doer in the early conservation field was Aldo Leopold. As a philosopher, Leopold summed up his feelings in *A Sand County Almanac:*

> Wilderness is the raw material out of which man has hammered the artifact called civilization. No living man will see again the long grass prairie, where a sea of prairie flowers lapped at the stirrups of the pioneer. No living man will see again the virgin pineries of the Lake States, or the flatwoods of the coastal plain, or the giant hardwoods.

Ralph Waldo Emerson

Henry David Thoreau

John Muir

Aldo Leopold

Leopold founded the field of game management. In the 1920s, while serving in the Forest Service, he worked for the development of a wilderness policy and pioneered his concepts of game management. He wrote extensively in the *Bulletin* of the American Game Association and stated that the amount of space and the type of forage of a wildlife habitat determine the number of animals

Rachel Carson

that can be supported in an area. Furthermore, he said that regulated hunting can maintain a proper balance of wildlife.

While most people talk about what's wrong with the way things are, few actually go ahead and change it. Rachel Carson ranks among those few. A distinguished naturalist and best-selling nature writer, Rachel Carson published a series of articles in the *New Yorker* in 1960, which generated widespread discussion about pesticides. In 1962, she published *Silent Spring,* which dramatized the potential dangers of pesticides to food, wildlife, and humans and eventually led to changes in pesticide use in the United States.

Although some technical details of her book have been shown to be in error by later research, her basic thesis that pesticides can contaminate and cause widespread damage to the ecosystem has been established. Unfortunately, Carson's early death from cancer came before her book was recognized as one of the most important events in the history of environmental awareness and action in this century.

Environmental Philosophy

Nature, growth, and progress are concepts that we all use, but which we seldom define either in discussion or to ourselves. We speak about environmental ethics, environmental philosophy, eco-philosophy, and so on, but what do we put into these concepts? We say that we have a responsibility for future generations, and that this is a question of morals, but how should questions about morals be decided?

We all have some kind of reasons for our opinions on these matters—whether we believe that nature shall serve man, that man shall serve nature, or something in between—but we seldom make them explicit or draw the conclusions from them. What are these reasons? Are they reasonable, rational, defensible, scientifically grounded, emotional, religious? And what do we mean by saying that a reason is "rational"? What value does it have that something is scientifically grounded? What weight should emotions carry in this context? Can we reason rationally about them and come to an agreement about them, or do we have to put them aside and just stick to facts? How can we know what is right?

These are all philosophical questions, and as such they may seem far removed from real, concrete environmental problems such as ozone holes and forest death. Can it really be meaningful to spend energy on a debate of these questions, when if not disaster then at least crisis stands at the doorway? Is it at all meaningful to philosophize about the environmental questions? Isn't it action that is needed?

Trying to answer the philosophical questions does not, of course, in itself solve any environmental problems, but on the other hand is it questionable whether we can solve these problems without discussing them on a philosophical level? Because whether we discuss them or not, we have ideas and conceptions which guide our way of thinking, what we see as a problem, what we see as causes of problems and what we see as possible, desirable, or necessary solutions. And, quite seriously, the problems which we stand before today are hardly founded on our having thought too much?

What are your thoughts on this question? Is there already too much talk or is it too little listening?

Source: Dept. of Philosophy, University of Gothenburg, Sweden. No copyright.

---

of most advanced societies. **Economic growth** and **resource exploitation** are attitudes shared by developing societies. We continue to consume natural resources as if the supplies were never ending. All of this is reflected in our increasingly unstable relationship with the environment, which grows out of our tendency to take from the "common good" without regard for the future.

This attitude is deeply embedded in the fabric of our society. Since the first settlers arrived in North America, nature has been considered an enemy. Frequently, the colonists expressed their relation to the wilderness in military terms. They viewed nature as an enemy to be "conquered," "subdued," or "vanquished" by a pioneer "army." Any qualms the pioneers may have felt about invading and exploiting the wilderness were justified by religious beliefs. This attitude toward nature is still popular today. Many view wilderness solely as underdeveloped land and see value in land only if it is farmed, built upon, or in some way developed. The notion that land and wilderness should be preserved is incomprehensible to some. The thought of purposely opting to not develop a resource is considered almost a sin by many.

CORPORATE ENVIRONMENTAL ETHICS

Many tasks of industry, such as procuring raw materials, manufacturing and marketing, and disposing of wastes, are in large part responsible for pollution. This is not because any industry or company has adopted pollution as a corporate policy. Industry is naturally dirty because it consumes energy and resources. When raw materials are processed, some waste (useless material) is inevitable. It is usually not possible to completely control the dispersal of all by-products of a manufacturing process. Also, some of the waste material may simply be useless.

For example, the food-service industry uses energy to prepare meals. Much of this energy is lost as waste heat. Smoke and odors are released into the atmosphere, and bone, fat, and discolored food items must be discarded.

The cost of controlling waste can be very important in determining a company's profit margin. **Corporations** are legal entities designed to operate at a profit,

which is not in itself harmful. The corporation has no ethics, but the people who make up the corporation are faced with ethical decisions. Ethics are involved when a corporation cuts corners in production quality or waste disposal to maximize profit. The cheaper it is to produce an item, the greater the possible profit. It is cheaper to dump wastes into a river than to install a wastewater treatment facility, and it is cheaper to release wastes into the air than it is to trap them in filters. Many people consider such pollution unethical and immoral, but corporations think of it as just one of the factors that determines **profitability.** (See figure 2.3.) Because stockholders expect an immediate return on their investment, corporations often make decisions based on short-term profitability rather than long-term benefit to society.

The amount of profit a corporation realizes determines how much it can expand. To expand continually, a corporation increases the demand for its products through advertising. The more it expands, the more power it attains. The more power it has, the greater its influence over decision makers who can create conditions favorable to its expansion plans. The process becomes a seemingly never-ending spiral.

A Corporate Perspective

Perhaps one of the most important and yet difficult lessons learned during the past quarter century is that the environment and the economy need not be viewed as opposites. On the contrary, we are beginning to understand that in order to have a healthy environment you must also have a healthy economy. The opposite is also equally true in that a healthy economy is dependent on a healthy environment. Many of the leading multinational corporations are adopting this new attitude as a philosophy of business. This is especially true in the area of sustainability. General Motors Corporation is one such company that has publicly stated its commitment to sustainability.

The corporate policy of General Motors has defined sustainability as a framework that drives them to seek continuous improvement in its operations and products in a way that integrates economic, environmental, and social objectives into daily business decisions. General Motors has defined sustainable development as an ongoing process of continuous improvement with discrete goals throughout. The baseline from which GM works regarding sustainability begins with the GM

Environmental Principles. This set of six principles, together with a preamble that establishes their global applicability, was adopted by the General Motors Board of Directors in March 1991.

Examples of some of GM's initiatives that are included in the corporate framework for sustainability include:

General Motors University (GMU) which was created in 1997 to align employees with the education and development needed to help GM succeed worldwide. Its focus is to bring about change through a network of learning opportunities designed to improve critical knowledge and skills. GMU could be instrumental in providing education about sustainability to employees because it offers a mechanism for communicating common systems and best practices, and for enhancing individual competencies. GMU will advance the concept that environmental protection, economic opportunities, and community objectives are interwoven in a common framework. This approach will be explained not only through traditional classroom learning, but through hands-on, functionally based training.

Sustainability Management Framework

Nations of the world must confront the problem of corporate irresponsibility toward the environment. In business, incorporation allows for the organization and concentration of wealth and power far surpassing that of individuals or partnerships. Some of the most important decisions affecting our environment are made not by governments or the public but by executives who wield massive corporate power. Often, these executives make only minimal concessions to the public interest, while they make every effort to maximize profits.

Business decisions and technological developments have increased the exploitation of natural resources. In addition, many political and legal institutions have generally supported the development of private enterprise. They have also defended and promoted private property rights rather than social and environmental concerns. Businesses and individuals typically use loopholes, political pressure, and the time-consuming nature of legal action to circumvent or delay compliance with social or environmental regulations.

Another GM initiative regarding sustainability is with the supplier community through the PICOS™ program. Established in 1989, this program sends teams of GM engineers to help suppliers identify faster, less costly, and more efficient ways of doing business. In 1996, PICOS™ was expanded to promote environmental excellence by helping suppliers identify and implement resource conservation and pollution-prevention (RCP2) projects that lower operating costs while reducing emissions. GM currently has over 200 supplier development engineers conducting over 2000 workshops per year at supplier facilities to reduce waste and improve productivity. The goal of PICOS™ is to promote eco-efficiency in the supplier community.

Internationally, GM is working with community groups to plan for sustainability. GM and its employees are involved in various community programs aimed at improving environmental quality and education. For example, Opel Taiwan has initiated the Opel Environmental

Award, which gives exposure to and awards individuals for exemplary environmental projects in Taiwan. It is judged by a panel of distinguished environmentalists, government officials, and industrialists. The mission of the program is to foster creative solutions for a better environment in Taiwan.

General Motors is not alone in the corporate world in its growing attention to the environment and sustainability. Most major corporations now publish an annual report on their environmental progress. Examples include DOW Chemical, which publishes its annual *Continuing the Responsible Care Journey: Steps Toward Sustainability;* Weyerhauser, which publishes its *Annual Environmental Performance Report;* and Boeing, which publishes its annual *Progress Report on Environmental Affairs.* Perhaps you could obtain similar reports from local companies and investigate whether they are living up to what they profess.

General Motors Environmental Principles

As a responsible corporate citizen, General Motors is dedicated to protecting human health, natural resources and the global environment.

This dedication reaches further than compliance with the law to encompass the integration of sound environmental practices into our business decisions.

The following environmental principles provide guidance to General Motors personnel worldwide in the conduct of their daily business practices.

1. We are committed to actions to restore and preserve the environment.

2. We are committed to reducing waste and pollutants, conserving resources and recycling materials at every stage of the products life cycle.

3. We will continue to participate actively in educating the public regarding environmental conservation.

4. We will continue to pursue vigorously the development and implementation of technologies for minimizing pollutant emissions.

5. We will continue to work with all governmental entities for the development of technically sound and financially responsible environmental laws and regulations.

6. We will continually assess the impact of our plants and products on the environment and the communities in which we live and operate with a goal of continuous improvement.

a.

b.

FIGURE 2.4   CERES Principles The 1989 oil spill in Alaska led to the development of the CERES Principles. (a) These waterfowl were victims of the spill. (b) The *Exxon Valdez.*

FIGURE 2.3   Corporate Decision Making Corporations must make a profit. When they look at pollution control, they view its cost like any other cost: any reductions in cost increase profits.

Is industry becoming more environmentally concerned? Corporations have certainly made more frequent references to worldwide environmental issues over the past several years. Is such concern only rhetoric and social marketing, or is it the beginning of a new corporate ethic? The "Valdez Principles," a tool that industry and the public can use to evaluate corporate environmental responsibility, were developed as a result of the 1989 oil spill in Alaska. (See figure 2.4.)

The Oil Protection Act of 1990, passed in the wake of the *Valdez* spill, was supposed to regulate supertankers and reduce the chances of supertanker oil spills. However, to get around the law many oil carriers have shifted their oil transport operations to lightly regulated oil barges pulled by tugboats. This reduction in oil spill safety has led to several barge oil spills, including one in January 1996 in Rhode Island's Moonstone Bay and

another in March 1997 in Texas's Galveston Bay.

In 1992, the Valdez Principles were proposed and adopted by the CERES Organization (Coalition for Environmentally Responsible Economics). This organization is an independent group of environmentalists and social investors. The Valdez Principles have been renamed the CERES Principles.

The CERES Principles are a set of codes that businesses may adopt voluntarily. The codes provide environmental standards against which all companies can be assessed and compared. The ten principles encompass a wide range of goals that include minimizing pollutants, making sustainable use of renewable resources, reducing health and safety risks for employees and communities, and representing environmental interests on corporate boards. The CERES Principles are looked upon as a guide for corporate environmentalism.

The goal, some argue, should be to make compliance with the CERES Principles a prerequisite for doing business.

Practicing an environmental ethic should not interfere with corporate and other social responsibilities or obligations, though this is not always the case. It must be integrated into overall systems of belief and coordinated with economic systems. Environmental advocates, in turn, need to consider others' objectives just as they demand that others consider the environmental consequences in decision making. It makes little sense to preserve the environment if that objective produces national economic collapse. Nor does it make sense to maintain stable industrial productivity at the cost of breathable air, drinkable water, wildlife species, parks, and wilderness. But to maintain profitability, influence, and freedom, businesses must be sensitive to their impact on current and future citizens, not just in terms of the price and quality of the goods they produce but also in terms of public

Chico Mendes and Extractive Reserves

Chico Mendes, whose real name was Francisco Alves Mendes Filho, was a Brazilian rubber tapper active in the rubber tappers' union. He and many other peasants made a living by extracting latex from rubber trees and selling it. Rubber tappers also collect and sell other natural products of the forest, such as Brazil nuts, fruits, and native medicines. Mendes was interested in preserving the portion of the Brazilian rainforest that provided their livelihood, and he supported the concept of "extractive reserves."

Extractive reserves involves setting aside land for rubber tappers, who would continue their traditional lifestyles and use the rainforest in its natural state for generations to come. This idea put Mendes in conflict with powerful people interested in clearing the rainforest to raise cattle. Most cattle-ranching operations show short-term economic gains but ultimately become uneconomical.

On 22 December 1988, as he walked from his home, Chico Mendes was shot by members of a vigilante group who supported local ranchers. In 1990, Darli Alves da Silva and his son were convicted of the murder.

Before his death, Mendes had said, "I want to live to defend the Amazon." His life and death appear to have made a difference in the way the Brazilian rainforest is being used. Due to the circumstances that surrounded his death and his role as a leader in the rubber tappers' union, his murder received international notice and caused many people to ask if the natural rainforest perhaps has as much to offer as ranches do.

During 1990, the Brazilian government established four refuges. The first, named the Chico Mendes Reserve in the Juruá River valley, covers about 6 percent of the state of Acre in northwest Brazil. Unfortunately, by 1993, economic realities were once again surfacing in the area. While confrontations between the rubber tappers and ranchers have decreased, there have been new problems relating to the mismanagement of funds in the Chico Mendes Foundation and mismanagement of the rubber tappers' cooperative that runs the Brazil-nut factory. Unfortunately by 1998, the ten-year anniversary of the murder of Chico Mendes, the future of both the cooperative and foundation were in doubt.

---

approval of their social and political influence. A 1997 Harris Poll, for example, found that 70 percent of Americans wanted increased government spending, either national or state, to address environmental problems, even if they had to pay higher taxes. In another poll, eight of ten Americans said they would be willing to pay extra for a product packaged with recyclable materials.

In the middle 1990s a concept emerged called industrial ecology that reflects the link between the economy and the environment. This concept argues that good ecology is also good economics and that alternatives exist for corporations to provide goods and services in ways that do not destroy the environment.

One of the most important elements of **industrial ecology** is that, as in biological systems, it accounts for waste. Dictionaries define waste as useless or worthless material. In nature, however, nothing is eternally discarded; in various ways, all materials are reused. In our industrial world, discarding materials taken from the Earth at great cost is also generally unwise. Perhaps materials and products that are obsolete should be termed *residues* rather than *wastes;* wastes are merely residues that our economy has not yet learned to use efficiently. A simpler way of saying this is to view a pollutant as a resource out of place. Such a statement forces us to view pollution and waste in a new way.

ENVIRONMENTAL JUSTICE

At its core, **environmental justice** means fairness. It speaks to the impartiality that should guide the application of laws designed to protect the health of human beings and the productivity of ecological systems on which all human activity, economic activity included, depends. It is emerging as an issue because studies show that certain groups of North Americans and citizens of other nations may suffer disproportionately from the effects of pollution.

Governments have established numerous laws, mandates, and directives to eliminate discrimination in housing, education, and employment, but few attempts have been made to address discriminatory environmental practices. In the United States, people of color have borne a disproportionate burden in the siting of municipal landfills, incinerators, and hazardous-waste treatment, storage, and disposal facilities.

Hazardous waste sites and incinerators are not randomly located. While waste generation is correlated directly with per capita income, few toxic waste sites are located in affluent suburbs. Waste facilities are often located in communities that have high percentages of poor, elderly, young, and minority residents. Often such facilities are deliberately sited in these communities because they are seen as providing the path of least resistance.

Questions of environmental justice extend beyond the siting of toxic waste sites. Exposure to harmful pesticides and other toxic agricultural substances is a major health issue among hired farm workers, the majority of whom are people of color. There is also concern that because some Native American communities consume much greater amounts of fish from certain areas such as the Great Lakes than does the general population, they are at greater risk for dietary exposure to toxic chemicals.

Historically, the environmental movement has been a concern of middle-class whites, but there is a growing level of activism by people of color. Minority participation has broadened the debate to include many issues that were being ignored. It has also forced a dialogue about race, class, discrimination, and equity. Minorities have pushed the plight of their communities to the forefront. They have also brought a new perspective to the environmental movement and will be a part of any future environmental agenda.

Environmental Justice Highlights

1979
Houston community group sues over landfill: While the black community group from Houston eventually lost its court case, its work produced some of the first research on environmental justice showing that most incinerators and landfills built in the city were in black neighborhoods.

1982
More than 500 arrested in dump fight: In mostly black Warren County, N.C., a battle over a landfill intended to hold PCB-contaminated dirt led to the arrest of hundreds including the Washington, D.C., delegate to Congress.

1983
Government report finds hazardous waste bias: An analysis by the Government Accounting Office found that three of the four largest hazardous waste sites in the Southeast were in black communities.

1987
Church finds environmental racism a national problem: The United Church of Christ published a report that showed that communities with hazardous waste facilities had higher percentages of minorities than those had had no such facilities.

1989
EPA takes up environmental justice: Bush administration EPA Administrator William Reilly establishes the Environmental Equity Work Group, marking the first official EPA response to the problem.

1993
EPA accepts first civil rights complaint: The Louisiana case was filed by the Tulane (university) Environmental Law Clinic. The legal group later would file another case over the Shintech Inc. PVC plant in rural Louisiana, which became a nationally important battle. At about the same time, the EPA formed the National Environmental Justice Advisory Council, a group of activists, local officials, and industry experts intended to serve as advisers.

1994
President Clinton signs environmental justice executive order: The order requires all federal agencies to begin taking the issue into account.

1998
EPA releases environmental justice "guidance": The new agency rules, produced with no input from states, cities, or industry, create an uproar.
The Detroit News, Sunday, June 28, 1998

INDIVIDUAL ENVIRONMENTAL ETHICS

The environmental movement has effectively influenced public opinion and moved the business community toward an environmental ethic. The result of this changing view of business's responsibilities will complicate business decision making into the next century. More complex demands by the public and a broadening of horizons on the part of business will be the dominant theme of corporate life during the next decade. As human populations and economic activity continue to grow, we are facing a number of environmental problems that threaten not only human health and the productivity of ecosystems, but in some cases the very habitability of the globe.

If we are to respond to those problems successfully, our environmental ethic must express itself in broader and more fundamental ways. We have to recognize that each of us is individually responsible for the quality of the environment we live in and that our personal actions affect environmental quality, for better or worse. The recognition of individual responsibility must then lead to changes in individual behavior. In other words, our environmental ethic must begin to express itself not only in national laws, but also in subtle but profound changes in the ways we all live our daily lives.

A Roper Poll in 1997 indicated that Americans think environmental problems can often be given a quick technological fix. Says the Roper organization, "They believe that cars, not drivers, pollute, so business should invent pollution-free autos. Coal utilities, not electricity consumers, pollute, so less environmentally dangerous generation methods should be found." It appears that many individuals want the environment cleaned up, but they do not want to make major lifestyle changes to make that happen.

Decisions and actions by individuals faced with ethical choices collectively determine the hopes and quality of life for everyone. As ecological knowledge and awareness begin to catch up with good intentions, people in all walks of life will need to live by an environmental ethic.

GLOBAL ENVIRONMENTAL ETHICS

In 1990, Noel Brown, the Director of the United Nations North American Environmental Programme, stated:

Suddenly and rather uniquely the world appears to be saying the same thing. We are approaching what I have termed a consensual moment in history, where suddenly from most quarters we get a sense that the world community is now agreeing that the environment has become a matter of global priority and action.

This new sense of urgency and common cause about the environment is leading to unprecedented cooperation in some

International Trade in Endangered Species

The United Nations Convention on International Trade in Endangered Species (CITES) continues to raise the issue of trade sanctions against China and Taiwan if they do not more rigorously attempt to halt trade in rhinoceros and tiger parts sold for medicinal purposes. The animals are killed to provide material for a growing international market. Rhino horns and tiger parts are still used frequently in traditional Chinese medicine to treat fevers, impotence, and joint and bone maladies such as rheumatism and arthritis. Recent investigations have shown that rhino horn is available for sale in pharmacies in Taiwan and through government-run companies in China and is a lucrative export commodity. In 1998, 110 shops in seven North American cities surveyed by Chinese investigators working for the World Wildlife Fund found that 50 percent of the shops offered for sale one or more medicines claiming to contain ingredients from tigers, rhinos, or leopards.

With one exception, the medicines discovered in North America had been manufactured in mainland China. China prohibited the manufacture and trade of medicines containing tiger, leopard, or rhino horn in 1993. Subsequent to the ban, however, medicines containing the above listed animals have repeatedly shown up in North America, New Zealand, and the European Union as well as China.

It is estimated that fewer than 7000 tigers and 10,000 rhinos remain in the world. Trade sanctions and the enactment of laws unfortunately is much easier to undertake than that of changing traditional customs and beliefs.

---

areas. Despite their political differences, Arab, Israeli, Russian, and American environmental professionals have been working together for several years. Ecological degradation in any nation almost inevitably impinges on the quality of life in others. For years, acid rain has been a major irritant in relations between the United States and Canada. Drought in Africa and deforestation in Haiti have resulted in waves of refugees. From the Nile to the Rio Grande, conflicts flare over water rights. The growing megacities of the Third World are time bombs of civil unrest.

Much of the current environmental crisis is rooted in and exacerbated by the widening gap between rich and poor nations. Industrialized countries contain only 20 percent of the world's population, yet they control 80 percent of the world's goods and create most of its pollution. The developing countries are hardest hit by overpopulation, malnutrition, and disease.

As these nations struggle to catch up with the developed world and improve the quality of life for their people, a vicious circle begins: Their efforts at rapid industrialization poison their cities, while their attempts to boost agricultural production often result in the destruction of their forests and the depletion of their soils, which lead to greater poverty. (See figure 2.5.)

Perhaps one of the most important questions for the future is, "Will the nations of the world be able to set aside their political differences to work toward a global environmental course of action?" The United Nations Conference on Human Environment held in Stockholm, Sweden, in 1972 was a step in the right direction. Out of that international conference was born the U.N. Environment Programme, a separate department of the United Nations that deals with environmental issues. A second world environmental conference was held in 1992 in

Brazil. It followed up the Stockholm conference with many new international initiatives. A major world conference on climate change was held in Kyoto, Japan in 1997. (See Global Perspective: Earth Summit and Global Perspective: The Kyoto Protocol.) Through organizations and conferences such as these, nations can work together to solve common environmental problems.

At the individual level, people have begun to respond to increased awareness of global environmental change by altering their values, beliefs, and actions. Changes in individual behavior are necessary but are not enough. As a global species, we are changing the planet. By pooling our knowledge, coordinating our actions, and sharing what the planet has to offer, we may achieve a global environmental ethic.

FIGURE 2.5 Lifestyle and Environmental Impact Significant differences in lifestyles and their environmental impact exist between the rich and poor nations of the world. What would be the environmental impact on the Earth if the citizens of China and India and other less developed countries enjoyed the standard of living of North Americans? Can we deny them that opportunity?

Few places on earth have not been exploited by humans. One such place is Antarctica. It is as close to an unpolluted environment as there is on earth, but it is not without its problems.

Seals and whales were the earliest exploited resource in Antarctica. There was money to be made, and this "opportunity" resulted in the near extinction of the southern fur seal, the elephant seal, and the blue whale.

By the 1950s, aboveground nuclear testing had spread radioactive particles around the planet, including Antarctica. Pesticides like DDT were turning up in the tissues and blood of certain Antarctic bird and marine mammal species. A growing hole in the ozone layer above the Antarctic continent is caused by the use of chlorofluorocarbons throughout the world. Fossil-fuel combustion contributes to the greenhouse effect, which in turn threatens to melt the ice in Antarctica's Western Peninsula.

During the past several decades, Antarctica has been the site of extensive scientific exploration. Much of this exploration has been economically motivated. For example, government scientists, with the aid of satellites, are advising oil and mineral prospectors. Much of the so-called scientific research is conducted with geopolitical or military objectives in mind.

Antarctica is also being proposed as a tourist attraction. Australia has suggested building a hotel, while Argentina is considering chartering a vessel to transport six hundred tourists from South America seven times a year. Several sites are also being viewed as potential ski resorts.

From an ecological perspective, Antarctica is fragile. The thin layer on the surface of the ocean, nourished by the sun, supports the tiny shrimp-like krill, which sustain fish, whales, seals, and penguins. These short, simple food chains are extremely sensitive to environmental insults.

In the mid-1970s, New Zealand proposed designating the continent an Antarctic World Park. This would turn Antarctica into an international wilderness area, a region on earth where we recognize that humanity does not belong.

In 1991, 24 countries signed an agreement to ban mineral and oil exploration in Antarctica for 50 years. The agreement, which was hailed as historic by governments and environmental groups, includes new regulations for wildlife protection, waste disposal, marine pollution, and continued monitoring of the Antarctic, which covers nearly one-tenth of the world's land surface. The signing of the agreement in Madrid, Spain, was the result of two years of negotiations. The protocol protects Antarctica's delicate flora and fauna and sets procedures to assess environmental effects of all human activities on the continent.

Should we turn a continent into a world park?

Would humanity be better served by developing the natural resources of Antarctica, such as oil and minerals?

Should the natural beauty of Antarctica be opened up to tourism so it can be enjoyed by many?

Is it possible to strike a balance between preservation and development in a fragile ecosystem? Can you give examples?

Source: United Nations Publications.

Should Antarctica be preserved or utilized for human use?

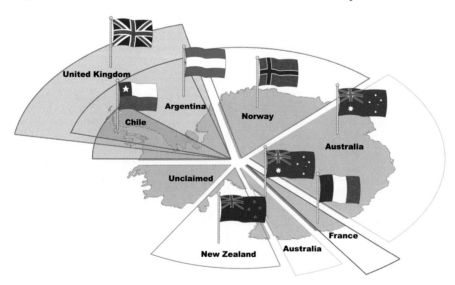

Territorial Claims

No country owns Antarctica, but seven countries have made claims to parts of it. (The United States, which has the largest research population on the continent, has not.) Overlapping claims led to the Antarctic Treaty, signed by 41 countries, which declares that "Antarctica shall be open to all nations to conduct scientific or other peaceful activities there."

Source: Composite drewing from United Nations' publications

GLOBAL PERSPECTIVE

Earth Summit

In June 1992, representatives from 178 countries, including 115 heads of state, met in Rio de Janeiro, Brazil, at the Earth Summit. Officially, the meeting was titled the United Nations Conference on Environment and Development (UNCED), and it was the largest gathering of world leaders ever held. The first Earth Summit had been held 20 years earlier in Stockholm, Sweden. At that time, the planet was divided into rival East and West blocs and was preoccupied with the perils of the nuclear arms race. With the collapse of the East bloc and the thawing of the cold war, a fundamental shift in the global base of power had occurred.

Today, the more important diversion, especially on environmental issues, is not between East and West but between "North" (Europe, North America, and Japan) and "South" (most of Asia, Africa, and Latin America). And, though the immediate threat of nuclear destruction has lifted, the planet is still at risk.

The idea behind the Earth Summit was that the relaxation of cold war tensions, combined with the growing awareness of ecological crises, offered a rare opportunity to persuade countries to look beyond their national interests and agree to some basic changes in the way they treat the environment. The major issues are clear: The developed countries of the North have grown accustomed to lifestyles that are consuming a disproportionate share of natural resources and generating the bulk of global pollution. Many of the developing countries of the South are consuming irreplaceable global resources to provide for their growing populations.

The Earth Summit was intended to promote better integration of nations' environmental goals with their economic aspirations. Although the hopes of some developing nations for large commitments of new foreign assistance did not fully materialize, much as accomplished during the Summit.

- The *Rio Declaration on Environment and Development* sets out 27 principles to guide the behavior of nations toward more environmentally sustainable patterns of development. The declaration, a compromise between developing and industrialized countries that was crafted at preparatory meetings, was adopted in Rio without negotiation due to fears that further debate would jeopardize any agreement.

- States at UNCED also adopted a voluntary action plan called *Agenda 21,* named because it is intended to provide an agenda for local, national, regional, and global action into the 21st century. UNCED Secretary General Maurice Strong called Agenda 21 "the most comprehensive, the most far-reaching and, if implemented, the most effective program of international action ever sanctioned by the international community." Agenda 21 includes hundreds of pages of recommended actions to address environmental problems and promote sustainable development. It also represents a process of building consensus on a "global work-plan" for the economic, social, and environmental tasks of the United Nations as they evolve over time.

- The third official product of UNCED was a *"non-legally binding authoritative statement of principles for a global consensus on the management, conservation, and sustainable development of all types of forests."* Negotiations on the forest statement, begun as negotiations for a legally binding convention on forests, were among the most difficult of the UNCED process. Many states and experts, dissatisfied with the end result, came away from UNCED seeking further negotiations toward agreement on a framework convention on forests.

Summary

People of different cultures view their place in the world from different perspectives. Among the things that shape their views are religious understandings, economic pressures, and fundamental knowledge of nature. Because of this diversity of backgrounds, different cultures put different values on the natural world and the individual organisms that compose it.

Three prevailing attitudes toward nature are the development ethic, which assumes that nature is for people to use for their own purposes; the preservation ethic, which assumes that nature has value in itself and should not be disturbed; and the conservation ethic, which recognizes that we will use nature but that it should be used in a sustainable manner.

Ethical issues can be examined at several levels. Growth and exploitation have been the prevailing priorities of our society for generations. This does not mean that everyone in society has the same opinions, but the general attitude has been one of the development rather than preservation. Most environmental decisions have really been economic decisions, and the rationale has been: If a resource is available for use, it should be used.

Corporate ethics are even more strongly influenced by economics. Corporations exist to make a profit. Any way that they can reduce costs makes them more profitable. Unfortunately, pollution and exploitation of rare resources may be costly to individuals or society while being profitable to corporations. In addition, corporations wield tremendous economic power and can sway public opinion and political will. Many corporations have begun to openly acknowledge their responsibilities to carefully examine their impact on the natural world.

Society and corporations are composed of individuals. An increasing sensitivity of individual citizens to environmental concerns can change the political and economic climate for society and corporations. However, people often do not have a clear idea

GLOBAL PERSPECTIVE

The Kyoto Protocol

On December 10, 1997, 160 nations reached agreement in Kyoto, Japan on limiting emissions of carbon dioxide and other greenhouse gases. The Kyoto Protocol was a significant agreement in focusing the attention of world leaders on the issues of climate change. The Kyoto Protocol called for the industrialized nations—the so-called Annex 1 countries—to reduce their average national emissions over the period 2008–2012 to about 5 percent below 1990 levels. The United States pledge is 7 percent below the 1990 level, slightly less than the European Union's pledge and slightly more than Japan's. None of the developing countries was required to set any limits.

The protocol initially covered only three greenhouse gases: carbon dioxide, methane, and nitrous oxide. Three more compounds—hydrofluorocarbons, perfluorocarbons, and sulfur hexafluoride—were to be added in subsequent years. The protocol also contains the elements of a program for international trading of greenhouse gas emissions. Such trading would employ market incentives to help ensure that the lowest-cost opportunities for reductions are pursued.

While the Kyoto climate agreement was important in bringing a new level of international attention to greenhouse gas emissions, there still remains many important issues to be resolved pertaining to the agreement. These include:

1. The rules and institutions that are to govern international trading of greenhouse gas emissions among Annex 1 countries must be better established.
2. The criteria used to judge compliance, and any penalties for non-compliance, must be clearly articulated.
3. To make longer-term objectives more credible, moderate but specific near-term goals should be set for Annex 1 countries and these

countries should be able to use early emissions reductions to meet longer-term requirements.

One thing is for sure, if the goals of the protocol are to be met by 2008 the developed nations of the world will be forced to focus attention on their energy policies and this in turn will affect all of us.

Key dates in the global warming story

1898—Swedish scientist Svante Ahrrenius warns that Industrial Revolution's carbon dioxide emissions, from coal and oil, could accumulate in atmosphere and lead to global warming.

1961—New observatory atop Hawaii's Mauna Loa volcano detects rise in atmospheric carbon dioxide.

1980s—Computer models of world climate project temperature rises.

1988—U.N. establishes authoritative network of climate scientists, Intergovernmental Panel on Climate Change.

1990—IPCC certifies scientific basis for "greenhouse effect" and global warming predictions.

1992—Climate change treaty signed, setting voluntary goals for industrial nations to lower greenhouse gas emissions to 1990 levels by 2000. Almost 170 nations eventually ratify.

October 1997—Negotiators end two years of preliminary tasks with major issues, including level of binding targets, unresolved and prepare for final conference in Kyoto.

Source: Data from "Cool Facts About Global Warming" USEPA. July 1998, Office of Policy, Planning and Evaluation.

---

of what should be done and often do not act in a way that supports their stated beliefs.

Global environmental concerns have become more important. The world is getting "smaller" and more interrelated. As more people are added to the world's population each year, there is increasing competition for the resources needed to live a decent life. An environmental disaster is no longer a local problem but affects us globally. The increasing economic difference between rich and poor nations affects the global environment, since the poor aspire to have what the rich take for granted. All peoples and nations need to work together to solve environmental problems.

Review Questions

1. How does personal wealth relate to ethics? Can you provide personal examples?
2. Why do industries pollute?
3. Why would normal economic forces work against pollution control? Do you feel that this is changing?

4. Is it reasonable to expect a totally unpolluted environment? Why or why not?
5. What has been the dominant societal attitude toward resource use?
6. Describe the differences between development, preservation, and conservation ethics. Must there always be conflict among these ethics?
7. What is a major motivating force of corporate management?
8. Why do decision makers view the actions of corporations differently from the way they view the actions of individuals?

Critical Thinking Questions

1. Using the definitions of moral and ethical judgment as presented in the text, identify at least two moral and ethical responses each to the issue of global climate change. What values, beliefs, and perspectives are at the root of these judgments?
2. What are our responsibilities to future generations regarding the environment? What values, beliefs, and perspectives lead you to think the way you do?
3. Compare and contrast the three approaches to environmental ethics outlined in the text. Which is closest to your own? Why? How does, and how could, your ethical stance influence your actions?

4. The text makes the point that up until recently humans have believed, almost universally, in unchecked growth as a positive good. Now, at the end of the 20th century, some are beginning to question this belief. What values, beliefs, and perspectives might these critics have? Describe some ways these critics might be received in a developing country. Why?

5. Imagine you are a business executive who wants to pursue an environmental policy for your company that limits pollution and uses fewer raw materials, but would cost more. What might be the discussion at your next board of directors meeting? How would you respond to your board of directors and shareholders? Why?

6. In 1997, Ojibwa Indians in northern Wisconsin sat on railroad tracks to block a shipment of sulfuric acid across their reservation that was headed for a controversial injection copper mine in northern Michigan. Try and put yourself in their position. What values, beliefs, and perspectives might have contributed to this action? Now put yourself in the position of the copper miners in northern Michigan. How might these copper miners have responded? What values, beliefs, and perspectives contribute to their action?

7. Read the Environmental Close-Up about environmental philosophy. Do you feel there is too much talk about environmental problems and not enough action? Or too little talk? Or some other problem? Please describe your position on this and your reasons for thinking the way you do.

8. Imagine yourself in the position of a person who lives on a poor Native American reservation that is contemplating building a storage facility for nuclear waste. What preconceptions, values, beliefs, or contextual perspectives might you bring to the issue? What might you propose as a course of action for yourself and for others? Why?

Key Terms

anthropocentric *17*
biocentric *18*
conservation ethic *20*
corporation *22*
development ethic *19*
ecocentrism *18*
economic growth *22*
environmental justice *26*

ethics *18*
industrial ecology *26*
morals *18*
preservation ethic *20*
profitability *22*
resource exploitation *22*

Related Internet Sites

Visit our website at http://www.mhhe.com/environmentalscience/ for specific information on each of the topics below.

Philosophical questions on environmental issues and ethics

EnviroLink Network on environmental resources

Indigenous treaties versus mining rights

Mining-related research, books, reports, and legislation

Environmental ethics

Philosophy and the environment

Wildlife ecologist Aldo Leopold

John Muir Education Project

PART TWO

Ecological Principles and Their Application

You can better understand many of the environmental problems that have surfaced over the past 25 years if you clearly understand how organisms interact with their physical environment and with each other. **Chapter 3** deals with energy principles and the structure of matter. **Chapters 4, 5,** and **6** present principles of ecology and population biology that can be applied to specific problems we face as a result of a growing world population putting demands on shrinking resources. **Chapter 7** focuses on human population issues. It specifically relates population growth to social and economic problems.

3
chapter

Interrelated Scientific Principles: Matter, Energy, and Environment

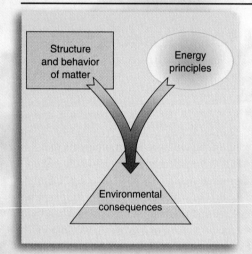

OBJECTIVES

After reading this chapter, you should be able to:

- Understand that science is exact because information is gathered in a manner that requires evaluation and revision.
- Understand that environmental science is a new discipline that includes both applied and theoretical aspects of traditional science, and that social, economic, and political aspects are also involved.
- Understand that matter is made up of atoms that have a specific subatomic structure of protons, neutrons, and electrons.
- Recognize that different elements have different atomic structures and that different atoms of the same element may differ in the number of neutrons present.
- Recognize that atoms may be combined and held together by chemical bonds to produce molecules.
- Understand that rearranging chemical bonds results in chemical reactions and that these reactions are associated with energy changes.
- Recognize that matter may be solid, liquid, or gas, depending on the amount of kinetic energy contained in the molecules.
- Realize that energy can be neither created nor destroyed, but when energy is converted from one form to another, some energy is converted into a useless form.
- Understand that energy can be of different qualities.

CHAPTER OUTLINE

Scientific Thinking

Environmental Close-Up: *Typical Household Chemicals*

Limitations of Science

The Structure of Matter
    Atomic Structure
    Molecules and Mixtures
    Acids, Bases, and pH
    Inorganic and Organic Matter
    Chemical Reactions
    Chemical Reactions in Living Things

Energy Principles
    States of Matter
    Kinds of Energy
    First and Second Laws of
        Thermodynamics
    Environmental Implications of
        Energy Flow

Issues and Analysis: *Improvements in Lighting Efficiency*

SCIENTIFIC THINKING

Since environmental science involves the analysis of data, it is useful to understand how scientists gather and evaluate information. It is also important to understand some chemical and physical principles as a background for evaluating environmental issues. An understanding of these scientific principles will also help you appreciate the ecological concepts in the chapters that follow.

The word *science* creates a variety of images in the mind. Some people feel that it is a powerful word and are threatened by it. Others are baffled by scientific topics and have developed an unrealistic belief that scientists are brilliant individuals who can solve any problem. For example, there are those who believe that the conservation of fossil fuels is unnecessary because scientists will soon "find" a replacement energy source. Similarly, many are convinced that if government really "wanted to" it would allocate sufficient funds to allow scientists to find a cure for AIDS. Such images do not accurately portray what science is really like. **Science** is a body of knowledge characterized by the requirement that information be gathered and evaluated by impartial testing of hypotheses and that it be shared so that it can be evaluated by others. It should also be remembered that science is based on a collection of information. The **scientific method** of gathering information generally involves observation, hypothesis formation, hypothesis testing, critical evaluation of results, and the publishing of findings. (See figure 3.1.) Underlying all of these activities is constant attention to accuracy and freedom from bias.

Observation simply means the ability to notice something. Sometimes, the observation is made with the unaided senses— we see, feel, or smell something. Often, machines such as microscopes, chemical analyzers, or radiation detectors may be used to extend our senses. Because these machines are complicated, we might get the feeling that science is incredibly complex, when in reality the questions being asked are relatively easy to understand.

A microscope has several knobs to turn and a specially designed light source.

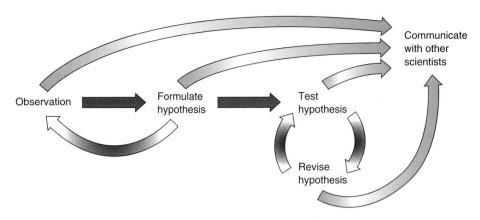

FIGURE 3.1 Elements of the Scientific Method The scientific method consists of several kinds of activities. Observation of a natural phenomenon is usually the first step. This is followed by the construction of a hypothesis that explains why the phenomenon occurred. The hypothesis is then tested to see if it is supported. Often this involves experimentation. If the hypothesis is not substantiated, it is modified and tested in its new form. It is important at all times that others in the scientific community be informed by publishing observations of unusual events, their probable cause, and the results of experiments that test hypotheses.

It requires considerable skill to use properly, but it is essentially a fancy magnifying glass that allows small objects to be seen more clearly. The microscope has enabled scientists to answer some relatively simple questions such as, Are there living things in pond water? and Are living things made up of smaller subunits? Similarly, a meter stick allows us to measure distance or a pH meter to measure a chemical property of a solution. Both are simple activities, but if we are not familiar with the units of measure, we might consider the processes hard to understand.

When you have formulated a question that needs scientific investigation, your first step is to form a hypothesis. A **hypothesis** is a logical statement that explains an event or answers a question. A good hypothesis should be as simple as possible, while taking all of the known facts into account. Furthermore, a hypothesis must be testable. In other words, you must be able to support it or prove it incorrect. The construction and testing of hypotheses is one of the most difficult (creative) aspects of the scientific method. Often, artificial situations must be constructed to test hypotheses. These are called **experiments.** A standard kind of experiment is one called a **controlled experiment.** Two groups are created that are identical in all respects except one. The control group has nothing out of the ordinary done to it. The experimental group has one thing different. If the

experimental group gives different results from the control group, those results must be caused by the single difference (variable) between the two groups.

The results of a well-designed experiment should be able to support or disprove a hypothesis. However, this does not always occur. Sometimes the results of an experiment are inconclusive. This means that a new experiment has to be conducted or that more information has to be collected. Often, it is necessary to have large amounts of information before a decision can be made about the validity of a hypothesis. The public often finds it difficult to understand why it is necessary to perform experiments on so many subjects, or why it is necessary to repeat experiments again and again.

The concept of **repeatability** is important to the scientific method. Because it is often not easy for scientists to eliminate unconscious bias, independent investigators must repeat the experiment to see if they get the same results. To do this, they must have a complete and accurate written document to work from. That means the scientists must publish the results of their experiment. This process of publishing results for others to examine and criticize is one of the most important steps in the process of scientific discovery. If a hypothesis is supported by many experiments and by different investigators, it is considered reliable.

ENVIRONMENTAL CLOSE-UP

Typical Household Chemicals

Modern society uses many different kinds of chemicals. A survey of a typical household would probably yield that following inorganic chemicals:

Common name	Chemical name	Use
Table salt	Sodium chloride, NaCl	Flavor
Saltpeter	Potassium nitrate, KNO_3	Preservative
Baking soda	Sodium bicarbonate, $NaHCO_3$	Leavening agent
Ammonia	Ammonia, NH_3	Disinfectant
Bleach	Sodium hypochlorite, NaHClO	Bleaching
Lye	Sodium hydroxide, NaOH	Drain cleaner

Other products we use contain mixtures of inorganic chemicals. Fertilizers are good examples. They usually contain a nitrate such as ammonium nitrate (NH_4NO_3), a phosphate compound (PO_4^{3+}), and potash, which is potassium oxide (K_2O).

In addition, we use a vast array of organic chemicals: ethyl alcohol in alcoholic beverages, acetic acid in vinegar, methyl alcohol for fuel, and cream of tartar (tartaric acid) for flavoring. We also use many complex mixtures of organic molecules in flavorings, pesticides, cleaners, and other applications.

Most of us know very little about the activities of the molecules we use. Many of them can be dangerous if used improperly. Fertilizer is poisonous, caustic soda can cause severe burns, and bleach or ammonia in high enough concentrations can damage skin or other tissues. The disposal of unused or unwanted household chemicals is a problem. Many of them should not just be dumped down the sink but should be disposed of in such a way that the material is converted to a harmless product or stored in a secure place. Unfortunately, most people do not know how to dispose of unwanted chemicals. For this reason, many manufacturers of household chemicals that have a potential to cause harm print statements on the containers explaining how to properly dispose of the unused product and the container. In addition, many communities have regular cleanup efforts for household hazardous waste, in which volunteers who know the contents of such products help determine how to dispose of them properly.

CARPET BEETLES—Thoroughly apply as a spot treatment. Spray along baseboards and edges of carpeting, under carpeting, rugs and furniture, in closets and on shelving, or wherever these insects are seen or suspected. **FLEAS, BROWN DOG TICKS**—Remove soiled bed bedding and clean thoroughly or destroy. Spray sleeping quarters of pets, along baseboards, windows, door frames, cracks and crevices, carpets, rugs, floors where these pests may be found. Put fresh bedding in pet quarters after spray has dried. **DO NOT SPRAY ANIMALS.** Pets should be treated with FLEA-B-GON® Flea Killer (aerosol) or ORTHO Pet Flea & Tick Spray Formula II (pump spray).
STORAGE: To store, rotate nozzle to closed position. Keep pesticide in original container. Do not put concentrate or dilute into food or drink containers. Avoid contamination of feed and foodstuffs. Store in a cool, dry place, preferably in a locked storage area.
DISPOSAL: PRODUCT—Partially filled bottle may be disposed of by securely wrapping original container in several layers of newspaper and discard in trash. **CONTAINER**—Do not reuse empty bottle. Rinse thoroughly before discarding in trash.

NOTICE: Buyer assumes all responsibility for safety and use not in accordance with directions.

Chevron Chemical Company © 1984
Ortho Consumer Products Division
P.O. Box 5047   San Ramon CA 94583-0947
Form 10152-N   Product 5466   Made in U.S.A.
EPA Reg. No. 239-2490-AA
EPA Est. 239-IA-3

0   8
7 1549 01980   C

A hypothesis that has survived repeated examination by many investigators over a long time and that has central importance to an area of science may become known as a **law.** A law also includes exact formulation. For example, the **biogenetic law** states that all living things are produced by previously living organisms. This has been tested repeatedly over hundreds of years and there have been no exceptions. Broadly written statements that cover large bodies of scientific knowledge are often called **theories.** A theory is well thought out and has lots of evidence for its support. For example, the **theory of evolution** holds that the characteristics of plants and animals and their kinds change over time. A theory is generally accepted by scientists to be true, but it cannot be proved true in *every* case because it is impossible to test *every* case. It is important to recognize that the word *theory* is often used in a much less restrictive sense. Often it is used to describe a vague idea or a hunch. This is not a theory in the scientific sense. So when you see or hear the word *theory* you must look at the context to see if the speaker or writer is referring to a theory in the scientific sense.

Now that we have some idea of how the scientific method works, let's look at an example. In many rivers in industrial parts of the world, it is possible to notice tumors of the skin and liver in the fish that live in the rivers (*observation*). This raises the question of what causes the tumors. Many people feel that the tumors are caused by the toxic chemicals that have been released into the rivers by industrial plants (*hypothesis*). Now, how could an experiment be conducted to test the hypothesis? If an industrial plant is suspected of releasing toxic chemicals that cause tumors, resident species of fish that do not migrate can be collected upstream and downstream from the plant's wastewater discharge pipes (outfall). Fish collected above the outfall constitute the control group, and those collected below the outfall constitute the experimental group. Large numbers of fish would have to be collected and examined. If the fish below the outfall have significantly more tumors than those above the outfall, it is because of where they live in the river and so the toxic chemicals from the industrial plants are probably the cause of the tumors. This is particularly true if the chemicals are already known to cause tumors. After the data were evaluated, the results of the experiment would be published. Certainly, the owners of the industrial plants would want to look at the data and might want to repeat the experiment to see if they get the same results.

LIMITATIONS OF SCIENCE

Science is a powerful tool for developing an understanding of the natural world, but it cannot analyze international politics, decide if family-planning programs should be instituted, or evaluate the significance of a beautiful landscape. These tasks are beyond the scope of scientific investigation. This does not mean that scientists cannot comment on such issues. They often do. But they should not be regarded as more knowledgeable on these issues just because they are scientists. Scientists may know more about the scientific aspects of these issues, but they struggle with the same moral and ethical questions that face all people, and their judgments on these matters can be just as faulty as anyone else's.

It is important to differentiate between the scientific data collected and the opinions scientists have about what the data mean. Scientists form and state opinions that may not always be supported by fact, just as other people do. Equally reputable

scientists commonly state opinions that are in direct contradiction. This is especially true in environmental science, where predictions about the future must be based on inadequate or fragmentary data. The issue of global warming (covered in chapter 17) is an example of this.

It is important to recognize that some knowledge can be used in both a scientific and a nonscientific manner. For example, it is a fact that many of the kinds of chemicals used in modern agriculture are toxic to humans and other animals. This does not mean that foods grown with the use of these chemicals are less nutritious or that "organically grown" foods are necessarily more nutritious because they have been grown without agricultural chemicals. The idea that something that is artificial is necessarily bad and something natural is necessarily good is false. After all, tobacco, poison ivy, and the prickly cactus are natural, while chemical fertilizers account for a large proportion of the food grown in the world. It is often very easy to jump to conclusions or confuse fact with hypotheis. This is particularly true when we generalize.

THE STRUCTURE OF MATTER

Now that we have an appreciation for the methods of science, it is time to explore some basic information and theories about the structure and function of various kinds of matter. **Matter** is anything that takes up space and has mass. Air, water, trees, cement, and gold are all examples of matter. A central theory that describes the structure and activity of matter is the **kinetic molecular theory.** This theory states that all matter is made up of tiny particles that are in constant motion. Although different kinds of matter have different properties, they are similar in one fundamental way. They are all made up of one or more kinds of smaller subunits called atoms.

Atomic Structure

Atoms are the fundamental subunits of matter. They are made up of protons, neutrons, and electrons. There are 92 kinds of atoms found in nature. Each kind forms a

specific type of matter known as an **element.** Gold (Au), oxygen (O), and mercury (Hg) are examples of elements. All atoms are composed of a central region known as a **nucleus,** which is composed of two kinds of relatively heavy particles: positively charged particles called **protons** and uncharged particles called **neutrons.** Surrounding the nucleus are clouds of relatively lightweight, fast-moving, negatively charged particles called **electrons.** As mentioned earlier, each kind of element is composed of a different kind of atom. The elements differ from one another in the number of protons, neutrons, and electrons in their atoms. For example, a typical atom of mercury contains 80 protons and 80 electrons; gold has 79, and oxygen only eight. (See figure 3.2.) All atoms of an element always have the same number of protons and electrons, but the number of neutrons may vary from one atom to the next. Atoms of the same element that differ from one another in the number of neutrons they contain are called **isotopes.**

Molecules and Mixtures

Atoms can be bonded to one another into stable units called **molecules.** When two or more different kinds of atoms are bonded to one another, the kind of matter formed is called a **compound.** While only

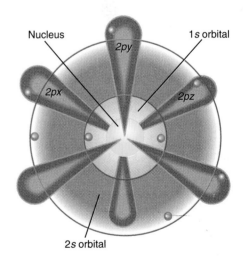

FIGURE 3.2 Diagrammatic Oxygen Atom
Most oxygen atoms are composed of a nucleus containing eight positively charged protons and eight neutrons without charges. Eight negatively charged electrons spin around the nucleus.

FIGURE 3.3 **Water Molecule** Water molecules are an atom of oxygen bonded to two atoms of hydrogen, with the hydrogen atoms located on one side of the oxygen.

92 kinds of atoms are commonly found, there are millions of ways atoms can be combined to form compounds. Water (H_2O), sugar ($C_6H_{12}O_6$), salt (NaCl), and methane gas (CH_4) are examples of compounds.

Many other kinds of matter are **mixtures,** variable combinations of atoms or molecules. Honey is a mixture of several sugars and water; concrete is a mixture of cement, sand, gravel, and reinforcing rod; and air is a mixture of several gases of which the most common are nitrogen and oxygen.

Acids, Bases, and pH

Acids and bases are two classes of compounds that are of special interest. Their characteristics are determined by the nature of their chemical bonds. When acids are dissolved in water, hydrogen ions (H^+) are set free. An **ion** is an atom or molecule that has gained or lost one or more electrons and, therefore, has either a positive charge or a negative charge. A *hydrogen ion* is positive because it has lost its electron and now has only the positive charge of the proton. An **acid** is any ionic compound that releases hydrogen ions in a solution. One other way of thinking of an acid is that it is a substance able to donate a proton to a solution. This is only part of the definition of an acid. We also think of acids as compounds that act like the hydrogen ion: they attract negatively charged particles. An example of a common acid is the sulfuric acid (H_2SO_4) in our automobile batteries.

A **base** is the opposite of an acid in that it is an ionic compound that releases a group known as a **hydroxyl ion,** or OH^- group. This group is composed of an oxygen atom and a hydrogen atom bonded together but with an additional electron. The hydroxyl ion is negatively charged. It is a base because it is able to donate electrons to the solution. A base can also be thought of as any substance able to attract positively charged particles. A very strong base used in oven cleaners is NaOH, or sodium hydroxide.

The strength of an acid or base is represented by a number called its **pH** number. The pH scale is a measure of hydrogen ion concentration. A pH of seven indicates that the solution is neutral and has an equal number of H^+ ions and OH^- ions to balance each other. As the number of hydrogen ions in the solution increases, the pH number gets smaller. The lower the number, the stronger the acid. A number higher than seven indicates that the solution has more OH^- than H^+. As the number of hydroxyl ions increases, the pH number gets larger. The higher the number, the stronger the base.

Inorganic and Organic Matter

Inorganic and organic matter are usually distinguished from one another by one fact: organic matter consists of molecules that contain carbon atoms bonded to form chains or rings. Consequently, organic molecules are usually large. Many different kinds of organic molecules are possible. Inorganic molecules are generally small and are of relatively few kinds. All living things contain organic molecules. They must either be able to manufacture organic molecules from inorganic molecules or be able to modify organic molecules they obtain from eating organic material. Typically, organic molecules contain a large amount of chemical energy that can be released when they are broken down to inorganic molecules. Salt, water, metals, sand, and oxygen are examples of inorganic matter. Sugars, proteins, fats, natural gas, gasoline, and coal are all examples of organic matter.

Chemical Reactions

The atoms within a molecule are held together by chemical bonds. (See figure 3.3.) **Chemical bonds** are physical attractions between atoms resulting from the interaction of their electrons. When chemical bonds are broken or formed, a chemical reaction occurs. When it does, the amount of energy within the chemical bonds changes, and some of the energy may be released as heat and light. A common example is the burning of natural gas. The primary ingredient in natural gas is the compound methane. When methane and oxygen are mixed together and a small amount of energy is used to start the reaction, the chemical bonds in the methane and oxygen (reactants) are rearranged to form two new compounds, carbon dioxide and water (products). In this kind of reaction, some chemical-bond energy is left over; it is released as light and heat. (See figure 3.4.) In every reaction, the amount of energy in the reactants and in the products can be compared and the differences accounted for by energy loss or gain. Even energy-yielding reactions usually need an input of energy to get the reaction started. This initial input of energy is called **activation energy.** In certain cases, the amount of activation energy required to start the reaction can be reduced by the use of a catalyst. A **catalyst** is a substance that alters the rate of a reaction, but the catalyst itself is not altered in the process. Catalysts are used in catalytic converters, which are attached to automobile exhaust systems. The purpose of the catalytic converter is to bring about more complete burning of the fuel, thus

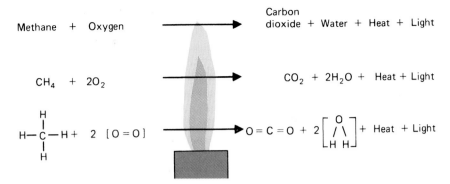

$$\text{Methane} + \text{Oxygen} \longrightarrow \text{Carbon dioxide} + \text{Water} + \text{Heat} + \text{Light}$$

$$CH_4 + 2O_2 \longrightarrow CO_2 + 2H_2O + \text{Heat} + \text{Light}$$

FIGURE 3.4 A Chemical Reaction When methane is burned, chemical bonds are changed, and the excess chemical bond energy is released as light and heat. The same atoms are present, but they are bonded in different ways, resulting in different molecules.

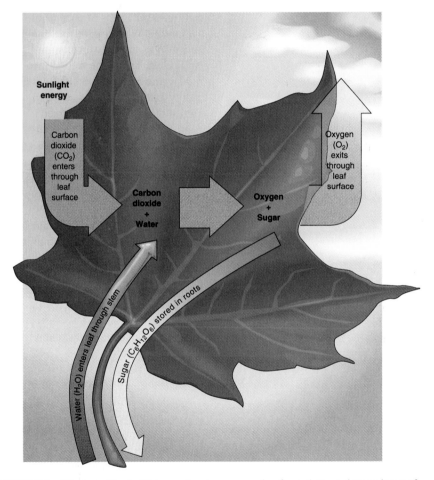

FIGURE 3.5 Photosynthesis This reaction is an example of one that requires an input of energy (sunlight) to combine low-energy molecules (CO_2 and H_2O) to form sugar with a greater amount of chemical bond energy. Oxygen is also produced.

resulting in less air pollution. Most of the materials that are not completely burned by the engine require high temperatures to react further; with the presence of catalysts, these reactions can occur at lower temperatures.

Chemical Reactions in Living Things

Living things contain catalysts called **enzymes,** which are composed of proteins. Most chemical reactions that occur in organisms are assisted by enzymes that reduce the activation energy needed to start the reactions. This is important since the high temperatures required to start these reactions without enzymes would destroy living organisms. Many of these enzymes are arranged in such a way that they cooperate in controlling a chain of reactions, as in photosynthesis and respiration.

Photosynthesis is the process plants use to convert inorganic material into organic matter, using the assistance of light energy. Light energy enables the smaller inorganic molecules (water and carbon dioxide) to be converted into organic sugar molecules. In the process, oxygen is released. Photosynthesis takes place in the green portions of the plant, usually the leaves. (See figure 3.5.) The organic molecules produced as a result of photosynthesis can be used as a source of energy by the plants and by organisms that eat the plants.

Respiration involves the use of oxygen to break down large, organic molecules (sugars, fats, and proteins) into smaller, inorganic molecules (carbon dioxide and water). This process releases energy the organisms can use. (See figure 3.6.) All organisms must carry on some form of respiration, since all need a source of energy to maintain life.

ENERGY PRINCIPLES

The previous section started out with a description of matter, yet it used the concepts of energy to describe chemical bonds, chemical reactions, and the movement of atoms and molecules. That is because energy and matter are inseparable. It is difficult to describe one without the other. **Energy** is the ability to do work, which typically results in matter being moved over a distance. This occurs even at the molecular level.

States of Matter

Depending on the amount of energy present, matter can occur in three states: solid, liquid, or gas. The physical nature of matter changes when the amount of kinetic energy its molecules contain changes, but the chemical nature of matter remains the

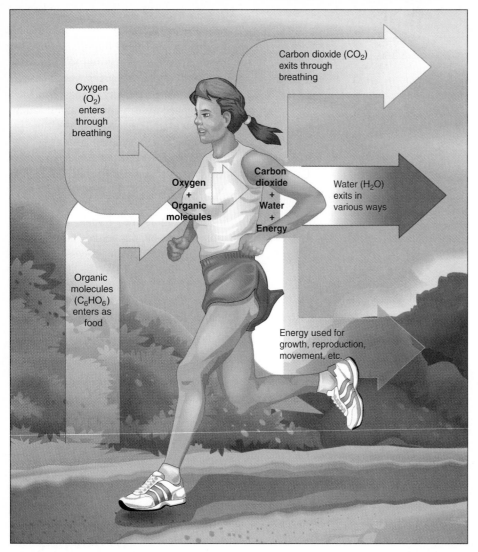

FIGURE 3.6 Respiration Respiration involves the release of energy from organic molecules when they react with oxygen. In addition to providing energy in a usable form, respiration produces carbon dioxide and water.

same. For example, water vapor, liquid water, and ice all have the same chemical composition but differ in the arrangement and activity of their molecules. The amount of kinetic energy molecules have determines how rapidly they move. (See figure 3.7.) In solids, the molecules have low amounts of energy, and they vibrate in place very close to one another. In liquids, the higher-energy molecules are farther apart and will roll, tumble, and flow over each other. The molecules of gases move very rapidly and are very far apart. All that is necessary to change the physical nature of a type of matter is an energy change. Heat energy must be added or removed.

Kinds of Energy

There are several kinds of energy. Heat, light, electricity, and chemical energy are common forms. The energy contained by moving objects is called **kinetic energy.** The moving molecules in air have kinetic energy, as does water running downhill or a dog chasing a ball. **Potential energy** is in a special category; it is the energy matter has because of its position. The water behind a dam has potential energy by virtue of its elevated position. (See figure 3.8.) An electron at some distance from the nucleus has potential energy due to the distance between the electron and the nucleus.

First and Second Laws of Thermodynamics

Energy can exist in several different forms, and it is possible to convert one kind of energy into another. However, the total amount of energy remains constant. The **first law of thermodynamics** states that energy can neither be created nor destroyed; it can only be changed from one form into another. From a human perspective, some forms of energy are more useful than others. We tend to make extensive use of electrical energy for a variety of purposes, but there is very little electrical energy present in nature. Therefore, we convert other forms of energy into electrical energy. When converting energy from one form to another, some of the useful energy is lost. This is the **second law of thermodynamics.** There is no loss of *total* energy, but there is a loss of *useful* energy. For example, coal can be burned in a power plant to produce electrical energy. However, large amounts of useless heat energy are also produced. Therefore, the amount of useful energy (electricity) is much less than the total amount of chemical energy present in the coal. (See figure 3.9.)

Energy is being converted from one form to another continuously within the universe. Stars are converting nuclear energy into heat and light. Animals are converting the chemical energy found in food into the kinetic energy of motion. Plants are converting sunlight energy into the chemical bond energy of sugar molecules. In each of these cases, some useless energy is produced, generally in the form of heat.

Environmental Implications of Energy Flow

The heat produced when energy conversions occur is dissipated throughout the universe. This is a common experience. Valuable things always disintegrate unless we work to maintain them. Houses fall into ruin, automobiles rust, and appliances wear out. In reality, all of these phenomena involve the loss of heat. The organisms that decompose the wood in our houses release heat. The chemical reaction that causes rust releases heat. Friction, caused by the movement of parts of a machine

FIGURE 3.7 States of Matter Matter exists in one of three states, depending on the amount of kinetic energy the molecules have. The higher the amount of energy, the greater the distance between molecules and the greater their degree of freedom of movement.

FIGURE 3.8 Kinetic and Potential Energy Kinetic and potential energy are interconvertible. The potential energy possessed by the water behind a dam is converted to kinetic energy as the water flows to a lower level.

against each other, generates heat and causes the parts to wear.

Orderly arrangements of matter, such as clothing, automobiles, or living organisms, always tend to become disordered. Eventually, nonliving objects wear out and living things die and decompose. This process of becoming more disordered coincides with the constant flow of energy toward a dilute form of heat. This dissipated, low-quality heat has little value to us.

It is important to understand that different energy forms are of different quality. Some are of high quality, such as electrical energy, which can be easily used to perform a variety of useful actions. Some are of low quality, such as the heat in the water of the ocean. Although the total *quantity* of heat energy in the ocean is much greater than the total amount of electrical energy in the world, little can be done with the heat energy in the ocean because it is of low *quality*. Therefore, it is not as valuable as other forms of energy that can be used to do work for us.

The reason the heat of the ocean is of little value involves the small temperature difference between two sources of heat. When two objects differ in temperature, heat will flow from the warmer to the cooler object. The greater the temperature difference, the more useful the work that can be done. For example, fossil-fuel power plants burn fuel to heat water and convert it to steam. Cold cooling-water provides a steep temperature gradient, and heat energy flows from the steam to the cold water. That causes a turbine to turn, which generates electricity. Because the average temperature of the ocean is not high, and it is difficult to find another object that has a greatly lower temperature than the ocean, the huge heat content of the ocean cannot do useful work for us.

These quantitative and qualitative factors are also evident in the energy expended by a stream as the water runs downhill. The steeper the slope, the greater the amount of energy expended per kilometer of its length. If there is no point along the stream where the slope is very steep, the stream has low-quality energy, because the energy is dissipated along the entire length of the stream. To make this a high-quality (concentrated) source of energy, the water must be dammed so that it will drop a long distance at one point. This means that it will give up much of its energy over a short distance. With damming, the *quantity* of energy has not changed but the *quality* has.

Because of the second law of thermodynamics, all organisms, including humans, are in the process of converting high-quality energy into low-quality energy. Waste heat is produced when the chemical-bond energy in food is converted into the energy needed to move, grow, or respond. The process of releasing chemical-bond energy from food by organisms is known as cellular respiration. From an energy point of view, it is similar to the process of **combustion,** which is the burning of fuel to obtain heat, light, or some other form of useful energy. The efficiency of cellular respiration is relatively high. About 40 percent of the energy contained in food is released in a useful form. The rest is dissipated as low-quality heat. Table 3.1 lists the efficiencies of many common energy conversion systems.

An unfortunate consequence of energy conversion is pollution. The heat lost from most energy conversions is a pollutant. The wear of the brakes used to stop cars results in pollution. The emissions from power plants pollute. All of these are examples of the effect of the second law of thermodynamics. If each individual on earth used less energy, there would be less waste heat and other forms of pollution that result from energy conversion. The amount of energy in the universe is limited. Only a small portion of that energy is of high quality. The use of high-quality energy decreases the amount of useful energy available, as more low-quality heat is generated. All life and all activities are subject to these important physical principles known as the first and second laws of thermodynamics.

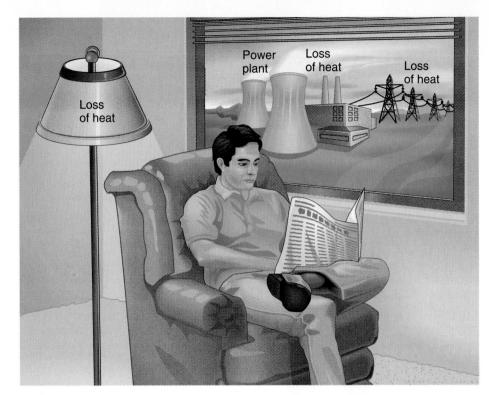

FIGURE 3.9 Second Law of Thermodynamics Whenever energy is converted from one form to another, some of the useful energy is lost, usually in the form of heat. The conversion of fuel to electricity produces heat, which is lost to the atmosphere. As the electricity moves through the wire, resistance generates some additional heat. When the electricity is converted to light in a light bulb, heat is produced as well. All of these steps produce useless heat as a result of the second law of thermodynamics.

TABLE 3.1	THE EFFICIENCY OF SOME ENERGY CONVERSION SYSTEMS	
Energy conversion system	**% efficiency***
Electric motor	93
Hydroelectric power plant	85
Home oil furnace	65
Fluorescent lamp	65
Steam-power plant	47
High-intensity lamp	32
Automobile engine	25
Incandescent lamp	4

*Efficiency with which the energy of the power or fuel source is converted to a useful form.
Source: Data from U.S. Department of Energy, 1998.

Summary

Science is a method of gathering and organizing information. It involves observation, hypothesis formation, the testing of hypotheses, and publication of the results for others to evaluate. A hypothesis is a logical prediction about how things work that must account for all the known information and be testable. The process of science attempts to be careful, unbiased, and reliable in the way information is collected and evaluated. This often involves conducting experiments to test the validity of a hypothesis. If a hypothesis is continually supported by the addition of new facts, it may become known as a law. A theory is a broadly written, widely accepted generalization that ties together large bodies of information.

In the United States, 36 percent of the energy consumed is electricity. Of that, nearly 40 percent is used for lighting. Improvements in the efficiency of lighting would significantly reduce the demand for electrical energy. All forms of lighting involve the conversion of electricity into light, but some systems are much more efficient than others. The incandescent light bulb is extremely inefficient yet is still used in many homes and commercial buildings. Compact fluorescent lights use about 25 percent of the energy of an incandescent bulb to produce the same amount of light and can be used in the standard incandescent light bulb socket. In many situations, such as commercial buildings, newer high-efficiency fluorescent lights with electronic ballasts can reduce energy consumption by about 15 percent over standard fluorescent bulbs.

In some cases, fluorescent lighting is not practical. It does not work well in the cold and in most situations cannot be used with dimmer switches. Other kinds of higher efficiency lighting are available, however. Halogen lights are incandescent lights with somewhat better efficiency than standard incandescent lights. Sodium vapor, mercury vapor, and metal halide lights are very efficient but produce a light of a different color than normal daylight. These lights also require several minutes to come up to full lighting power. They are used in places where color is not important and where they are not turned on and off repeatedly, such as exterior lighting in parking lots. The U.S. Department of Energy has helped develop a new sulfur lamp that is even more efficient than fluorescent lighting and has better color than other high-efficiency lamps.

Any improvement in the efficiency with which electricity is converted to light also reduces the amount of waste heat produced, which reduces the amount of electricity needed to cool buildings. Often the cost of replacing inefficient incandescent light bulbs is offset by subsidies from local electric utilities, since increased lighting efficiency reduces the demand for electricity and allows utilities to put off building expensive new power plants.

How many incandescent light bulbs do you have in your home? Why haven't they been replaced?

The fundamental unit of matter is the atom, which is made up of protons and neutrons in the nucleus and electrons circling the nucleus. The number of protons for any one type of atom is constant, but the number of neutrons in different atoms of the same type may vary. The number of electrons is equal to the number of protons. Protons have a positive charge, neutrons lack a charge, and electrons have a negative charge.

When two or more atoms combine with one another, they form stable units known as molecules. Chemical bonds are physical attractions between atoms resulting from the interaction of their electrons. When chemical bonds are broken or formed, a chemical reaction occurs, and the amount of energy within the chemical bonds is changed. Chemical reactions require activation energy to get the reaction started.

Matter that is composed of only one kind of atom is known as an element. Matter that is composed of molecules containing atoms bonded in specific ratios is known as a compound. Atoms or molecules that have gained or lost electrons so that they have an electric charge are known as ions.

Matter can occur in three states: solid, liquid, and gas. These three differ in the amount of energy the molecules contain and the distance between the molecules. Kinetic energy is the energy contained by moving objects. Potential energy is the energy an object has because of its position.

The first law of thermodynamics states that the amount of energy in the universe is constant, that energy can neither be created nor destroyed. The second law of thermodynamics states that when energy is converted from one form to another, some of the useful energy is lost. Some forms of energy are more useful than others. The quality of the energy determines how much useful work can be accomplished by expending the energy. Low-temperature heat sources are of poor quality, since they cannot be used to do useful work.

Review Questions

1. How do scientific disciplines differ from nonscientific disciplines?
2. What is a hypothesis? Why is it an important part of the way scientists think?
3. Why are events that happen only once difficult to analyze from a scientific point of view?
4. What is the scientific method, and what processes does it involve?
5. How are the second law of thermodynamics and pollution related?
6. Diagram an atom of oxygen and label its parts.
7. What happens to atoms during a chemical reaction?
8. State the first and second laws of thermodynamics.
9. How do solids, liquids, and gases differ from one another at the molecular level?
10. List five kinds of energy.
11. Are all kinds of energy equal in their capacity to bring about changes? Why or why not?

Critical Thinking Questions

1. You observe that a high percentage of frogs, especially sensitive to environmental poisons, in small ponds in your agricultural region have birth defects. Suspecting agricultural chemicals present in runoff to be the culprit, state the hypothesis in your own words. Next devise an experiment that might help you support or reject your hypothesis.
2. Given the experiment you proposed in Question 12, imagine some results that would support that hypothesis. Now imagine you are a different scientist that is very skeptical of the initial hypothesis. How convincing do you find this data? What other possible explanations (hypotheses) might there be to explain the results? Devise a different experiment to test this new hypothesis.
3. Increasingly, environmental issues like global climate change are moving to the forefront of world concern. What role should science play in public policy decisions? How should we decide between competing scientific explanations about an environmental concern

like global climate change? What might be some of the criteria for deciding what is "good science" and what is "bad science"?

4. How important are the first and second laws of thermodynamics to explain environmental issues? Using the concepts in these laws of thermodynamics, try and explain a particular environmental issue. How does an understanding of thermodynamics change your conceptual framework regarding this issue?

5. The text points out that incandescent light bulbs are only 4% efficient at using energy to accomplish their task, while new, initially more expensive, compact fluorescent lighting uses significantly less electricity to provide the same quantity of light. Examine the contextual framework of those who advocate for new lighting methods and the contextual framework of those who continue to design and build using the old methods. What are the major differences in perspective? What could you suggest be done to help bring these different perspectives closer together?

6. Some scientists argue that living organisms constantly battle against the principles of the second law of thermodynamics using the principles of the first law of thermodynamics. What might they mean by this? Do you think this is accurate? What might be some of the implications of this for living organisms?

Key Terms

acid *40*

activation energy *40*

atom *39*

base *40*

biogenetic law *38*

catalyst *40*

chemical bond *40*

combustion *43*

compound *39*

controlled experiment *37*

electron *39*

element *39*

energy *41*

enzyme *41*

experiment *37*

first law of thermodynamics *42*

hydroxyl ion *40*

hypothesis *37*

ion *40*

isotope *39*

kinetic energy *42*

kinetic molecular theory *31*

law *38*

matter *39*

mixture *40*

molecule *39*

neutron *39*

nucleus *39*

observation *37*

pH *40*

photosynthesis *41*

potential energy *42*

proton *39*

repeatability *37*

respiration *41*

science *37*

scientific method *37*

second law of thermodynamics *42*

theory *38*

theory of evolution *38*

Related Internet Sites

Visit our website at http://www.mhhe.com/environmentalscience/ for specific information on each of the topics below.

Department of Energy's home page

Department of Energy's Alternative Fuels Data Center

Energy issues dealing with water, sun, wind, plant, and geothermal power

Efficiency and renewable energy networks

Database on chemical compounds

Chemical Science Technology Division at Los Alamos Laboratory

Renewable energy sources

Educational resources on all forms of energy

Interactive periodic table

Educational materials and programs about energy, natural resources, and the environment

Horticulture in a virtual perspective

Interactions: Environment and Organisms

4

chapter

OBJECTIVES

After reading this chapter, you should be able to:

- Identify the abiotic and biotic factors in an ecosystem.
- Define niche.
- Describe the process of natural selection as it operates to refine the fit between organism, habitat, and niche.
- Describe predator–prey, parasite–host, competitive, mutualistic, and commensalistic relationships.
- Differentiate between a community and an ecosystem.
- List some of the components of an ecosystem.
- Define the roles of producer, herbivore, carnivore, omnivore, scavenger, parasite, and decomposer.
- Describe energy flow in an ecosystem.
- Relate the concepts of food webs and food chains to trophic levels.
- Explain the cycling of nutrients such as nitrogen, carbon, and phosphorus through an ecosystem.

CHAPTER OUTLINE

Ecological Concepts
  Environment
  Limiting Factors
  Habitat and Niche

Environmental Close-up: *Habitat Conservation Plans: Tool or Token?*

The Role of Natural Selection and Evolution
  Species Definition
  The Mechanism of Natural Selection

Kinds of Organism Interactions
  Predation
  Competition
  Symbiotic Relationships

Environmental Close-up: *Human Interaction—A Different Look*

Community and Ecosystem Interactions
  Major Roles of Organisms
  Energy Flow Through Ecosystems
  Food Chains and Food Webs
  Nutrient Cycles in Ecosystems

Environmental Close-Up: *Name That Relationship*

Global Perspective: *Human Impact on Nutrient Cycles*

Environmental Close-Up: *Colorado River Restoration*

Environmental Close-Up: *Organic Contaminants in Great Lakes Fish*

Issues and Analysis: *Reintroducing Wolves to Yellowstone*

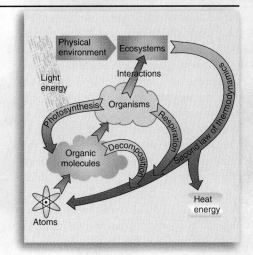

ECOLOGICAL CONCEPTS

The science of **ecology** is the study of the way organisms interact with each other and with their nonliving surroundings. It deals with the ways in which organisms are molded by their surroundings, how they make use of these surroundings, and how an area is altered by the presence and activities of organisms. These interactions involve energy and matter. Living things require a constant flow of energy and matter to assure their survival. If the flow of energy and matter ceases, the organisms die.

All organisms are dependent on other organisms in some way. One organism may eat another and use it for energy and raw materials. One organism may temporarily use another without harming it. One organism may provide a service for another, such as when animals distribute plant seeds or bacteria break down dead organic matter for reuse. The study of ecology can be divided into many specialties and be looked at from several levels of organization. (See figure 4.1.) Before we can explore the field of ecology in greater depth, we must become familiar with some of the standard vocabulary used.

Environment

Everything that affects an organism during its lifetime is collectively known as its **environment.** Environment is a very broad concept. For example, during its lifetime an animal such as a raccoon is likely to interact with millions of other organisms (bacteria, food organisms, parasites, mates, predators), drink copious amounts of water, breathe huge quantities of air, and respond to daily changes in temperature and humidity. This list only begins to describe the various components that make up the raccoon's environment. Because of this complexity, it is useful to subdivide the concept of environment into **abiotic** (nonliving) and **biotic** (living) **factors.**

Abiotic factors include the flow of energy necessary to maintain any organism, the physical factors that affect it, and the supply of molecules required for its life functions. The ultimate source of energy

FIGURE 4.1 **Ecology** The study of ecology is pursued at various levels, ranging from atoms and molecules to individual organisms to groups of interacting organisms to ecosystems.

for almost all organisms is the sun; in the case of plants, the sun directly supplies the energy necessary for them to maintain themselves. Animals obtain their energy by eating plants or other animals that eat plants. Ultimately, the amount of living material that can exist in an area is determined by the plants and the amount of energy they can trap.

Other physical factors include such things as climate (average weather patterns over a number of years); temperature (average annual temperature as well as daily variations); precipitation, including its type (rain, snow, hail), amount, and seasonal distribution; type of soil present (sandy or clay, dry or wet, fertile or infertile); and even the three-dimensional shape of the space the organism inhabits.

All forms of life require atoms such as carbon, nitrogen, and phosphorus, and molecules such as water, to construct and maintain themselves. Organisms constantly obtain these materials from their

environment by eating food or taking them up through the process of photosynthesis. The atoms are used for a period of time as part of the organism's body structure, and eventually are returned to the environment through respiration, excretion, or death and decay.

The biotic factors influencing an organism include all forms of life with which it interacts. Plants that carry on photosynthesis; animals that eat other organisms; bacteria and fungi that cause decay; and bacteria, viruses, and other parasitic organisms that cause disease are all part of an organism's biotic environment.

Limiting Factors

Although organisms interact with their surroundings in many ways, certain factors may be recognized as key to a particular species' success. A shortage or absence of a key factor restricts the success of the species; thus the factor is

FIGURE 4.2 **Limiting Factors** In aquatic habitats, the amount of oxygen dissolved in the water is often a limiting factor for many species of fish. Cool, highly oxygenated water, which is typical of the rapidly flowing upper sections of river system, supports trout, but warmer, less oxygenated water is unsuited for trout. Other fish, which are more tolerant of low levels of oxygen, such as bass, catfish, bullheads, and carp, occupy the lower sections of the river, where the water is warmer, there is less oxygen, and the river contains much silt and other soil particles.

known as a **limiting factor.** Limiting factors may be either abiotic or biotic and can be quite different from one species to another. Many plants are limited by scarcity of water or specific soil nutrients. Animals may be limited by climate or the availability of a specific food. For example, many snakes and lizards are limited to the warmer parts of the world, and Monarch butterflies require milkweed plants as food for their developing caterpillars.

The limiting factor for many species of fish is the amount of dissolved oxygen in the water. In a swiftly flowing, tree-lined mountain stream, the level of dissolved oxygen is high and so provides a favorable environment for trout. (See figure 4.2.) As the stream continues down the mountain, the steepness of the slope decreases, which results in fewer rapids where the water tumbles over rocks and becomes oxygenated. In addition, the canopy of trees over the stream usually is

thinner, allowing more sunlight to reach the stream and warm the water. Warm water cannot hold as much dissolved oxygen as cool water. Therefore, slower-flowing, warm-water streams contain less oxygen than rapidly moving, cool streams. Fish such as black bass and walleye are adapted to such areas, since they are able to tolerate lower oxygen concentrations and higher water temperatures. Trout are not able to survive under such conditions and are not found in warm, less well-oxygenated water. Each of these species has a specific **range of tolerance** to oxygen concentration and water temperature. Thus, low levels of oxygen and high water temperatures are limiting factors for the distribution of trout.

Other factors such as the abundance of silt may influence the ability of water to support certain species of fish. Silt reduces visibility, making it difficult for fish to find food, and covers gravel beds fish

need for spawning. Under these conditions, the bass and walleye may be replaced by such species as carp and catfish, which have an even greater ability to withstand high temperatures and low oxygen concentrations and are better able to survive in water with a high amount of silt.

The amount of grass available for food is a limiting factor for grazing animals; the saltiness (salinity) of water is limiting for many kinds of ocean animals and prevents their migration into freshwater; and the amount of available sunlight is a limiting factor for many kinds of plants.

Habitat and Niche

As we have just seen, it is impossible to understand an organism apart from its environment. The environment influences the organism, and organisms affect the environment. To focus attention on specific elements of this interaction, ecologists

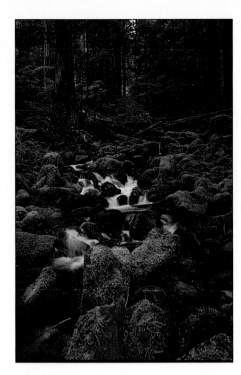

FIGURE 4.3 Moss Habitat The habitat of mosses must be cool and shady, since mosses tend to dry out and die if they are exposed to the drying action of the sun.

have developed two concepts that need careful explanation: habitat and niche.

The **habitat** of an organism is the space that the organism inhabits, the place where it lives (its address). We tend to identify an organism's habitat with particular physical environmental characteristics like soil type, availability of water, or climatic conditions, or we identify it with the predominant plant species that exist in the area. For example, mosses are small plants that dry out and die if they are exposed to sunlight, wind, and drought. Therefore, the habitat of moss is likely to be cool, moist, and shady. (See figure 4.3.) Likewise, a rapidly flowing, cool, well-oxygenated stream with many bottom-dwelling insects is good trout habitat, while open prairie with lots of grass is preferred by bison, prairie dogs, and many kinds of hawks and falcons. Elm bark beetles will reside only in areas where elm trees are found. The particular biological requirements of an organism determine the kind of habitat in which it is likely to be found.

The **niche** of an organism is the functional role it has in its surroundings (its profession). An organism's niche includes

FIGURE 4.4 Ecological Niche The ecological niche of an organism involves everything that affects it during its lifetime as well as all the effects it has. The ecological niche of beaver includes the need for streams with aspen trees nearby, the building of dams and flooding of forested areas, the killing of trees, the providing of ponds for ducks and other animals, and many other effects.

everything that affects the organism and everything affected by the organism during its lifetime. For example, beavers frequently flood areas by building dams of mud and sticks across streams. (See figure 4.4.) The flooding has several effects. It provides beavers with a larger area of deep water, which they need for protection; it provides a pond habitat for many other species of animals like ducks and fish; and it kills trees that cannot live in saturated soil. The animals attracted to the pond and the beavers often fall to

predators. After the beavers have eaten all the suitable food, such as aspen, they will abandon the pond and migrate to other areas along the stream and begin the whole process over again.

In this recitation of beaver characteristics, we have listed several effects that the animal has. It changes the physical environment by flooding, it kills trees, it enhances the environment for other animals, and it is a food source for predators. This is only a superficial glimpse of the many aspects of the beaver's interaction with its

ENVIRONMENTAL CLOSE-UP

Habitat Conservation Plans: Tool or Token?

The Endangered Species Act places strict regulations on the destruction of the habitat of an endangered species. Since endangered species are typically restricted to specific habitats and have very local distribution, specific patches of habitat are critical to their survival. Landowners feel that the presence of an endangered species on their land unfairly deprives the landowners of the use of their land, since they may not "incidentally take" (accidentally kill) members of the endangered species. Destruction of the habitat would constitute incidental take. Many landowners have argued that, since they have lost the use of their land due to the presence of an endangered species, they should be compensated by the government for the value they have lost. The cost to the government would be enormous. However, the Endangered Species Act allows for habitat conservation plans to be put in place which would allow the landowner some limited use of the land while assuring the protection of the endangered species.

The process of developing a habitat conservation plan is really a negotiated settlement that many conservationists consider totally inadequate. They argue that it is very difficult to anticipate all the critical features of the habitat since many of the species being "protected" are poorly understood. Therefore, habitat conservation plans are very likely to be inadequate to protect the species from extinction. Habitat conservation plans become binding documents and allow landowners to continue to operate under the habitat conservation plan even if new scientific information becomes available that demonstrates that the plan is inadequate. Landowners feel this is necessary because they should not be held responsible for unforeseen consequences of the plan. Critics feel that most plans are political compromises that do not protect the endangered species and that the plans make no provisions for the recovery of the species. And since the habitat is being affected, the ultimate fate of most species the plans are supposed to protect is likely to be extinction.

FIGURE 4.5 The Niche of a Dandelion A dandelion is a common plant that does well in disturbed sites with lots of sunlight. It is able to invade these areas easily because it produces many seeds that are blown easily to new areas. Lawn mowing benefits dandelions since it prevents shading by tall grasses and other plants.

environment. A complete catalog of all aspects of its niche would make up a separate book.

Another familiar organism is the dandelion. (See figure 4.5.) It is an opportunistic plant that rapidly becomes established in sunny, disturbed sites. In a few days it can produce thousands of parachutelike seeds that are easily carried by the wind over long distances. (You have probably helped this process by blowing on the fluffy white head of a mature dandelion flower.) Furthermore, it

often produces several sets of flowers per year. Since there are so many seeds and they are so easily distributed, the plant can easily establish itself in any sunny, disturbed site, including lawns. Since it is a plant, one major aspect of its niche is the ability to carry on photosynthesis and grow. Dandelions need a large amount of sunlight to grow successfully and do not do well in shady areas. Mowing lawns helps provide just the right conditions for dandelions, since the vegetation is never allowed to get so tall that dandelions are shaded. Many kinds of animals, including humans, use the plant for food. The young leaves may be eaten in a salad, and the blossoms can be used to make dandelion wine. Bees visit the flowers regularly to obtain nectar and pollen.

THE ROLE OF NATURAL SELECTION AND EVOLUTION

As we have seen, each organism is finely tuned to a particular habitat and has a very specific role (niche) within its habitat. But how is it that each plant, animal, fungus, or bacterium fits into its environment in such a precise way? The process that leads

to this close fit between organisms and their environment is known as natural selection. **Natural selection** is the process of more successful individuals surviving and reproducing larger numbers of offspring than those that are less successful. Those that reproduce more successfully pass on to the next generation the characteristics that made them successful. Thus each kind of organism is continually refined as each generation is subjected to the same environmental conditions.

Species Definition

It is important to recognize that natural selection happens only within a **species,** which is a population of organisms in which the individuals are potentially able to interbreed and produce fertile offspring. Individual organisms are members of a species. This definition of species contains two points that require explanation. Obviously, there are individuals in any population that never reproduce, and many pairs of individuals that will never meet one another. However, they still have the potential to interbreed and are considered members of the same species. The second point is the ability to produce fertile offspring. In some instances, two kinds of organisms may interbreed and produce offspring, but the offspring are sterile and never reproduce. For example, horses and donkeys can breed and produce offspring called mules, but since the mules are sterile, the horse and donkey are considered separate species.

Some species are easy to recognize. We easily recognize humans as a distinct species. Most people recognize a dandelion when they see it and do not confuse it with other kinds of plants that have yellow flowers. Other species are not as easy to recognize. Most of us cannot tell one species of mosquito from another or identify different species of grasses. Because of this we tend to lump organisms into large categories and do not recognize the many subtle niche differences that exist among the similar-appearing species. However, different species of mosquitoes are quite distinct from one another. Only certain ones carry the human disease malaria. Other species carry dog heartworm. Each species is active during certain portions of the day or night. And each species requires specific conditions to reproduce.

The Mechanism of Natural Selection

In 1859, Charles Darwin, a British naturalist, introduced the theory of natural selection in his book *On the Origin of Species by Means of Natural Selection, or the Preservation of Favored Races in the Struggle for Life.* As a young man Darwin had the opportunity to travel around the world as the naturalist on a British survey ship. Wherever Darwin visited, he saw many different kinds of plants and animals and noted how well they were adapted to their particular situations. This caused him to wonder about the mechanism that caused this high degree of adaptation. He concluded that several factors could interact to allow for natural selection:

1. Individuals within a species showed variation; some of the variations were very useful and others were not. For example, individual animals that are part of the same species show color variations. Some colors make the animal more conspicuous while others make it less conspicuous.

2. Organisms within a species typically produce many more offspring than can survive. This means that there is not enough suitable habitat for all of the offspring to grow to maturity. One apple tree may produce hundreds of apples with several seeds in each apple, or a pair of rabbits may have three to four litters of offspring each summer, with several young in each litter.

3. The excess number of individuals results in a struggle for survival. Individuals within the population must compete with each other for food, space, mates, or other requirements that are in limited supply. If you plant 100 bean seeds in a pot, many of them will begin to grow, but eventually some will become taller and shade the remaining plants. Great horned owls typically produce two young at a time, but if food is in short supply, the larger of the two young will get the majority of the food.

4. Because of variation among individuals, some have a greater chance of surviving and reproducing than others. The competition for resources often results in the less fit individuals dying; therefore, they do not get a chance to reproduce themselves. Even if they do not die, they may mature more slowly and not be able to reproduce as many times as the more fit members of the species.

5. As time passes and each generation is subjected to the same process, the percentage of individuals showing favorable variations will increase and those having unfavorable variations will decrease. Thus, the species will become better and better adapted to its environment.

Many people have tested the theory of natural selection, and it continues to be tested today. In a classic experiment, H.B.D. Kettlewell studied the processes that resulted in some areas of England having mostly dark-colored forms of the peppered moth, while other areas had mostly light-colored forms. He hypothesized that the color of tree trunks was a

FIGURE 4.6   Natural Selection in Peppered Moths Air pollution in industrial regions of England resulted in blackened tree trunks. The dark form of peppered moths was less conspicuous and, therefore, they were not seen as readily by their bird predators. The light form was more easily seen and more frequently eaten. The number of dark moths increased and the number of light moths decreased, because the predators ate more of the light form of the moth, preventing them from breeding and passing on their genes for light color.

major influence. Peppered moths normally rest on tree trunks during the day. If they are noticed by birds, they are eaten. In industrialized regions of England, the trees were blackened by the presence of soot from the smokestacks of coal-burning industrial plants. Kettlewell discovered that, in these areas, the light-colored moths were more conspicuous and were more frequently located and eaten by birds. (See figure 4.6.) In areas with less air pollution the trees were light colored and were covered with light-colored lichens. Birds had greater difficulty locating the light-colored moths, and the more conspicuous dark-colored moths were rare. In these two different locations, natural selection, through the act of birds killing and eating moths, was selecting for those moths that were able to blend in with their surroundings. Each generation of moths had a greater proportion of individuals that were able to blend in than the previous generation had.

Since Darwin's time, many discoveries have helped to explain how the process of natural selection works. Today, we know that each organism has DNA that determines the characteristics it will develop. We also know that the genes contained in DNA control, in addition to structural characteristics, behavior and how organs and body parts function. We also have a better appreciation for how subtle competition can be and how the behavior of organisms contributes to their successful reproduction. We better understand the age of the Earth and appreciate how it has changed over millions of years. In fact, studies of recent fossils and other geologic features show that only thousands of years ago, huge glaciers covered much of Europe and the northern parts of North America, affecting the lives of humans who ate mammoths and encountered saber-toothed tigers and giant cave bears. As the climate became warmer and the glaciers receded, new pressures affected the world's organisms. Some, including the mammoths, saber-toothed tigers, and giant cave bears, did not adapt and became extinct. Others, such as humans, horses, and most kinds of plants, adapted to the new conditions and so survive to the present. Natural selection is constantly at work shaping organisms to fit a changing environment.

Scientists have continuously shown that this theory of natural selection can explain the development of most aspects of the structure, function, and behavior of organisms. It is the central idea that helps explain how species adapt to their surroundings. When we discuss environmental problems, it is helpful to understand that species change, and that as the environment is changed either naturally or by human action, some species will adapt to the new conditions while others will not.

When natural selection occurs over long periods of time (thousands to millions of years), considerable change can be seen in the nature of a species. These changes in the kinds of organisms that exist and in their characteristics is called **evolution.** During the process of evolution, which occurs as a result of natural selection, new species can be produced. The production of new species from previously existing species is known as **speciation** and is thought to occur as a result of a species dividing into two isolated subpopulations. If this separation occurs and the two subpopulations must adapt to somewhat different environments, they will probably develop different characteristics. Eventually the differences may be so great that the two subpopulations are

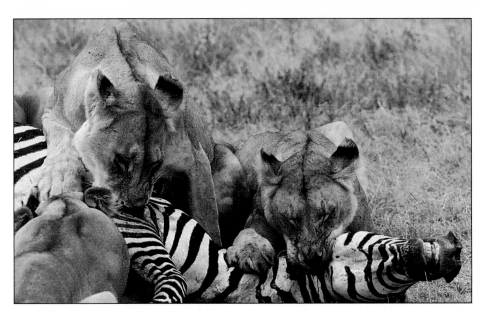

FIGURE 4.7 Predator/Prey Relationship Lions are predators on zebras. The quicker lions are more likely to get food, and the slower, sickly, or weaker zebras are more likely to become prey.

not able to interbreed. At this point they are two different species.

As natural selection occurs it is possible that some species will not be able to adjust to the changes that occur in their environment. If this occurs the species may go extinct. **Extinction** is the loss of an entire species and is a common feature of the evolution of organisms. Most of the species that have ever existed are extinct. It is clear that humans have had a significant impact on the extinction of many kinds of species. Wherever humans have modified the environment for their purposes (farming, forestry, cities), species were displaced from the area. If large areas are modified, the entire species may have been displaced. Ultimately humans are also subject to evolution and the possibility of extinction as well.

KINDS OF ORGANISM INTERACTIONS

Ecology looks at organisms and how they interact with their surroundings, including other organisms. We will discuss several kinds of interactions recognized as distinct types. The concept of natural selection allows us to see how interacting organisms can result in populations that are better adapted to their environment. As you read

this section, notice how different species have characteristics that fit them for their specific roles (niches).

Predation

One common kind of interaction occurs when one animal, known as a **predator,** kills and eats another, know as the **prey.** (See figure 4.7.) Some examples of predator–prey relationships are lions and zebras, birds and worms, wolves and moose, and frogs and insects. Predator–prey relationships are often thought to be one-sided. The predator seems to reap all the benefits. It has the meal and goes on living after killing and eating the prey. However, this relationships does have value for the prey as well. Prey species have a higher reproductive rate than predator species. For example, field mice may have 10 to 20 offspring per year, while hawks typically have two to three. Because of this high reproductive rate, prey species can endure a high mortality rate. Certainly, the individual organism that has been killed and eaten does not benefit, but the species does since the prey organisms that die are likely to be the old, the slow, the sick, and the less fit members of the population. The healthier, quicker, and more fit individuals are more likely to survive. When these survivors reproduce, their

FIGURE 4.8 Competition Whenever a needed resource is in limited supply, organisms compete for it. This competition may be between members of the same species and is called intraspecific competition, or it may be between different species and is called interspecific competition. This photograph shows several vultures competing for a food source.

a slightly different niche, migrate to a different geographic area, or become extinct. As with intraspecific competition, one of the effects of interspecific competition is that the successful species emerges from the interaction better adapted to its environment.

Symbiotic Relationships

Symbiosis is a close, long-lasting, physical relationship between two different species. In other words, the two species are usually in physical contact and at least one of them derives some sort of benefit from this contact. There are three different categories of symbiotic relationships: parasitism, commensalism, and mutualism.

Parasitism

Parasitism is a relationship in which one organism, known as the **parasite,** lives in or on another organism, known as the **host,** from which it derives nourishment. Generally, the parasite is much smaller than the host. Although the host is harmed by the interaction, it is generally not killed by the parasite. Because the evolution of a parasitic way of life involves a long-standing interaction between two species—the parasite and the host—the two species generally evolve in such a way that they can accommodate one another. It is not in the parasite's best interest to kill its host. If it does, it must find another. Likewise, the host evolves defenses against the parasite, often reducing the harm done by the parasite to something the host can tolerate.

Parasites that live on the surface of their hosts are known as **ectoparasites.** Fleas, lice, and some molds and mildews are examples of ectoparasites. (See figure 4.9.) Many other parasites, like tapeworms, malaria parasites, many kinds of bacteria, and some fungi, are called **endoparasites.** They live inside the bodies of their hosts. The tapeworm lives in the intestines of its host where it is able to resist being digested and makes use of the nutrients in the intestine. If the host has only one or two tapeworms, it can live for some time with little discomfort, supporting itself and its parasites. If the number of parasites is large, the host may die.

offspring are more likely to have characteristics that help them survive; they are better adapted to their environment. At the same time, a similar process is taking place in the predator population. Since poorly adapted individuals are less likely to capture prey, they are less likely to survive and reproduce. The predator is a participant in the natural selection process and so is the prey. This dynamic relationship between predator and prey species is a complex one that continues to intrigue ecologists.

Competition

A second type of interaction between species is **competition,** in which two organisms strive to obtain the same limited resource. In the process, both organisms are harmed to some extent. (See figure 4.8.) However, this does not mean that there is no winner. If two robins are competing for the same worm, only one gets it. Both organisms were harmed because they had to expend energy in fighting for the worm, but one got some food and was harmed less than the one that fought and got nothing. This example of competition, which is between members of the same species, is known as **intraspecific competition.** Other examples of intraspecific competition include corn plants in a field competing for water and nutrients, male elk competing with one another for the right to mate with the females, and certain species of woodpeckers competing for the holes in dead trees to use for nesting sites.

Competition among members of the same species does not have only a negative impact. As was the case with predation, competition has its good side when seen from the point of view of the species as a whole. When resources are limited, less fit organisms are more likely to die or be denied mating privileges. Consequently, the next generation of organisms will contain more of the genetic characteristics that are favorable for survival in that particular environment. Since individuals of the same species have similar needs, competition among them is usually very intense. A slight advantage on the part of one individual may mean the difference between survival and death.

Competition may also occur between organisms of different species. This is called **interspecific competition.** In a forest, very little light reaches the forest floor. The different species of plants must compete for the available light. Mosses and ferns can tolerate shade while grasses cannot; therefore, the grasses lose in this competitive interaction. In grasslands the same kind of competition occurs. The tall species shade the short species which are able to live under lower light conditions.

The more similar two species are, the more intense the competition between them. If one of the two competing species is better adapted to live in the area than the other, the less fit species must evolve into

Flea (external parasite)

FIGURE 4.10 **Commensalism** Remoras hitchhike a ride on sharks and feed on the scraps of food lost by the sharks. This is a benefit to the remoras. The sharks do not appear to be affected by the presence of the remoras.

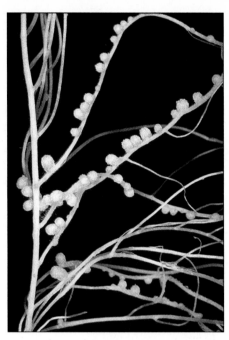

FIGURE 4.11 **Mutualism** The growths on the roots of this plant contain bacteria that are beneficial to the plant because they make nitrogen available to it. The relationship is also beneficial to the bacteria because the plant makes necessary raw materials available to the bacteria. It is a mutually beneficial relationship.

Tapeworm (internal parasite)

FIGURE 4.9 **Parasitism** Fleas are small insects that live in the feathers of birds or the fur of mammals, where they bite their hosts to obtain blood. Since they live on the outside of their hosts, they are called ectoparasites. Tapeworms live inside the intestines of their hosts, where they absorb food from their hosts' intestines. Since they live inside their hosts, they are called endoparasites.

Even plants can be parasites. Mistletoe is a flowering plant that is parasitic on trees. It establishes itself on the surface of a tree when a bird transfers the seed to the tree. It then grows down into the tissues of the tree and uses it as a source of nutrients.

Parasitism is a very common niche. If we were to categorize all the organisms in the world, we would find many more parasitic species than nonparasitic species. Each organism, including you, has many others that use it as a host.

Commensalism

If the relationship between organisms is one in which one organism benefits while the other is not affected, it is called **commensalism.** It is possible to visualize a parasitic relationship evolving into a commensal one. Since parasites generally evolve to do as little harm to their host as possible and the host is combating the negative effects of the parasite, they might eventually evolve to the point where the host is not harmed at all. There are many examples of commensal relationships. Many orchids use trees as a surface upon which to grow. The tree is not harmed or helped, but the orchid needs a surface upon which to establish itself and also benefits by being close to the top of the tree, where it can get more sunlight and rain. Some mosses, ferns, and many vines also make use of the surfaces of trees in this way.

In the ocean, many sharks have a smaller fish known as a remora attached to them. Remoras have a sucker on the top of their heads that they can use to attach to the shark. In this way, they can hitchhike a ride as the shark swims along. When the shark feeds, the remora obtains small bits of food that the shark misses. The shark does not appear to be positively or negatively affected by the remoras. (See figure 4.10.)

Mutualism

Some symbiotic relationships are actually beneficial to both species involved. This kind of a relationship is called **mutualism.** In many mutualistic relationships, the relationship is obligatory. Flowers of the yucca plant are pollinated by a specific insect known as the yucca moth. The moth, in turn, lays its eggs in the flower, where the immature moth larvae feed on the developing seeds. The yucca is dependent on the moth for pollination, and the moth is dependent on the seeds as a source of food for its larvae. Neither organism can exist without the other. The number of eggs laid by the moth is small enough that some of the yucca seeds grow to maturity to provide new yucca plants for future generations of yucca moths. Many other flowering plants are also dependent on specific insects as pollinators.

One plant nutrient that is usually in short supply in soil is nitrogen. Many kinds of plants, such as beans, clover, and alder trees, have bacteria that live in their roots in little nodules. The roots form these nodules when they are infected with certain kinds of bacteria. The bacteria do not cause disease but provide the plants with nitrogen-containing molecules that the plants can use for growth. The nitrogen-fixing bacteria benefit from the living site and nutrients that the plants provide, and the plants benefit from the nitrogen they receive. (See figure 4.11.)

Human Interaction—A Different Look

The niche of humans involves a very complicated set of interactions with other kinds of organisms. We can look at these interactions in the same way we examine the relationships between nonhuman organisms.

Predator — Humans throughout the world use animals as food. Some actually kill animals themselves while others rely on employees of slaughterhouses to do the killing.

Herbivore — Humans rely on many kinds of plants as a source of food. In many respects we graze lettuce, broccoli, celery, and many other kinds of plants. We also eat the fruits and seeds of many plants.

Scavenger — Hunter/gatherer peoples are basically scavengers who look for food of various kinds wherever they happen to be. Even more sophisticated cultures engage in scavenging activities when they pick wild berries, mushrooms, or asparagus.

Commensal — Humans find themselves on both sides of the commensal relationship. Many kinds of organisms live off our leavings. Rats, opossums, raccoons, and many other animals regularly raid our garbage, deriving benefit from something we

have no use for. Other animals benefit from the animals we accidentally kill along our highways. We also derive benefit from organisms without affecting them such as when we are able to get out of the hot sun by sitting under the shade of a tree.

Parasite — Humans are often parasites on other organisms. Maple syrup is made by "bleeding" maple trees. Similarly humans tap rubber trees to obtain sap that is used to make rubber. Many other fragrances and flavors are extracted from plants without killing them. Many human activities regularly rob animals. Honeybees are robbed of their honey and chickens are robbed of their eggs.

Mutualism — Humans have many mutualistic relationships with plants and animals. Our domesticated plants and animals rely on us for support and nutrition and we extract from them payment in the form of companionship, food, or other valuable resources.

Competition — Humans are in competition with all other organisms on Earth. As we convert land and aquatic resources to our uses we deprive other organisms of what they need to survive. We usually win in the game of competition.

COMMUNITY AND ECOSYSTEM INTERACTIONS

Thus far, we have discussed specific ways in which individual organisms interact with one another and with their physical surroundings. We have come to think of interacting groups of species as **communities** in which each kind of organism has a specific niche or role to play. Some organisms play minor roles while others play major roles, but all are part of the community. For example, the grasses of the prairie have a major role since, without them, there would be no prairie. Grasshoppers, prairie dogs, and bison are important consumers of grass. A meadowlark, though a conspicuous and colorful part of the prairie scene, has a relatively minor role and has little to do with maintaining a prairie community.

Communities consist of interacting species, but these species do not interact in a vacuum. They must interact with the physical world around them as well. The physical world has a major impact on

what kinds of plants and animals can live in an area. We do not expect to see a banana tree in the arctic or a walrus in the Mississippi River. Banana trees are adapted to warm, moist, tropical areas, and walruses require cold ocean waters. Periodic fires are often important to maintain grasslands and certain kinds of forests. The kind of soil and the amount of moisture also influence the kinds of organisms found in an area.

Organisms can also impact their physical surroundings. Trees break the force of the wind, grazing animals form paths, and earthworms create holes in the soil. This system of interacting organisms and their nonliving surroundings is frequently referred to as an **ecosystem.** While the concepts of community and ecosystem are closely related, an ecosystem is a broader concept because it involves the physical as well as the biological realm.

Every system has parts that are related to one another in specific ways. A bicycle has wheels, a frame, handlebars, brakes, pedals, and a seat. These parts must be organized in a certain way or the system known as a bicycle will not function. Similarly, ecosystems have parts that

must be organized in specific ways or the systems will not operate. In order to more fully develop the concept of ecosystem, we will look at ecosystems from three points of view: the major roles played by organisms, the way energy is utilized within ecosystems, and the way atoms are cycled from one organism to another.

Major Roles of Organisms

Several categories of organisms are found in any ecosystem. **Producers** are able to make new, complex, organic material from the atoms in their environment. To do this, they must have a source of energy. In nearly all ecosystems, this energy is supplied by the sun, and the organisms that utilize this energy are plants, algae, and tiny aquatic organisms called phytoplankton that carry on photosynthesis. Since producers are the only organisms in an ecosystem that can trap energy and make new organic material from inorganic material, all other organisms rely on producers as a source of food, either directly or indirectly. These other organisms are called **consumers** because they consume organic matter to provide themselves with

TABLE 4.1   ROLES IN AN ECOSYSTEM

Category	Major role or action	Examples
Producer	Converts simple inorganic molecules into organic molecules by the process of photosynthesis	Trees, flowers, grasses, ferns, mosses algae
Consumer	Uses organic matter as a source of food	Animals, fungi, bacteria
Herbivore	Eats plants directly	Grasshopper, elk, human vegetarian
Carnivore	Kills and eats animals	Wolf, pike, dragonfly
Omnivore	Eats both plants and animals	Rats, raccoons, most humans
Scavenger	Eats meat, but often gets it from animals that died by accident or illness, or that were killed by other animals	Coyote, vulture, blowflies
Parasite	Lives in or on another living organism and gets food from it	Tapeworm, many bacteria, some insects
Decomposer	Returns organic material to inorganic material; completes recycling of atoms	Fungi, bacteria, some insects and worms

energy and the organic molecules necessary to build their own bodies. **Primary consumers** eat producers (plants or phytoplankton) as a source of food. They are also known as **herbivores. Secondary consumers** or **carnivores** eat other animals. Some carnivores eat herbivores, while others may eat other carnivores. In addition, many animals, called **omnivores,** have mixed diets that include both plants and animals.

A final category of consumer is the decomposer. **Decomposers** use nonliving organic matter as a source of food. Many small animals, fungi, and bacteria fill this niche. (See table 4.1.)

Energy Flow Through Ecosystems

An ecosystem is a stable, self-regulating unit. To maintain itself, it must have a continuous input of energy. The only significant source of energy for most ecosystems is sunlight energy. Producers are the only organisms that are capable of trapping this energy through the process of photosynthesis and making it available to the ecosystem. The energy is stored in the form of chemical bonds in large organic molecules such as carbohydrates (sugars, starches), fats, and proteins. The energy stored in the molecules of plants can be transferred to other organisms when they consume plants. Each step in the flow of energy through an ecosystem is known as a **trophic level.** Producers (plants) constitute the first trophic level, and herbivores constitute the second trophic level. Carnivores that eat herbivores are the third trophic level, and carnivores that eat other carnivores are the fourth trophic level. Omnivores, parasites, and scavengers occupy different trophic levels, depending on what they happen to be eating at the time. If we eat a piece of steak, we are at the third trophic level; if we eat celery, we are at the second trophic level. (See figure 4.12.)

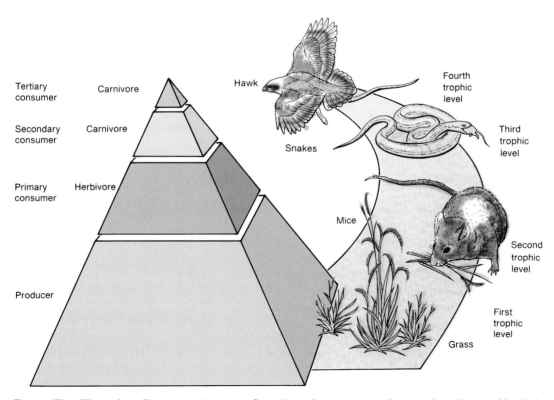

FIGURE 4.12   Energy Flow Through an Ecosystem As energy flows through an ecosystem, it passes through several levels known as trophic levels. Each time energy moves to a new trophic level, approximately 90 percent of the useful energy is lost. Therefore, higher trophic levels contain less energy and fewer organisms in most ecosystems.

As energy passes from one trophic level to the next, some of the useful energy is lost due to the second law of thermodynamics. Much of this loss is in the form of low-quality heat, which is dissipated to the surroundings and warms the air, water, or soil. In addition to this loss of heat, organisms must expend energy to maintain their own life processes. It takes energy to chew food, defend nests, walk to waterholes, or raise offspring. Therefore, the amount of energy contained in higher trophic levels is considerably less than that at lower levels. Approximately 90 percent of the useful energy is lost with each transfer to the next highest trophic level. So in any ecosystem, the amount of energy contained in the herbivore trophic level is only about 10 percent of the energy contained in the producer trophic level. The amount of energy at the third trophic level is approximately 1 percent of that found in the first trophic level.

Because it is difficult to actually measure the amount of energy contained in each trophic level, ecologists often use other measures to approximate the relationship between the amounts of energy at each level. One of these is the biomass. The **biomass** is the weight of living material in a trophic level. It is often possible in a simple ecosystem to collect and weigh all the producers, herbivores, and carnivores. The weights often show the same 90 percent loss from one trophic level to the next as happens with the amount of energy.

Food Chains and Food Webs

The passage of energy from one trophic level to the next as a result of one organism consuming another is known as a **food chain.** For example, willow trees grow well in very moist soil, perhaps near a pond. The trees' leaves capture sunlight and convert carbon dioxide and water into sugars and other organic molecules. The leaves serve as a food source for insects, such as caterpillars and leaf beetles, that have chewing mouth parts and a digestive system adapted to plant food. Some of these insects are eaten by spiders, which fall from the trees into the pond below, where they are consumed by a frog. As the frog swims from one lily pad to another, a large bass consumes the frog. A human may use

an artificial frog as a lure to entice the bass from its hiding place. A fish dinner is the final step in this chain of events that began with the leaves of a willow tree. (See figure 4.13.) This food chain has five trophic levels. Each organism occupies a specific niche and has special abilities that fit it for its niche, and each organism in the food chain is involved in converting energy and matter from one form to another.

When a plant or animal dies, the chemical energy contained within its body is ultimately released as heat by organisms that decompose the body into smaller molecules, such as water and carbon dioxide. These decomposers—bacteria, fungi, and some small consumers, such as insects and worms—play a significant role in the ultimate recycling of the atoms that make up the bodies of living things. They are involved in the breakdown of small bits of nonliving organic material called **detritus.** Detritus food chains rely on a constant supply of dead organic material being supplied from situations where photosynthesis is taking place. Detritus food chains are found in a variety of situations. The bottoms of the deep lakes and oceans are too dark for photosynthesis. The animals and decomposers that live there rely on a steady rain of small bits of organic matter from the upper layers of the water where photosynthesis does take place. Similarly, in most streams, leaves and other organic debris serve as the major source of organic material and energy. A sewage treatment plant is also a detritus food chain in which particles and dissolved organic matter are constantly supplied to a series of bacteria and protozoa that use this material for food.

In another example, the soil on a forest floor receives leaves, which fuel a detritus food chain. In detritus food chains, a mixture of insects, crustaceans, worms, bacteria, and fungi cooperate in the breakdown of the large pieces of organic matter, while at the same time feeding on one another. When a leaf dies and falls to the forest floor it is colonized by bacteria and fungi, which begin the breakdown process. An earthworm will also feed on the leaf and at the same time consume the bacteria and fungi. If that earthworm is eaten by a bird, it becomes part of a larger food chain that includes material from both a detritus food chain and a photosynthesis-driven

FIGURE 4.13 Food Chain As one organism feeds on another organism, energy flows through the series. This is called a food chain.

food chain. When several food chains overlap and intersect, they make up a **food web.** (See figure 4.14.)

Notice in the upper-left-hand corner of figure 4.14 that the Cooper's and sharp-shinned hawks use many different kinds of birds as a source of food. These hawks fit into several food chains. If one source of prey is in short supply, they can switch to something else without too much trouble.

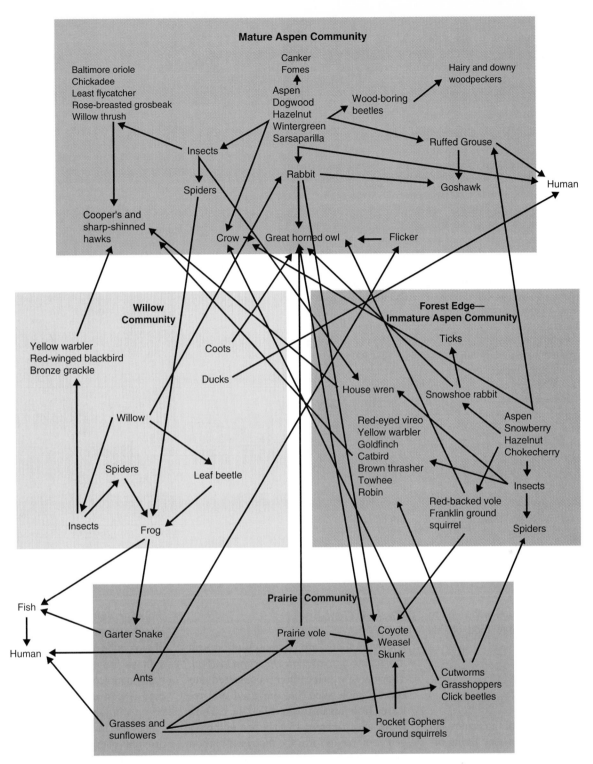

FIGURE 4.14 **Food Web** The many kinds of interactions among organisms in an ecosystem constitute a food web. In this network of interactions, several organisms would be affected if one key organism were reduced in number. Look at the rabbit in the mature aspen community and note how many organisms use it as food.

Source: R.D. Bird, "Biotic Communities of the Aspen Parkland of Central Canada" in *Ecology*, 11:410, April 1930.

Name That Relationship

Sometimes it is not easy to categorize the relationships that organisms have with each other. For example, it is not always easy to say whether a relationship is a predator–prey relationship or a host–parasite relationship. How would you classify a mosquito or a tick? Both of these animals require blood meals to live and reproduce. They don't kill and eat their prey. Neither do they live in or on a host. This question points out the difficulty encountered when we try to place all organisms into a few categories.

Another relationship that doesn't fit well is the relationship that certain birds like cowbirds and cuckoos have with other birds. Cowbirds and cuckoos do not build nests but lay their eggs in the nests of other species of birds, who are left to care for a foster nestling at the expense of their own nestlings, who generally die. This could be called parasitism, but it doesn't fit the standard textbook definition.

What about grazing animals? Are they predators on the plants that they eat? And what kind of relationships do humans have with their domesticated plants and animals? Do we have a mutualistic relationship with cows and chickens? We certainly derive benefit from the milk and eggs we get, and cows and chickens benefit from the care they are given. Or are we primarily parasites on these species? Certainly, we are sometimes predators when we kill and eat them.

It would be convenient to unambiguously classify every kind of relationship into a clear category. But this is not possible. Moreover, each relationship may change in character over time.

Is this cowbird nestling being fed by a red-eyed vireo a parasite?

Is this mosquito a parasite?

Are bison grass predators?

These kinds of complex food webs tend to be more stable than simple food chains with few cross-links.

Nutrient Cycles in Ecosystems

Organisms are composed of molecules and atoms that are cycled from one living organism to another. Some atoms are more common in living things than others. Carbon, nitrogen, oxygen, hydrogen, and phosphorus are found in important molecules like proteins, DNA, carbohydrates, and fats, which are found in all kinds of living things. These atoms are passed from one organism to another when one organism consumes another, and they are recycled when an organism dies.

Carbon Cycle

All living things are composed of organic molecules that contain the atom carbon. The way in which carbon is converted into organic molecules that are used and released from organisms is called the **carbon cycle.** (See figure 4.15.) Carbon and oxygen combine to form the molecule carbon dioxide (CO_2), which is present in small quantities as a gas in the atmosphere and dissolved in water. During photosynthesis, carbon dioxide from the atmosphere is taken into plant leaves and combined with hydrogen from water molecules, (H_2O) which are absorbed from the soil by the roots and transported to the leaves. Complex organic molecules such as carbohydrates (sugars) are formed. Oxygen molecules (O_2) are released from the leaves into the atmosphere during the process of photosynthesis when water molecules are split to provide hydrogen atoms for the organic molecules. In this process, light energy is converted to chemical-bound energy in organic molecules such as sugar. Plants use these sugars for growth and to provide energy for other processes. In aquatic ecosystems, plants, algae, and phytoplankton absorb carbon dioxide from the water and release oxygen into the water during photosynthsis.

Herbivores can use these complex organic molecules as food. When an herbivore eats plants or algae, it breaks down the complex organic molecules into simpler molecules, which can be incorporated into the chemical structure of the

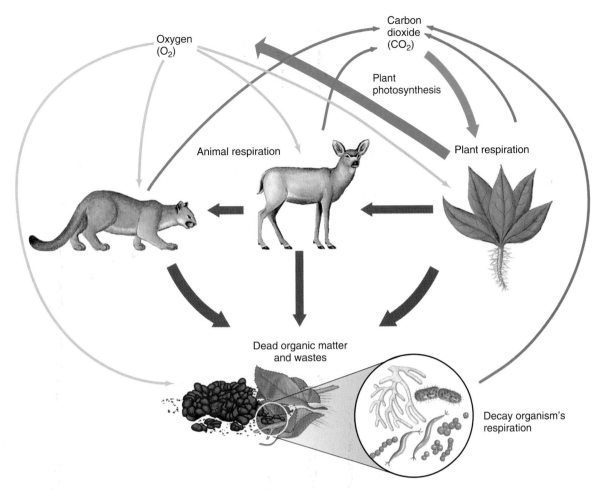

FIGURE 4.15 Carbon Cycle Carbon atoms are cycled through ecosystems. Plants can incorporate carbon atoms from carbon dioxide into organic molecules when they carry on photosynthesis. The carbon-containing organic molecules are passed to animals when they eat plants or other animals. Organic wastes or dead organisms are consumed by decay organisms. All organisms, plants, animals, and decomposers return carbon atoms to the atmosphere when they carry on respiration. Oxygen atoms are being cycled at the same time that carbon atoms are being cycled.

animal. The carbon atom, which was once part of the producer, is now part of the herbivore. All organisms also carry on the process of respiration, in which oxygen from the atmosphere is used to break down large organic molecules into carbon dioxide and water. Much of the chemical-bond energy is released by respiration and is lost as heat, but the remainder is used by the herbivore for movement, growth, and other activities.

When an herbivore is eaten by a carnivore, the carbon-containing molecules of the herbivore become incorporated into the body of the carnivore. Carnivores also carry on respiration and release carbon dioxide and water in the process of obtaining energy from organic molecules.

The waste products of all kinds of organisms, and any organisms that die, are acted upon by decomposers so that even the naturally occurring organic molecules in nonliving matter are used by decomposers to provide energy. Carbon dioxide is produced in the process. The carbon atom begins as part of a carbon dioxide molecule in the atmosphere, becomes part of an organic molecule in one or more organisms, and then is released back into the atmosphere by the process of respiration.

Nitrogen Cycle

Another very important cycle, the **nitrogen cycle,** involves the flow of nitrogen atoms through organisms in an ecosystem. Seventy-eight percent of the gas in the air we breathe is made up of molecules of nitrogen gas (N_2). However, very few organisms are able to use it in this form. Since plants and other producers are at the base of nearly all food chains, they must make new nitrogen-containing molecules, such as proteins and DNA. Plants and other producers are unable to use the nitrogen in the atmosphere and must get it in the form of nitrate (NO_3^-) or ammonia (NH_3). The amount of nitrogen available to plants controls their growth and is often a limiting factor. The primary way in which plants obtain nitrogen compounds they can use is with the help of bacteria.

Some bacteria are able to convert the nitrogen gas (N_2) in the atmosphere into forms that plants can use. These bacteria are called **nitrogen-fixing bacteria.** Some kinds of these bacteria live in the soil and are called **free-living nitrogen-fixing bacteria.** Others live in nodules in the roots of plants known as legumes (peas,

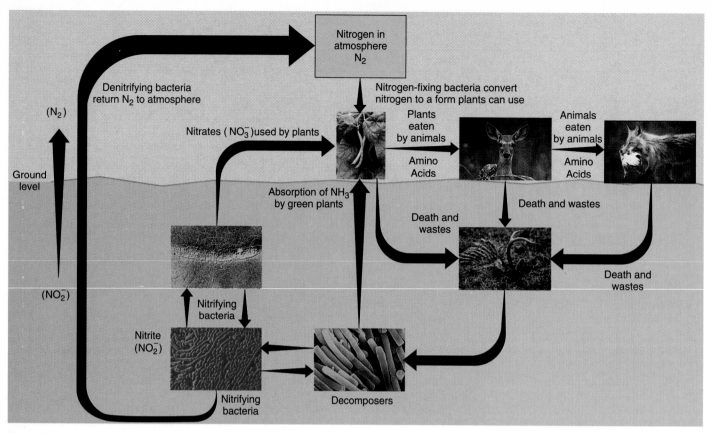

FIGURE 4.16 **Nitrogen Cycle** Nitrogen atoms are cycled in ecosystems. Atmospheric nitrogen is converted by nitrogen-fixing bacteria to a form that plants can use to make protein and other compounds. Proteins are passed to other organisms when one organism is eaten by another. Dead organisms and waste products are acted on by decay organisms to form ammonia, which may be reused by plants or converted to other nitrogen compounds by other kinds of bacteria. Denitrifying bacteria are able to convert inorganic compounds into atmospheric nitrogen.

beans, and clover) and are known as **symbiotic nitrogen-fixing bacteria.** Some grasses and evergreen trees appear to have a similar relationship with certain root fungi that seem to improve the nitrogen-fixing capacity of the plant.

Other kinds of bacteria can also make nitrogen available to plants. Dead organisms and their waste products contain molecules that contain nitrogen. Many kinds of soil bacteria decompose these large nitrogen-containing organic molecules, releasing ammonia, which can be used directly by many kinds of plants or be converted to nitrate by other kinds of bacteria. Additional kinds of bacteria, known as **denitrifying bacteria,** are, under certain conditions, able to convert a molecule known as nitrite to atmospheric nitrogen gas. The nitrogen atoms in the gas can enter the cycle again with the aid of nitrogen-fixing bacteria. (See figure 4.16.)

In naturally occurring soil, nitrogen is often a limiting factor of plant growth. To increase yields, farmers often provide extra sources of nitrogen in several ways. Inorganic fertilizers are a primary method of increasing the nitrogen available. These fertilizers may contain ammonia, nitrate, or both.

Since the manufacture of nitrogen fertilizer takes a lot of energy, fertilizer is expensive. Farmers have looked for alternative methods to reduce this cost. A farmer might alternate nitrogen-demanding crops like corn with nitrogen-yielding crops like soybeans. Soybeans are legumes that have symbiotic nitrogen-fixing bacteria in their roots. If soybeans are planted one year, the excess nitrogen left in the soil can be used by the corn plants grown the next year. A slightly different technique involves growing a nitrogen-fixing crop for a short period of time and then plowing the crop under and letting the organic matter decompose. Farmers can also add nitrogen to the soil by spreading manure on the field and relying on the soil bacteria to decompose the organic matter and release the nitrogen for plant use.

Phosphorus Cycle

Phosphorus is another kind of atom common in the structure of living things. It is present in many important biological molecules such as DNA and in the membrane structure of cells. In addition, the bones and teeth of animals contain significant quantities of phosphorus. The ultimate source of phosphorus atoms is rock. In nature, new phosphorus compounds are released by the erosion of rock. Plants use the dissolved phosphorus compounds to construct the molecules they need. Animals consume plants or other animals and obtain needed phosphorus. When an organism dies or excretes waste products, decomposer organisms recycle the phosphorus compounds back into the soil. (See figure 4.17.) In many soils, phosphorus is in short supply and must be provided to crop plants

FIGURE 4.17 Phosphorus Cycle The source of phosphorus is rock that, when dissolved, provides a source of phosphate used by plants and animals.

GLOBAL PERSPECTIVE

Human Impact on Nutrient Cycles

The three nutrient cycles discussed in this chapter (carbon, nitrogen, and phosphorus) are important to understanding how ecosystems function. However, it is also important to understand that humans have significantly altered these cycles in many ways. The carbon cycle has been affected by humans in two primary ways. The burning of fossil fuels (coal, oil, and natural gas) has released large amounts of carbon dioxide into the atmosphere. The burning of fossil fuels is in effect releasing carbon that was stored in a form that was not accessible to living things. Furthermore, the conversion of natural ecosystems that tend to store carbon for long periods, such as forests, with agricultural ecosystems that only store carbon temporarily has also disrupted the natural carbon cycle. The use of large amounts of nitrogen and phosphorus as fertilizer has also caused major changes in the amount of these nutrients that are cycled through ecosystems. These nutrients are introduced into the bodies of the plants and animals that we raise for food. A major impact of the widespread use of nitrogen and phosphorus in agriculture is the impact on aquatic ecosystems. Normally aquatic ecosystems are starved for these nutrients. The presence of large amounts of these nutrients in either freshwater or saltwater results in increased rates of growth of bacteria, algae, and aquatic plants. Increases in these organisms can have many different effects. Many algal blooms are toxic, resulting in fish kills and human poisoning. Large amounts of plants and algae in aquatic ecosystems will also deplete the water of oxygen when they die. Each summer a major "dead zone" (18,000 km^2) develops in the Gulf of Mexico off the mouth of the Mississippi River. This "dead zone" contains few fish and bottom-dwelling organisms. It is caused by low oxygen levels brought about by the rapid growth of algae and bacteria in the nutrient-rich waters. The nutrients can be traced to the extensive use of fertilizer in the major farming areas of the central United States.

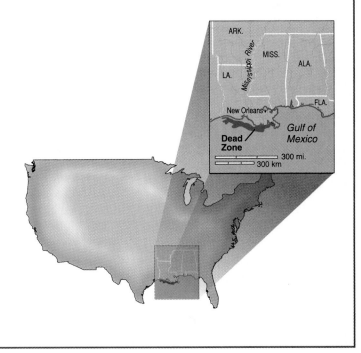

Colorado River Restoration

The Colorado River originally was a warm-water river that fluctuated widely in flow and delivered large amounts of sediment to the river system. The building of dams on the river changed its nature significantly. The sediments were deposited above the dams, and since water was discharged from the bottom of the dam, the water was colder. Furthermore, since the river was dammed to provide water resources for cities and agriculture, the amount of water flowing through the river system was reduced.

In order to restore some of the original character to the portion of the river that flows through the Grand Canyon, the U.S. Department of the Interior in March 1996 released large amounts of water from the Glen Canyon Dam. This action was intended to simulate the normal flooding that used to occur before the dam was built. The flooding restored beaches, marshes, and backwaters that are important for many species native to the Grand Canyon portion of the Colorado River. The humpback chub, for example, requires backwaters for spawning. Other fish species benefit from the restoration as well. The flooding increased the amount of organic material in the water. This is important as a source of food for insects that are in turn eaten by fish and other organisms.

Information gathered from this test release will be important in assessing future management of the Colorado River ecosystem and will have application to other river systems where dams alter flow.

Organic Contaminants in Great Lakes Fish

The Great Lakes area has a concentration of industry because of the availability of water for manufacturing processes, and because water transportation is an efficient way to move raw materials and products. In the past, many of these industries released heavy metals and organic molecules into the water as an accidental by-product of the manufacturing process, or because it was a cheap way to get rid of unwanted material. Many of the organic molecules are products of modern organic synthesis and, therefore, are not something that bacteria and fungi are able to decay. Since these molecules are of recent origin, the decomposers have not had time to evolve mechanisms to break them down.

Approximately five hundred different organic compounds that scientists think are contaminants have been identified in the bodies of Great Lake's fish. Most are present in extremely small amounts and probably do not represent a serious hazard, but others are present in high enough concentrations to cause public health officials to be concerned. The contaminants are a problem because they cannot be broken down and eliminated. They persist. Fish that are carnivores tend to accumulate more of these toxic materials in their bodies as they get older. It just so happens that most of the fish that people catch and eat are carnivores. Furthermore, the older a fish is, the greater can be its concentration of contaminants.

It is not possible to check every fish caught to see if it is fit to eat. The cost of doing so would be on the order of several hundred to several thousands of dollars per individual fish, depending on which contaminants are looked for. Therefore, the states and provinces surrounding the Great Lakes have developed advisory statements to help people avoid unsafe fish.

1. Some species of fish, such as carp and catfish, feed on the bottom and tend to accumulate contaminants more quickly than others. In many areas of the Great Lakes, people are advised not to eat these fish.

2. Larger fish have generally consumed more food and have had an opportunity to accumulate more contaminants. Therefore, people are advised to eat only smaller fish.

3. Since many of the organic contaminants are fat soluble, removal of the fat or cooking in such a way that the fat is allowed to separate from the flesh is also advised in many areas.

4. Because the amount of contamination a person is exposed to is directly related to the number of fish a person eats, people are advised to limit the number of fish consumed, and women of childbearing age and young children are often advised to not eat the fish at all.

The lip cancer on this bullhead indicates carcinogens in Wisconsin's Fox River. In some tributaries of the Great Lakes, fish cancer rates may reach 84 percent.

At one time large predators such as wolves, mountain lions, and grizzly bears were an important part of the balance of organisms in the region of Yellowstone National Park. For generations ranchers in the area had reduced the numbers of these large predators because they often killed livestock. By the early 1900s the wolf was extinct in the U.S. Rocky Mountains and was listed as an endangered species in 1967. Without large predators the populations of large grazing animals in the Park (elk, moose, and bison) increased to levels that were not natural. In the early 1990s the U.S. Fish and Wildlife Service reintroduced wolves into the Park to provide a more natural mix of species and control the burgeoning grazer populations. The wolves were successful and are reducing the elk population. The wolves kill coyotes and reduce their populations as well.

However, many of the ranchers in the area were not pleased. It is highly likely that as the population of wolves increases they will migrate from the Park and prey on livestock. This has already happened in a few instances. The losses of livestock to wolves had been anticipated and a fund was set up to reimburse ranchers for livestock they lost to wolf predation. However, many ranchers were not satisfied with this mechanism and filed a lawsuit to have the wolves removed from the Park. The lawsuit is based on a technicality of the Endangered Species Act. The wolves in Yellowstone received a "nonessential experimental" designation under the Endangered Species Act. Because the wolves were migrating from Canada into the area, the reintroduction program simply accelerated the process. The special designation would allow ranchers to kill reintroduced wolves that attack their livestock; however, wolves that migrate naturally from Canada would not be allowed to be killed. The ranchers argued that they would not be able to tell reintroduced "nonessential experimental" wolves from those that arrived by migration from other areas. In December of 1997 a U.S. District Judge ordered the removal of the wolves from Yellowstone. The ruling has been appealed.

- Is it important to return species to areas that were part of their former range?
- Should the economic interests of ranchers be more important than the biological interests of managing Yellowstone National Park as an example of a natural ecosystem?
- Should special interests be allowed to use "technicalities" to circumvent the generally accepted policy?

to get maximum yields. Phosphorus is also in short supply in aquatic ecosystems.

Fertilizers usually contain nitrogen, phosphorus, and potassium compounds. The numbers on a fertilizer bag indicate the percentage of each in the fertilizer. For example, a 6–24–24 fertilizer has 6 percent nitrogen, 24 percent phosphorus, and 24 percent potassium compounds. In addition to carbon, nitrogen, and phosphorus, potassium and other elements are cycled within ecosystems. In an agriculture ecosystem, these elements are removed when the crop is harvested. Farmers must return these elements to the soil with fertilizer. In aquaculture such as that used to raise catfish fertilizer is added to the body of water to stimulate the production of algae which is the base of many aquatic food chains.

Summary

Everything that affects an organism during its lifetime is collectively known as its environment. The environment of an organism can be divided into biotic (living) and abiotic (nonliving) components.

The space an organism occupies is known as its habitat, and the role it plays in its environment is known as its niche. The niche of an organism is the result of natural selection directing the adaptation of the organism to a specific set of environmental conditions.

Organisms interact with one another in a variety of ways. Predators kill and eat prey. Organisms that have the same needs compete with one another and do mutual harm, but one is usually harmed less and survives. Symbiotic relationships are those in which organisms live in physical contact with one another. Parasites live in or on another organism and derive benefit from the relationship, harming the host in the process. Commensal organisms derive benefit from another organism but do not harm the host. Mutualistic organisms both derive benefit from their relationship.

A community is a set of interacting groups of organisms. Those organisms and their abiotic environment constitute an ecosystem. In an ecosystem, energy flows from producers through various trophic levels of consumers (herbivores, carnivores, omnivores, and decomposers). About 90 percent of the energy is lost as it passes from one trophic level to the next. This means that the amount of biomass at higher trophic levels is usually much less than that at lower trophic levels. The sequence of organisms through which energy flows is known as a food chain. Several interconnecting food chains constitute a food web.

The flow of atoms through an ecosystem involves all the organisms in the community. The carbon, nitrogen, and phosphorus cycles are examples of how these materials are cycled in ecosystems.

Review Questions

1. Define environment.
2. Describe, in detail, the niche of a human.
3. How is natural selection related to the concept of niche?
4. List five predators and their prey organisms.

5. How is an ecosystem different from a community?
6. Humans raising cattle for food is what kind of relationship?
7. Give examples of organisms that are herbivores, carnivores, and omnivores.
8. What are some different trophic levels in an ecosystem?
9. Describe the carbon cycle, the nitrogen cycle, and the phosphorus cycle.
10. Analyze an aquarium as an ecosystem. Identify the major abiotic and biotic factors. List members of the producer, primary consumer, secondary consumer, and decomposer trophic levels.

Critical Thinking Questions

1. Ecologists and political scientists look at habitat destruction differently. Consider the Environmental Close-Up about conservation plans and the political/economic and scientific issues that surround it. Identify some perspectives each discipline has to contribute to our understanding of habitat destruction? What values does each place on the ideas of the other discipline? What do you think about the issue of creating conservation plans? Protecting habitat from destruction? Why?

2. Even before humans entered the scene, many species of plants and animals were extinct and new ones had developed. Why are we even concerned about endangered species given the fact that species have always come and gone?

3. Humans have had a major impact on many ecosystems. Consider the questions in the Issues and Analysis section about reintroducing wolves to Yellowstone Park. Is this the appropriate way for humans to try to repair the environmental damage they have caused? Identify some of the values and beliefs that make up these different perspectives: ranchers and conservationists. Which do you find the most compelling? Why?

4. Would you establish any rules about which animals, or plants, should be reintroduced? How would you establish those rules?

5. Concentrations of industrial chemicals are high in some species of fish, high enough to call for an advisory to limit the number of fish a person should eat within a given period of time. Many of these chemicals are thought to cause cancer, but cancer is an effect that is often not felt for decades after exposure. How do scientists decide how many fish can be safely eaten? Is there any "safe" level? What evidence would convince you that there is danger? How could you tell?

6. Humans move around a lot, and, intentionally or not, introduce species to new habitats that these species had not formerly inhabited. Using the concepts and terms from this chapter, describe what you think the effect of introduced species might be on the new ecosystems. Why?

7. You notice that after using pesticides on your farm field that the number of insects declines for a year. The next year, though, they come back and you need to reapply the pesticide. This time, though, there is less of an effect on the insect population. A third application in another year has even less of an effect. What is your hypothesis about what is happening here? Design an experiment that tests your hypothesis.

Key Terms

abiotic factors 48	host 54
biomass 58	interspecific competition 54
biotic factors 48	intraspecific competition 54
carbon cycle 60	limiting factor 49
carnivore 57	mutualism 55
commensalism 55	natural selection 51
community 56	niche 50
competition 54	nitrogen cycle 61
consumer 56	nitrogen-fixing bacteria 61
decomposer 57	omnivore 57
denitrifying bacteria 62	parasite 54
detritus 58	parasitism 54
ecology 48	predator 53
ecosystem 56	prey 53
ectoparasite 54	primary consumer 57
endoparasite 54	producer 56
environment 48	range of tolerance 49
evolution 53	secondary consumer 57
extinction 53	speciation 53
food chain 58	species 51
food web 58	symbiosis 54
free-living nitrogen-fixing bacteria 61	symbiotic nitrogen-fixing bacteria 62
habitat 50	trophic level 57
herbivore 57	

Related Internet Sites

Visit our website at http://www.mhhe.com/environmentalscience/ for specific information on each of the topics below.

Living in harmony with nature

Atmosphere, hydrosphere, and biosphere

Global environmental change

Animal diversity

Conservation, preservation, and restoration

Conserving endangered species

Mollusks, marine mammals, and birds

Protecting endangered ecosystems

5
chapter

Kinds of Ecosystems and Communities

OBJECTIVES

After reading this chapter, you should be able to:

- Recognize the difference between primary and secondary succession.
- Describe the process of succession from pioneer to climax community in both terrestrial and aquatic situations.
- Associate typical plants and animals with the various terrestrial biomes.
- Recognize the physical environmental factors that determine the kind of climax community that will develop.
- Differentiate the forest biomes that develop based on temperature and rainfall.
- Describe the various kinds of aquatic ecosystems and the factors that determine their characteristics.

CHAPTER OUTLINE

SUCCESSION

Ecosystems are dynamic, changing units. On a daily basis, plants grow and die, animals feed on plants and on one another, and decomposers recycle the chemical elements that make up the biotic portion of any ecosystem. Since all organisms are linked together in a community, any change in the community affects many organisms within it. Certain conditions within a community are keys to the kinds of organisms that are found associated with one another. Grasshoppers need grass for food, robins need trees to build nests, and herons need shallow water to find food. Each organism has specific requirements that must be met in a community, or it will not survive.

Over long periods of time, it is possible to see trends in the way the structure of a community changes. This series of regular, predictable changes in the structure of a community over time is called **succession.** Succession occurs because organisms cause changes in their surroundings that make the environment less suitable for themselves and more suitable for other kinds of organisms. One community of organisms is replaced by a slightly changed community, which itself is replaced by a subsequent collection of organisms. Ecologists recognize two different kinds of succession: **primary succession,** which begins with bare mineral surfaces or water, and **secondary succession,** which begins with the destruction or disturbance of an existing ecosystem.

Primary Succession

Primary succession can begin on a bare rock surface, pure sand, or standing water. Since succession on rock and sand is somewhat different from that which occurs with watery situations, we deal with them separately. We discuss terrestrial succession first.

Terrestrial Primary Succession

Bare rock is a very inhospitable place for organisms to live. The temperature changes drastically, there is little moisture, the organisms are exposed to the damaging effect of the wind, few nutrients are available, and few places are available for organisms to attach themselves or hide. Primarily what is lacking is soil; however, a few kinds of organisms can survive in even this inhospitable environment. This collection of organisms is known as the **pioneer community** because it is the first to colonize bare rock. (See figure 5.1.)

The dominant organism in this initial community is something called a lichen. Lichens are actually mutualistic relationships between two kinds of organisms: algae that are green and carry on photosynthesis and fungi that retain moisture and attach to the rock surface. The growth and development of lichens is a slow process. It may take lichens one hundred years to grow as large as a dinner plate. A lichen serves as a producer in this simple ecosystem, and many tiny organisms may be found associated with a lichen. They may feed on it and use it as a place of shelter, since even a drizzle is like a torrential rain for a microscopic animal. Since lichens are firmly attached to rock surfaces, they also tend to accumulate bits of airborne debris and store small amounts of water that would otherwise blow away or run off the rock surface. Acids produced by the lichen tend to cause the breakdown of the rock substrate into smaller particles. This fragmentation of rock, aided by physical and chemical weathering processes, along with the trapping of debris and the contribution of organic matter by the death of lichens and other organisms, ultimately leads to the accumulation of a very thin layer of soil.

This thin layer of soil is the key to the next stage in the successional process. The layer can support some fungi, certain small worms, insects, bacteria, protozoa, and perhaps a few tiny annual plants. As these organisms grow, reproduce, and die, they contribute additional organic material for the soil-building process. This stage eliminates the lichen community and is then replaced by a community of small perennial plants. The perennial grasses and herbs are eventually replaced by shrubs, which are often replaced by trees that require lots of sunlight, which are replaced by trees that can tolerate shade. Eventually, a relatively stable, long-lasting, more complex, and interrelated community of plants, animals, fungi, and bacteria is

FIGURE 5.1 Pioneer Organism The lichen growing on this rock is able to accumulate bits of debris, carry on photosynthesis, and aid in breaking down the rock. All of these activities contribute to the formation of a thin layer of soil, which is necessary for plant growth in the early stages of succession.

produced, known as the **climax community.** Each step in this process is called a **successional stage,** or **seral stage,** and the entire sequence of stages—from pioneer community to climax community—is called a **sere.** (See figure 5.2.) The specific kind of climax community produced depends on such things as climate and soil type, which are discussed in greater detail later in this chapter. However, certain characteristics are typical of climax communities regardless of the type.

1. Climax communities are able to maintain their mix of species.

2. They are in energy balance, while successional communities show changes in species and tend to accumulate large amounts of new material (gain energy).

3. They tend to have many more kinds of organisms and kinds of interactions among organisms than does a successional community.

The general trend in climax communities is toward increasing complexity and energy efficiency, compared to the successional communities that preceded them.

Aquatic Primary Succession

The principal concepts of land succession can be applied to aquatic ecosystems. Except for the oceans, most aquatic ecosystems are considered temporary. Certainly,

| Bare rock | Lichens | Small annual plants, lichens | Perennial herbs, grasses | Grasses, shrubs, shade-intolerant trees | Shade-tolerant trees |

Pioneer stages | Intermediate stages | Climax community

Hundreds of years

FIGURE 5.2 **Primary Succession on Land** The formation of soil is a major step in primary succession. Until soil is formed, the area is unable to support large amounts of vegetation, which modify the harsh environment. Once soil formation begins, the site proceeds through an orderly series of stages toward a climax community.

some are going to be around for thousands of years, but eventually they will disappear and be replaced by terrestrial ecosystems as a result of normal successional processes. All aquatic ecosystems receive a continuous input of soil particles and organic matter from surrounding land, which results in the gradual filling in of shallow bodies of water like ponds and lakes.

In deep portions of lakes and ponds, only floating plants and algae can exist, but as the amount of sediment accumulates, it becomes possible for submerged plants to become established on the bottom of shallow bodies of water. They carry on photosynthesis, resulting in a further accumulation of organic matter, along with the trapping of sediments that flow into the pond or lake from streams or rivers. Eventually, as the water becomes shallower, emergent plants become established. They have leaves that float on the surface or project into the air. The network of roots and stems below the surface of the water results in the accumulation of more material, and the water becomes even shallower as material accumulates on the bottom. As the process continues, more sediment accumulates, the wet soil thus formed begins to dry out, and grasses and other plants that can live in wet soil

become established. This is often called a wet meadow. Once this occurs, the stage is set for a typical terrestrial successional series of changes, eventually resulting in a climax community. (See figure 5.3.)

Since the shallower portions of most lakes and ponds are at the shore, it is often possible to see the various stages in aquatic succession from the shore. In the central, deeper portions of the lake, there are only floating plants and algae. As we approach the shore, we first find submerged plants like *Elodea* and algal mats, then emergent vegetation like water lilies and cattails, then grasses and sedges that can tolerate wet soil, and on the shore the beginnings of a typical terrestrial succession resulting in the climax community typical for the area.

In many northern ponds and lakes, sphagnum moss forms thick, floating mats. These mats may allow certain plants that can tolerate wet soil conditions to become established. The roots of the plants bind the mat together and establish a floating bog, which may contain small trees and shrubs as well as many other smaller, flowering plants. (See figure 5.4.) You can recognize that the entire system is floating only when you jump on it and the trees sway or when you step through a weak

zone in the mat and sink to your hips in water. Eventually these bogs will become increasingly dry and the normal climax vegetation for the area will succeed the more temporary bog stage.

Secondary Succession

The same processes and activities that drive primary succession result in secondary succession. The major difference is that secondary succession occurs when an existing community is destroyed in some way. A forest fire, a flood, or the conversion of a natural ecosystem to agriculture may be the cause. Usually, the destroyed ecosystem is not completely returned to bare rock. Much of the soil may remain, and many of the nutrients necessary for plant growth may be available for the reestablishment of the previously existing ecosystem. Consequently, secondary succession tends to be more rapid than primary succession. Figure 5.5 shows the typical secondary succession found on abandoned farmland in the southeastern United States.

Similarly, when beavers flood an area, the existing trees die and an aquatic ecosystem is established. As the area behind the dam fills in with sediment and organic

FIGURE 5.3 Succession from a Pond to a Wet Meadow A shallow pond will fill slowly with organic matter from producers in the pond. Eventually, a floating mat will form over the pond and grasses will become established. In many areas this will be succeeded by a climax forest.

FIGURE 5.4 Floating Bog In many northern regions, sphagnum moss forms a floating mat that can support the growth of other plants. If a person were to walk on this mat, it would bounce up and down, because it is floating on water.

matter, it goes through a series of stages that eventually return the area to the typical climax community. (See figure 5.6.)

MAJOR TYPES OF CLIMAX COMMUNITIES: BIOMES

Biomes are terrestrial climax communities with wide geographic distribution. (See figure 5.7.) The species involved may not be identical in all parts of the world, but the general structure of the ecosystem and the kinds of niches are similar in broad terms. Two primary nonbiological factors have major impacts on the kind of climax community that develops: precipitation and temperature. Several aspects of precipitation are important: the total amount of precipitation per year, the form in which it arrives (rain, snow, sleet), and the seasonal distribution of the precipitation. Is it evenly spaced throughout the year? Are there wet and dry seasons?

The temperature can vary considerably. Tropical areas have warm, relatively unchanging temperatures throughout the year. Areas near the poles have long winters with extremely cold temperatures and relatively short, cool summers. Other areas are more evenly divided between cold and warm periods of the year. The interplay between these two major parameters can account for many of the differences seen in climax communities. (See figure 5.8.)

In addition to temperature and precipitation, several other factors may influence the kind of climax community present. Some climax communities seem to rely on periodic fires to prevent the establishment of larger, woody species. Similarly, severe wind may prevent the establishment of trees and cause rapid drying of the soil. Sandy soils tend to dry out quickly and may not allow the establishment of more water-demanding species like trees, while extremely wet soils may allow only certain species of trees to grow. Obviously, the kinds of organisms currently living in the area are also important, since their offspring will be the ones available to colonize a new area. Let's look at some of the major biomes and the factors that shape their character.

Desert

Deserts are areas that generally receive fewer than 25 centimeters of precipitation per year. A lack of water is the primary factor that determines that an area will be a desert. (See figure 5.9.) Rain comes in the form of thundershowers at infrequent intervals. Much of the water runs off into gullies and, since the rate of evaporation is high, the rains provide only brief periods of rapid growth. Deserts are also likely to be windy. We often think of deserts as hot, dry wastelands devoid of life. However, many deserts are quite cool during a major part of the year. Certainly, the Sahara Desert and the deserts of the southwestern United States and Mexico are hot during much of the year, but the desert areas of the northwestern United States and the Gobi Desert in Central Asia can be extremely cold during winter months and have relatively cool summers. Furthermore, the temperature can vary greatly

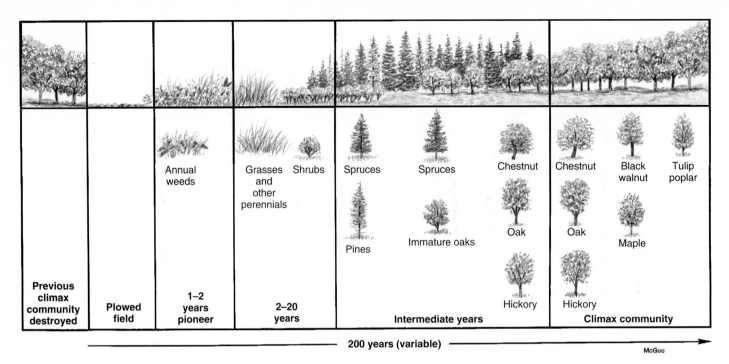

Annual weeds | Grasses and other perennials | Shrubs | Spruces | Spruces | Chestnut | Chestnut | Black walnut | Tulip poplar

Pines | Immature oaks | Oak | Oak | Maple

Hickory | Hickory

| Previous climax community destroyed | Plowed field | 1–2 years pioneer | 2–20 years | Intermediate years | Climax community |

← 200 years (variable) →

McGee

FIGURE 5.5 Secondary Succession on Land A plowed field in the southeastern United States shows a parade of changes over time involving plant and animal associations. The general pattern is for annual weeds to be replaced by grasses and other perennial herbs, which are replaced by shrubs, which are replaced by trees. As the plant species change, so do the animal species.

FIGURE 5.6 Beaver Pond Succession—Secondary Succession from Water A colony of beavers can dam up streams and kill trees by the flooding that occurs and by using trees for food. Once the site is abandoned, it will slowly return to the original forest community by a process of succession.

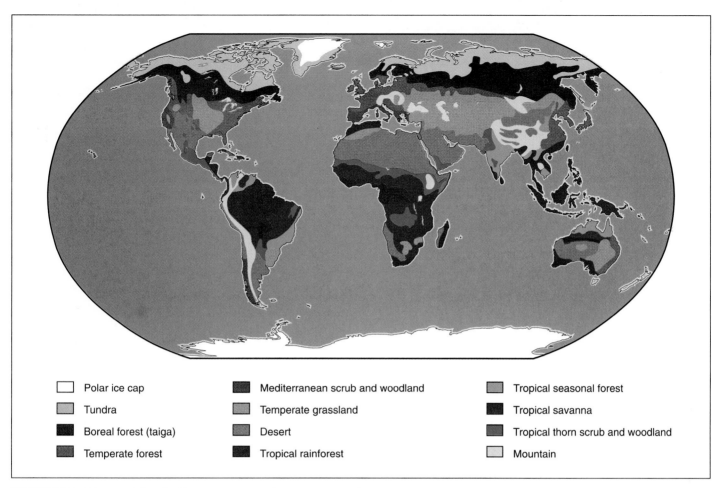

FIGURE 5.7 **Biomes of the World** Although most biomes are named for a major type of vegetation, each includes a specialized group of animals adapted to the plants and the biome's climatic conditions.

Legend:
- Polar ice cap
- Tundra
- Boreal forest (taiga)
- Temperate forest
- Mediterranean scrub and woodland
- Temperate grassland
- Desert
- Tropical rainforest
- Tropical seasonal forest
- Tropical savanna
- Tropical thorn scrub and woodland
- Mountain

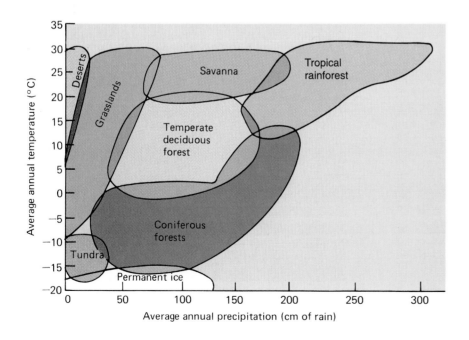

FIGURE 5.8 **Influence of Precipitation and Temperature on Vegetation** Temperature and moisture are two major factors that influence the kind of vegetation that can occur in an area. Areas with low moisture and low temperatures produce tundra; areas with high moisture and freezing temperatures during part of the year produce deciduous or coniferous forests; dry areas produce deserts; moderate amounts of rainfall or seasonal rainfall support grasslands or savannas; and areas with high rainfall and high temperatures support tropical rainforests.

City: Cairo, Egypt
Latitude: 29°52'N
Altitude: 381 ft. (116m)
Yearly precipitation: 0.7 in. (1.8cm)

Climate name: Hot desert
Other cities with similar climates:
  Mecca, Karachi

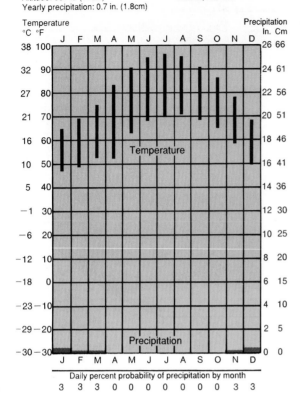

Daily percent probability of precipitation by month
3 3 3 0 0 0 0 0 0 0 3 3

a.

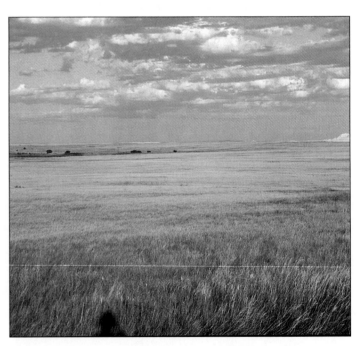

b.

FIGURE 5.9 Desert (*a*) Climagraph for Cairo, Egypt. (*b*) The desert receives less than 25 centimeters of precipitation per year, yet it teems with life. Cactus, sagebrush, lichens, snakes, small mammals, birds, and insects inhabit the desert. Because daytime temperatures are often high, most animals are active only at night, when the air temperature drops significantly. Cool deserts also exist in many parts of the world, where rainfall is low but temperatures are not high.

City: Tehran, Iran
Latitude: 35° 41'N
Altitude: 4002 ft. (1220m)
Yearly precipitation: 10.1 in. (26cm)

Climate name: Midlatitude dryland
Other cities with similar climates:
  Salt Lake City, Ankara

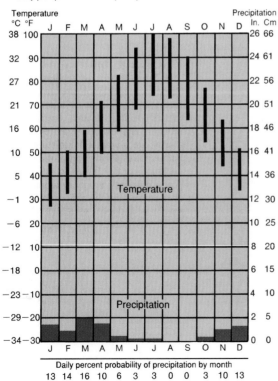

Daily percent probability of precipitation by month
13 14 16 10 6 3 3 0 0 3 10 13

a.

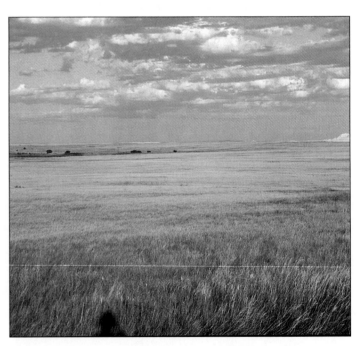

b.

FIGURE 5.10 Grassland (*a*) Climagraph for Tehran, Iran. (*b*) Grasses are better able to withstand low water levels than are trees. Therefore, in areas that have moderate rainfall, grasses are the dominant plants.

The Changing Nature of the Climax Concept

When European explorers traveled across the North American continent they saw huge expanses of land covered by the same kinds of organisms. Deciduous forests in the East, coniferous forests in the North, grasslands in central North America and deserts in the Southwest. These collections came to be considered the steady-state or normal situation for those parts of the world. When ecologists began to explore the way in which ecosystems developed over time they began to think of these ecosystems as the end point or climax of a long journey beginning with the formation of soil and its colonization by a variety of plants and other organisms.

As settlers removed the original forests or grasslands and converted the land to farming, the original "climax" community was destroyed. Eventually, as poor farming practices destroyed the soil, the farms were abandoned and the land was allowed to return to its "original" condition. This secondary succession often resulted in forests that resembled those that had been destroyed. However, in most cases these successional forests contained fewer species and in some cases were entirely different kinds of communities from the originals. These new stable communities were also called climax communities, but they were not the same as the originals.

Ecologists began to recognize that there was not a fixed, predetermined community for each part of the world and began to modify the way they looked at the concept of climax communities. The concept today is a more plastic one. It is still used to talk about a stable stage following a period of change, but ecologists no longer feel that land will eventually return to a "preordained" climax condition. They have also recognized in recent years that the type of climax community that develops depends on many other factors than simply climate. One of these is the availability of seeds to colonize new areas. Two areas with very similar climate and soil characteristics may contain different species because of the seeds available when the lands were released from agriculture. Furthermore, we need to recognize that the only thing that differentiates a "climax" community from a successional one is the time scale over which change occurs. "Climax" communities do not change as rapidly as successional ones. However, all communities are eventually replaced, as were the swamps that produced coal deposits, the preglacial forests of Europe and North America, and the pine forests of the Northeastern U.S.

So what should we do with this concept? Although the climax concept embraces a false notion that there is a specific end point to succession, it is still important to recognize that there is a predictable pattern of change during succession and that later stages in succession are more stable and longer lasting than early stages. Whether we call it a climax community is not really important.

---

during a 24-hour period. Since deserts receive little rainfall, it is logical that they have infrequent cloud cover. Since no clouds block out the sun, throughout the day the soil surface and the air above it tend to heat up rapidly. After the sun has set, the absence of clouds allows heat energy to be reradiated from the earth, and the area cools off rapidly. Cool to cold nights are typical even in "hot" deserts, especially during the winter months.

Another misconception about deserts is that they have limited species diversity. Many types of organisms live in deserts. They are specially adapted to survive in dry, often hot environments. For example, since water evaporates from the surfaces of leaves, many plants have very small leaves that allow them to conserve water. Some even lose their leaves entirely during the driest part of the year. Many desert plants are spiny. Some, like cactus, have the ability to store water in their spongy bodies for use during drier periods. Other plants have parts or seeds that lie dormant until the rains come. Then they grow rapidly, reproduce, and die, or become dormant until the next rains. Even the perennial plants are tied to the infrequent rains. During these times, the plants are most likely to produce flowers and reproduce.

The desert has many kinds of animals, often inconspicuous and overlooked because they are small or are inactive during the hot part of the day. They also aren't seen in large, conspicuous groups. Many lizards, snakes, small mammals, grazing mammals, carnivorous mammals, and birds are common in desert areas. All of the animals that live in deserts are able to survive with a minimal amount of water. Some get nearly all of their water from the moisture in the food they eat. They generally have a waterproof outer skin or cuticle, so they lose little water by evaporation. They often limit their activities to the cooler part of the day (the evening) and may spend considerable amounts of time in underground burrows during the day, which allows them to avoid extreme temperatures and to conserve water.

Grassland

Grasslands, also known as **prairies** or **steppes,** are widely distributed over the world. As with deserts, the major factor that contributes to the establishment of a grassland is the amount of available moisture. Grasslands generally receive between 25 and 75 centimeters of precipitation per year. These areas are windy with hot summers and cold to mild winters. Fire is an important force in this biome. Trees, which generally require greater amounts of water, are rare in these areas except along watercourses. (See figure 5.10.) Grasses make up 60 to 90 percent of the vegetation. Many other kinds of flowering plants are interspersed with the grasses.

Characteristic of these areas are large herds of migratory, grazing mammals like bison, wildebeests, wild horses, and various kinds of sheep, cattle, and goats. Most of the grasslands of the world have been converted to agriculture, since the rich, deep soil that developed as a result of the

City: Rangoon, Burma
Latitude: 16°46'N
Altitude: 18 ft. (5.5m)
Yearly precipitation: 99.2 in.
(250cm)

Climate name: Savanna (monsoon type)
Other cities with similar climates:
Bombay, Calcutta, Miami

Daily percent probability of precipitation by month
0 0 3 7 45 77 84 81 67 32 10 3

a.

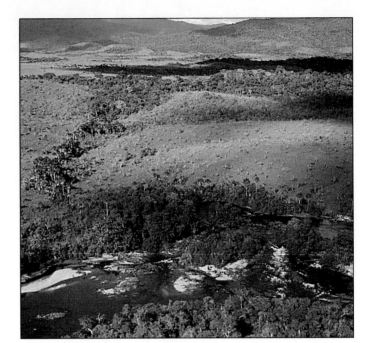

b.

FIGURE 5.11 Savanna (*a*) Climagraph for Rangoon, Burma. (*b*) Savannas develop in tropical areas that have seasonal rainfall. They typically have grasses as the dominant vegetation with drought and fire-resistant trees scattered through the area.

City: Singapore
Latitude: 1°20'N
Altitude: 33 ft. (11m)
Yearly precipitation: 100.7 in. (250cm)

Climate name: Tropical rainforest
Other cities with similar climates:
Colombo, Panama City, Jakarta, Lagos

Daily percent probability of precipitation by month
52 46 42 50 45 43 39 42 47 48 60 58

a.

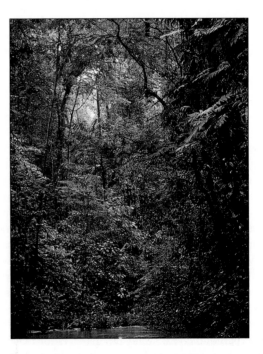

b.

FIGURE 5.12 Tropical Rainforest (*a*) Climagraph for Singapore. (*b*) Tropical rainforests develop in areas with high rainfall and warm temperatures. They have an extremely diverse mixture of plants and animals.

Grassland Succession

Because there are many kinds of grasslands, it is difficult to generalize about how succession takes place in these areas. Most grasslands in North America have been heavily influenced by agriculture and the grazing of domesticated animals. The grasslands reestablished in these areas may be somewhat different from the original ecosystem. However, there appear to be several stages typically involved in grassland succession.

After land is abandoned from cultivation, a short period of time, perhaps one to three years, elapses in which the field is dominated by annual weeds. In this respect, grassland succession is like deciduous forest succession. The next stage varies in length (10 or more years) and is dominated by annual grasses. Usually, in these early stages, the soil is in poor condition, lacking organic matter and nutrients. After several years, the soil fertility increases as organic material accumulates from the death and decay of annual grasses. This leads to the next stage in development, perennial grasses. Eventually, a mature grassland develops as prairie flowers invade the area and become interspersed with the grasses. In general, throughout this sequence, the soil becomes more fertile and of higher quality.

Because so much of the original North American grassland has been used for agriculture, when the land is allowed to return to a prairie, there may not be seeds of all of the original plants native to the area. Thus the grassland that results from secondary succession may not be exactly like the original; some species may be missing. Consequently, in many managed restorations of prairies, seeds that are no longer available in the local soils are introduced from other sources.

The low amount of rainfall and the fires typical of grasslands generally cause the successional process to stop at this point. However, if more water becomes available or if fire is prevented, woody trees may invade moist sites.

Actively farmed

Recently abandoned

Several years of succession

activities of centuries of soil building is useful for growing cultivated grasses like corn (maize) and wheat. The drier grasslands have been converted to the raising of domesticated grazers like cattle, sheep, and horses.

In addition to grazing mammals, many kinds of insects, including grasshoppers and other herbivorous insects, dung beetles (which feed on the dung of grazing animals), and flies (which bite the large mammals), are common. Small herbivorous mammals, such as mice and ground squirrels, are also common. Birds are often associated with grazing mammals. They eat the insects stirred up by the mammals or feed on the insects that bite them. Other birds feed on seeds and other plant parts. Reptiles (snakes and lizards) often feed on small mammals and insects.

Savanna

Parts of Africa, South America, and Australia have extensive grasslands spotted with occasional trees or patches of trees. (See figure 5.11.) This kind of a biome is often called a **savanna.** These areas of the world are typically tropical, with 50 to 150 centimeters of rain per year. However, the rain is not distributed evenly throughout the year. Typically, a period of heavy rainfall is followed by a prolonged drought. This results in a very seasonally structured ecosystem. The plants and animals time their reproductive activities to coincide with the rainy period, when limiting factors are least confining. The predominant plants are grasses, but many drought-resistant, flat-topped, thorny trees are common. The trees are also resistant to fire damage. Many of these trees are particularly important because they are involved in nitrogen fixation. They also provide shade and nesting sites for animals. As with grasslands, the predominant mammals are the grazers. Kangaroos in Australia, various species of antelope in Africa, and llamas in South America are examples. Many kinds of rodents, birds, insects, and reptiles are associated with this biome. Among the insects, mound-building termites are particularly common.

Tropical Rainforest

Tropical rainforests are located near the equator in Central and South America, Africa, Southeast Asia, and some islands in the Caribbean Sea and Pacific Ocean. (See figure 5.12.) The temperature is normally warm and relatively constant. There

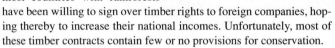

Destruction of the Rainforests

Two hundred years ago rainforests covered about 1500 million hectares, an area the size of Europe, but today only 900 million hectares remain. Modern technology makes short work of clearing the rainforest, taking less than an hour to clear one hectare. Exactly how much rainforest is disappearing is not known but it seems likely that over 20 million hectares are destroyed each year. At this rate there will be no rainforests remaining in 40 years.

Ranching

Large areas of rainforest are being burned down and converted into ranchland even though this is the worst possible use of the land. Central America has already lost two-thirds of its rainforest to cattle ranching. Where tropical rainforests have been cleared for pasture and cattle ranches established, the production of meat hardly reaches 50 kilograms of meat/hectare/year whereas North American farms produce more than 600 kilograms of meat/hectare/year. The soil quickly loses its scarce nutrients and becomes useless. When this happens the land is abandoned and more rainforest is cleared. The quickest way to clear an area of rainforest is by burning, often by aircraft dropping firebombs.

Logging

Rainforests were spared from exploitation in earlier years because of their inaccessibility, the relative low value of most of the trees for timber purposes, and the limited world demand. Recently this situation has changed and a wide variety of woods previously considered worthless are now used for pulp or as cellulose for the production of plastics. With new machines and better transportation it has become profitable to remove trees from previously remote areas. Logging companies may be interested in only one or two hardwood species, such as teak or mahogany. Solid wood is still used for furniture but more often a thin sheet of hardwood, called a veneer, is applied to a cheaper material. As there may be only three suitable trees in a hectare, to remove them may not seem a threat. However, heavy machinery is needed to clear a path and when the tree is felled and dragged through the forest, the amount of damage to other trees can be enormous. Faced with a high demand for their forest products, most countries with rainforests have been willing to sign over timber rights to foreign companies, hoping thereby to increase their national incomes. Unfortunately, most of these timber contracts contain few or no provisions for conservation.

Farming

Settlers make use of roads built by logging and mining companies to gain access to previously remote areas. Government policies often encourage people to settle in logged areas, however inappropriate they are for growing crops. The settlers move in and carry out shifting cultivation or slash-and-burn, cutting down the trees and other unwanted plants. They grow vegetables in the clearing for a few years, then abandon it and move on. After two or three crops have been taken up from a plot of land the store of nutrients is used up and crop yields become very low. When the tree canopy is removed the fierce heat of the sun dries and hardens the soil and the torrential rain washes it away.

Mining

Many rainforests are rich in oil deposits and mineral reserves such as bauxite, coal, copper, diamonds, gold, iron ore, nickel, tin, and uranium. While mining is a minor cause of deforestation, land is cleared for access and the mines are often in areas which are the only habitats of certain plant and animal species. The small amount of deforestation can cause a disproportionate amount of damage and the mining often releases toxic wastes into rivers destroying the animal and plant life.

is no frost, and it rains nearly every day. Most areas receive in excess of 200 centimeters of rain per year. Some receive 500 centimeters or more. Because of the warm temperatures and abundant rainfall, most plants grow very rapidly; however, soils are often poor in nutrients because water tends to carry away any nutrients not immediately taken up by plants. Many of the trees have extensive root networks, associated with fungi, near the surface of the soil that allow them to capture nutrients from decaying vegetation before they can be carried away. Most of the nutrients in a tropical rainforest are tied up in the biomass, not in the soil, and, therefore, most of these areas do not make good farmland.

These forests have a very large number of species. A small area of a few square kilometers is likely to have hundreds of species of trees. Balsa, teakwood, and many other ornamental woods are from tropical trees. The trees are usually in two or three layers, with some standing taller than their neighbors. Since most of the sunlight is captured by the trees, many shade-tolerant plants live beneath the trees' canopy. Each tree serves as a surface for the growth of ferns, mosses, orchids, and vines. Associated with this variety of plants is an equally large variety of animals. Insects, such as ants, termites, moths, butterflies, and beetles, are particularly abundant. Birds also are extremely common, as are many climbing mammals, lizards, and tree frogs. Since flowers and fruits are available throughout the year, there are many kinds of nectar- and fruit-feeding birds and mammals. Because of the low light levels, many of the animals communicate by making noise.

More species are found in the tropical rainforests of the world than in the rest of the world combined. Recently, biologists discovered a whole new community of organisms that live in the canopy (leafy tops) of these forests. The dense vegetation reduces the amount of light on the forest floor. This, coupled with the bewildering array of plants and animals, makes study of this kind of ecosystem difficult.

Rainforest Products

It might be assumed that the rainforest is not important in our everyday lives. This assumption is far from the truth since much of the food we eat and materials we use originally came from the rainforest. Every time we drink coffee, eat chocolate, bananas, or nuts or use anything made of rubber, we are enjoying products from the rainforests. Many rainforest plants have been collected as seeds or plant cuttings and then bred for cultivation. Sometimes these cultivated varieties are attacked by new diseases, yields fall, and plants die. Natural or "wild" strains are needed to crossbreed to produce new disease-resistant plants.

Below is a list of some products that originated in the rainforests, and their uses:

Fruit: pineapples, bananas, oranges, lemons, limes, grapefruit, and tangerines.

Vegetables: tomatoes, avocados, and many types of beans.

Nuts: brazil nuts, cashew nuts, peanuts, sesame seeds, and coconuts.

Spices: chilies, pepper, cloves, nutmeg, vanilla, cinnamon, and turmeric.

Drinks: tea, coffee, cocoa, chocolate, kola nut extract used in Coca-Cola.

Gums and resins: rubber for household and industrial goods, copals for paints and varnishes, chicle gum used for chewing gum, balata used in golf balls.

Perhaps the greatest benefits within the rainforests lie in the realm of pharmacology. Plants contain many substances known as phytochemicals which enable them to survive in the rainforest's unique environment. There is a one-in-four chance that the next time you enter a pharmacy for a prescription you will leave with a product derived from the rainforest. Over 2000 plants that promise to be valuable in treating cancer have been discovered but species are being eliminated faster than they can be identified and researched. Seventy percent of plants identified as having anti-cancer properties come from the rainforests. In fact, almost 50 percent of our medications are derived from plants yet less than 1 percent of tropical rainforest species have been examined for their possible value in medicine.

ENVIRONMENTAL CLOSE-UP

Forest Canopy Studies

In the past few years, a new frontier of ecological study has developed. Scientists have traditionally looked at forest ecosystems at ground level. Trees were identified, understory species were categorized, and the animals that live on or near the forest floor were studied. Gradually, scientists began to realize that many kinds of plants and animals that are important parts of forest ecosystems rarely descend from the tops of trees to the forest floor. They began to devise methods for studying these canopy-dwelling organisms. Initially, they relied on techniques and gear used by mountain climbers, but that approach was labor intensive and dangerous. Recently, ecologists have established several sites where large construction cranes have been built in forests. These cranes allow researchers to study the chemistry, climate, and organisms found in the canopy.

Several surprising discoveries have been made, including the discovery of new species of insects. A canopy study in an old-growth temperate rainforest on Vancouver Island, British Columbia, identified more than 60 new species of insects. Researchers estimate that hundreds of new species will eventually be identified. Similar studies in tropical forests have identified hundreds more new species of insects. Studies in Panama recognized that vines (lianas) can constitute up to 70

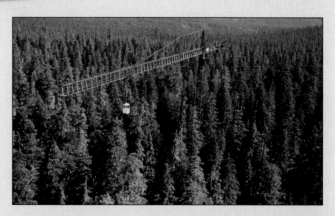

percent of the forest canopy and that these vines compete for sunlight with the trees that support them.

The way in which animals use the canopy is another interesting area of research. Birds, bats, monkeys, squirrels, and insects use specific parts of the forest canopy for food, nesting, and travel routes. Researchers are learning much as they spend time in the forest canopy with the animals.

City: Chicago, Illinois
Latitude: 41°52'N
Altitude: 595 ft. (181m)
Yearly precipitation: 33.3 in. (85cm)

Climate name: Humid continental (warm summer)
Other cities with similar climates:
New York, Berlin, Warsaw

Daily percent probability of precipitation by month
35   36   39   37   39   37   29   29   30   29   33   35

a.

b.

FIGURE 5.13   Temperate Deciduous Forest   (*a*) Climagraph for Chicago, Illinois. (*b*) A temperate deciduous forest develops in areas that have significant amounts of moisture throughout the year, but where the temperature falls below freezing for parts of the year. During this time the trees lose their leaves. This kind of forest once dominated the eastern half of the United States and southeastern Canada.

City: Moscow, Russia
Latitude: 55°46'N
Altitude: 505 ft. (154m)
Yearly precipitation: 21.8 in. (55cm)

Climate name: Humid continental (cool summer)
Other cities with similar climates:
Montreal, Winnipeg, Leningrad

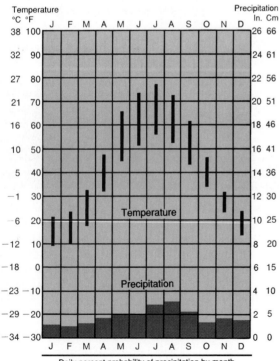

Daily percent probability of precipitation by month
35   32   26   30   29   33   39   39   30   35   33   29

a.

b.

FIGURE 5.14   Taiga, Northern Coniferous Forest, or Boreal Forest   (*a*) Climagraph for Moscow. (*b*) The taiga, northern coniferous forest, or boreal forest occurs in areas with long winters and heavy snowfall. The trees have adapted to these conditions and provide food and shelter for many of the animals that live there.

Old-Growth Temperate Rainforests of the Pacific Northwest

The coastal areas of northern California, Oregon, Washington, British Columbia, and southern Alaska have an unusual set of environmental conditions that support a special kind of forest, a temperate rainforest. The prevailing winds from the west bring moisture-laden air to the coast. As this air meets the coastal mountains and is forced to rise, it cools and the moisture falls as rain or snow. Most of these areas receive 200 or more centimeters of precipitation per year. This abundance of water, along with fertile soil and mild temperatures, results in a luxuriant growth of plants.

Sitka spruce, Douglas fir, and western hemlock are typical evergreen coniferous trees. Undisturbed (old-growth) forests of this region have trees as old as 800 years that are almost as tall as the length of a football field. Deciduous trees of various kinds (red alder, bigleaf maple, black cottonwood) grow in places where they can get enough light. All trees are covered with mosses, ferns, and other plants that grow on their surface. The dominant color is green, since most surfaces have something growing on them.

When a tree dies and falls to the ground, it rots in place and often serves as a site for the establishment of new trees. This is such a common feature of the forest that the fallen, rotting trees are called nurse trees. The fallen trees also serve as a food source for a variety of insects, which are food for a variety of other animals. Some animals, such as the northern spotted owl, the marbled murrelet (a seabird), and the Roosevelt elk seem to be dependent on undisturbed forest.

Because of the rich resource of trees, 90 percent of the original temperate rainforest has been logged. What remains has become a source of controversy. Some maintain that it should be protected as a remnant of the original forest of the region, just as small patches of prairie and eastern woodland have been preserved in other parts of North America. Others maintain that the trees are a resource to be used and that they should be harvested for lumber since they are old and will die in the near future. Both sides in the controversy have taken firm positions so that the issue is not likely to be resolved without legislation or legal action.

Temperate Deciduous Forest

Forests in temperate areas of the world that have a winter–summer change of seasons typically have trees that lose their leaves during the winter and replace them the following spring. This kind of forest is called a **temperate deciduous forest** and is typical of the eastern half of the United States, parts of south central and southeastern Canada, southern Africa, and many areas of Europe and Asia.

These areas generally receive 100 or more centimeters of relatively evenly distributed precipitation per year. Each area of the world has certain species of trees that are the major producers for the biome. In contrast to tropical rainforests, where individuals of a tree species are scattered throughout the forest, temperate deciduous forests generally have many fewer species, and many forests may consist of two or three dominant tree species. In deciduous forests of North America and Europe, common species are maples, aspen, birch, beech, oaks, and hickories. These tall trees shade the forest floor, where many small flowering plants bloom in the spring. These spring wildflowers store food in underground structures. In the spring, before the leaves come out on the trees, the wildflowers can capture sunlight and reproduce before they are shaded. Many smaller shrubs also may be found in these forests. (See figure 5.13.)

These forests are home to a great variety of insects, many of which use the leaves and wood of trees as food. Beetles, moth larvae, wasps, and ants are examples. The birds that live in these forests are primarily migrants that arrive in the spring of the year, raise their young during the summer, and leave in the fall. Many of these birds rely on the large summer insect population for their food. A few kinds of birds, including woodpeckers, grouse, turkeys, and some of the finches, are year-round residents. Several kinds of small and large mammals inhabit these areas. Mice, squirrels, deer, moles, and rabbits are common examples. Major predators are foxes, badgers, weasels, coyotes, and birds of prey.

Taiga, Northern Coniferous Forest, or Boreal Forest

Throughout the southern half of Canada, parts of northern Europe, and much of Russia, there is a coniferous forest known as the **taiga, northern coniferous forest,** or **boreal forest.** Spruces, firs, and larches are the trees most common in these areas. (See figure 5.14.) The climate is one of short, cool summers and long winters with abundant snowfall. The winters are extremely harsh and can last as long as six months. Typically, the soil freezes during the winter. Precipitation ranges between 25 and 100 centimeters per year, but the climate is humid because the generally low temperatures during all parts of the year reduce evaporation. The landscape is typically dotted with lakes, ponds, and bogs.

The trees are specifically adapted to winter conditions. Winter is relatively dry

City: Fairbanks, Alaska
Latitude: 64°51'N
Altitude: 440 ft. (134m)
Yearly precipitation: 12.4 in. (31.5cm)

Climate name: Subarctic tundra
Other cities with similar climates:
Yellowknife, Yakutsk

Daily percent probability of precipitation by month
32 21 19 13 29 33 42 48 33 35 33 23

a.

b.

FIGURE 5.15   Tundra   (*a*) Climagraph for Fairbanks, Alaska. (*b*) In the northern latitudes and on the tops of some mountains, the growing season is short and plants grow very slowly. Trees are unable to live in these extremely cold areas, in part because there is a permanently frozen layer of soil beneath the surface, known as the permafrost. Because growth is so slow, damage to the tundra can still be seen generations later.

as far as the trees are concerned because the moisture falls as snow and stays above the soil until it melts in the spring. The needle-shaped leaves are adapted to prevent water loss; the larch loses its needles in the fall. The branches of these trees are flexible, allowing the snow to slide off the pyramid-shaped trees without greatly damaging them. As with the temperate deciduous forest, many of the inhabitants of this biome are temporarily active during the summer. Most birds are migratory and feed on the abundant insect population, which becomes inactive during the long, cold winter. A few woodpeckers, owls, and grouse are permanent residents. Typical mammals are deer, caribou, moose, wolves, weasels, mice, and squirrels. Because of the cold, few reptiles and amphibians live in this biome.

Tundra

North of the taiga is a biome that lacks trees and has a permanently frozen soil layer or **permafrost.** It is known as the **tundra.** (See figure 5.15.) Scattered patches of tundralike communities also are found on mountaintops throughout the world. These are known as **alpine tundra.** Because of the permanently frozen soil and extremely cold, windy climate (up to 10 months of winter), no trees can live in the area. During the brief summer, the top layers of the soil thaw, and many plants and lichens (reindeer moss) grow. The plants are short, usually less than 20 centimeters. A few hardy mammals like musk oxen, caribou or reindeer, arctic hare, and lemmings can survive by feeding on the grasses and other plants that grow during the short, cool summer. Although the amount of precipitation is similar to that in some deserts (25 cm per year), the short summer is generally moist because of the melting of the snow that falls during the winter. In addition, the permafrost does not let the water sink into the soil, resulting in waterlogged soils and many shallow ponds and pools. Consequently, many waterfowl like ducks and geese migrate to the tundra, where they mate and raise their young. Insects are common during the summer and serve as food for migratory birds. Permanent resident birds are the ptarmigan and snowy owl. No reptiles or amphibians survive in this extreme climate. Because of the very short growing season, damage to this kind of ecosystem is slow to heal, so the land must be handled with a great deal of care.

Altitude and Latitude

The distribution of terrestrial ecosystems is primarily related to precipitation and temperature. The temperature is warmest near the equator and becomes cooler toward the poles. Similarly, as the altitude increases, the average temperature decreases. This means that even at the equator it is possible to have cold temperatures on the peaks of tall mountains. As one proceeds from sea level to the tops of mountains, it is possible to pass through a series of biomes that are similar to what would be encountered as one traveled from the equator to the North Pole. (See figure 5.16.)

MAJOR AQUATIC ECOSYSTEMS

Terrestrial biomes are determined by the amount and kind of precipitation and by temperatures. Other factors, such as soil type and wind, also play a part. Aquatic ecosystems also are shaped by primary determiners. Four such factors are the ability of the sun's rays to penetrate the water, the nature of the bottom substrate, the water temperature, and the amount of dissolved materials.

Marine Ecosystems

An important determiner of the nature of aquatic ecosystems is the amount of salt dissolved in the water. Those that have little dissolved salt are called **fresh water ecosystems** and those that have a high salt content are called **marine ecosystems.**

Pelagic Marine Ecosystems

In the open ocean, many kinds of crustaceans, fish, and whales swim actively as they pursue food. These kinds of animals are called **pelagic,** and the ecosystem they are a part of is called a **pelagic ecosystem.** As with all ecosystems, the organisms at the bottom of the energy pyramid carry on photosynthesis. In the ocean, a majority of these organisms are small, microscopic, floating algae called **phytoplankton.** Since sunlight cannot penetrate to great depths in water, these phytoplankton are located in the upper regions. This region where the sun's rays penetrate is known as the **euphotic zone.** Small, weakly swimming animals of many kinds, known as **zooplankton,** feed on the phytoplankton. Zooplankton are often located at a greater depth in the ocean than the phytoplankton but migrate upward at night and feed on the large population of phytoplankton. The zooplankton are in turn eaten by larger animals like fish and larger shrimp, which are eaten by larger fish like salmon, tuna, sharks, and mackerel. (See figure 5.17.)

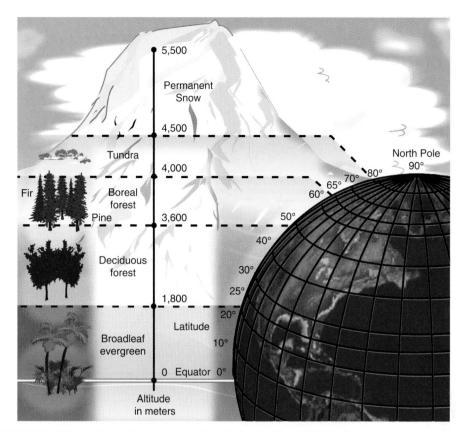

FIGURE 5.16 Relationship Between Altitude, Latitude, and Vegetation As one travels up a mountain, the climate changes. The higher the elevation, the cooler the climate. Even in the tropics tall mountains can have snow on the top. Thus it is possible to experience the same change in vegetation by traveling up a mountain as one would experience traveling from the equator to the North Pole.

A major factor that influences the nature of a marine community is the kind and amount of material dissolved in the water. Probably more important is the amount of nutrients available to the organisms carrying on photosynthesis. Phosphorus, nitrogen, and carbon are all required for the construction of new living material. In water, these are often in short supply. Therefore, the most productive aquatic ecosystems are those in which these essential nutrients are most common. These areas include places in oceans where currents bring up nutrients that have settled to the bottom and areas where rivers deposit their load of suspended and dissolved materials.

Benthic Marine Ecosystems

Organisms that live on the ocean bottom, whether attached or not, are known as **benthic** organisms, and the kind of ecosystem that consists of these organisms is called a **benthic ecosystem.** Some fish, clams, oysters, various crustaceans, sponges, sea anemones, and many other kinds of organisms live on the bottom. In shallow water, sunlight can penetrate to the bottom, and a variety of attached photosynthetic organisms like kelp (commonly called seaweeds) are common. Since they are attached and can grow to very large size, many other bottom-dwelling organisms are associated with them.

The substrate is very important in determining the kind of benthic community that develops. Sand tends to shift and move, making it difficult for large plants or algae to become established, although some clams, burrowing worms, and small crustaceans find sand to be a suitable habitat. Clams filter water for plankton and detritus, or burrow through the sand, feeding on other inhabitants. Muds may provide suitable habitats for some kinds of rooted plants. Muds generally have little oxygen in them but still may be inhabited by a

FIGURE 5.17 Marine Ecosystems   All of the photosynthetic activity of the ocean occurs in shallow water called the euphotic zone, either by attached algae near the shore or by minute phytoplankton in the upper levels of the open ocean. Consumers are either free-swimming pelagic organisms or benthic organisms that live on the bottom. Small animals that feed on phytoplankton are known as zooplankton.

variety of burrowing organisms that feed by filtering the water above them or feed on other animals in the mud. Rocky surfaces in the ocean provide a good substrate for many kinds of large algae. Associated with this profuse growth of algae is a large variety of animals. (See figure 5.18.)

At great depths in the ocean is an ecosystem that must rely on a continuous rain of organic matter from the euphotic zone. These areas are known as abyssal areas, and the ecosystem is known as an **abyssal ecosystem.** Essentially, all of the organisms in this environment are scavengers that feed on whatever drifts their way.

Temperature also has an impact on the kind of benthic community established. Some communities, such as coral reefs or mangrove swamps, are found only in areas where the water is warm. **Coral reef ecosystems** are the result of large numbers of small animals that build cupshaped external skeletons around themselves. They are able to protrude from their skeletons to capture food and expose themselves to the sun. This is important because corals contain single-celled algae within their bodies. These algae carry on photosynthesis and provide both themselves and the coral with the nutrients necessary for growth. Because they require warm water, coral ecosystems are found only near the equator. Coral ecosystems also require shallow, clear water since the algae must have ample sunlight to carry on photosynthesis. This mutualistic relationship between algae and coral is the basis for a very productive community of organisms. The skeletons of the corals provide a surface upon which many other kinds of animals live. Some of these animals feed on corals directly, while others feed on small plankton and bits of algae that establish themselves among the coral organisms. Many kinds of fish, crustaceans, sponges, clams, and snails are members of coral reef ecosystems. Coral reefs are considered one of the most productive ecosystems on earth. (See figure 5.19.)

Mangrove swamp ecosystems occupy a region near the shore. The dominant organisms are special kinds of trees that are able to tolerate the high salt content of the ocean. In areas where the water is shallow and wave action is not too great, the trees can become established. They have long seeds that float in the water. When the seeds become trapped in mud, they take root. The trees can excrete salt from their leaves. They also have extensively developed roots that are above the water, where they can obtain oxygen and prop up the plant. The trees trap sediment and provide places for crabs, jellyfish, sponges, and fish to live. The trapping of sediment and the continual movement of mangroves into shallow areas result in the development of a terrestrial ecosystem in what was once shallow ocean. Mangroves are found in South Florida, the Caribbean, Southeast Asia, Africa, and other parts of the world where conditions are suitable. (See figure 5.20.)

Estuaries

Estuaries, a special category of marine ecosystem, consist of shallow, partially enclosed areas where freshwater enters the ocean. The saltiness of the water in the estuary changes with tides and the flow of water from rivers. The organisms that live here are specially adapted to this set of physical conditions, and the number of species is less than in the ocean or in freshwater. Estuaries are particularly productive ecosystems because of the large amounts of nutrients introduced into the basin from the rivers that run into them. This is further enhanced by the fact that the shallow water allows light to penetrate to most of the water in the basin. Phytoplankton and attached algae and plants are able to use the sunlight and the nutrients for rapid growth. This photosynthetic activity supports many kinds of organisms in the estuary. Estuaries are especially important as nursery sites for fish and crustaceans like flounder and shrimp. The adults enter these productive, sheltered areas to reproduce and then return to the ocean. The young spend their early life in the estuary and eventually leave as they get larger and are more able to survive in the ocean. Estuaries also trap sediment. This activity tends to prevent many kinds of pollutants from reaching the ocean and also results in the gradual filling in of the estuary, which may eventually become a salt marsh and then part of a terrestrial ecosystem.

Freshwater Ecosystems

Freshwater ecosystems differ from marine ecosystems in several ways. The amount of salt present is much less, the temperature of the water can change greatly, the water is in the process of moving downhill, oxygen can often be in short supply, and the organisms that inhabit freshwater systems are different.

Freshwater ecosystems can be divided into two categories: those in which the water is relatively stationary, such as lakes, ponds, and reservoirs, and those in which the water is running downhill, such as streams and rivers.

Lakes and Ponds

Large lakes have many of the same characteristics as the ocean. If the lake is deep, there is a euphotic zone at the top, with many kinds of phytoplankton, and zooplankton that feed on the phytoplankton. Small fish feed on the zooplankton, which are in turn eaten by larger fish. The species of organisms found in freshwater lakes are different from those found in the ocean, but the roles played are similar, so the same terminology is used.

Along the shore and in the shallower parts of lakes, the euphotic zone contains many kinds of flowering plants. They are rooted in the bottom and some have leaves that float on the surface or protrude above the water. These are called **emergent plants.** Cattails, bulrushes, arrowhead plants, and water lilies are examples. Rooted plants that stay submerged below the surface of the water are called **submerged plants.** *Elodea* and *Chara* are examples.

Many kinds of freshwater algae also grow in the shallow water, where they may appear as mats on the bottom or attached to vegetation and other objects in the water. Associated with the plants and algae are a large number of different kinds of animals. Fish, crayfish, clams, and many kinds of aquatic insects are common inhabitants of this mixture of plants and algae. This region, with rooted vegetation, is known as the **littoral zone,** and the portion of the lake that does not have rooted vegetation is called the **limnetic zone.** (See figure 5.21.)

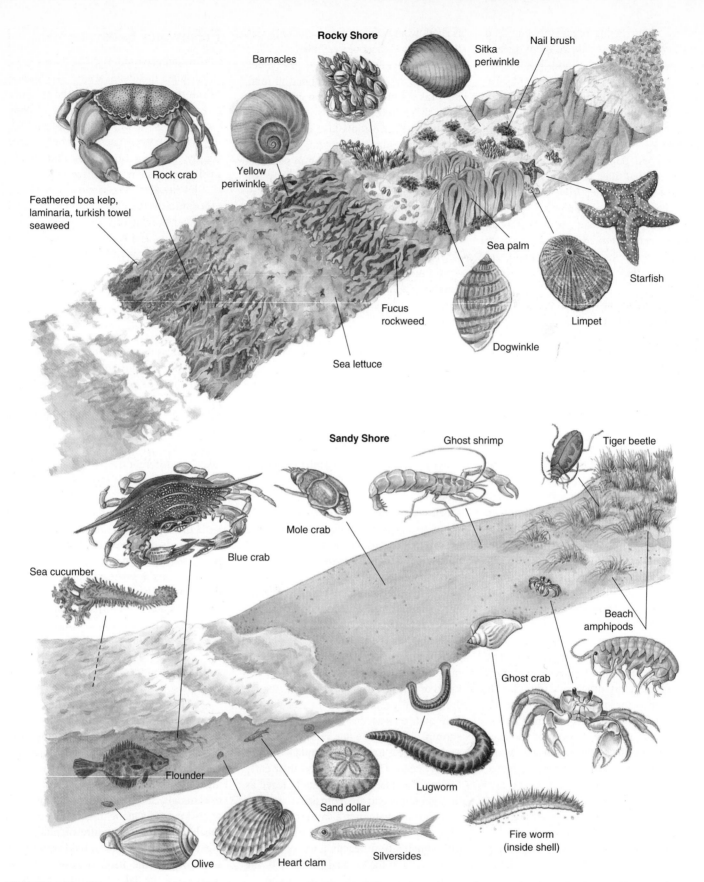

FIGURE 5.18 Types of Shores The kind of substrate determines the kind of organisms that can live near the shore. Rocks provide areas for attachment that sands do not, since sands are constantly shifting. Muds usually have little oxygen in them; therefore, the organisms that live there must be adapted to those kinds of conditions.

FIGURE 5.19   Coral Reef   Corals are small sea animals that secrete external skeletons. They have a mutualistic relationship with certain algae, which allows both kinds of organisms to be very successful. The skeletal material serves as a substrate for many other kinds of organisms to live.

FIGURE 5.20   Mangrove Swamp   Mangroves are tropical trees that are able to live in very wet, salty muds found along the ocean shore. Since they are able to trap additional sediment, they tend to extend farther seaward as they reproduce.

The productivity of the lake is determined by several factors. Temperature is important, since cold temperatures tend to reduce the amount of photosynthesis. Shallow lakes tend to be more productive because most of the water has sunlight penetrating it and, therefore, photosynthesis can occur to the bottom of the lake. Shallow lakes also tend to be warmer as a result of the warming effects of the sun's rays. A third factor that influences the productivity of lakes is the amount of nutrients present. This is primarily determined by the rivers and streams that carry nutrients to the lake. River systems that run through areas that donate many nutrients

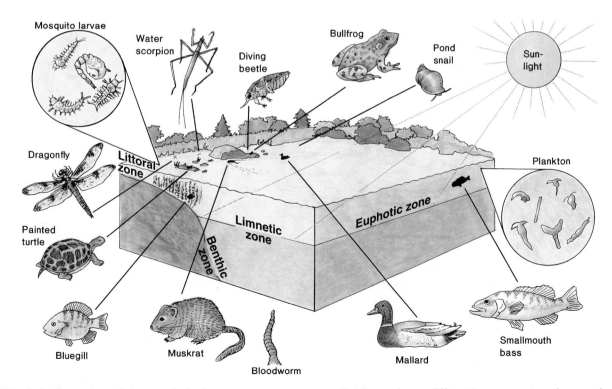

FIGURE 5.21   Lake Ecosystem   Lakes are similar in structure to oceans except that the species are different because most marine organisms cannot live in fresh water. Insects are common organisms in freshwater lakes, as are many kinds of fish, zooplankton, and phytoplankton,

Wherever humans have had major influence on ecosystems, they have changed them substantially. In most of the world, all the land that can be used for agriculture has been converted to that use. In the process, entire ecosystems have been destroyed and the original complex mixtures of organisms have been replaced by more simple, managed agricultural ecosystems. In North America, much of the original, temperate deciduous forest has been cleared of trees to provide farmland. Similarly, most of the prairie ecosystems have been plowed to produce corn, wheat, and other grains. In other cases, human activity has accidentally altered ecosystems. For example, rivers have been dammed or controlled to provide protection from floods or to facilitate river navigation.

Today, there is interest in restoring some of these altered ecosystems. This is not as easy as it sounds. To restore the Everglades, for example, would require that water currently being diverted for municipal and agricultural use be returned to the Everglades. Restoring a prairie often involves planting seeds of native plants, since farmers would have systematically removed native prairie plants as weeds, thus eliminating the seeds. These seeds must be returned to the soil to reestablish a prairie. Furthermore, since many prairie species are adapted to periodic fires, it may be necessary to burn the prairie as a normal management practice. The absence of fires allows fire-adapted species to be crowded out of the ecosystem.

All of these restorations require an investment of time and money to preserve some small part of the original ecosystem.

- What values are there in protecting special ecosystems?
- Should society bear the cost of removing land from cultivation and replanting it to prairie or temperate forest?
- Who would be inconvenienced if water were returned to the Everglades?

---

will carry the nutrients to the lakes. Exposed soil and farmland tend to release nutrients, as do other human activities like depositing sewage into streams and lakes. Deep, cold, nutrient-poor lakes are low in productivity and are called **oligotrophic lakes.** Shallow, warm, nutrient-rich lakes are called **eutrophic lakes.**

Although water (H_2O) has oxygen as part of its molecular structure, this oxygen is not available to organisms. The oxygen that they need is molecular oxygen (O_2), which enters water from the air or as a result of photosynthesis by aquatic plants. Wave action as wind blows across the surface tends to mix air and water and allows more oxygen to dissolve in the water.

Dissolved oxygen content in the water is an important factor since the quantity determines the kinds of organisms that can inhabit the lake. Organic materials in the lake result from the metabolic wastes of organisms that live in and around the lake, plants and animals that die, and parts of organisms, such as leaves from trees, that fall into the lake. The breakdown of these organic materials by bacteria and fungi uses oxygen from the water. This is called the **biochemical oxygen demand** or **BOD.** The amount and kinds of organic matter determine, in part, how much oxygen is left to be used by other organisms, such as fish, crustaceans, and snails. Many lakes may experience periods when oxygen is low, resulting in the death of fish and other organisms. These topics are discussed in greater depth in chapter 15.

Streams and Rivers

Streams and rivers are a second category of freshwater ecosystem. Since the water is moving, plankton organisms are less important than are attached organisms. Most algae grow attached to rocks and other objects on the bottom. This collection of algae, animals, and fungi is called the **periphyton.** Since the water is shallow, light can penetrate easily to the bottom (except for large or extremely muddy rivers). Even so, it is difficult for photosynthetic organisms to accumulate the nutrients necessary for growth, and most clear-water streams are not very productive. As a matter of fact, the major input of nutrients is from organic matter that falls into the stream from terrestrial sources. These are primarily the leaves from trees and other vegetation, as well as the bodies of living and dead insects. Within the stream is a community of organisms that are specifically adapted to use the debris as a source of food. Bacteria and fungi colonize the organic matter, and many kinds of insects shred the material as they eat it and the fungi and bacteria living on it. The feces (intestinal wastes) of these insects and the tiny particles produced during the eating process become food for other insects that build nets to capture the tiny bits of organic matter that drift their way. These insects are in turn eaten by carnivorous insects and fish.

Organisms in larger rivers and muddy streams, which have less light penetration, rely on the accumulation of food that drifts their way from the many streams that empty into the river. These larger rivers tend to be warmer and to have slower-moving water. Consequently, the amount of oxygen is usually less, and the species of plants and animals change. Any additional organic matter added to the river system adds to the BOD, further reducing the oxygen in the water. Plants may become established along the shore and contribute to the ecosystem by carrying on photosynthesis and providing hiding places for animals.

Just as estuaries are a bridge between freshwater and marine ecosystems, swamps and marshes are a transition between aquatic and terrestrial ecosystems. **Swamps** contain trees that are able to live in places that are either permanently flooded or flooded for a major part of the year. **Marshes** are dominated by grasses and reeds. Many swamps and marshes are successional states that eventually become totally terrestrial communities.

Summary

Ecosystems change as one kind of organism replaces another in a process called succession. Ultimately, a relatively stable stage is reached, called the climax community. Succession may begin with bare rock or water, in which case it is called primary succession, or may occur when the original ecosystem is destroyed, in which case it is called secondary succession. The stages that lead to the climax are called successional stages.

Major regional terrestrial climax communities are called biomes. The primary determiners of the kinds of biomes that develop are the amount and yearly distribution of rainfall and the yearly temperature cycle. Major biomes are desert, grassland, savanna, tropical rainforest, temperate deciduous forest, taiga, and tundra. Each has a particular set of organisms that is adapted to the climatic conditions typical for the area. As one proceeds up a mountainside, it is possible to witness the same kind of change in biomes that occurs if one were to travel from the equator to the North Pole.

Aquatic ecosystems can be divided into marine (saltwater) and freshwater ecosystems. In the ocean, some organisms live in open water and are called pelagic organisms. Light penetrates only the upper few meters of water; therefore, this region is called the euphotic zone. Tiny photosynthetic organisms that float near the surface are called phytoplankton. They are eaten by small animals known as zooplankton, which in turn are eaten by fish and other larger organisms.

The kind of material that makes up the shore determines the mixture of organisms that lives there. Rocky shores provide surfaces for organisms to attach; sandy shores do not. Muddy shores are often poor in oxygen, but marshes and swamps may develop in these areas. Coral reefs are tropical marine ecosystems dominated by coral animals. Mangrove swamps are tropical marine shoreline ecosystems dominated by trees. Estuaries occur where freshwater streams enter the ocean. They are usually shallow, very productive areas. Many marine organisms use estuaries for reproduction.

Insects are common in freshwater and are absent in marine systems. Lakes show similar structure to the ocean, but the species are different. Deep, cold-water lakes with poor productivity are called oligotrophic, while shallow, warm-water, highly productive lakes are called eutrophic. Streams differ from lakes in that most of the organic matter present in streams falls into it from the surrounding land. Thus, organisms in streams are highly sensitive to the land uses that occur near the streams.

Review Questions

1. Describe the process of succession. How does primary succession differ from secondary succession?
2. How does a climax community differ from a successional community?
3. List three characteristics typical of each of the following biomes: tropical rainforest, desert, tundra, taiga, savanna, grassland, and temperate deciduous forest.
4. What two primary factors determine the kind of terrestrial biome that will develop in an area?
5. How does altitude affect the kind of biome present?
6. What areas of the ocean are the most productive?
7. How does the nature of the substrate affect the kinds of organisms found at the shore?
8. What is the role of each of the following organisms in a marine ecosystem: phytoplankton, zooplankton, algae, coral organisms, and fish?
9. List three differences between freshwater and marine ecosystems.
10. What is an estuary?

Critical Thinking Questions

1. Does the concept of a "climax community" make sense? Why or why not?
2. What do you think about restoring ecosystems that have been degraded by human activity? Should it be done or not? Why? Who should pay for this reconstruction?
3. Since ecosystems change as environments change, how do you decide what form of the ecosystem to use as an example of what to return to?
4. We humans seem to often deny our place in the "natural world." Do you see human-made changes as part of the natural world? How do you make distinctions between human-made changes and natural changes? What lines do you draw to separate acceptable human-made changes from unacceptable human ones?
5. Identify the biome in which you live. What environmental factors help make this a biome? What is the current health of your biome? What are the current threats to its health? How might your biome have looked 100, 1000, 10,000 years ago?
6. Using the thinking you have done in question 15, what would it take to restore your biome to its previous state? Would you favor a change in this direction? What are some of the implications for not restoring? For restoring?
7. Imagine you are a conservation biologist who is being asked by local residents what the likely environmental outcomes of development would be in the tropical rainforest in which they live. What would you tell them? Why do you give them this evaluation? What evidence can you cite for your claims?
8. The text says that 90% of the old growth temperate rainforest in the Pacific Northwest has been logged. What to do with the remaining 10% is still a question. Some say it should be logged and others say it should be preserved. What values, beliefs, and perspectives are held by each side? What is your ethic regarding logging old growth in this area? What values, beliefs, and perspectives do you hold regarding this issue?
9. Much of the old growth forest in the United States has been logged, economic gains have been realized, and secondary growth forests have become established. This is not the case in the tropical rainforests, although they are being lost at alarming rates. Should developed countries, who have already "cashed in" on their resources, have anything to say about what is happening in developing countries? Why do you think the way you do?

Key Terms

abyssal ecosystem *85*
alpine tundra *82*
benthic *83*
benthic ecosystem *83*
biochemical oxygen demand
  (BOD) *88*
biome *71*
boreal forest *81*
climax community *69*
coral reef ecosystem *85*
desert *71*
emergent plants *85*
estuary *85*
euphotic zone *83*
eutrophic lake *88*
freshwater ecosystem *83*
grassland *75*
limnetic zone *85*
littoral zone *85*
mangrove swamp ecosystem *85*
marine ecosystem *83*
marsh *88*
northern coniferous forest *81*

oligotrophic lake *88*
pelagic *83*
pelagic ecosystem *83*
periphyton *88*
permafrost *82*
phytoplankton *83*
pioneer community *69*
prairie *75*
primary succession *69*
savanna *77*
secondary succession *69*
seral stage *69*
sere *69*
steppe *75*
submerged plants *85*
succession *69*
successional stage *69*
swamp *88*
taiga *81*
temperate deciduous forest *81*
tropical rainforest *77*
tundra *82*
zooplankton *83*

Related Internet Sites

Visit our website at http://www.mhhe.com/environmentalscience/ for specific information on each of the topics below.

The state of the world's rainforests

Rainforest preservation

Sustainable forests directory

Forestry information

American forests and policies surrounding them

Coral forests and how to save the reefs

Coastal management and resource center

Great Lakes region of North America

Dams, logging, and other environmental issues in the Northwest

Hubbard Brook Ecosystem Study. Forest site information, research, and experimental data

Scientists with an interest in ecology

The world's drylands

Population Principles

OBJECTIVES

After reading this chapter, you should be able to:

- Understand that birthrate and death rate together determine population growth rate.
- Define the following characteristics of a population: natality, mortality, sex ratio, age distribution, reproductive potential, and spatial distribution.
- Explain why the biotic potential for most populations is much greater than needed.
- Describe how a growing population goes through lag, exponential growth, and stable equilibrium phases.
- Describe how the carrying capacity for a population in a region is determined by environmental resistance.
- List the four categories under which limiting factors can be classified.
- Describe how some populations enter a death phase after the stable equilibrium phase.
- Explain why humans are subject to the same forces of environmental resistance as are other organisms.
- Understand the implications of overreproduction.
- Understand that the human population is still growing rapidly.
- Explain how human population growth is influenced by social, theological, philosophical, and political thinking.

CHAPTER OUTLINE

POPULATION CHARACTERISTICS

A population can be defined as a group of individuals of the same species inhabiting an area. Just as individuals within a population are recognizable, populations have specific characteristics that distinguish them from one another. Some of these characteristics are natality (birthrate), mortality (death rate), sex ratio, age distribution, growth rates, density, and spatial distribution.

Natality and Mortality

Natality refers to the number of individuals added to the population through reproduction. Bacteria and other tiny organisms are able to divide to form new individuals that are identical to the original organism. In plant populations, reproduction results in the production of numerous seeds, but the seeds must land in appropriate soil conditions before they will germinate to produce a plant. Animal species also typically produce large numbers of offspring as a result of sexual reproduction. In human populations, natality is usually described in terms of the **birthrate,** the number of individuals born per one thousand individuals per year. For example, if a population of 2000 individuals produced 20 offspring during one year, the birthrate would be 10 per thousand per year. The natality for most species is typically quite high. Most organisms produce many more offspring than are needed to replace the parents.

It is important to recognize that the growth of a population is not determined by the birthrate (natality) alone. **Mortality,** the number of deaths per year, is also important. For most organisms, mortality rates are very high. Of all the seeds that plants produce, very few will result in a mature plant that itself will produce offspring. Many seeds are eaten by animals, some never find proper soil conditions, and those that germinate must compete with other organisms for nutrients and sunlight. In animals, most immature organisms die before they have an opportunity to reproduce. In human population studies, mortality is usually discussed in terms of the **death rate,** the number of people who die per one thousand individuals per year. Compared to the high mortality of the young of most species, the infant death rate of long-lived animals like humans is relatively low. In order for the size of a population to grow, the number of individuals added by reproduction must be greater than the number leaving it by dying. (See figure 6.1)

Sex Ratio and Age Distribution

A population's ability to grow by reproduction depends on its sex ratio and age distribution. The **sex ratio** refers to the

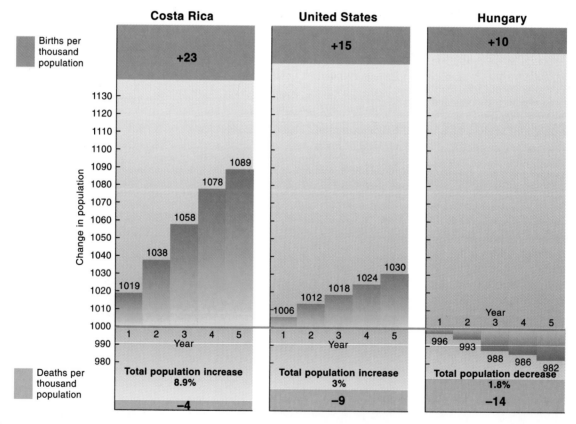

FIGURE 6.1 **Effect of Birthrate and Death Rate on Population Size** For a population to grow, the birthrate must exceed the death rate for a period of time. These three human populations illustrate how the combined effects of births and deaths would change population size if birthrates and death rates were maintained for a five-year period.

Source: Data from World Population Data Sheet 1998, Population Reference Bureau, Inc., Washington, D.C.

relative numbers of males and females. (Many kinds of organisms, such as earthworms and most plants, have both kinds of sex organs in the same body; sex ratio has no meaning for these species.) Many kinds of organisms have about equal numbers of males and females; however, even minor differences can be important since in all species the number of offspring produced is tied more closely to the number of females than to the number of males. In many game animal populations, where the males are shot and the females are not, there may be an uneven sex ratio in which the females outnumber the males. In many social insect populations (bees, ants, and wasps), the number of females greatly exceeds the number of males at all times, though most of the females are sterile. In humans, about 106 males are born for every 100 females. However, in the United States, by the time people reach their mid-twenties, a higher death rate for males has equalized the sex ratio. The higher male death rate continues into old age, when women outnumber men.

As you can see, populations can differ in **age distribution,** the number of individuals of each age in the population. Some are prereproductive juveniles, some are reproducing adults, and some are postreproductive adults. Many kinds of organisms, particularly those that have short life spans, have age distributions that change significantly during the course of a year.

If the majority of a population is made up of reproducing adults, a rapid increase in the number of prereproductive individuals will probably result. Most species of plants and animals typically produce a large number of prereproductive individuals at a specific time of the year (spring, following rain, or when food is plentiful) but have a very high mortality rate among the young. Thus the reproduction of the young declines sharply as time passes. A good example of this type of population is the cottontail rabbit. Many young are produced in the spring, but most of them die, a few reach sexual maturity, and very few live to old age. (See figure 6.2.)

In species that live a long time, it is possible for a stable population to have a balanced age distribution in which the number of individuals in these three categories is relatively constant. For most long-lived plant and animal populations, this means more prereproductive individuals than reproductive individuals, and more reproductive individuals than postreproductive individuals. If the majority of a population is postreproductive, the population declines. This age structure develops in many insect populations in the fall after eggs have been laid.

Human populations exhibit several types of age distribution. (See figure 6.3.) Kenya's population has a large prereproductive and reproductive component. This means that it will continue to increase rapidly for some time. Germany has a relatively even age distribution and will maintain a nearly constant population size, although it will probably have a declining population in the future. The United States has a very large reproductive component with a declining number of prereproductive individuals. Eventually, the U.S. population will begin to decline if current trends in birthrates and death rates continue.

Population Density and Spatial Distribution

Because of such factors as soil type, quality of habitat, and availability of water, organisms normally are distributed unevenly. Some populations have many individuals clustered into a small space while other populations of the same species may be widely dispersed. **Population density** is the number of organisms per unit area. For example, fruitfly populations are very dense around a source of rotting fruit, while they are rare in other places. Similarly, humans are often clustered into dense concentrations we call cities, with lower densities in rural areas. When the population density is too great, all organisms are injured because they compete severely with each other for necessary resources. Plant populations may compete for water, soil nutrients, or sunlight.

In animal populations, overcrowding might cause exploration and migration into new areas. This **dispersal** of organisms relieves the overcrowded conditions in the home area and, at the same time, leads to the establishment of new populations. Often, juvenile individuals relieve the overcrowding by leaving. The pressure

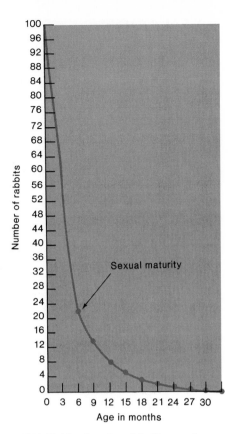

FIGURE 6.2   Survivorship Curve of Cottontail Rabbit In most natural populations, mortality rates are so high that very few individuals reach sexual maturity, and even fewer reach old age.

for out-migration **(emigration)** may be a result of seasonal reproduction leading to a rapid increase in population size, or environmental changes that intensify competition among members of the same species. For example, as water holes dry up, competition for water increases, and many desert birds emigrate to areas where water is still available.

The organisms that leave one population often become members of a different population. This in-migration **(immigration)** may introduce characteristics that were not in the population originally. When Europeans immigrated to North America, they brought genetic and cultural characteristics that had a tremendous impact on the existing Native American population. Among other things, Europeans brought diseases that were foreign to the Native Americans. These diseases increased their death rate and lowered the birthrate.

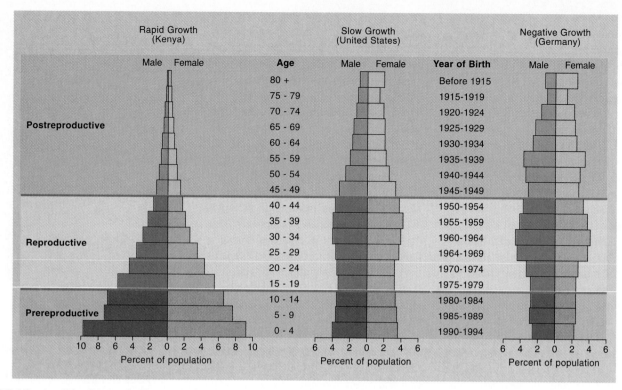

FIGURE 6.3   Age Distribution in Human Population   The relative numbers of individuals in each of the three categories (prereproductive, reproductive, and postreproductive) are good clues to the future of the population. Kenya has a large number of young individuals who will become reproducing adults. Therefore, this population is likely to grow rapidly. The United States has a declining proportion of prereproductive individuals, and Germany has a relatively balanced population. Therefore, at present birthrates and death rates, they will probably have declining populations sometime in the future, with Germany's.

Source: Data from Population Reference Bureau.

Droughts, wars, and political persecution have caused people to emigrate from their native lands to other countries. In 1997, the United Nations High Commissioner for Refugees estimated that there were about 18 million refugees worldwide. Often, the receiving countries are unable to cope with the large influx of new inhabitants.

Summary of Factors That Influence Population Growth Rates

Populations have an inherent tendency to increase in size. However, as we have just seen, many factors influence the rate at which a population can grow. At the simplest level, the rate of increase is determined by subtracting the number of individuals leaving the population from the number entering. Individuals leave the population either by death or emigration. They enter the population by birth or immigration. Birthrates and death rates are influenced by several factors, including the number of females in the population and their age. In addition, the density of a population may encourage individuals to leave because of intense competition for a limited supply of resources.

A POPULATION GROWTH CURVE

Sex ratios and age distributions directly influence the rate of reproduction within a population. Each species has an inherent reproductive capacity, or **biotic potential,** which is its ability to produce offspring. Generally, this biotic potential is many times greater than the number of offspring needed to replace the parents when they die. (See figure 6.4.) Consequently, most of the young die; only a few survive to become reproductive adults.

This high reproductive potential results in a natural tendency for populations to increase. For example, two mice produce four offspring, which, if they live, will produce offspring while their parents are also reproducing. Therefore, the population will grow exponentially (2, 4, 8, 16, 32, etc.).

Population growth tends to follow a particular pattern, consisting of a lag phase, an exponential growth phase, and a stable equilibrium phase. Figure 6.5 shows a typical population growth curve. During the first portion of the curve, known as the **lag phase,** the population grows very slowly because the process of reproduction and growth of offspring takes time. Organisms must mature into adults before they can reproduce. This period is followed by mating and the development of the young into independent organisms. By the time the first batch of young have reached sexual maturity, the parents may be producing a second set of offspring. Since more organisms now are reproducing, the population begins to increase at an exponential rate. This stage is known as the **exponential growth phase.**

Apples

Ducks

Pigs

FIGURE 6.4 **Biotic Potential** The ability of a population to reproduce greatly exceeds the number necessary to replace those who die. Here are some examples of the prodigious reproductive abilities of some species.

This growth will continue for as long as the birthrate exceeds the death rate. Eventually, however, the death rate and the birthrate will come to equal one another, and the population will stop growing and reach a relatively stable population size. This stage is known as the **stable equilibrium phase.**

It is important to recognize that although the size of the population may not be changing, the individuals are changing. As new individuals enter by birth or immigration, others leave by death or emigration. For most organisms, the first indication that a population is entering a stable equilibrium phase is an increase in the death rate. A decline in the birthrate may also contribute to the stabilizing of population size. Usually, this occurs after an increase in the death rate. To understand why populations cannot grow continuously, it is necessary to discuss the concept of carrying capacity.

CARRYING CAPACITY

The **carrying capacity** of an area is the number of individuals of a species that can survive in that area over time. For most populations, four factors interact to set the carrying capacity. These are: (1) the availability of raw materials, (2) the availability of energy, (3) the accumulation of waste products and their means of disposal, and (4) interactions among organisms. All of these forces acting together to

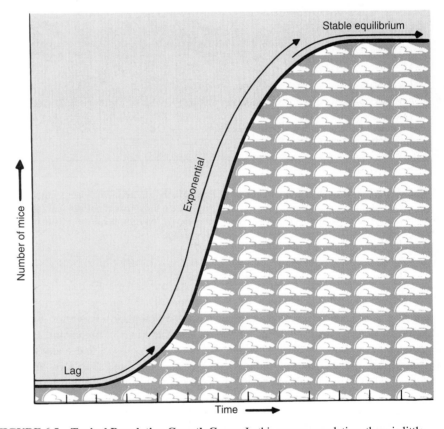

FIGURE 6.5 **Typical Population Growth Curve** In this mouse population, there is little growth during the lag phase. During the exponential growth phase, the population increases rapidly as the offspring reach reproductive age. Eventually the population reaches a stable equilibrium phase, during which the birthrate equals the death rate.

limit population size is known as **environmental resistance.**

Certain **limiting factors** have a primary role in limiting the size of a population. In some cases, these limiting factors are easy to identify. Lack of food, lack of

oxygen, competition with other species, or disease are examples. In other cases the limiting factors may be less obvious. (See figure 6.6.) For example, in grass plants, nitrogen and magnesium in the soil are necessary raw materials for the manufacture of

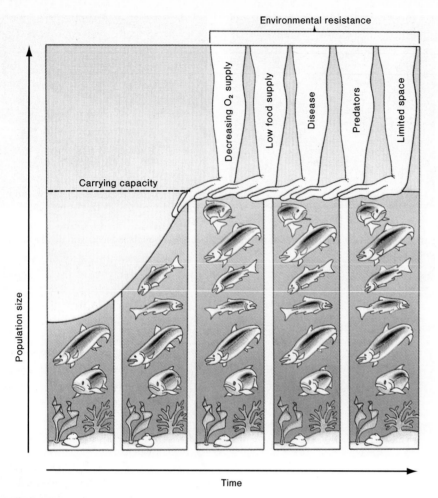

Environmental resistance

FIGURE 6.6 Carrying Capacity A number of factors in the environment, such as food, oxygen supply, diseases, predators, and space, determine the number of organisms that can survive in a given area—the carrying capacity of that area. The environmental factors that limit populations are known collectively as environmental resistance.

This decline in population size is known as the **death phase.** When a population decreases rapidly, it is said to crash.

Interactions among organisms are also important in controlling population size. Since many birds rely on grass seeds for food, birds have a limiting effect on the size of grass populations. Decomposer organisms may allow for increased populations because they prevent the buildup of toxic wastes. Parasites and predators may cause the premature death of individuals, thus limiting the size of the population. A good example of predator–prey interaction is the relationship between the cat known as the Canada lynx and a member of the rabbit family known as the varying hare. (See figure 6.8.) The varying hare has a high biotic potential that the lynx helps to control by using the hare as food. In turn, the size of the varying hare population determines the size of the lynx population.

Some studies indicate that populations can be controlled by interaction among individuals within the population. A study of laboratory rats shows that crowding causes a breakdown in normal social behavior, which leads to fewer births and increased deaths. The changes observed include abnormal mating behavior, decreased litter size, fewer litters per year, lack of maternal care, and increased aggression in some rats or withdrawal in others. Thus, limiting factors can reduce birthrates as well as increase death rates.

REPRODUCTIVE STRATEGIES AND POPULATION FLUCTUATIONS

So far, we have talked about population growth as if all organisms reach a stable population when they reach the carrying capacity. That is an appropriate way to begin to understand population changes, but the real world is much more complicated. Species can be divided into two broad categories based on their reproductive strategies. **K-strategists** are usually large organisms that have relatively long lives, produce few offspring, and provide care for their offspring. Their populations

chlorophyll. If these minerals are not present in sufficient quantities, the grass population cannot increase. The application of fertilizers containing these minerals removes this limiting factor, and the individual grass plants grow and reproduce, resulting in a larger population. In effect, the carrying capacity has been increased because this limiting factor has been removed. The carrying capacity has not been eliminated because now some new limiting factor will predominate. Perhaps it will be the amount of water, the number of insects that feed on the grass, or competition for sunlight. Because plants require energy in the form of sunlight for photosynthesis, the amount of light can be a limiting factor for many plants. When small plants are in the shade of trees, they often do not grow well and have small populations.

Accumulation of waste products is not normally a limiting factor for plants, since they produce few wastes, but it can be for other kinds of organisms. Bacteria, other tiny organisms, and many kinds of aquatic organisms that live in small ecosystems like puddles, pools, or aquariums may be limited by wastes. When a small number of a species of bacterium are placed on a petri plate with nutrient agar (a jellylike material containing food substances), the population growth follows a curve shown in figure 6.7. As expected, it begins with a lag phase, continues through an exponential growth phase, and eventually levels off in a stable equilibrium phase. However, in this small, enclosed space, there is no way to get rid of the toxic waste products, which accumulate, eventually killing the bacteria.

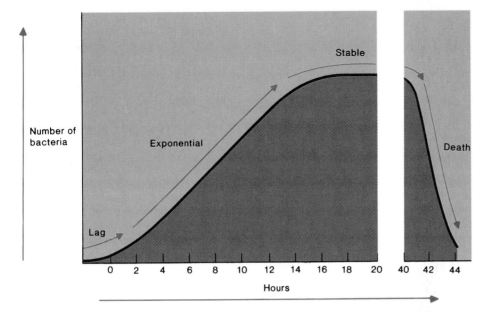

FIGURE 6.7 **Bacterial Growth Curve** The change in population size follows a typical population growth curve until waste products become lethal. The buildup of waste products lowers the carrying capacity. When a population begins to decline, it enters the death phase.

FIGURE 6.8 **Predator/Prey Interaction** Interaction between predator and prey species is complex and often difficult to interpret. These data were collected from the records of the number of pelts purchased by the Hudson Bay Company. They show that the two populations fluctuate, with change in the lynx population usually following changes in the varying hare population.

Source: Data from D.A. MacLulich, "Fluctuations in the Numbers of the Varying Hare (*Lepus americanus*)," in *University of Toronto Studies, Biology Series #43* 1937 (reprinted 1974), University of Toronto Press, Toronto.

typically stabilize at the carrying capacity. Their reproductive strategy is to invest a great deal of energy in producing a few offspring that have a good chance of living to reproduce. Deer, lions, and swans are examples of this kind of organism. Humans generally produce single offspring, and even in countries with high infant mortality, 80 percent of the children survive. Generally, populations of K-strategists are limited by density-dependent limiting factors. **Density-dependent limiting factors**

are those that become more severe as the size of the population increases. For example, as a population of hawks increases, the hawks begin to compete for available food, such as mice, snakes, and small birds. When food is in short supply, many of the young in the nests die, and population growth slows. They have reached the carrying capacity.

The **r-strategist** is typically a small organism that has a short life, produces many offspring, and does not reach a

carrying capacity. Examples are grasshoppers, gypsy moths, and some mice. The reproductive strategy of r-strategists is to expend large amounts of energy producing many offspring but to provide limited care (often none) for them. Consequently, there is high mortality among the young. For example, one female oyster may produce a million eggs, but few of them ever find suitable places to attach themselves and grow. Typically, these populations are limited by **density-independent limiting factors** in which the size of the population has nothing to do with the limiting factor. Typical density-independent limiting factors are changing weather conditions that kill large numbers of organisms, the drying up of a small pond, or the death of entire populations due to the destruction of their food source. The population size of r-strategists is likely to fluctuate wildly. They reproduce rapidly, population size increases until some factor causes the population to crash, and then they begin the whole process all over again.

Since humans are K-strategists, it may be difficult for us to appreciate that the r-strategy can be viable from an evolutionary point of view. There is no carrying capacity for temporary resources. Resources that are present only for a short time can be exploited most effectively if many individuals of one species monopolize the resource, while denying other species access to it. Rapid reproduction can place a species in a position to compete against other species that are not able to increase numbers as rapidly. Obviously, most of the individuals will die, but not before they have left some offspring or resistant stage that will be capable of exploiting the resource should it become available again.

Even K-strategists, however, have population fluctuations for a variety of reasons. One reason is that the environment is not constant from year to year. Floods, droughts, fires, extreme cold, and similar events may affect the carrying capacity of an area, thus causing fluctuations in population size. Epidemic disease or increased predation may also lead to populations that vary from year to year. Figure 6.8 shows rather substantial changes in the populations of lynx and varying hare. The size of the lynx population seems to be

Managing Elephant Populations—Harvest or Birth Control?

During the 1980s, uncontrolled hunting of elephants for ivory and food had reduced the African elephant, *Loxodonta africana,* population from 1.3 million to 650,000. Many conservation organizations became alarmed by the rapid decline in the numbers of elephants and supported a ban on the export of ivory. In 1989, the Convention on International Trade in Endangered Species of Wild Fauna and Flora (CITES) banned all international trade in elephant products. The ban worked. The price of ivory fell and poaching became unprofitable. Since then, some southern African countries have experienced an elephant population explosion that threatens to destroy the limited habitat available to elephants and increasingly results in conflicts between farmers and elephants.

The African elephant requires huge amounts of food (150–250 kg/day) to sustain its large body. The animals strip bark from trees, uproot the trees, and eat large quantities of grass. Where humans and elephants share the same habitat, elephants can do great damage to crops. Because they have been increasingly confined to national parks and nature preserves, two natural options for relieving population pressure (migration or starvation) are not available. Migration from the park results in increased agricultural damage and risk of injury to farmers, and increasing populations can cause irreparable damage to the protected habitat they currently occupy as they seek food. If they are allowed to starve there would be enormous public pressure from wildlife groups, and the park managers would be condemned. If the populations are to remain healthy, not destroy their habitat, and not infringe on farming communities, the size of the population must be controlled. There are only two ways to do that—increase the death rate (culling) or decrease the birthrate (sterilization or birth control).

Several southern African countries have argued that they should be allowed to manage their elephant populations as a natural resource in the same way that other large game animals (deer, elk, caribou) are managed. The sustainable harvest of surplus elephants would be wise use of a natural resource that would provide income to the local people as well as a much needed source of protein in this region. Furthermore, the income from the harvesting of elephants could be used to provide additional funding for park management.

The Humane Society of the United States has advocated the use of birth control to reduce the birthrate and solve the elephant population problem. Several methods have been tested and they can work. However, critics of this approach suggest that preventing female elephants from conceiving for long periods of time may disrupt the normal social structure of the herd. Under most circumstances a female that is pregnant will not come into heat again for two years. If females do not conceive they will come into heat approximately every 15 weeks and there will be more females in heat than is normal throughout the year. Since the mating activity of elephants is disruptive of their usual pattern of activity, some people are concerned that the normal social structure of the herd will be disrupted. Furthermore a herd usually includes various ages of immature individuals. The younger individuals learn from older siblings or cousins. The use of birth control could disrupt the typical spacing of pregnancies and interfere with the normal process of "educating" young elephants. What looks like a simple problem to solve has resulted in two diametrically opposed camps: those that advocate increased deaths by harvesting and those that advocate decreased births by manipulating births. Neither is "natural" but something must be done or the survival of elephants could again become a problem.

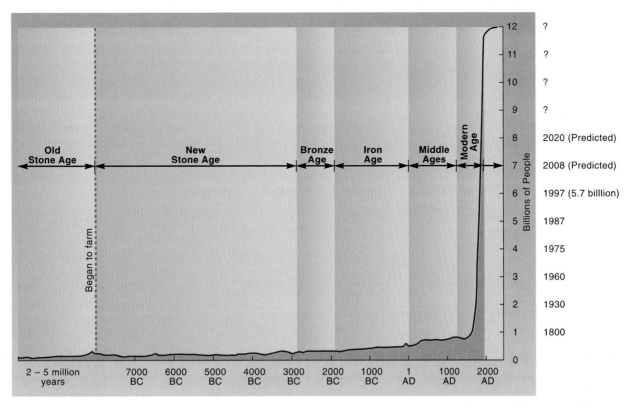

FIGURE 6.9 Human Population Curve From A.D. 1800 to A.D. 1930, the number of humans doubled (from one billion to two billion) and then doubled again by 1975 (four billion) and could double again (eight billion) by the year 2020. How long can this pattern continue before the earth's ultimate carrying capacity is reached?

Source: Data from Jean Van Det Tak, et al., "Our Population Predicament: A New Look" in *Population Bulletin,* Vol. 34, No. 5, December 1979, Population Reference Bureau, Washington, D.C.; and more recent data taken from Population Reference Bureau.

tied to the size of the hare population, which is logical since lynx eat hares. However, the causes of fluctuations in the varying hare population are unclear. One possibility is that periodic epidemic disease may cause dense populations of organisms like hares to crash, leading to the crash of their predators' populations.

Although local human populations often show fluctuations, the worldwide human population has increased continually for the past several hundred years. Humans have been able to reduce environmental resistance by eliminating competing organisms, increasing food production, and controlling disease organisms.

HUMAN POPULATION GROWTH

The human population growth curve has a long lag phase followed by a sharply rising exponential growth phase that is still rapidly increasing. (See figure 6.9.) A major reason for the increase is that the human species has lowered its death rate. When various countries reduce environmental resistance by increasing food production or controlling disease, they share this technology throughout the world. Developed countries send health care personnel to all parts of the globe to improve the quality of life for people in less-developed countries. Physicians offer advice on nutrition, and engineers develop wastewater treatment systems. Improved sanitary facilities in India and Indonesia, for example, decreased deaths caused by cholera. These advancements tend to reduce mortality and directly increase the *quantity* of life, because birthrates remain high.

Let us examine this situation from a different perspective. The world population is currently increasing at an annual rate of 1.4 percent. That may not seem like much, but even at 1.4 percent the population is growing rapidly. It can be difficult to comprehend the impact of a 1 or 2 percent annual increase. Remember that a growth rate in any population compounds itself, since the additional individuals eventually reproduce, thus adding more individuals. One way to look at this growth is to determine how much time is needed to double the population. This is a valuable method because most of us can appreciate what life would be like if the number of people in our locality were doubled, particularly if the number were to occur within our lifetime.

Figure 6.10 shows the relationship between the rate of annual increase for the human population and the number of years it would take to double the population if that rate were to continue. At a 1 percent rate of annual increase, the population will double in approximately 70 years. At a 2 percent rate of annual increase, the population will double in about 35 years. The current worldwide rate of annual increase of about 1.4 percent will double the world population in about 49 years.

What does this very rapid rate of growth mean to the human species? As a species, humans are subject to the same

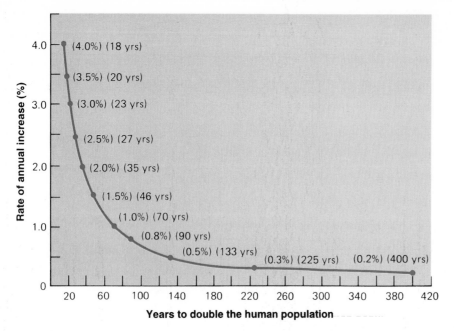

FIGURE 6.10 Doubling Time for the Human Population This graph shows the relationship between the rate of annual increase in percent and doubling time. A population growth rate of 1 percent per year would result in the doubling of the population in about seventy years. A population growth rate of 3 percent per year would result in a population doubling in about twenty-three years.

limiting factors as all other species. We cannot increase beyond our ability to acquire raw materials and energy and safely dispose of our wastes. We also must remember that interactions with other species and with other humans will help determine our carrying capacity.

Let us look at these four factors in more detail. Many of us think of raw materials simply as the amount of food available. However, we have become increasingly dependent on technology, and our lifestyles are directly tied to our use of other kinds of resources, such as irrigation water, genetic research, and antibiotics. Food production is becoming a limiting factor for some segments of the world's human population. Malnutrition is a serious problem in many parts of the world because sufficient food is not available. Chapter 7 deals with the problems of food production and distribution and their relationship to human population growth.

The second factor, available energy, involves problems similar to those of raw materials. Essentially all species on earth are ultimately dependent on sunlight for their energy. New, less disruptive methods of harnessing this energy must be developed to support an increasing population.

Currently, the world population depends on fossil fuels to raise food, modify the environment, and move from place to place. When energy prices increase, much of the world's population is placed in jeopardy because incomes are not sufficient to pay the increased costs for food and other essentials.

Waste disposal is the third factor determining the carrying capacity for humans. Most pollution is, in reality, the waste product of human activity. Some people are convinced that disregard for the quality of our environment will be a major limiting factor. In any case, it makes good sense to control pollution and to work toward cleaning our environment.

The fourth factor that determines the carrying capacity of a species is interaction with other organisms. We need to become aware that we are not the only species of importance. When we convert land to meet our needs, we displace other species from their habitats. Many of these displaced organisms are not able to compete with us successfully and must migrate or become extinct. Unfortunately, as humans expand their domain, the areas available to these displaced organisms become more rare. Parks and natural areas

have become tiny refuges for the plants and animals that once occupied vast expanses of land. If these refuges fall to the developer's bulldozer or are converted to agricultural use, many organisms will become extinct. What today seems like an unimportant organism, one that we could easily do without, may someday be seen as an important link to our very survival.

Humans Are Social Animals

Human survival depends upon interaction and cooperation with other humans. Current technology and medical knowledge are available to control human population growth and to improve the health of the people of the world. Why then does the population continue to increase, and why does poverty become worse every year? Humans are social animals who have freedom of choice and frequently do not do what is considered "best" from an unemotional, uninvolved, biological point of view. People make decisions based on history, social situations, ethical and religious considerations, and personal desires. The biggest obstacle to controlling human population is not biological but falls under the province of philosophers, theologians, politicians, and sociologists. People in all fields need to understand the cause of the population problem if they are to successfully deal with every aspect of it.

Ultimate Size Limitation

The human population cannot increase indefinitely, because, eventually, the weight of the human tissue would equal the weight of the earth. We can say with certainty that the population will ultimately reach its carrying capacity and level off long before this absurd condition would occur. Many people speculate about the limits of human population growth. The United Nations estimates that a moderate estimate of future human population is that the world will contain 10 billion people by the year 2050 and will continue to grow. But as the population grows, limiting factors will become more effective and the human population will ultimately be forced to level off. Some people suggest that a lack of food, a lack of water, or increased waste heat will ultimately

ENVIRONMENTAL
CLOSE-UP

Population Growth of Invading Species

When a new species is introduced into an area suitable for its survival, it has great potential to increase its population size, since it will not have natural enemies to keep mortality high. New species may also be able to compete favorably for available resources and lower the populations of native species. A typical population growth curve starts with a few individuals being released. Once established, the population will increase in number and expand its range. The zebra mussel, for example, is thought to have entered the Great Lakes about 1985. Today, it is found throughout all the Great Lakes and has been introduced into the Mississippi River and its tributaries, where it has been discovered as far south as New Orleans. Similarly, the gypsy moth has spread from its place of original release to much of the forestland of the Midwest, and is still expanding its range. Another invading species is the kudzu vine, which has become a pest in many areas of the southern United States.

Eventually, the invading species occupies all the habitat suitable to it and the population stabilizes. Dandelions and starlings are introduced species that are no longer expanding their range. They have simply become a normal part of the biology of North America.

Dandelion

Zebra mussels

Starlings

Gypsy moths

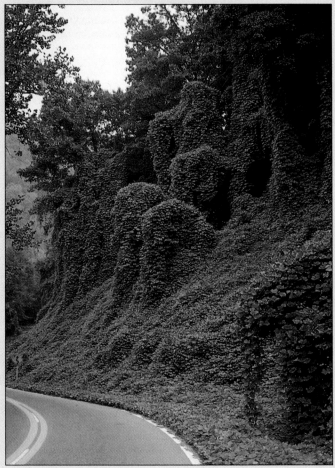
Kudzu vine

Isle Royale in Lake Superior has been the site of a long-term study of the relationship between moose and wolf populations. Wolves (probably a single pair) reached Isle Royale by crossing the ice from Canada during the winter of 1947–48 and had reached a population of about 20 by 1958 when a succession of biologists began a long-term study of the wolves and their relationships with moose and other organisms on the island. In over forty years they have learned a great deal about the food habits of wolves, pack behavior, the relationships among various wolf packs, and the relationship between the predator (wolf) and prey (moose) populations. Although there are several factors that affect population size for wolves and moose, an examination of the fluctuations in moose and wolf populations suggests that they have an effect on one another. When wolf populations decline, moose populations rise and when wolf populations rise, moose populations fall. The major changes in moose and wolf populations since 1980 appear to have had triggering events that resulted in major reductions in one of the two populations that had a subsequent effect on the other. For example, the major decline in wolves in the early 1980s from about 50 to 13 wolves was probably initiated by the introduction of a viral disease of domesticated dogs (parvovirus). The reduction in the number of wolves allowed a greater number of new young moose to survive and the moose population increased rapidly.

The rapid decline in the moose population during the mid-1990s was probably initiated by a combination of factors. The rapid growth in the years prior to 1995 had reduced the amount of food available and resulted in malnutrition, slow growth, and poor reproduction. A very severe winter and a tick infestation probably contributed to the die-off as well. During the winter of 1995–96, the moose population suffered a 50 percent reduction as it fell to about 1200. This decline continued as the population fell to about 500 individuals in 1997, an 80 percent reduction from its maximum. However it rebounded to about 700 indi-

viduals in 1998. What was a bad time for moose was a good time for wolves, since weak and dying moose provided an abundant food supply, and the wolf population rose to 24 individuals, its highest population since the early 1980s. However, this abundant food supply was short-lived and the remaining moose were younger, healthy individuals that are harder to take as prey. In 1998 the wolf population fell to 14, raising concern that the population may not be able to sustain itself. On the positive side, a smaller wolf population reduces competition for food. In addition, most of the remaining wolves are young and have the potential to produce many sets of pups over the next several years. On the negative side, recent survival of pups has been low. This could be due to several factors. Perhaps the canine parvovirus is affecting the survival of the pups. Lack of genetic variety may also be contributing to the low survival, since the entire population is thought to be descended from a single mated pair. Observations over the next few years may identify causes for the decline in the wolf population. For example blood samples taken from captured wolves could clarify questions about the presence of parvovirus and concerns about genetic variety. It is also possible that the wide fluctuations seen in these two populations may be normal for isolated, island populations in which prey and predator populations have significant influences on one another.

- When populations of animals become endangered, efforts are often made to preserve them. Should special efforts be made to preserve this wolf population?

- If lack of genetic variety is shown to be a contributing factor to the decline of the wolf population, should efforts be made to introduce genetic variety?

- In what ways is this wolf population different from that of populations of California condors or whooping cranes?

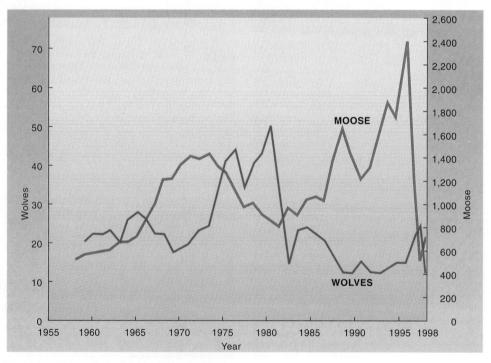

Source: Data from Rolf O. Peterson, "Ecological Studies of Wolves on Isle Royale," Annual Report 1995–96 and subsequent reports.

control the size of the human population. Still others suggest that, in the future, social controls will limit population growth. It is also important to consider the age structure of the world population. In most of the world, there are many reproductive and prereproductive individuals. Since most of these individuals are currently reproducing or will reproduce in the near future, even if they reduce their rate of reproduction, there will be sharp increase in the number of people in the world in the next few years.

Others are concerned that politically related destruction of humanity will occur. No one knows what the ultimate human population size will be or what the most potent limiting factors will be, but most agree that we are approaching the maximum sustainable human population. If the human population continues to increase, the amount of agricultural land available will not be enough to satisfy the demand for food.

Summary

A population is a group of organisms of the same species that inhabits an area. The birthrate (natality) is the number of individuals entering the population by reproduction during a certain period. The death rate (mortality) measures the number of individuals that die in a population during a certain period. Population growth is determined by the combined effects of the birthrate and death rate.

The sex ratio of a population is a way of stating the relative number of males and females. Age distribution and the sex ratio have a profound impact on population growth. Most organisms have a biotic potential much greater than that needed to replace dying organisms.

Interactions among individuals in a population, such as competition, predation, and parasitism, are also important in determining population size. Organisms may migrate into (immigrate) or migrate out of (emigrate) an area as a result of competitive pressure.

A typical population growth curve shows a lag phase followed by an exponential growth phase and a stable equilibrium phase at the carrying capacity. The carrying capacity is determined by many limiting factors that are collectively known as environmental resistance. The four major categories of environmental resistance are available raw materials, available energy, disposal of wastes, and interactions among organisms. Some populations experience a death phase following the stable equilibrium phase.

K-strategists typically are large, long-living organisms that reach a stable carrying capacity. Their population size is usually controlled by density-dependent limiting factors. r-strategists are generally small, short-living organisms that reproduce very quickly. Their populations do not generally reach a carrying capacity but crash because of some density-independent limiting factor.

The human population is increasing at a rapid rate. The earth's ultimate carrying capacity for humans is not known. The causes for human population growth are not just biological but also social, political, philosophical, and theological.

Review Questions

1. How are biotic potential and age distribution interrelated?
2. List three characteristics populations might have.
3. Why do some populations grow? What factors help to determine the rate of this growth?
4. Under what conditions might a death phase occur?
5. List four factors that could determine the carrying capacity of an animal species.
6. How do the concepts of birthrate and population growth differ?
7. How does the population growth curve of humans compare with that of bacteria on a petri dish?
8. How do r-strategists and K-strategists differ?
9. As the human population continues to increase, what might happen to other species?
10. All successful organisms overproduce. What advantage does this provide for the species? What disadvantages may occur?

Critical Thinking Questions

1. Why do you suppose some organisms display high natality and others display lower natality? For example, why do cottontail rabbits show high natality and wolves relatively low natality? Why wouldn't all organisms display high natality?
2. Do you think African elephants should be managed by southern African countries like deer are managed in the United States? What values, beliefs, and perspectives lead you to your conclusion?
3. Where do you stand on the issue of birth control for elephants or culling of the herd? Why do you think the way you do? What is the position of the Humane Society in the United States? Why do you think they hold the position they do? Do you think conservationists in the United States should have a say over what is done with the elephants in Africa? Should Africans have a say in what is done in the United States? Why?
4. Consider the differences between r-strategists and K-strategists. What costs are incurred by adopting either strategy? What evolutionary benefits does each strategy enjoy?
5. Why do invading species, which can survive in a new environment, often show exponential growth rates soon after they are introduced to that environment?
6. If humans see that a population of organisms, like the population of wolves on Isle Royale, is declining for "natural" reasons, should humans intervene and try to preserve these populations? Why?

Key Terms

age distribution *93*

biotic potential *94*

birthrate *92*

carrying capacity *95*

death phase *96*

death rate *92*

density-dependent limiting
  factors *97*

density-independent limiting
  factors *97*

dispersal *93*

emigration *93*

environmental resistance *95*

exponential growth phase *94*

immigration *93*

K-strategists *96*

lag phase *94*

limiting factors *95*

mortality *92*

natality *92*

population density *93*

r-strategists *97*

sex ratio *92*

stable equilibrium phase *95*

Related Internet Sites

Visit our website at http://www.mhhe.com/environmentalscience/
for specific information on each of the topics below.

Zebra mussels in the western part of North America

The world, humans, and the environment

Comprehensive access to population data

U.S. Census Bureau's official statistics

Population analysis of wildlife in the northwestern U.S.

Carrying capacity of regions in the U.S.

Animal populations

Global population distribution database

Endangered and extinct species lists from around the world

Ecological urban living

Human Population Issues

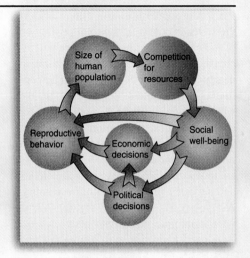

OBJECTIVES

After reading this chapter, you should be able to:

- Apply some of the principles discussed in chapter 6 to the human population.
- Differentiate between birthrate and population growth rate.
- Explain the current population situation in the United States.
- Explain how age distribution and women's roles affect population projections.
- Recognize that U.S. society is adjusting as the average age increases.
- Recognize that most of the world still has a rapidly growing population.
- Describe the implications of the hypothesis that a demographic transition occurs.
- Understand how an increasing world population will alter the worldwide ecosystem.
- Understand the implications of the fact that most of the countries of the world are not able to produce enough food to feed their people.
- Explain why less-developed nations have high birthrates and why they will continue to have a low standard of living.
- Recognize that the developed nations of the world will be under greater pressure to share their abundance.

CHAPTER OUTLINE

Current Population Trends

Population and Standard of Living

The Human Population Issue

Causes of Population Growth
  Biological Reasons for Population Growth
  Social Reasons for Population Growth

Global Perspective: *The Impact of AIDS on Populations*

Environmental Close-Up: *Control of Births*

Global Perspective: *Governmental Policy and Population Control*

Global Perspective: *Thomas Malthus and His Essay on Population*
  Political Factors That Affect Population Growth

The Demographic Transition Concept

The U.S. Population Picture

Global Perspective: *The Urbanization of the World's Population*

Hunger, Food Production, and Environmental Degradation

Global Perspective: *Population and Poverty: A Vicious Cycle*

Anticipated Changes with Continued Population Growth

Global Perspective: *Canadian Population Overview*

Issues and Analysis: *Population Growth in Mexico*

CURRENT POPULATION TRENDS

The human population predicament is very complex. In order to appreciate it, we must understand current population trends and how they are related to economic conditions. Currently, the world population is over 5.9 billion people. By the year 2025, this is expected to increase to just over 8 billion. Much of this increase is expected to occur in Africa, Asia, and Latin America, which already have over 82 percent of the world population. (See figure 7.1.) If these trends continue, the total population of Africa, Asia, and Latin America will increase from the current 4.87 billion to nearly 7 billion by 2025, when these continents will contain 86 percent of the world's people. These regions have not only the highest population growth rates but also the lowest per capita **gross national product (GNP).** The GNP is an index that measures the total goods and services generated within a country. As you can see from figure 7.2, a wide economic gap exists between less-developed countries with low GNPs and the developed countries with high GNPs. This large difference in economic well-being is reflected in a dissimilarity in the **standard of living,** an abstract measure of the degree to which necessities and comforts of daily life are met. Yet the people of less-developed countries aspire to the same standard of living enjoyed by people in the developed world.

POPULATION AND STANDARD OF LIVING

Standard of living is a difficult concept to quantify since various cultures have different attitudes and feelings about what is desirable. Here, we compare averages of several aspects of the cultures in three countries: (1) the United States, which is an example of a highly developed industrialized country; (2) Argentina, which is a moderately developed country; and (3) Zimbabwe, which is less developed. Figure 7.3 shows several statistics that relate to the standard of living in these three countries. The United States produces approximately three times more goods and services per person than does Argentina and about forty-five times more goods and services than Zimbabwe. U.S. citizens consume five times more energy than the average Argentinean and about sixty times more than the average Zimbabwean. Infant death is high in Zimbabwe, intermediate in Argentina, and low in the United States. In addition, people live longer and go to school longer in the United States than in the other two countries. Zimbabwe is more densely populated than Argentina. The average Zimbabwean is less well fed, a factor related to high mortality, low life expectancy, and underweight children, whereas the average citizen of the United States or Argentina has more food than needed.

Obviously, tremendous differences exist in the standard of living among these three countries. What the average U.S. citizen would consider poverty level would be considered a luxurious life for the average person in a poorly developed country. Standard of living seems to be closely tied to energy consumption. In general, the higher the per capita energy consumption, the higher the standard of living of the people. This is related to the degree of industrialization or development. However,

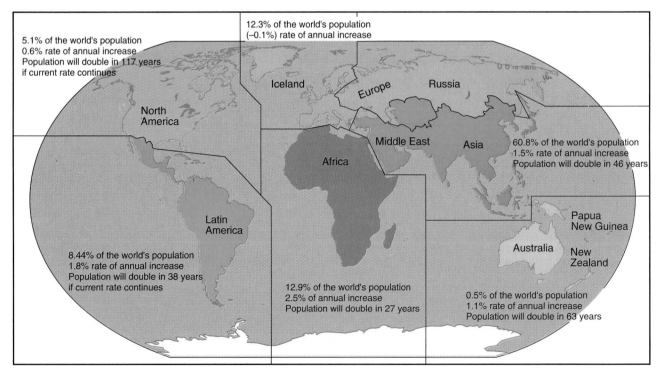

FIGURE 7.1 **Population Growth in the World (1998)** The population of the world is not evenly distributed. Currently over 82 percent of the world's population is in Latin America, Africa, and Asia. These areas also have the highest rates of increase and are generally considered less developed. Because of the high birthrates, they are likely to remain less developed and will constitute nearly 84 percent of the world's population by the year 2010.

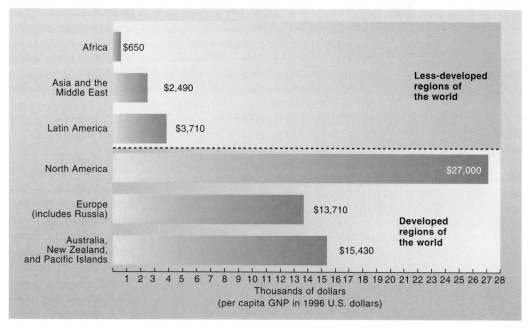

FIGURE 7.2 Per Capita Gross National Product of Nations The world can be divided into developed and less-developed regions. One measure of the degree to which a country is developed is its per capita gross national product. A less-developed country produces very little for each person in the country, while a developed country has high productivity and, therefore, also has goods and services available for consumption by its people.

Source: Data from 1998 World Population Data Sheet.

with industrialization comes pollution, as huge amounts of fossil fuels are consumed to provide the goods and services that contribute to the standard of living.

THE HUMAN POPULATION ISSUE

There are many ways to show that world population growth is a contributing factor in nearly all world problems. Current population growth has led to famine in areas where food production cannot keep pace with increasing numbers of people; political unrest in areas with great disparities in availability of resources (jobs, goods, food); environmental degradation by poor agricultural practices (erosion, desertification); water pollution by human and industrial waste; air pollution caused by the human need to use energy for personal use and for industrial applications; extinctions caused by people converting natural ecosystems to managed agricultural ecosystems; and destructive effects of exploitation of natural resources (strip mining, oil spills, groundwater mining).

Several factors interact to determine the impact of a society on the resources of the country.

$$\begin{matrix} \text{Impact} \\ \text{of a} \\ \text{population} \end{matrix} = \cfrac{\begin{matrix}\text{Population}\\\text{size}\end{matrix}}{\begin{matrix}\text{Area}\\\text{occupied}\end{matrix}} \times \begin{matrix}\text{Degree of}\\\text{technological}\\\text{development}\end{matrix}$$

The larger the size of a population, the greater the demand on the resources of the country. However, population size is not the only important factor. **Population density,** the number of people per unit of land area, is also important. A million people spread out over the huge area of the Amazon have much less impact on resources than that same million people in a small island country, because the impact is distributed over a greater land surface.

The degree of technological development is also important. The environmental impact of the developed world is often underestimated because the population in these countries is relatively stable and local environmental conditions are good. However, highly developed countries consume huge amounts of resources. Citizens of these countries eat more food, particularly animal protein, which requires larger agricultural inputs than does a vegetarian diet. They have more material possessions, and consume vast amounts of energy. These developed countries purchase goods and services from other parts of the world, often degrading environmental

conditions in less-developed countries. The environmental impact of highly developed regions like North America, Japan, and Europe is often felt in distant places, while the impact on resources in the home country may be minimal.

While controlling world population would not eliminate the problems, it could reduce the rate at which environmental degradation is occurring. The quality of life for many people in the world would improve. Why, then, does the human population continue to grow at such a rapid rate?

CAUSES OF POPULATION GROWTH

In chapter 6, we examined populations from a biological point of view. We looked at their characteristics, the causes of growth, and the forces that cause populations to stabilize. All of these biological factors apply to human as well as nonhuman populations. There is an ultimate carrying capacity for the human population. Eventually, limiting factors will cause human populations to stabilize. However, unlike other kinds of organisms, humans are also influenced by social, political, economic, and ethical factors. We

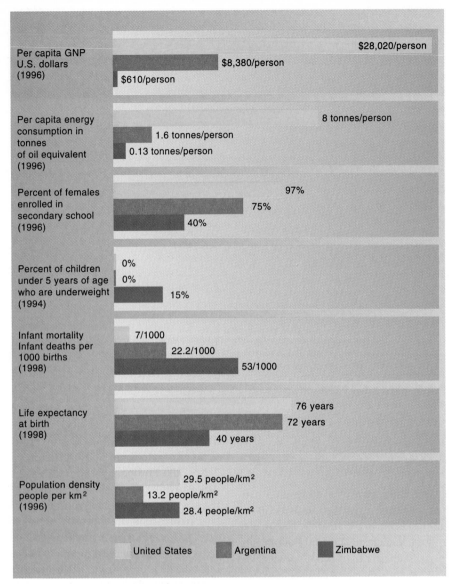

FIGURE 7.3 **Standard of Living in Three Countries** Standard of living is a measure of how well one lives. It is not possible to get a precise definition, but when we compare the United States, Argentina, and Zimbabwe, it is obvious that there are great differences in how the people in these countries live. Zimbabwe has a low life expectancy, a high death rate, and low productivity, and the people are often starving. The United States has a high life expectancy, a low death rate, and high productivity, and many people eat too much. Argentina is intermediate in all of these characteristics.

have accumulated knowledge that allows us to predict the future. We can make conscious decisions based on the likely course of events and adjust our lives accordingly. Part of our knowledge is the certainty that as populations continue to increase, death rates and birthrates will become equal. This can happen by allowing the death rate to rise or by choosing to limit the birthrate. Controlling human population would seem to be a simple process. Once everyone understands that lowering the birthrate is more humane than allowing

the death rate to rise, most people should make the "correct" decision; however, it is not quite that simple.

Biological Reasons for Population Growth

The study of human populations, their characteristics, and what happens to them is known as **demography.** Demographers can predict the future growth of a population by looking at several biological indicators. Currently, in almost all

countries of the world, the birthrate exceeds the death rate. Therefore, the size of the population must increase. (See table 7.1.) Some countries that have high birthrates and high death rates, with birthrates greatly exceeding the death rates, will grow rapidly (Afghanistan and Ethiopia). Such countries usually have an extremely high mortality rate among children because of disease and malnutrition.

Some countries have high birthrates and low death rates and will grow extremely rapidly (Guatemala and Syria). Infant mortality rates are moderately high in these countries. Other countries have low birthrates, and death rates that closely match the birthrates; they will grow slowly (Sweden and the United Kingdom). These and other more-developed countries typically have very low infant mortality rates. The disruption caused by the political upheaval in the former Soviet Union and Eastern Europe has resulted in several countries (e.g., Russia and Germany) with death rates that are equal to or exceed birthrates, causing their populations to decline. Because of these countries and the generally low rates of growth in the rest of Europe, the European region as a whole has a declining population.

The most important determinant of the rate at which human populations grow is related to how many women in the population are having children and the number of children each woman will have. The **total fertility rate** of a population is the number of children born per woman per lifetime. A total fertility rate of 2.1 is known as **replacement fertility,** since in the long run, if the total fertility rate is 2.1, population growth will stabilize. A rate of 2.1 is used rather than 2.0 because some children do not live very long after birth and therefore will not contribute to the population for very long. When population is not growing, and the number of births equals the number of deaths, it is said to exhibit **zero population growth.**

A total fertility rate of 2.1 will not necessarily immediately result in a stable population with zero growth, for several reasons. First, the death rate may fall as living conditions improve and people live longer. If the death rate falls faster than the birthrate, there will still be an increase in the population even though it is reproducing at the replacement rate.

TABLE 7.1 POPULATION GROWTH RATES IN SELECTED COUNTRIES (1998)

Country	Current population (millions) (1998)	Births per 1000 individuals	Deaths per 1000 individuals	Infant mortality rate (deaths per 1000 live births)	Rate of natural increase (annual %)	Time needed to double population (years)
Russia	146.9	9	14	17	(−0.5)	—
Germany	82.3	10	10	4.9	(−0.1)	—
Sweden	8.9	10	11	3.9	(−0.0)	—
Belgium	10.2	12	10	5.8	0.1	529
United Kingdom	59.1	13	11	6.1	0.2	433
Japan	126.4	10	7	3.8	0.2	330
Canada	30.6	12	7	6.3	0.5	136
United States	270.2	15	9	7.0	0.6	116
China	1,242.5	17	7	31	1.0	69
Argentina	36.1	19	8	22.2	1.1	62
Zimbabwe	11.0	35	20	53	1.5	46
Turkey	64.8	22	7	42	1.6	45
India	988.7	27	9	72	1.9	37
Uzbekistan	24.1	26	6	26	2.0	35
Mexico	97.5	27	5	28	2.2	32
Afghanistan	24.8	43	18	150	2.5	28
Ethiopia	58.4	46	21	128	2.5	28
Syria	15.6	33	6	35	2.8	25
Togo	4.9	46	11	84	3.6	19

Source: Data from *World Population Data Sheet,* 1998, Population Reference Bureau, Washington, D.C.

The age structure of a population also has a great deal to do with the rate of population growth. If a population has many young people who are raising families or who will be raising families in the near future, the population will continue to increase even if the families limit themselves to two children. Depending on the number of young people in a population, it may take 20 years to a century for the population of a country to stabilize so that there is no net growth.

Social Reasons for Population Growth

It is clear that the economically developed countries of the world have low fertility rates and low rates of population growth and that the less-developed countries have high fertility rates and high population growth rates. It also appears obvious that reducing fertility rates would be to everyone's advantage; however, not everyone in the world feels that way. Several factors influence a person's desired family size. Some are religious, some are traditional, some are social, and some are economic.

The major social factors that determine family size are the role and desires of women in the culture. In many male-dominated cultures, the traditional role of women is to marry and raise children. Often this role is coupled with strong religious input as well. Typically, little value is placed on educating women, and early marriage is encouraged. In these cultures, women are totally dependent on their husbands and children in old age. Because early marriage is encouraged, fertility rates are high, since women are exposed to the probability of pregnancy for more of their fertile years. Lack of education reduces options for women in these cultures. They do not have the option to not marry or to delay marriage and thus reduce the number of children they will bear. By contrast, in much of the developed world women are educated, delay marriage, and have fewer children. It has been said that the single most important activity needed to reduce the world population growth rate is to educate women. Whenever the educational level of women increases, fertility rates fall.

Data on the age of mothers giving birth indicates that women in the 15 to 19-year-old range are responsible for 12 percent of births in Africa, 6 percent of births in Asia (excluding China) and 8 percent of births in Latin America. Their total fertility rates are 5.6, 3.3, and 3.0 children

per woman per lifetime, respectively. In the developed world, the average age of first marriage is much higher, between age 25 and 27; early marriages are rare; about 3 percent of mothers are between 15 to 19 years of age, and the total fertility rate is about 1.6 children per woman per lifetime.

As women become better educated and obtain higher paying jobs, they become financially independent and can afford to marry later and consequently have fewer children. Better-educated women are also more likely to have access to and use birth control. In economically advanced countries, about 70 percent of women typically use contraception. In the less-developed countries, contraceptive use is much lower—less than 20 percent in Africa, about 45 percent in Asia (except China), and about 60 percent in Latin America. Changing attitudes toward divorce may also be tied to reduced fertility.

It is important to recognize that access to birth control will not solve the population problem, What is most important is the desire of women to limit the size of their families. In developed countries, use of birth control is extremely important in regulating the birthrate. This is true regardless of religion and previous historical birthrates. Italy and France, both

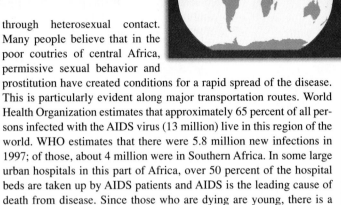

The Impact of AIDS on Populations

The AIDS (*acquired immunodeficiency syndrome*) virus has caused a worldwide epidemic, which can be called a pandemic because it continues to spread to all countries of the world. The disease is spread through direct transfer of body fluids containing the virus into the bloodstream of another person. Sharing of contaminated needles among intravenous drug users and sexual contact are the most likely methods of passage. In the United States, the disease was once considered a problem only for the homosexual community and those who use intravenous drugs. This perception is rapidly changing. Many of the new cases of AIDS are being found in women infected by male sex partners and in children born to infected mothers.

The World Health Organization (WHO) estimated that 30 million adults and children were infected with the AIDS virus by 1997 and that 40 million people will be infected worldwide by the year 2000. WHO estimates that 1.8 million people died of AIDS in 1997 and that 12 million have died since the pandemic began. The distribution of the virus is lowest in the economically developed countries and highest in the developing countries. The figure shows estimates by World Health Organization of the numbers of HIV-infected people in various parts of the world. In the less-developed world there is little medical care to treat AIDS and a lack of resources to identify those who have HIV. Many people do not know they are infected and will continue to pass the disease to others.

Sub-Saharan Africa has been hit hardest by this disease. In Africa, AIDS has always been primarily a sexually transmitted disease spread through heterosexual contact. Many people believe that in the poor coutries of central Africa, permissive sexual behavior and prostitution have created conditions for a rapid spread of the disease. This is particularly evident along major transportation routes. World Health Organization estimates that approximately 65 percent of all persons infected with the AIDS virus (13 million) live in this region of the world. WHO estimates that there were 5.8 million new infections in 1997; of those, about 4 million were in Southern Africa. In some large urban hospitals in this part of Africa, over 50 percent of the hospital beds are taken up by AIDS patients and AIDS is the leading cause of death from disease. Since those who are dying are young, there is a change in the age structure of the population of heavily infected countries. This has resulted in large numbers of AIDS orphans—over 8 million AIDS orphans in the world at the end of 1997. Some villages are already beginning to notice a change in the structure of their populations. With the death of young infected adults, villages are composed primarily of older people and children.

The economic burden on these countries is tremendous. Those with AIDS symptoms are unable to work and need medical care and medication. Because of poverty there is often little medical care available. Also the millions of orphaned children have resulted in an additional economic burden on relatives.

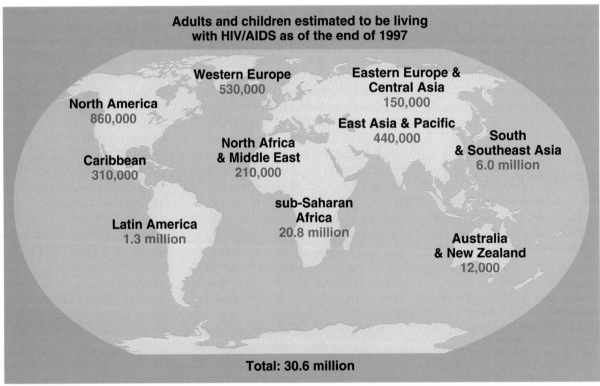

Adults and children estimated to be living with HIV/AIDS as of the end of 1997

Western Europe 530,000

Eastern Europe & Central Asia 150,000

North America 860,000

East Asia & Pacific 440,000

North Africa & Middle East 210,000

South & Southeast Asia 6.0 million

Caribbean 310,000

Latin America 1.3 million

sub-Saharan Africa 20.8 million

Australia & New Zealand 12,000

Total: 30.6 million

Source: Data from World Health Organization of the United Nations.

ENVIRONMENTAL CLOSE-UP

Control of Births

The use of technology to control disease and famine has greatly reduced the death rate of the human population. Technological developments can also be used to control the birthrate. A variety of contraceptive methods are available to help people regulate their fertility. Research is continuing to develop more effective, more acceptable, and less expensive methods of controlling conception. Because of cultural and religious differences, some forms of contraception may be more acceptable to one segment of the world's population than another.

The most common methods of contraception are oral contraceptive pills, diaphragms and spermicidal jelly, intrauterine contraceptive devices, spermicidal vaginal foam, condoms, vasectomy, and tubal ligation. The range of effectiveness of these methods, shown in the table, is the result of individual fertility differences and the degree of care employed in the use of each method.

In addition to various methods of conception control, abortion can terminate unwanted conceptions early in pregnancy. Most countries with low birthrates, such as the United States, Japan, and many European countries, allow abortions.

EFFECTIVENESS OF VARIOUS METHODS OF CONTRACEPTION

Method	Percent of women experiencing an unintended pregnancy within the first year of use	
	Typical use	Perfect use
No contraceptive method used	85	85
Spermicidal foams, creams, gels, suppositories, and vaginal films	26	6
Cervical cap		
Women who have had children	40	26
Women who have not had children	20	9
Sponge		
Women who have had children	40	20
Women who have not had children	20	9
Female condom	21	5
Diaphragm with spermicide	20	6
Withdrawal	19	4
Male condom	14	3
Periodic abstinence (natural family planning)		
Calendar method		9
Ovulation method		3
Temperature method		2
Post-ovulation method		1
Intrauterine device (IUD)	2	1.5
Female sterilization (tubal ligation)	0.5	0.5
Contraceptive pill		0.5
Contraceptive injection (Depo-Provera)	0.3	0.3
Male sterilization (vasectomy)	0.15	0.10
Contraceptive implant (Norplant)	0.05	0.05

Source: Data from J. Trussel, "Contraceptive Efficacy" in R. A. Hatcher, et al., *Contraceptive Technology*, 17th Revised Edition, 1998, Irvington Publishers.

traditionally Catholic countries, have birthrates of 9/1000 and 12/1000, respectively. The average for the developed countries of the world is 11/1000. Obviously, women in these countries make use of birth control to help them regulate the size of their families. (See Environmental Close-Up: Control of Births.) By contrast, Mexico, which is also a traditionally Catholic country, has a birthrate of 27/1000, which is typical of birthrates in the less-developed world regardless of religious tradition.

When women in the less-developed world are asked what family size they desire, they typically suggest four or more children. However, the desired number is usually fewer than the number they actually have, and both are much higher than the replacement fertility rate of 2.1 children. Access to birth control will allow them to limit the number of children they actually have to their desired number and will allow them to space their children at more convenient intervals, but they still desire more children than the 2.1 needed for replacement. Why do they desire large families? There are several reasons. In areas where infant mortality is high, it is traditional to have large families since several of a woman's children may die before they reach adulthood. This is particularly important in the less-developed world where there is no government program of social security. Parents are more secure in old age if they have several children to contribute to their needs when they can no longer work.

In less-developed countries, the economic benefits of children are extremely important. Even young children can be given jobs that contribute to the family economy. They can protect livestock from predators, gather firewood, cook, or carry water. In the developed world, large numbers of children are an economic drain.

Governmental Policy and Population Control

Governments can have a significant impact on the population growth patterns of nations. Some countries have policies that encourage couples to have children. Canada pays a bonus to couples on the birth of a child. The U.S. tax code indirectly encourages more children by providing exemptions for them. Some countries in Europe are concerned about a future lack of working-age people and are considering programs to encourage births. Many countries have no official population policy, but the position of most of those that do is to reduce the birthrate.

China has long been the most populated nation of the world. It now contains over 1.2 billion people, about 21 percent of the world's people. Its history of population control is an interesting study of how government policy affects reproductive activity among its citizens. When the People's Republic of China was established in 1949, the population was about 540 million. The official policy of the government was to encourage births because more Chinese would be able to produce more goods and services, and production was the key to economic prosperity.

Because of China's high birthrate and falling death rate, its population increased to 614 million in 1955. This rapid increase and lack of economic growth led government officials to make changes that would lead to population control. Abortions became legal in 1953, and the first family-planning program began in 1955, as a means of improving maternal and child health. Birthrates fell. (See graph.) The failure of government economic programs during the Great Leap Forward, and the establishment of communes to produce food and other products, resulted in widespread famine and increased death rates and low birthrates in the late 1950s and early 1960s. A second family-planning program began in 1962 but was not particularly effective.

The present family-planning policy began in 1971 with the launching of the *wan xi shao* campaign. Translated, the phrase means "later" (marriages), "longer" (intervals between births), and "fewer" (children). This program raised the legal ages for marriage. For women and men in rural areas, the ages were raised to 23 and 25, respectively; for women and men in urban areas, the ages were raised to 25 and 28, respectively. These policies resulted in a reduction of birthrates by nearly 50 percent between 1970 and 1979.

An even more restrictive, one-child campaign was begun in 1978–79. The program offered incentives for couples to restrict their family size to one child. Couples enrolled in the program would receive free medical care, cash bonuses for their work, special housing treatment, and extra old-age benefits. Those who broke their pledge were penalized by the loss of these benefits as well as other economic penalties. By the mid-1980s, less than 20 percent of the eligible couples were signing up for the program. Rural couples particularly desired more than one child. In fact, in a country where about 70 percent of the population is rural, the rural total fertility rate was 2.5 children per woman. In 1988, a second child was sanctioned for rural couples if their first child was a girl, which legalized what had been happening anyway.

The current total fertility rate is 1.8 children per woman. Over 80 percent of couples use contraception; the most commonly used forms are male and female sterilization and the intrauterine device. Abortion is also an important aspect of this program, with a ratio of over 600 abortions per 1,000 live births. However, because a large proportion of the population is under 30 years of age, the population is still expected to double in about 69 years. It appears that China has undergone a demographic transition of sorts. Birthrates and death rates were high in 1949, death rates fell steadily (except during the famine years), and birthrates have fallen to an all-time low. However, this demographic transition occurred as a result of government family-planning policies. Living standards are still low, with a per capita gross national product of US$750, and the population has not yet become stable. Compare the graph here with figure 7.4 on the demographic transition.

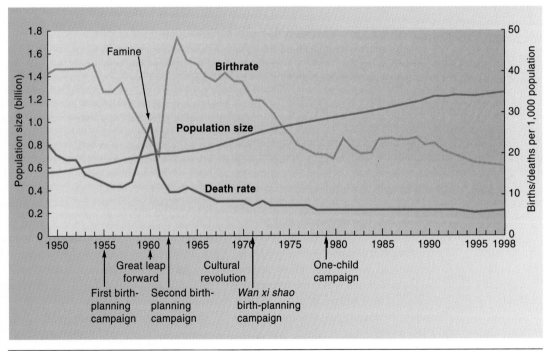

Source: Data from H. Yuan Tien, "China's Demographic Dilemmas" in *Population Bulletin, 1992,* Population Reference Bureau, Inc., Washington, D.C. and National Family Planning Commission of China; and more recent data taken from Population Reference Bureau.

GLOBAL PERSPECTIVE

Thomas Malthus and His Essay on Population

In 1798 Thomas Robert Malthus, an Englishman, published an essay on human population. It presented an idea that was contrary to popular opinion. His basic thesis was that human population increased in a geometric or exponential manner (2, 4, 8, 16, 32, 64, etc.), while the ability to produce food increased only in an arithmetic manner (1, 2, 3, 4, 5, 6, etc.). The ultimate outcome of these different rates would be that population would outgrow the ability of the land to produce food. He concluded that wars, famines, plagues, and natural disasters would be the means of controlling the size of the human population. His predictions were hotly debated by the intellectual community of his day. His assumptions and conclusions were attacked as erroneous and against the best interest of society. At the time he wrote the essay, the popular opinion was that human knowledge and "moral constraint" would be able to create a world that would supply all human needs in abundance. One of Malthus's basic postulates was that "commerce between the sexes" (sexual intercourse) would continue unchanged, while other philosophers of the day believed that sexual behavior would take

less procreative forms and human population would be limited. Only within the past fifty years, however, have really effective conception-control mechanisms become widely accepted and used, and they are used primarily in developed countries.

Malthus did not foresee the use of contraception, major changes in agricultural production techniques, or the exporting of excess people to colonies in the Americas, Australia,, and other parts of the world. These factors, as well as high death rates, prevented the most devastating of his predictions from coming true. However, in many parts of the world today, people are experiencing the forms of population control (famine, epidemic disease, wars, and natural disasters) predicted by Malthus in 1798. Many people feel that his original predictions were valid—only his time scale was not correct—and that we are seeing his predictions come true today.

They are prevented by law from working, they must be sent to school at great expense, and they consume large amounts of the family income. Parents in the developed world make an economic decision about having children in the same way they buy a house or car: "We are not having children right away. We are going to wait until we are better off financially."

Political Factors That Affect Population Growth

Two other factors that influence population growth are political pressures and immigration. Governments often encourage or discourage population growth in several ways. (See Global Perspective: Governmental Policy and Population Control.) They can have stated policies that describe population goals and pass laws that penalize those who fail to meet the goals, or reward those who meet them. Governments can also subtly encourage people to regulate their reproductive activities by providing tax advantages or low-cost access to birth control.

The immigration policies of a country also have a significant impact on the rate at which the population grows. Birthrates are currently so low in several European countries, Japan, and China that they will

likely have a shortage of those of working age in the near future. One way to solve this problem is to encourage immigration from other parts of the world.

The developed countries are under tremendous pressure to accept immigrants. Their standard of living is a tremendous magnet for refugees or people who seek a better life than is possible where they currently live. The continuing economic and political reorganization in central Europe is causing significant increases in the number of immigrants and placing considerable strain on the social systems of Germany, Austria, and other countries of the region. In the United States, approximately one-third of the population increase experienced each year is the result of immigration.

THE DEMOGRAPHIC TRANSITION CONCEPT

The relationship between the standard of living and the population growth rate seems to be that countries with the highest standard of living have the lowest population growth rate, and those with the lowest standard of living have the highest population growth rate. This has led many

people to suggest that countries naturally go through a series of stages called **demographic transition.** This model is based on the historical, social, and economic development of Europe and North America. In a demographic transition, the following four stages occur (see figure 7.4):

1. Initially, countries have a stable population with a high birthrate and a high death rate. Death rates often vary because of famine and epidemic disease.

2. Improved economic and social conditions (control of disease and increased food availability) bring about a period of rapid population growth as death rates fall. Birthrates remain high.

3. As countries become industrialized, the birthrates begin to drop because people desire smaller families and make use of contraceptives.

4. Eventually, birthrates and death rates again become balanced, with low birthrates and low death rates.

This is a very comfortable model because it suggests that, if a country can become industrialized, then social, political, and economic processes will naturally cause its population to stabilize.

However, the model leads to some serious questions. Can the historical pattern

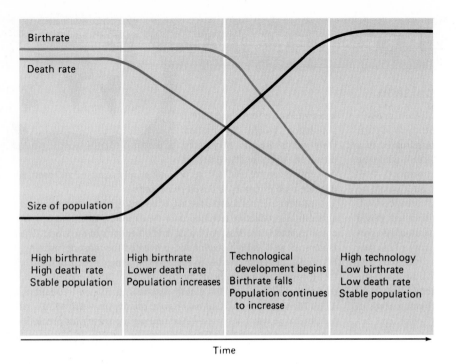

High birthrate	High birthrate	Technological	High technology
High death rate	Lower death rate	development begins	Low birthrate
Stable population	Population increases	Birthrate falls	Low death rate
		Population continues	Stable population
		to increase	

Time

FIGURE 7.4   Demographic Transition The demographic transition model suggests that, as a country develops technologically, it automatically experiences a drop in the birthrate. This certainly has been the experience of the developed countries of the world. However, the developed countries make up about 20 percent of the world's population. It is doubtful whether the less-developed countries can achieve the kind of technological advances experienced in the developed world.

exhibited by Europe and North America be repeated in the less-developed countries of today? Europe, North America, Japan, and Australia passed through this transition period when world population was lower and when energy and natural resources were still abundant. It is doubtful whether these supplies are adequate to allow for the industrialization of the major portion of the world currently classified as less-developed.

A second concern is the time element. With the world population increasing as rapidly as it is, industrialization probably cannot occur fast enough to have a significant impact on population growth. As long as people in less-developed countries are poor, there is a strong incentive to have large numbers of children. Children are a form of social security because they take care of their elderly parents. Only people in developed countries can save money for their old age. They can choose to have children, who are expensive to raise, or to invest money in some other way.

When the countries of Europe and North America passed through the demographic transition, they had access to large expanses of unexploited lands, either within their boundaries or in their colonies. This provided a safety valve for expanding populations during the early stages of the transition. Without this safety valve, it would have been impossible to deal adequately with the population while simultaneously encouraging economic development. Today, less-developed countries may be unable to accumulate the necessary capital to develop economically, since an ever-increasing population is a severe economic drain.

THE U.S. POPULATION PICTURE

As a result of the 1990 census in the United States, several changes have occurred in how demographers view the structure of and future trends in the U.S. population. In many ways, the U.S. population is similar to those of other developed countries of the world with low birthrates and slow population growth. However, the U.S. population includes a **postwar baby boom** component, which has significantly affected population trends. These baby boomers were born during an approximately 20-year period following World War II, when birthrates were much higher than today, and constitute a bulge in the age distribution profile. (See figure 7.5.) As members of this group have raised families, they have had a significant influence on how the U.S. population has grown. Some of the older persons in this group are beginning to retire, and as more of them do, the population will gradually age. By 2030, about 20 percent of the population will be 65 years of age or older.

A changing age structure will lead to social changes as well. The baby boom of the late 1940s and the 1950s encouraged growth in service industries needed by young families. Maternity wards had to be expanded, schools could not be built fast enough, baby-care companies saw unprecedented sales, and the toy industry flourished. Today, these "babies" are in their forties and fifties. They are buying homes, cars, and appliances, but are raising fewer children than did their parents. However, because baby boomers are such a large segment of the population, they have contributed a large number of children to the population. The children of the baby boomers now constitutes a "baby boomlet." In many parts of the country schools are seeing increased numbers of children in the early grades. At the same time, there are more elderly, because people are living longer. That creates a need for additional services for the elderly. This trend toward an aging population will be accentuated as the baby boomers retire. What will the social needs be in the year 2020 when many of these baby boomers will have retired?

Even with the current lower total fertility rate, the population is still growing by about 1.1 percent per year. About 0.6 percent is the result of natural increases owing to the difference between birthrates and death rates. The remainder is the result of immigration into the United States, The U.S. Census Bureau projects that immigration will increase significantly and account for 50 percent of population growth by the year 2050.

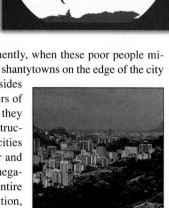

GLOBAL PERSPECTIVE

The Urbanization of the World's Population

The United Nations estimates that currently about 45 percent of the world's population lives in cities. This is expected to increase to over 50 percent by 2005 and reach 60 percent by 2025. The economically developed world is currently about 75 percent urban, and the less-developed world is about 37 percent urban. Therefore, the increase in the urban population will occur primarily in the less-developed world where resources to deal with urban problems are least available.

Urbanization is not necessarily bad. The economic activity of the developed world takes place primarily in cities. Cities can be planned and managed to be healthy, interesting places to live. Cities offer jobs, health care, schools, and other services that are usually lacking in rural areas. Often the rural economy of a country is considered to be of low status, and young people wish to move to cities where higher status jobs and desirable cultural amenities are available.

Dense populations of people have a concentrated impact on local resources. Often water must be transported long distances, wastes are difficult to get rid of, and air quality drops as each additional person places a burden on the local environment. Unfortunately, in the less-developed world it is impossible to plan for growth and provide basic resources such as drinking water, sewers, transportation, and housing as fast as the population is growing. New migrants to the cities often lack education or training for the high wage jobs available, have lost the opportunity to raise their own food, and lack the support of their family units. Consequently, when these poor people migrate to the cities, they often live in shantytowns on the edge of the city or form squatter settlements on hillsides and other unused land. These centers of poverty lack basic services since they are often outside the bureaucratic structure of the city. The rich in such cities live a short distance from the poor and may be relatively isolated, but the negative environmental impact of the entire urban region (air and water pollution, transportation problems, etc.) affects all the residents. Even the long-established urban centers of the developed world have major problems. Many contain pockets of poverty that have led to problems of social unrest and crime.

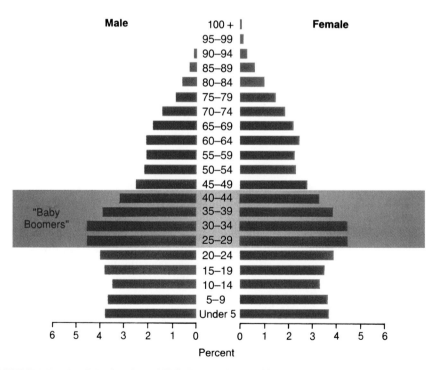

FIGURE 7.5 **Age Distribution of U.S. Population (1990)** This graph shows the number of people at each age level. Notice that at age 40 to 44, a bulge begins that ends at about age 25 to 29. These people represent the "baby boom" that followed World War II. In recent years, births have increased.

Source: Data from the U.S. Department of Commerce, Bureau of the Census.

Current immigration policy in the United States is difficult to characterize. Strong measures are being taken to reduce illegal immigration across the southern border. This is in part due to pressures placed on Congress by states that receive large numbers of illegal immigrants. Illegal immigrants add to the education and health care that states must fund. At the same time, some segments of the U.S. economy maintain that they are unable to find workers to do certain kinds of work. Consequently, special guest workers are allowed to enter the country for limited periods to serve the needs of these segments of the economy. There is also a consistent policy of allowing immigration that reunites families of U.S. residents. Obviously, the families that fall into this category are likely to be U.S. citizens who were recent immigrants themselves. It is obvious that most immigration policy is the result of political decisions rather than decisions that relate to population policy or a concern about the rate at which the U.S. population is growing.

Projections based on the 1990 census indicate that the population will grow more

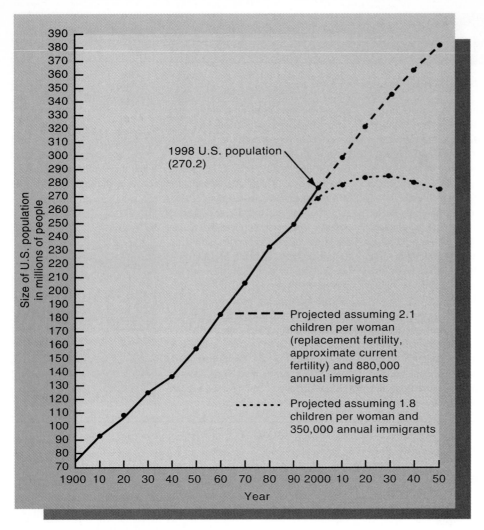

FIGURE 7.6 Population Growth The population of the United States has grown continuously until the present. The graph indicates that the size of the population was about 230 million people in 1980. The U.S. Census Bureau has made projections based on birthrate and presumed immigration. The ultimate size of the U.S. population differs considerably, depending on the estimates used.

Source: Data from the U.S. Department of Commerce, Bureau of the Census.

rapidly than previously projected, and it does not seem to be moving toward zero growth. Indeed, it appears that the total fertility rate, which is currently about 2 children per woman per lifetime, will increase to 2.12 by the year 2050. This would result in an increase from 248.7 million people in 1990 to about 275 million by 2000 and a further increase to about 383 million by 2050. (See figure 7.6.)

These increases are partly the result of higher fertility rates among some minority populations. The Hispanic and Asian American portions of the population will grow rapidly, and the Caucasian portion will decline. Thus, the future U.S. population will be much more ethnically diverse.

HUNGER, FOOD PRODUCTION, AND ENVIRONMENTAL DEGRADATION

As the human population increases, there is an increased demand for food. People must either grow food themselves or purchase it. Most people in the developed world purchase what they need and have more than enough food to eat. Most people in the less-developed world must grow their own food and have very little money to purchase additional food. Typically, these farmers have very little surplus. If crops fail, people starve. Even in countries with the highest population (China and India), the majority of the people live on the land and farm.

The human population can increase only if the population of other kinds of plants and animals decrease. Each ecosystem has a maximum biomass that can exist within it. There can be shifts within ecosystems to allow an increase in the population of one species, but this always adversely affects certain other populations because they are competing for the same basic resources.

When humans need food, they convert natural ecosystems to artificially maintained agricultural ecosystems. The natural mix of plants and animals is destroyed and replaced with species useful to humans. If these agricultural ecosystems are mismanaged, the region's total productivity may fall below that of the original ecosystem. The dust bowl of North America, desertification in Africa, and destruction of tropical rainforests are well-known examples. In countries where food is in short supply and the population is growing, there is intense pressure to convert remaining natural ecosystems to agriculture. Typically, these areas are the least desirable for agriculture and will not be productive. However, to a starving population, the short-term gain is all that matters. The long-term health of the environment is sacrificed for the immediate needs of the population.

A consequence of the basic need for food is that people in less-developed countries generally feed at lower trophic levels than do those in the developed world. (See figure 7.7.) Converting the less concentrated carbohydrates of plants into more nutritionally valuable animal protein and fat is an expensive process. During the process of feeding plants to animals and harvesting animal products, approximately 90 percent of the energy in the original plants is lost. Therefore, in terms of economics and energy, people in less-developed countries must consume the plants themselves rather than feed the plants to animals and then consume the

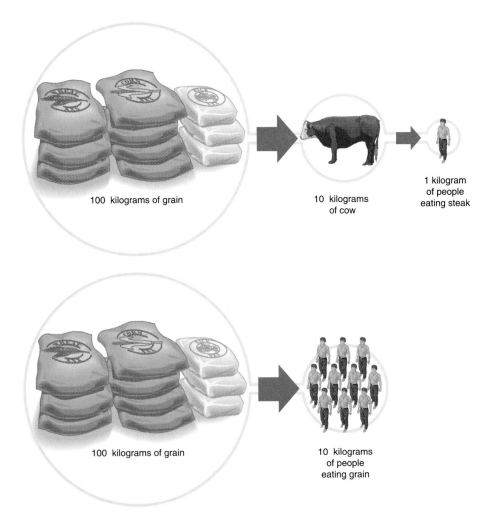

100 kilograms of grain

10 kilograms of cow

1 kilogram of people eating steak

100 kilograms of grain

10 kilograms of people eating grain

FIGURE 7.7 Population and Trophic Levels The larger a population, the more energy it takes to sustain the population. Every time one organism is eaten by another organism, approximately 90 percent of the energy is lost. Therefore, when populations are very large, they usually feed at the herbivore trophic level because they cannot afford the 90 percent energy loss that occurs when plants are fed to animals. The same amount of grain can support ten times more people at the herbivore level than at the carnivore level.

animals. In most cases, if the plants were fed to animals, many people would starve to death. On the other hand, a lack of protein in diets that consist primarily of plants can lead to malnutrition. Many people in the less-developed world suffer from a lack of adequate protein, which stunts their physical and mental development.

In contrast, in most of the developed world, meat and other animal protein sources are important parts of the diet. Many people suffer from overnutrition (they eat too much); they are "malnourished" in a different sense. The ecological impact of one person eating at the carnivore level is about ten times that of a person eating at the herbivore level. If people in the developed world were to reduce

their animal protein intake, they would significantly reduce their demands on world resources. Almost all of the corn and soybeans grown in the United States is used as animal feed. If these grains were used to feed people rather than animals, less grains would have to be grown and the impact on farmland would be less.

In countries where food is in short supply, agricultural land is already being exploited to its limit, and there is still a need for more food. This makes the United States, Canada, Australia, Argentina, New Zealand, and the European Economic Community net food exporters. Many countries, like India and China, are able to grow enough food for their people but do not have any left for export. Others,

including many nations of the former Soviet Union, are not able to grow enough to meet their own needs and, therefore, must import food.

A country that is a net food importer is not necessarily destitute. Japan and some European countries are net food importers but have enough economic assets to purchase what they need. Hunger occurs when countries do not produce enough food to feed their people and cannot obtain food through purchase or humanitarian aid.

The current situation with respect to world food production and hunger is very complicated. It involves the resources needed to produce food, such as arable land, labor, and machines; appropriate crop selection; and economic incentives. It also involves the maldistribution of food within countries. This is often an economic problem, since the poorest in most countries have difficulty finding the basic necessities of life, while the rich have an excess of food and other resources. In addition, political activities often determine food availability. War, payment of foreign debt, and poor management often contribute to hunger and malnutrition.

Improved plant varieties, irrigation, and improved agricultural methods have dramatically increased food production in some parts of the world. In recent years, India, China, and much of southern Asia have moved from being food importers to being self-sufficient, and in some cases, food exporters.

The areas of greatest need are in sub-Saharan Africa. Africa is the only major region of the world where per capita grain production has decreased over the past few decades. People in these regions are trying to use marginal lands for food production, as forests, scrubland, and grasslands are converted to agriculture. Often, this land is not able to support continued agricultural production. This leads to erosion and desertification.

What should be done about countries that are unable to raise enough food for their people and are unable to buy the food they need? This is not an easy question. A simple humanitarian solution to the problem is for the developed countries to supply food. Many religious and humanitarian organizations do an excellent

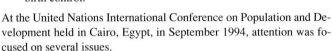

GLOBAL PERSPECTIVE

Population and Poverty: A Vicious Cycle

It is clear that the areas of the world where the human population is growing most rapidly are those that have the lowest standard of living. The developed countries of Europe and North America have abundant wealth and have relatively slowly growing populations. Countries in the southern hemisphere are generally poor and have high population growth rates. What forces are involved? Although not all cases are the same, the following factors seem to be interrelated.

Let us look at why some of these factors might be interrelated.

1. Poor people cannot afford birth control and, since they are often poorly educated, may not be able to read the directions on how to use various birth control mechanisms. Therefore, they have more children than they may wish to have.

2. Poor people need to obtain income in many ways. Often this includes taking children out of school so that they are able to work on the farm or in other jobs to provide income for the family. Poorly educated people cannot get high-paying jobs and so remain in poverty.

3. Women in poor countries are usually poorly educated and do not have disposable income. Therefore, they are dependent on their husbands or the family unit for their livelihood. Women who do not have independent income are more likely to have children they do not want because they cannot afford birth control.

At the United Nations International Conference on Population and Development held in Cairo, Egypt, in September 1994, attention was focused on several issues.

1. Improving the educational status of women was promoted. This would lead to improved financial standing for women, which could allow them to have fewer children.

2. Access to birth control was advanced as a goal, but was attacked by fundamentalist Muslims and Catholics. The final document was approved by both groups.

3. There was recognition that economic well-being is tied to solving the population problem. However, the huge, rapidly growing population of the poor countries of the world cannot hope to consume at the rate the rich countries do. Furthermore, the rich countries of the world need to reduce their rate of consumption.

	Countries with low population growth rates	Countries with high population growth rates
Wealth	Rich	Poor
Educational level	Nearly all people can read and write.	Many people cannot read and write.
Status of women	Educated	Uneducated
	Financially independent	Financially dependent on husband or family unit
Status of children	Laws prevent children from working.	Many children required to work.

---

service by taking food to those who need it, and save many lives. However, the aim should always be to provide temporary help and insist that the people of the country develop mechanisms for solving their own problem. Often, emergency food programs result in large numbers of people migrating from their rural (agricultural) areas to cities, where they are unable to support themselves. They become dependent on the food aid and stop working to raise their own food, not because they do not want to work, but because they need to leave their fields to go to the food distribution centers. Many humanitarian organizations now recognize the futility of trying to feed people with gifts from the developed world. The emphasis must be on self-sufficiency.

ANTICIPATED CHANGES WITH CONTINUED POPULATION GROWTH

As the world human population continues to increase, pressure for the necessities of life will become greater. Differences in the standard of living between developed and less-developed countries will remain significant because population will increase most in less-developed countries. The supply of fuel and other resources is dwindling. Pressure for these resources will intensify as industrialized countries seek to maintain their current standard of living. People in less-developed countries will seek more land to raise crops and feed themselves unless major increases in food production per hectare occur. Since most of these people live in tropical areas, tropical forests will be cleared for farmland. The resulting erosion or alteration of the soil will make it no longer suitable for forest or crops. This conversion of natural ecosystems to agricultural ecosystems could cause profound changes in the world ecosystem.

Developed countries may have to choose between two alternatives: helping the less-developed countries, thus maintaining their friendship, or isolating themselves from the problems of the less- developed nations. Neither of these policies will be able to prevent a change in lifestyle as world population increases. The resources

GLOBAL PERSPECTIVE

Canadian Population Overview

Although the total Canadian population is slightly more than 11 percent of the U.S. population (30.6 million versus 270.2 million), the structure of the Canadian population is very similar to that of the United States. The chart shows several comparisons between the two populations for 1998.

These data are typical for populations in developed nations where birthrates and death rates are low and a relatively large proportion of the population is made up of older members. The relatively small, young proportion of the population, coupled with low total fertility rates, leads to relatively slow population growth. In both countries, a significant amount of the population growth is attributable to immigration. The high standard of living in both countries attracts people who desire a better life.

	Population size 1998 (millions)	Birth rate per 1000	Death rate per 1000	Rate of natural increase	Doubling time (years)	Total fertility rate (children/ woman)	Life expectancy (years)	Infant morality rate (deaths/1000 births)	Per capita GNP 1996 US$	% under 15 years of age	% over 65 years of age
Canada	30.6	12	7	0.5	136	1.6	78	6.2	19,020	20	12
United States	270.2	15	9	0.6	116	2.0	76	7.0	28,020	22	13

Source: Population Reference Bureau "1998 Population Data Sheet"

ISSUES & ANALYSIS — Population Growth in Mexico

Mexico currently has a population growth rate of 2.2 percent per year, which will result in a doubling of its population in 32 years. This is an improvement over 1970, when the population was growing at a rate of 3.5 percent. The Mexican government's goal is to reduce the population growth rate to 1 percent. The program includes family planning, sex education, improving the status of women, and educational reform.

The Mexican population is very young; 36 percent are under 15 years of age. As these children mature, they will have children and will need to find jobs. There may not be enough jobs to go around, since current unemployment is high. The government is seeking to create labor-intensive industries to absorb this growing labor force.

The United States has made it more difficult for Mexican nationals to find work north of the border. The U.S. government has set penalties for employers who use illegal immigrant labor and has reduced the number of documented Mexicans who can legally work in the United States. This could have a major impact on the Mexican economy, since 21 percent of Mexicans rely on income from the United States for their support.

Many U.S. and European businesses have built assembly plants in Mexico to make use of the abundant inexpensive labor. The North American Free Trade Agreement (NAFTA) will allow more fluid exchange of goods and services between Canada, the United States, and Mexico. It is unclear what the eventual outcome of free trade will be, but labor leaders in Canada and the United States are concerned about the effect access to low-cost labor will have on their membership.

- Should the U.S. government be involved in setting Mexico's population policy?

- Why can't the Mexican population be limited as easily as European or North American populations have been?

- Should the United States enact policies that make the Mexican population problem worse?

of the world are finite. Even if industrialized countries continue to use a disproportionate share of the world's resources, the amount available per person will decline as population rises. As world population increases, the less-developed areas will probably maintain their low standard of living. It is difficult to see how that standard of living could get much lower, since some people in these countries are already starving to death. Lifestyles in developed nations will probably change very little but may become less consumption-oriented. What some people currently view as necessities (meals in restaurants, vacations to remote sites, and two cars per family) will probably become luxuries. Many people who enjoy the freedom of mobility associated with the automobile will have their travel limited as public transportation replaces private transportation. Recreation may cease to involve expensive, energy-demanding machines (powerboats, motorcycles, and electric-powered toys) and emphasize such activities as hiking, bicycling, and reading instead. These changes will not come quickly, unless some catastrophic political or economic force causes major world-wide adjustments. Most likely, changes will occur only as economic pressures affect families. Many economists and political thinkers feel that as the economics of the world become more linked, and jobs flow more freely from country to country, wealth will be redistributed from the former rich countries to emerging economies in developing countries. The possibility of such a redistribution has become a major political issue in many of the developed countries of the world.

Summary

Many of the problems of the world are caused by or made worse by an increasing human population. Currently, the world's population is growing very rapidly. Most of the growth is occurring in the less-developed nations of the world (Africa, Asia, and Latin America) where people have a low standard of living. The developed regions of the world, with their high standard of living, have relatively slow population growth and in some instances have declining populations.

Demography is the study of human populations and the things that affect them. Demographers study the sex ratio and age distribution within a population to predict future growth. Population growth rates are determined by biological factors such as birthrate, which is determined by the number of women in the population and the age of the women, and death rate. Sociological and economic conditions are also important, since they affect the number of children desired by women, which helps set the population growth rate. In developed countries, women usually have access to jobs. Couples marry later, and they make decisions about the number of children they will have based on the economic cost of raising children. In the less-developed world, women marry earlier, and children have economic value as additional workers, future caregivers for the parents, and as status for either or both parents.

The current U.S. population will continue to grow. The average age of the population will increase and the racial mix of the population will change significantly.

The demographic transition model suggests that, as a country becomes industrialized, its population begins to stabilize, but there is little hope that the earth can support the entire world in the style of the industrialized nations. It is doubtful whether there are enough energy and other natural resources to develop the less-developed countries or time to change trends of population growth. Highly developed nations should anticipate increased pressure in the future to share their wealth with less-developed countries.

Review Questions

1. What is demography?
2. What is demographic transition? What is it based upon?
3. What is a baby boom?
4. What does age distribution of a population mean?
5. List 10 differences between your standard of living and that of someone in a less-developed country?
6. Why do people who live in overpopulated countries use plants as their main source of food?
7. Although predicting the future is difficult, what do you think your life will be like in 10 years? Why.?
8. List five changes you might anticipate if world population were to double in the next 50 years.
9. Which three areas of the world have the highest population growth rate? Which three areas of the world have the lowest standard of living?
10. How many children per woman would lead to a stable U.S. population?
11. What role does the status of women play in determining population growth rates?
12. Describe three reasons why women in the less-developed world might desire more than two children.

Critical Thinking Questions

1. Do you think it is appropriate for developed countries to persuade less developed countries to limit their population growth? What would be appropriate and inappropriate interventions, according to your ethics? Now imagine you are a citizen of a less developed country. What might be your reply to those who live in more developed countries? Why?
2. The Chinese government has been very involved in regulating population growth in China. What do you think about this kind of government intervention in China's population problem? What values, beliefs, and perspectives do you hold that lead you to think the way you do?
3. Population growth causes many environmental problems. Identify some of these problems. What role do you think technology will

play in solving these problems? Are you optimistic or pessimistic about these problems being solved through technology? Why?

4. Do you think that demographic transition will be a viable option for world development? What evidence leads you to your conclusions? What role should the developed countries play in the current demographic transition of developing countries? Why?

5. By the year 2050, the United States Census Bureau projects that immigration will account for 50% of the population growth in the United States. What values and perspectives should guide our immigration policy? Why?

6. Imagine a debate between an American and a Sudanese about human population and the scarcity of resources. What perspectives do you think the American might bring to the debate? What perspectives do you think the Sudanese would bring? What might be their points of common ground? On what might they differ?

7. Many people in developing countries hope to achieve the standard of living of those in the developed world. What might be the effect of this pressure on the environment in developing countries? The political relationship between developing countries and already developed countries? What ethical perspective do you think should guide this changing relationship?

8. The demographic changes occurring in Mexico have an influence on the United States. What problems does Mexico face regarding its demographics? Should the United States be involved in Mexican population policy?

Key Terms

demographic transition *113-114*
demography *108*
gross national product (GNP) *106*
population density *107*

postwar baby boom *114*
replacement fertility *108*
standard of living *106*
total fertility rate *108*
zero population growth *108*

Related Internet Sites

Visit our website at http://www.mhhe.com/environmentalscience/ for specific information on each of the topics below.

Population programs and activities in Africa

Human dimensions of global environmental change

Searchable database of the "Population" index

The impact of rapid population growth

Demography and ecology

World population clock from the U.S. Census Bureau

Carolina population center

Online resource for population information

Population trends in the U.S.

Statistics on Canadian land, people, economy, and the state

Information facilities in demography

PART THREE

Energy

All living systems can be described by the flow of energy through them: Energy enables simple forms of matter to be changed into more complex forms and energy is needed to maintain this complexity. Energy sustains the complex technical and social units typical of human populations.

All living things (including humans) rely on the sun as a source of energy. Coal, oil, and natural gas are energy sources available today because organisms in the past captured sunlight energy and stored it in the complex organic molecules that made up their bodies, which were then compressed and concentrated.

Technological development and fossil-fuel exploitations are directly related to one another and have allowed us an increasingly higher standard of living. **Chapter 8** traces the development of energy consumption, its interrelationships with economic development and lifestyles, and current uses and demands. **Chapter 9** discusses the sources of energy currently being used, as well as the significance of each source and its impact. **Chapter 10** discusses nuclear energy, other applications of radioactive substances, and the concerns their use generates.

8
chapter

Energy and Civilization: Patterns of Consumption

OBJECTIVES

After reading this chapter, you should be able to:

- Explain why all organisms require a constant input of energy.
- Describe how per capita energy consumption increased as civilization developed from hunting and gathering to primitive agriculture to advanced cultures.
- Describe how advanced modern civilizations developed as new fuels were used to run machines.
- Recognize that coal deposits are not uniformly distributed throughout the world.
- Correlate the Industrial Revolution with social and economic changes.
- Explain how cheap oil and natural gas led to a consumption-oriented society.
- Explain how the automobile changed people's lifestyles.
- Explain why overall energy use in the United States declined during the 1970s and 1980s.

HISTORY OF ENERGY CONSUMPTION

Every form of life and all societies require a constant input of energy. If the flow of energy through organisms or societies ceases, they stop functioning and begin to disintegrate. Some organisms and societies are more energy efficient than others. In general, history shows that complex industrial societies use the most energy. If societies are to survive, they must continue to expend energy. However, they may need to change their pattern of energy consumption as traditional sources become limited.

Biological Energy Sources

Energy is essential to maintain life. In every ecosystem, the sun provides that energy. (See chapter 4.) The first transfer of energy occurs during photosynthesis, when plants convert light energy into chemical energy in the production of food. Herbivorous animals utilize the food energy in the plants. The herbivores, in turn, are a source of energy for carnivores. Because nearly all of their energy requirements were supplied by food, primitive humans were no different from other animals in their ecosystems. In such hunter-gatherer cultures, nearly all human energy needs were met by using plants and animals as food, tools, and fuel. (See figure 8.1.)

Early in human history, people began to use additional sources of energy to make their lives more comfortable. They domesticated plants and animals to provide a more dependable supply of food. They no longer needed to depend solely upon gathering wild plants and hunting wild animals for sustenance. Domesticated animals also furnished a source of energy for transportation, farming, and other tasks. (See figure 8.2.) Wood provided a source of fuel for heating and cooking. Eventually this biomass energy was used in simple technologies, such as shaping tools and extracting metals.

Increased Use of Wood

Early civilizations, such as the Aztecs, Greeks, Egyptians, Romans, and Chinese,

FIGURE 8.1 Hunter-Gatherer Society In this type of society people obtain nearly all of their energy from the collection of wild plants and the hunting of animals. These societies do not make large demands on fossil fuels.

FIGURE 8.2 Animal Power This bas-relief panel from an Egyptian tomb depicts an important accomplishment in the development of human civilization. With the use of domesticated animals, people had a source of power other than their own muscles.

were culturally advanced, but their societies used human muscle, animal muscle, and fire as sources of energy. Except for limited use of some wind-powered and water-powered devices such as ships and canoes, the controlled use of fire was the first use of energy in a form other than food. Wood was the primary fuel. (Wood was also used for building materials and

other cultural uses.) The energy provided by wood enabled people to cook their food, heat their dwellings, and develop a primitive form of metallurgy. Such advances separated humans from other animals. When dense populations of humans made heavy use of wood for fuel and building materials, they eventually used up the readily available sources and had to import wood or seek alternative forms of fuel.

Because of a long history of high population density, India and some other parts of the world experienced a wood shortage hundreds of years before Europe and North America did. In many of these areas, animal dung replaced wood as a fuel source. It is still used today in some parts of the world.

Western Europe and North America were able to use wood as a fuel for a longer period of time. The forests of Europe supplied sufficient fuel until the thirteenth century. In North America, vast expanses of virgin forests supplied adequate fuel until the late nineteenth century. Fortunately, when local supplies of wood declined in Europe and North America, coal, formed from fossilized plant remains, was available as an alternative energy source. By 1880 coal had replaced wood as the primary energy source.

Fossil Fuels and the Industrial Revolution

Fossil fuels are the remains of plants, animals, and microorganisms that lived millions of years ago. (The energy in these fuels is stored sunlight, just as the biomass of wood represents stored sunlight.) During the Carboniferous period, 275 to 350 million years ago, conditions in the world were conducive to the formation of large deposits of fossil fuels. (See figure 8.3.) Ever since machines replaced muscle power, the major energy sources for the world have been fossil remains from the distant past.

Historically, the first fossil fuel to be used extensively was coal. In the early eighteenth century, regions of the world that had readily available coal deposits were able to switch to this new fuel and participate in a major cultural change known as the **Industrial Revolution.** The Industrial Revolution began in England and spread to much of Europe and North America. It involved the invention of machines that replaced human and animal labor in manufacturing and transporting goods. Central to this change was the invention of the steam engine, which could convert heat energy into the energy of motion. The source of energy for steam engines was either wood or coal; wood

a.   b.

FIGURE 8.3   Carboniferous Period Approximately 300 million years ago, this kind of ecosystem was common throughout the world. Plant material accumulated in these swamps and was ultimately converted to coal.

was quickly replaced by coal in most cases. Nations without a source of coal, or those possessing coal reserves that were not easily exploited, did not participate in the Industrial Revolution.

Prior to the Industrial Revolution, Europe and North America were predominately rural. Goods were manufactured on a small scale in the home. As machines and the coal to power them became increasingly available, the factory system of manufacturing products replaced the small home-based operation. Because expanding factories required a constantly increasing labor supply, people left the farms and congregated in areas surrounding the factories. Villages became towns, and towns became cities. Widespread use of coal in cities resulted in increased air pollution. In spite of these changes, the Industrial Revolution was viewed as progress. Energy consumption increased, economies grew, and people prospered. Within a span of two hundred years, the daily per capita energy consumption of industrialized nations

increased eightfold. This energy was furnished primarily by coal, but a new source of energy was about to he discovered: oil.

The Chinese used some gas and oil as early as 1000 B.C., yet these resources remained virtually untapped until fairly recently. The oil well that Edwin L. Drake, an early oil prospector, drilled in Pennsylvania in 1859 was not the world's first oil well, but it was the beginning of the modern petroleum era. By 1870, oil production in the United States had reached over four million barrels a year and supplied 1 percent of the nation's energy requirements. It grew to nearly 50 percent by 1970 and currently contributes just over 40 percent. (See figure 8.4.)

For the first 60 years of production, the principal use of oil was to make kerosene, a fuel for lamps. The gasoline produced was discarded as a waste product. During this time, the supply of oil exceeded the demand. However, the automobile dramatically increased the demand. In 1900, the United States had only 8000 automobiles.

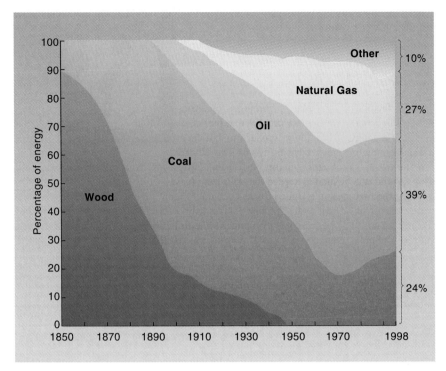

FIGURE 8.4 Oil Replaces Coal Just as wood was replaced by coal, coal was later replaced by oil. This graph represents the production of energy by various sources in the United States. Oil has remained a dominant energy source for the past 40 years, coal has decreased slightly, natural gas has increased, and other sources such as nuclear, hydroelectric, wind, and solar power have been increasing.

Source: Data from Annual Energy Review, 1998, Energy Information Administration, and more recent data from *BP Statistical Review of World Energy,* 1998 and pervious editions.

By 1920, it had 8 million cars, and by 1998, 160 million. More oil was needed to make automobile fuel and lubricants.

The use of natural gas did not increase as rapidly as the use of oil. It primarily was used for home heating. In the early 1900s, 90 percent of the natural gas was "flared," that is, burned as a waste product at oil wells.

A series of events involving the U.S. government and the need for more oil was ultimately responsible for increased use of natural gas in the United States. World War II greatly increased the energy demand for manufacturing and transportation. In 1943, a federally financed pipeline was constructed to transport oil within the United States. This 2000-kilometer pipeline transported oil more efficiently from wells in Texas, Louisiana, and Oklahoma to refineries and factories in the eastern section of the country. In 1944, a longer (2400 kilometers) federally financed pipeline was built to increase the flow of oil to the country's eastern and midwestern regions.

After the war, the federal government sold these pipelines to private corporations.

The corporations converted the pipelines to transport natural gas. Thus, a direct link was established between the natural gas fields in the Southwest and the markets in the Midwest and East. By 1971, there were 400,000 kilometers of long-range transmission to pipelines and 986,000 kilometers of distribution pipelines in the United States. Approximately 1,613,000 kilometers of natural gas distribution pipelines are used today. Natural gas is used in many parts of the developed world both for home heating and for industrial purposes.

ENERGY AND ECONOMICS

Most industrial societies want to ensure a continuous supply of inexpensive energy. The higher the price of energy the more expensive goods and services become. To keep costs down, many countries have subsidized their energy industries and maintained energy prices artificially low. International trade in fossil fuels has a major influence on the world economy and politics. The emphasis on low-priced fuels has encouraged high rates of consumption.

Economic Growth and Energy Consumption

There is a direct link between economic growth and the availability of inexpensive energy. The replacement of human and animal energy with fossil fuels began with the Industrial Revolution and was greatly accelerated by the supply of cheap, easy-to-handle, and highly efficient fuels. Because the use of inexpensive fossil fuels allows each worker to produce more goods and services, productivity increased. The result was unprecedented economic growth in Europe, North America, and the rest of the industrialized world.

In North America, World War II was a prime factor in ending the economic depression of the 1930s. Military activities created millions of defense jobs. Almost everyone was employed, but there was a scarcity of consumer goods. After World War II, consumer goods that had been unavailable during the war were in great demand. Industries set up to produce military goods turned to the production of consumer goods. High employment, a rapidly expanding population, and a supply of inexpensive energy encouraged a period of rapid economic growth.

In Europe and Japan, where much of the industrial base was destroyed by the war, the recovery was slow. However, foreign aid and an intense rebuilding effort on the part of the people resulted in a reindustrialization of these countries. Because they had to build new facilities, often their technology was better than that of North America where few factories were affected by the war.

The Role of the Automobile

The cheap, abundant energy that fueled industries produced an ever-increasing amount and array of consumer goods. One product was the automobile. The growth of the automobile industry, first in the United States and then in other industrialized countries, led to roadway construction, which required energy. Thus, the energy costs of driving a car were greater

GLOBAL PERSPECTIVE

Gasoline Prices and Government Policy

The price of a liter of gasoline is determined by two major factors: (1) the cost of purchasing and processing crude oil into gasoline and (2) various taxes. Most of the differences in gasoline prices among countries are a result of taxes and reflect differences in government policy toward motor vehicle transportation.

A major objective of governments is to collect money to build and repair roads. Governments often charge road users by taxing the fuel their cars or trucks run on. Governments can also discourage the use of automobiles by increasing the cost of fuel. An increase in fuel costs also creates a demand for increased fuel efficiency in all forms of motor transport.

Many European countries raise more money from fuel taxes than they spend on building and repairing roads. The United States, on the other hand, raises approximately 60 percent of the monies needed for roads from fuel taxes. The relatively low cost of fuel in the United States encourages more travel, which increases road repair costs. The cost of taxes to the U.S. consumer is about 28 percent of the retail gasoline price, while in Japan and many European countries, the cost is 47 to 77 percent. (See table.)

WORLD GASOLINE PRICES AND TAXATION

Country	Total price per gallon (U.S. dollars)	Total price per liter (U.S. dollars)	Total price per liter (local currency)	Tax % of price
France	$4,08	$1.20	6 francs	77%
Italy	$5.50	$1.48	1600 liras	75%
Germany	$4.04	$1.60	1.64 marks	71%
Spain	$4.01	$1.11	103 pesetas	69%
Britain	$3.38	$0.97	.48 pound	69%
Japan	$5.35	$1.41	175 yen	47%
United States	$1.19	$0.31	$0.31	28%

Source: Runzheimer International, 1998.

than just the fuel consumed in travel. As roads improved, higher speeds were possible. People demanded faster cars, and automobile companies were quick to build them. Bigger and faster cars required more fuel and even better roads. So roads were continually being improved, and better cars were being produced. A cycle of *more chasing more* had begun. In North America and much of Europe, the convenience of the automobile encouraged two-car families, which created a demand for more energy. It requires energy to mine ore, process it into metals, form the metals into automobile components, and transport all the materials. As the economy grew, so did energy needs.

More cars meant more jobs in the automobile industry, the steel industry, the glass industry, and hundreds of other industries. Constructing thousands of kilometers of roads created additional jobs. From their beginnings as suppliers

of lamp oil, the oil companies grew into one of the largest industries in the world. Thus, the automobile industry played a major role in the economic development of the industrialized world. All this wealth gave people more money for cars and other necessities of life. The car, originally a luxury, was now considered a necessity.

The car not only created new jobs but also altered people's lifestyles. Vacationers could travel greater distances. New resorts and chains of motels, restaurants, and other businesses developed to serve the motoring public, creating thousands of new jobs. Because people could live farther from work, they began to move to the suburbs. (See figure 8.5.) Large shopping centers in suburban areas hastened the decline of central business districts. Today, fewer than 50 percent of retail sales are made in central business districts of North American cities, which has resulted in a

loss of jobs in these areas. In Philadelphia, 79 percent of all retail jobs were located in the city in 1930; by 1970, they had declined to 43 percent.

As people moved to the suburbs, they also changed their buying habits. Labor-saving, energy-consuming devices became essential in the home. The vacuum cleaner, dishwasher, garbage disposal, and automatic garage door opener are only a few of the ways human power has been replaced with electrical power. Eleven percent of the electrical energy in North America is used to operate home appliances. Other aspects of our lifestyles illustrate our energy dependence. The small, horse-powered farm of yesterday has grown into the huge, diesel-powered farm of today. Regardless of where we live, we expect Central American bananas, Florida oranges, California lettuce, Texas beef, Hawaiian pineapples, Ontario fruit, and Nova Scotia lobsters to be readily available at all seasons. What we often fail to consider is the amount of energy required to process, refrigerate, and transport these items. The car, the modern home, the farm, and the variety of items on our grocery shelves are only a few indications of how our lifestyles are based on cheap, abundant energy.

HOW ENERGY IS USED

The amount of energy consumed by countries of the world varies widely. (See table 8.1.) The highly industrialized countries use most of the world's energy; less-developed countries use much less. Even countries with the same level of development vary in the amount of energy they use as well as how they use it. To maintain their style of living, individuals in Canada and the United States use about twice as much energy as people in France or Japan and about 25 times as much energy as people in Africa.

Countries also use energy in different ways. Industrialized nations use energy about equally for three purposes: (1) residential and commercial uses, (2) industrial uses, and (3) transportation. Less-developed nations with little industry use most of their energy for residential purposes (cooking and heating). Developing countries use much of their energy to develop their industrial base.

TABLE 8.1 ENERGY CONSUMPTION 1998

Region	Energy consumption per capita per year (in tonnes of oil equivalent)
Africa	0.33
Latin America	0.67
Japan	3.72
France	4.05
Germany	4.11
Canada	7.61
United States	7.86

Source: Data from *BP Statistical Review of World Energy,* June 1998 and Population Reference Bureau *1998 World Population Data Sheet.*

FIGURE 8.5 Energy-Demanding Lifestyle Building private homes on large individual lots some distance from shopping areas and places of employment is directly related to the heavy use of the automobile as a mode of transportation. Heating and cooling a large enclosed shopping mall, along with the gasoline consumed in driving to the shopping center, increase the demand for energy.

Residential and Commercial Energy Use

The amount of energy required for residential and commercial use varies greatly throughout the world. Although a country with a high gross domestic product (GDP) uses a large amount of energy, it uses a lower percentage of its energy per capita for residential and commercial needs than

TABLE 8.2 PER CAPITA ENERGY USE FOR TRANSPORTATION, 1998

Country	Energy use in gigajoules/capita
India	2
Zimbabwe	4
Mexico	17
Argentina	18
Japan	28
Russia	26
Netherlands	41
Denmark	42
Australia	86
United States	106

FIGURE 8.6 Open Fire Cooking About half of the energy demand in Africa is for cooking. Using a stove instead of an open fire could reduce this energy need by nearly 50 percent.

does a less-developed country. For example, about 30 percent of the energy used in North America is for residential and commercial purposes, while in India, 90 percent of the energy is for residential uses. The ways residential and commercial energy is used also vary widely. In North America, 75 percent is used for air conditioning, refrigeration, water heating, and space heating. In India, almost all of the energy is used in the home for cooking since the scarcity and high cost of fuel precludes other uses.

The current pattern of residential and commercial energy use in each region of the world determines what conservation methods will be effective. In Canada, which has a cold climate, 40 percent of the residential energy is used for heating.

Proper conservation practices could reduce this by 50 percent. In Africa, almost half of the energy used in the home is for cooking. (See figure 8.6.) Using fuel-efficient stoves instead of open fires could reduce these energy requirements by 50 percent.

Industrial Energy Use

The amount of energy countries use for industrial processes varies considerably. Nonindustrial countries use little energy for industry. Countries that are developing new industries dedicate a high percentage of their energy use to them. They divert energy to the developing industries at the expense of other sectors of their economy. Highly industrialized countries use a significant amount of their energy in industry,

but their energy use is high in other sectors as well. In the United States, industry claims about 30 percent of the energy used.

The amount of energy required in a country's industrial sector depends on the types of industrial processes used. Many countries use inefficient processes and could reduce their energy consumption by converting to more energy-efficient ones. However, they need capital investment to upgrade their industries and reduce energy consumption. Some countries cannot afford the upgrade. For example, India, a nation with few coal deposits, still uses outdated open-hearth furnaces to produce steel. These furnaces require nearly double the worldwide energy average to produce a metric ton of steel. The high cost of converting forces India to continue to use this energy-expensive method rather than convert to the more efficient electric-hearth method.

Transportation Energy Use

As with residential, commercial, and industrial uses, the amount of energy used for transportation varies widely throughout the world. In some of the less-developed nations, transportation uses are very small. Per capita energy use for transportation is larger in developing countries, and highest in highly developed countries. (See table 8.2.)

Once a country's state of development has been taken into account, the mix of bus, rail, water, and private automobiles is the main factor in determining a country's energy use for transportation. In Europe, Latin America, and many other parts of the world, rail and bus transport are widely used because they are more efficient than private automobile travel, governments support these transportation methods, or a large part of the populace is unable to afford an automobile. In countries with high population densities, rail and bus transport is particularly efficient. (See figure 8.7.) In these countries, automobiles require about four times more energy per passenger kilometer than bus or rail transport require. In addition, most of these countries have high taxes on fuel, which raise the cost to the consumer and encourage the use of public transport. In North America, the situation is different. Government policy has kept the cost of

ENVIRONMENTAL CLOSE-UP

Why Do Gasoline Prices Vary from Day to Day?

The price of gasoline at your local station is driven by the market forces of supply and demand. There are many external factors that influence the price you pay for gasoline including the world price of crude oil, the cost of refining and delivering the gasoline to your local station, and the international gasoline market. Each of these factors can and does vary from day to day.

The market forces that help determine the price of gasoline vary from the state of the local economy to the marketing strategies of the companies that produce the gasoline. Customer buying preferences and the number of stations in any given area are also factors. In addition to the above, the wholesale price of gasoline at any given time is also a factor. These factors can all affect the price of the gasoline you buy.

You have also no doubt witnessed what are referred to as price wars over gasoline at your local stations. Price wars can begin when a single dealer may use gasoline as the "loss leader" to attract you to the station to purchase other items such as car parts or groceries. The evolution from traditional gas stations to "mini-marts" has increased price wars and "loss leader" marketing.

Gasoline is a very competitive market and the visible pricing of the cost of a gallon or liter will attract customers. Pricing, however, in the long run will always stabilize in order to provide the station owner a reasonable profit after accounting for wholesale delivery and operating costs.

Governmental bodies at both the federal and state level closely monitor gasoline price swings. Such swings are indicative of free market competition and often benefit the consumer by offering lower prices.

---

energy low and supported the automobile industry while removing support for bus and rail transport. Consequently, the automobile plays a dominant rote, and public transport is primarily used only in metropolitan areas. Rail and bus transport are about twice as energy efficient as private automobiles. Private automobiles in North America consume over 15 percent of the world's oil production, while the rest of the automobiles in the world consume 7 percent. Air travel is relatively expensive in terms of energy, although it is slightly more efficient than private automobiles. Passengers, however, are paying for the convenience of rapid travel over long distances.

ELECTRICAL ENERGY

Electrical energy is such a large proportion of energy consumed in most countries that it deserves special comment. Electricity is both a way that energy is consumed and a way that it is supplied. Almost all electric energy is produced as a result of burning fossil fuels. Thus we can look at electrical energy as a use to which fossil fuel energy is put. In the same way we use natural gas to heat homes, we can use natural gas to produce electricity. Because the transportation of electrical energy is so simple and the uses to which it can be put

FIGURE 8.7 Public Transportation In regions of the world where energy is expensive, such as in Germany, people make maximum use of cheaper public transportation.

are so varied, electricity is a major energy source for many people of the world.

As with other forms of energy use, electrical consumption in different regions of the world varies widely. The amount of electricity used by all of the less-developed nations of the world, which have about 80 percent of the world's population, is only 66 percent of that used by the United States alone. The per capita use of electricity in North America is 25 times greater than average per capita use in the less-developed countries. In Nepal, the annual per capita use of electricity is 23 kilowatt hours, which is enough to light a 100-watt lightbulb for one week. The per capita consumption of electricity in North America is 270 times greater than in Nepal. There is also a wide variation in electrical consumption among developed countries. For example, the per capita use in Europe is about half that in North America.

The production and distribution of electricity is a major step in the economic

Electric Car Development

Clean air laws and regulations in countries around the world are providing a significant new push for the commercial development of electric vehicles. The Air Resources Board of the California Environmental Protection Agency voted in early 1996 to reaffirm a requirement of zero-emission vehicles. Although they voted to reduce the number of electric vehicles required in the late 1990s, they reaffirmed the requirement that 10 percent of all cars and light duty trucks sold in California in 2003 and thereafter be zero-emission vehicles. Electric cars are the only zero-emission vehicles currently available. Based on the number of automobiles sold in California, the following companies must comply with the requirement: General Motors, Ford, Chrysler, Mazda, Toyota, Honda, and Nissan. They must provide cars with a 125-mile range between charges, and they must participate in research leading to more efficient batteries. All seven manufacturers currently produce electric vehicles.

Electric car development received a boost following the 1997 Kyoto Global Climate Agreement. In early 1998 General Motors announced the phasing in of nickel-metal-hydride batteries to a new series of electric vehicles, the EV1 and S-10 electric truck. General Motors plans to have a production-ready hybrid electric vehicle by 2001 and a fuel cell electric vehicle by 2004. Nickel-metal-hydride (NiMH) batteries are capable of increasing the range to 160 miles. NiMH batteries have little in common with lead-acid batteries. NiMH batteries work by storing and releasing hydrogen during the charge and discharge processes. These batteries are completely sealed, require no maintenance and can be safely disposed of at the end of their life. All materials used in the battery are recyclable.

EV1

S-10
Electric Pickup

GM Ovonic Nickel-Metal-Hydride Batteries

Electric car

development of a country. In developed nations, about a quarter of the electricity is used by industry. The remainder is used primarily for residential and commercial purposes. In nations that are developing their industrial base, over half of the electricity is used by industry. For example, industries use 55 percent of the electricity used in Mexico and 70 percent of that used in South Korea.

ENERGY CONSUMPTION TRENDS

From a historical point of view, it is possible to plot changes in energy consumption. Economics, politics, public attitudes, and many other factors must be incorporated into an analysis of energy use trends. (See table 8.3 on page 135.)

In 1998, world energy consumption was around 24 million metric tons of oil equivalent per day, an increase of 18 percent since 1985. Of this total, conventional fossil fuels—oil, natural gas, and coal—accounted for about 90 percent.

Over half of world energy is consumed by the 25 countries that are

ENVIRONMENTAL CLOSE-UP

Alternative-Fuel Vehicles

Electricity is the only zero-emission option now available for vehicles, but many experts believe that other fuels will prove viable in *reduced-emission* vehicles. Here are some of the options.

Compressed Natural Gas (CNG)

Status:
Fuels thousands of vehicles already on U.S. highways. Mass production is still a couple of years away, at the earliest.

Advantages:
- Substantial U.S. supply of natural gas.
- Easy—though expensive—to convert conventional gasoline cars to CNG.
- Cheaper: the equivalent of buying gasoline at 70 cents a gallon.

Disadvantages:
- Requires heavy, bulky fuel tanks.
- Public refueling stations expensive to build.
- Lower performance and shorter range than gasoline vehicles.

Emissions Benefits:
- Burns 80 percent cleaner than gasoline.

Methanol/Flex-Fuel

Status:
Some heavy vehicles—including 354 buses operated by Southern California's Metropolitan Transportation Authority—run on pure methanol. More than 10,000 flex-fuel vehicles—running on methanol-gasoline combinations—now on U.S. roads.

Advantages:
- Can be made from natural gas (98% of current production) and a range of renewable sources.
- Requires only modest changes in gasoline engines and fueling infrastructure.
- Higher octane than gasoline, giving 5 percent better performance.

Disadvantages:
- Highly corrosive.
- Lower energy content than gasoline, so requires bigger fuel tanks.
- Fuel cost slightly higher than gasoline.

Emissions Benefits:
- An 85 percent methanol mix burns 30 to 50 percent cleaner than pure gasoline; pure methanol vehicles potentially could be much cleaner.

Hydrogen

Status:
Still largely experimental. Vehicles could be powered by hydrogen fuel cells or by engines that burn the gas. Practical, wide-scale use is at least a decade away.

Advantages:
- A high-energy fuel, boosting vehicle range.
- Potentially unlimited fuel supply if produced from water; currently made from natural gas.
- Produces no carbon dioxide, the gas blamed for possible global warming.

Disadvantages:
- Highly explosive, though ultimately could be less dangerous than gasoline.
- Costly and difficult to produce.
- Costly and difficult to store.

Emissions Benefits:
- If produced from water using solar energy—and then used in a fuel cell—would be a zero-emissions technology.

Propane

Status:
Broadly used in transportation worldwide. Environmentalists, regulators, and automakers contend that propane makes the most sense for heavy vehicles as a replacement for diesel.

Advantages:
- A proven, low-cost fuel.
- Public refueling infrastructure in place.
- Nontoxic, so storage tanks are exempt from environmental regulations.

Disadvantages:
- Supplies are limited, compared with other alternatives. At most could replace 10 percent of gasoline.
- Highly flammable.

Emissions Benefits:
- At least 50 percent cleaner than conventional gasoline.

Sources: Data from California Air Resources Board, California Energy Commission, South Coast Air Quality Management District, Ward's Communications, Ford Motor Co., General Motors Corp., Chrysler Corp., and LP Gas Coalition.

GLOBAL PERSPECTIVE

OPEC

OPEC, the Organization of the Petroleum Exporting Countries, began in September 1960 when the governments of five of the world's leading oil exporting countries agreed to form a cartel. Three of the original members—Saudi Arabia, Iraq, and Kuwait—were Arab countries, while Venezuela and Iran were not. Today, twelve countries belong to OPEC. These include seven Arab states—Saudi Arabia, Kuwait, Libya, Algeria, Iraq, Qatar, and United Arab Emirates—and five non-Arab members—Iran, Indonesia, Nigeria, Gabon, and Venezuela. (Ecuador was a member but withdrew from OPEC in 1992. See map.) OPEC nations control over 75 percent of the world's estimated oil reserves of 1200 billion barrels of oil. Middle Eastern OPEC countries control over 60 percent of this total, which makes OPEC and the Middle East important world influences.

During the Arab-Israeli War of 1973 OPEC was supplying 55 percent of the world's oil. To protest the war, seven Arab members of OPEC reduced their production of oil. This resulted in a worldwide oil shortage, which caused all oil companies, in OPEC and non-OPEC countries, to increase their prices. In the United States, the price of oil skyrocketed from $3.39 per barrel to $13.93 per barrel resulting in oil shortages, long lines at gas stations, and increased domestic exploration. The unity of OPEC, however, was not to last into the 1980s.

During the 1980s OPEC members had important differences concerning oil pricing and production rules. These differences weakened the cartel. Then the 1990 invasion of Kuwait by Iraq deeply divided many of the OPEC countries.

Owing in part to friction caused by the Kuwait conflict and the decline in world oil prices during the middle and late 1990s, OPEC's power has continued to slide. This situation could change rapidly, however, in the event of renewed hostilities in the Middle East or a dramatic rise in the cost of oil worldwide.

OPEC OIL PRODUCTION, 1990–2010

Year	(Million barrels per day)		
	Ref. case	Sensitivity range	
History			
1990	25.1	—	—
Projections			
1995	29.5	29.3	30.3
2000	34.8	33.4	38.9
2005	41.7	39.7	48.3
2010	46.2	43.5	55.0

Note: Includes the production of crude oil, natural gas plant liquids, refinery gain, and other liquid fuels.

Sources: History: Energy Information Administration (EIA); Data from California Energy Commission, 1998.

TABLE 8.3 WORLD TOTAL ENERGY CONSUMPTION BY REGION AND FUEL, 1990–2010

(quadrillion Btu)

Region/Fuel	History		Projections	
	1995	2000	2005	2010
OECD	195.1	214.4	227.4	239.0
Oil	83.2	89.6	94.7	98.2
Natural gas	40.0	45.6	49.5	53.9
Coal	37.0	40.5	42.4	44.0
Nuclear	17.8	18.8	19.2	18.7
Renewables	16.8	19.8	21.5	24.0
EE/FSU	62.1	63.7	69.4	74.7
Oil	16.4	12.8	15.7	19.2
Natural gas	26.0	29.1	31.6	33.3
Coal	17.1	15.4	15.4	15.0
Nuclear	2.7	3.4	3.3	3.3
Renewables	2.6	3.1	3.4	3.8
Non-OECD Asia	59.2	79.4	92.1	104.2
Oil	19.1	29.5	32.7	34.6
Natural gas	3.8	4.7	5.9	7.2
Coal	30.1	39.1	45.7	52.8
Nuclear	0.9	1.3	1.8	1.9
Renewables	3.4	4.8	6.0	7.5
Middle East	12.4	15.2	16.9	18.4
Oil	8.1	9.4	10.4	11.4
Natural gas	4.8	5.4	6.0	6.3
Coal	0.2	0.2	0.2	0.3
Nuclear	0.0	0.0	0.0	0.0
Renewables	0.1	0.3	0.4	0.4
Africa	11.1	12.6	13.6	14.6
Oil	4.9	5.8	6.3	6.9
Natural gas	1.7	1.9	2.1	2.4
Coal	3.9	4.1	4.2	4.4
Nuclear	0.1	0.1	0.1	0.1
Renewables	0.5	0.7	0.8	0.9
Central and South America	15.1	17.2	19.1	20.8
Oil	7.6	9.4	10.2	10.9
Natural gas	2.3	2.5	3.1	3.7
Coal	0.9	1.1	1.4	1.6
Nuclear	0.1	0.2	0.3	0.3
Renewables	4.1	4.0	4.2	4.4
World total	355.0	402.6	438.6	471.7
Oil	139.3	156.5	170.0	181.3
Natural gas	78.6	89.2	98.1	106.8
Coal	89.1	100.5	109.4	118.0
Nuclear	21.6	23.7	24.6	24.4
Renewables	27.5	28.1	36.4	41.1

Notes: OECD = Organization for Economic Cooperation and Development. EE/FSU = Eastern Europe/Former Soviet Union.

Sources: History data from Energy Information Administration (EIA), *International Energy Annual*, DOE/EIA-0219(92), January 1995, Washington, D.C.; Projections from EIA, World Energy Projections 1998.

members of the Organization for Economic Cooperation and Development (OECD). These countries (Australia, New Zealand, Japan, Canada, Mexico, the United States, and the countries of Europe) are the developed nations of the world. In the last decade, per capita energy consumption in OECD countries has risen moderately while economic growth has continued. There has also been a shift toward service-based economies, with energy-intensive industries moving to non-OECD countries. In contrast, in countries that are becoming more economically advanced energy consumption is increasing at a faster rate.

Oil remains the world's major source of energy, accounting for about 39 percent of primary energy demand. Coal accounts for 24 percent, and natural gas for 27 percent; the remainder is supplied mainly by nuclear energy and hydropower.

Since 1973, the year of the first "oil shock," demand for natural gas has increased faster than the demand for other fossil fuels. From 1973 to 1995, consumption of natural gas increased by about 80 percent, coal consumption increased by about 40 percent, and oil consumption by about 15 percent.

Since 1950, North American energy consumption has steadily increased. However, there were two periods of decline, one beginning in 1973 and the other in 1979. Both of these episodes were the result of political turmoil in the Middle East and the increasing influence of OPEC. (See Global Perspective: OPEC.) (See figure 8.8.) The same trend occurred in Western Europe, Japan, and Australia, resulting in a slowing of the world's growth in energy consumption. (See figure 8.9.) Several factors contributed to these declines in energy consumption. Increased prices for oil and all other forms of energy forced businesses and individuals to become more energy conscious and to expand efforts to conserve energy. From 1970 to 1983, the amount of energy used for heat per dwelling in the United States declined by 20 percent. Comparable reductions were made in Denmark, West Germany, and Sweden.

However, since 1970, energy consumption has increased in North America and Europe. Like the stabilization of energy use in 1973 and 1979, the increase since 1980 is due to the price of oil. In 1979, oil was selling for about forty dollars a barrel. Beginning in 1980, the price began to drop, and in 1998, oil was selling for less than 20 dollars a barrel, When Iraq invaded Kuwait in August 1990, prices rose dramatically, but dropped back at the end of hostilities.

During the 1980s, with energy costs declining, people in North America and Europe became less concerned about their energy consumption. They used more energy to heat and cool their homes and buildings, used more home appliances, and bought bigger cars. Obviously, the two primary factors that determine energy use are political stability in parts of the world that supply oil, and the price of that oil. Insecurity leads to a price rise, but

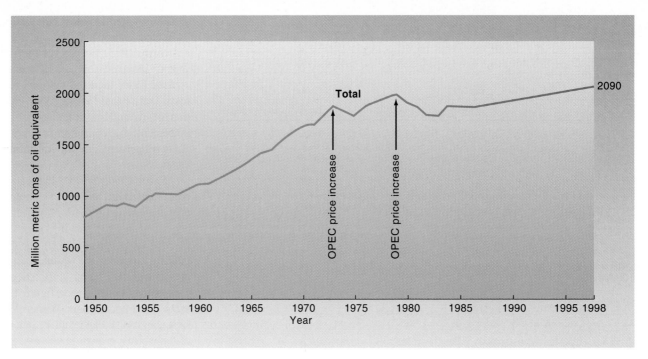

FIGURE 8.8   Changes in U.S. Energy Consumption   Energy consumption in the United States experienced a gradual increase until 1973 when the OPEC countries increased the price of oil. The increase in price reduced consumption. A similar action in 1979 resulted in another decrease. However, since 1983 consumption has steadily increased.

Source: Data from *BP Statistical Review of World Energy*, 1998.

FIGURE 8.9   World Energy Consumption   Energy consumption has increased steadily for the past 25 years. The slowdown in growth beginning in 1973 and the decline in the early 1980s in the result of price increases instituted by OPEC. The slight decline noted in 1991 and 1992 is the result of decreased energy consumption in the former Soviet Union and Eastern Europe, and is related to economic problems in those countries.

Source: Data from *BP Statistical Review of World Energy*, June 1998.

Energy Development in China

China is currently the second largest energy consumer in the world, following the United States (about 37 quadrillion Btu in 1997 versus 89 quadrillion Btu in the United States). China's rapidly growing economy will drive energy demand growth of about four to five percent annually through 2015 (compared with growth of about 1 percent in the industrialized countries). China currently consumes about 10 percent of world's energy, and also accounts for about 10 percent of the world energy production.

China became a net importer of energy in 1995, and is expected to become increasingly dependent on imports; however, it is expected to remain a net exporter of coal. China has been a net importer of oil since 1993.

In 1997, coal accounted for 75 percent of China's primary energy production. In the same year, petroleum accounted for 19 percent, hydroelectricity 6 percent, natural gas 2 percent, and nuclear power 0.5 percent. China's electricity is generated overwhelmingly by coal (about 75%) followed by hydroelectricity at about 20 percent. Coal is expected to retain its importance in China's fuel production mix, increasing its share to 78 percent by 2015. Hydroelectricity, natural gas, and nuclear energy will also become increasingly important, while the petroleum share declines.

One of the biggest concerns regarding energy consumption in China is that of carbon emissions and the threat of increasing global warming. China's carbon emissions are expected to increase 4 percent annually through 2015, driven by rapid economic growth and a rapid increase in coal use. The country's total carbon emissions should exceed 2 billion metric tons shortly after 2015, over three times the 1986 emissions and more than double the 1995 emissions.

Source: Data from EIA, U.S. Department of Energy, International Energy Outlook, 1997.

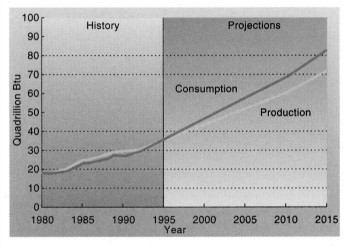

a. China's energy production and comsumption.

b. Energy production by fuel type, 1980–2015.

Source: History data from Energy Information Administration (EIA), International Petroleum Statistics Report, DOE/EIA–520 (92/08), August 1992, Washington, D.C. Projections from EIA, National Energy Modeling System, Internatoinal Energy MOdule, 1995.

governments can also manipulate prices by changing taxes, granting subsidies, and using other means. The energy consumption behavior of most people is motivated by economics rather than a desire to wisely use energy resources.

When Will We Run Out of Oil?

Currently, world oil prices are extremely low, which gives the impression that there is a vast supply of oil available for use. Is it true? Yes and No. There is a vast amount of oil available but the low price is contributing to its being used very rapidly. Oil is also a finite, nonrenewable resource that will eventually be exhausted. The primary questions about when we will run out of oil revolve around price, rates of use, and rates at which new discoveries are made. In the early 1950s the late M. King Hubbert, a petroleum geologist, developed a mechanism for projecting when oil production in a region of the world would begin to decline. This method involved looking at the rate at which new oil discoveries were made. In 1956 he predicted that oil production in the 48 contiguous states of the United States would peak and begin to decline between 1965 and 1970. Although many laughed at his prediction, oil production in the lower 48 states did in fact begin to decline in 1970 and has been declining ever since. (See figure *a*.)

By using the same kind of information many respected organizations are predicting world oil production will peak somewhere between 2010 and 2020. (See figure *b*.) The current low price for oil is partly the result of the exploitation of recently discovered oil resources that are not controlled by the Organization of Petroleum Exporting Countries (OPEC). When production from these sources begins to fall (about 2000), the price of oil will rise significantly. Furthermore, the Middle Eastern countries of OPEC will control about 65 percent of the known supply of oil unless major new oil deposits are discovered. When this happens OPEC countries will again be able to regulate the price of this very important resource. Economists argue that as oil production begins to fall the market place will adjust by finding substitutes and changing the way we use energy. Many scientists argue there are no equivalents for petroleum that could serve as substitutes. Nevertheless, we should look for major changes in energy policy, international relations, and energy use early in the next century.

a.

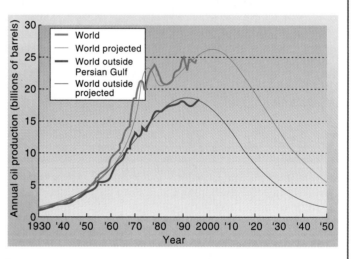

b.

Sources: History data from Energy Information Administration (EIA), International Petroleum Statistics Report, *DOE/EIA-520 (92/08), August 1992, Washington, D.C.; Projections from EIA, National Energy Modeling System, International Energy Module, 1995.*

Summary

A constant supply of energy is required by all living things. Energy has a major influence on society. A direct correlation exists between the amount of energy used and the complexity of civilizations.

Wood furnished most of the energy and construction materials for early civilizations. Heavy use of wood in densely populated areas eventually resulted in shortages, so fossil fuels replaced wood as a prime source of energy. Fossil fuels were formed from the remains of plants, animals, and microorganisms that lived about 300 million years ago. Fossil-fuel consumption in conjunction with the invention of labor-saving machines resulted in the Industrial Revolution, which led to the development of technology-oriented societies today in the developed world.

Throughout the world, residential and commercial uses, industries, transportation, and electrical utilities require energy. Because of financial, political, and other factors, nations vary in the amount of energy they use as well as in how they use it.

Analysts expect the worldwide demand for energy to increase steadily.

Review Questions

1. Why was the sun able to provide all energy requirements for human needs before the Industrial Revolution?
2. In addition to food, what energy requirements does a civilization have?
3. Why were some countries unable to use the technologies developed during the Industrial Revolution?
4. What factors caused a shift from wood to coal as a source of energy?
5. How were energy needs in World War II responsible for the subsequent increased consumption of natural gas?
6. What part does government regulation play in changing the consumption of natural gas and oil?
7. Why was much of the natural gas that was first produced wasted?
8. What was the initial use of oil? What single factor was responsible for a rapid increase in oil consumption?
9. List the three purposes for which a civilization uses energy.
10. Why is OPEC important in the world's economy?

Critical Thinking Questions

1. Imagine you are an historian writing about the history of the Industrial Revolution, but you also have your new knowledge and perspective of the environmental science you have learned in this course. What kind of a story would you tell about the development of industry in Europe and the United States? Would it be a story of triumph, tragedy, or some other story? Why?
2. What might be some of the effects of raising gasoline taxes in the United States to the rate that most Europeans pay for gasoline? Why? What do you think about this possibility?
3. Some argue that the price of gasoline in the United States is artificially low because it does not take into account all of the costs of producing and using gasoline. If you were to figure out the "true" cost of gasoline, what kinds of factors would you want to take into account to get an accounting of the actual cost of gasoline?
4. How has the ubiquitous nature of automobiles changed the United States? Do you feel these changes are, in balance, positive or negative? What should the future look like regarding automobile use in the United States? How can this be accomplished?
5. Some energy experts predict that Organization of Petroleum Exporting Countries (OPEC) will control about 65% of the known oil reserves by the year 2010. What political and economic effects do you think this will this have? Will this have any effect on energy use?

6. How do you think projected energy consumption will affect world politics and economics given current concerns about global warming?

Key Terms

fossil fuels *126*
Industrial Revolution *126*

Related Internet Sites

Visit our website at *http://www.mhhe.com/environmentalscience/* for specific information on each of the topics below.

Energy use, economics, and the environmental implications of energy use

The process of photosynthesis

Air pollution emissions from motor vehicles

Environmental issues affecting the oil and gas industry

Production, recovery, and conversion of energy crops to environmentally beneficial fuels and power generation

Efficient electric motor systems

Global energy production and consumption

Problems and possible solutions of transportation and vehicle emissions

U.S. Department of Energy—Office of Fossil Energy

The potential of hydrogen as an energy source

Links to alternative fuel web sites

chapter

Energy Sources

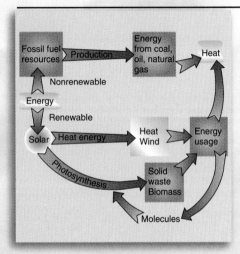

OBJECTIVES

After reading this chapter, you should be able to:

- Differentiate between resources and reserves.
- Identify peat, lignite, bituminous coal, and anthracite coal as steps in the process of coal formation.
- Recognize that natural gas and oil are formed from ancient marine deposits.
- Explain how various methods of coal mining can have negative environmental impacts.
- Explain why surface mining of coal is used in some areas and underground mining in other areas.
- Explain why it is more expensive to find and produce oil today than it was in the past.
- Recognize that secondary recovery methods have been developed to increase the proportion of oil and natural gas obtained from deposits.

- Recognize that transport of natural gas is still a problem in some areas of the world.
- Explain why the amount of energy supplied by hydroelectric power is limited.
- Describe how wind, geothermal, and tidal energy are used to produce electricity.
- Recognize that wind, geothermal, and tidal energy can be developed only on areas with the proper geologic or geographical features.
- Describe the use of solar energy in passive heating systems, active heating systems, and the generation of electricity.
- Recognize that fuelwood is a major source of energy in many parts of the less-developed world and that fuelwood shortages are common.
- Describe the potential and limitations of biomass conversion and waste incineration as a source of energy.
- Recognize that energy conservation can significantly reduce our need for additional energy sources.

ENERGY SOURCES

Chapter 8 outlined the historical development of energy consumption and how advances in civilizations were closely linked to the availability and exploitation of energy. New manufacturing processes relied on dependable sources of energy. Technology accelerated in the twentieth century. Between 1900 and 2000, world energy consumption increased by a factor of fourteen, the quantity of products manufactured increased by over thirtyfold (actually 35), but population increased only threefold. As societies used more energy, they produced goods and services more efficiently. (See table 9.1.)

The energy sources most commonly used by industrialized nations are the fossil fuels: oil, coal, and natural gas, which supply about 90 percent of the world's commercially traded energy. Fossil fuels were formed hundreds of millions of years ago. They are the accumulation of energy-rich organic molecules produced by organisms as a result of photosynthesis over millions of years. We can think of fossil

fuels as concentrated, stored solar energy. The rate of formation of fossil fuels is so slow that no significant amount of fossil fuels will be formed over the course of human history. Since we are using these resources much faster than they can be produced and the amount of these materials is finite, they are known as **nonrenewable energy sources.** (See figure 9.1.) Eventually, human demands will exhaust the supplies of coal, oil, and natural gas.

In addition to nonrenewable fossil fuels, there are several renewable energy sources. **Renewable energy sources**

replace themselves or are continuously present as a feature of the solar system. For example, in plants, photosynthesis converts light energy into chemical energy.

$$\text{Carbon dioxide} + \text{water} + \frac{\text{Sunlight}}{\text{Energy}} \rightarrow \frac{\text{Biomass}}{(\text{Chemical Energy})} + \text{Oxygen}$$

This energy is stored in the organic molecules of the plant as wood, starch, oils, or other compounds. Any form of biomass—plant, animal, alga, or fungus—can be traced back to the energy of the sun. Since biomass is constantly being produced, it is

TABLE 9.1 WORLD POPULATION, ECONOMIC OUTPUT, AND FOSSIL-FUEL CONSUMPTION

	Population (billions)	Gross world product (trillion 1995 dollars)	Fossil-fuel consumption (billion tons coal equivalent)
1900	1.6	0.6	1
1950	2.5	2.9	3
1998	5.9	32.0	14

Source: Data from *CIA World Fact Book*.

Predicting Energy Needs

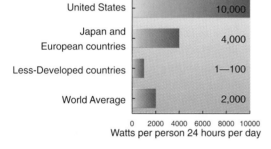

Rate of Average Energy Use

- United States: 10,000
- Japan and European countries: 4,000
- Less-Developed countries: 1—100
- World Average: 2,000

0 2000 4000 6000 8000 10000
Watts per person 24 hours per day

Annual Total World Energy Consumption

10 TWyr 5 billion people x 2000 watts per person = 10 trillion watts/year (TWyr)

Future Energy Needs in the Lifetime of Today's Children

30 TWyr

Assuming that there will be 10 billion people and that each person can reach a satisfactory living standard with less than one-third the energy an average American uses now: the world will consume

(10 billion x 3000 watts per person) = 30 trillion watts/year

Predicting Energy Resources

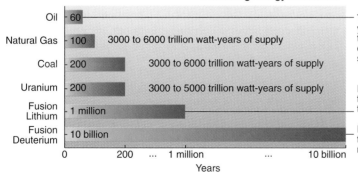

- Oil: 60
- Natural Gas: 100 — 3000 to 6000 trillion watt-years of supply
- Coal: 200 — 3000 to 6000 trillion watt-years of supply
- Uranium: 200 — 3000 to 5000 trillion watt-years of supply
- Fusion Lithium: 1 million
- Fusion Deuterium: 10 billion

0   200   ...   1 million   ...   10 billion
Years

The world has 300 to 400 billion watt years of known oil reserves. In addition there may be decades or centuries of energy available from unconventional oil supplies (tar sands and shale oil).

If fusion power plants use deuterium-tritium reaction, the lithium (for making tritium) will last more than 1 million years.

If deuterium-deuterium reactions are used, the deuterium (from heavy water) will last more than 10 billion years.

FIGURE 9.1   Predicting Energy Resources and Energy Needs

Source: Princeton Plasma Physics Laboratory, U.S. Department of Energy, Washington, D.C., 1997.

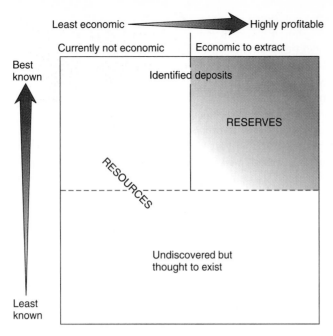

FIGURE 9.2 Resources and Reserves Each term describes the amount of a natural resource present. Reserves are those known deposits that can be profitably obtained using current technology under current economic conditions. Reserves are shown in the box in the upper right-hand corner in this diagram. The darker the color, the more valuable the reserve. Resources are a much larger quantity that includes undiscovered deposits and deposits that currently cannot be profitably used, although it might be feasible to do so if technology or market conditions change.

Source: Adapted from the U.S. Bureau of Mines.

a form of renewable energy. Solar, geothermal, and tidal energy are renewable energy sources because they are continuously available. Anyone who has ever lain in the sun, seen a geyser or hot springs, or been swimming in the surf has experienced these forms of energy. However, many technical problems must be solved before these renewable energy forms can contribute significantly to meeting humans' energy demands.

RESOURCES AND RESERVES

When discussing deposits of nonrenewable resources, such as fossil fuels, we must differentiate between deposits that can be exploited and those that cannot. From a technical point of view, a **resource** is a naturally occurring substance of use to humans that can *potentially* be extracted using current technology. **Reserves** are

known deposits from which materials *can* be extracted profitably with existing technology *under present economic conditions.* It is important to recognize that *reserves* is an economic concept and is only loosely tied to the total quantity of a material present in the world. Therefore, reserves are smaller than resources. (See figure 9.2.) Both terms are used when discussing the amount of mineral or fossil-fuel deposits a country has at its disposal. This can cause considerable confusion if the difference between these concepts is not understood. The total amount of a resource such as coal or oil changes only by the amount used each year. The amount of a reserve changes as technology advances, new deposits are discovered, and economic conditions vary. There can be large changes in the amount of reserves, while the resource remains almost constant. (See figure 9.3.)

When we read about the availability of fossil fuels, we must remember that energy is needed to extract the wanted

material from the earth. If the material is dispersed or hard to reach, it is expensive to extract. If the cost of removing and processing a fuel is greater than the fuel's market value, no one is going to produce it. Also, if the amount of energy used to produce, refine, and transport a fuel is greater than its potential energy, the fuel will not be produced. A net useful energy yield is necessary to exploit the resource. However, in the future, new technology or changing prices may permit the profitable removal of some fossil fuels that currently are not profitable. If so, those resources will be reclassified as reserves.

To further illustrate the concept of reserves and how technology and economics influence their magnitude, let us look at the history of oil. The ancient Chinese are said to have been the first to use oil as a fuel. The only oil available to them was the small amount that naturally seeped out of the ground. These seepages represented the known oil reserves as well as the known oil resources at that time.

Nearly two thousand years passed before oil reserves increased significantly. When the first oil well in North America was drilled in Pennsylvania in 1859, it greatly expanded the estimate of the amount of oil in the earth. There was a sudden increase in the known oil reserves. In the years that followed, new deposits were discovered. Better drilling techniques led to the discovery of deeper oil deposits, and offshore drilling established the location of oil under the ocean floor. At the time of their initial discovery, these deep deposits and the offshore deposits added to the estimated size of the world's oil resources. But they did not necessarily add to the reserves because it was not always profitable to extract the oil. With advances in drilling and pumping methods and increases in oil prices, it eventually became profitable to obtain oil from many of these deposits. As it became economical to extract them, they were reclassified as reserves.

The amount of fossil fuel reserves is in a constant state of flux. For example, prior to 1973, many oil wells were capped because it was not profitable to remove the oil. The oil still in the ground at these wells was not economic to produce and was not included in the reserves category. During the oil embargoes of 1973–1974 and

FIGURE 9.3 **Changes in Oil Reserves** The figure shows the changes in oil reserves over a 28-year period. The changes that occurred in 1987 and 1989 are the result of changes in reporting, not the result of new discoveries. Since 1989 new discoveries and revisions have matched consumption. Thus, reserves have remained nearly constant.

Source: Data from *BP Statistical Review of World Energy,* June 1998.

1979–1980, when some of the OPEC countries reduced oil production, the price of oil increased. With the increase in oil prices, these wells became profitable, and the oil was reclassified as reserves. Then, in the 1980s, an oil surplus developed because the high oil prices resulting from the embargo caused increased production worldwide. Prices fell. Many of these same wells were capped again, and the oil was no longer classified as part of the reserve. With the exception of 1990 and 1991 when Iraq invaded Kuwait, the price of oil in the 1990s has been relatively stable.

FOSSIL-FUEL FORMATION

Fossil fuels are the remains of once living organisms that were preserved and altered as a result of geologic forces. There are significant differences in the formation of coal from that of oil and natural gas.

Coal Formation

Tropical freshwater swamps covered many regions of the earth 300 million years ago. Conditions in these swamps favored extremely rapid plant growth, resulting in large accumulations of plant material. Because this plant material collected under water, decay was inhibited, and a spongy mass of organic material formed, called peat. Peat deposits are 90 percent water, 5 percent carbon, and 5 percent volatile materials. In some parts of the world, peat is cut, dried, and used as fuel. However, because of its high water content, it is regarded as a low grade of fuel.

Due to geological changes in the earth, some of the swamps containing peat were submerged by seas. The plant material that had collected in the swamps was now covered by sediment. The weight of the plant material plus the weight of the sediment on top of it compressed it into a harder form of low-grade coal known as

lignite, which contains less water and a higher proportion of burnable materials.

If the weight of the sediment was great enough, the heat from the earth high enough, and the length of time long enough, the lignite was changed into bituminous (soft) coal. The major change from lignite to bituminous coal is a reduction in the water content from over 40 percent to about 3 percent. If the heat and pressure continued over time, some of the bituminous coal was changed to anthracite (hard) coal, which is about 96 percent carbon. Through this combination of events, which occurred over hundreds of millions of years, present-day coal deposits were created. Most parts of the world have significant coal deposits. (See figure 9.4.)

Oil and Natural Gas Formation

Oil and natural gas, like coal, are products from the past. They probably originated

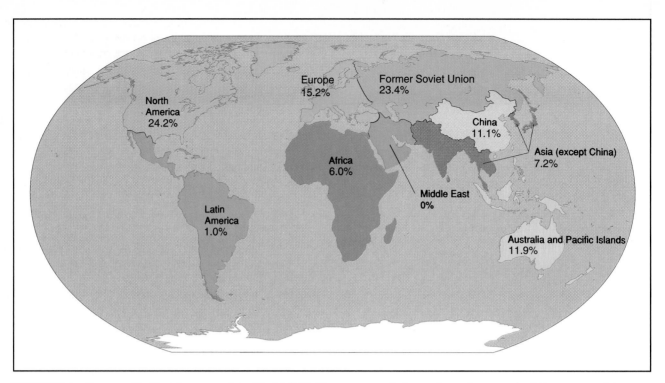

FIGURE 9.4 Recoverable Coal Reserves of the World 1997 The percentage indicates the coal reserves in different parts of the world. This coal can be recovered under present local economic conditions using available technology.

Source: Data from *BP Statistical Review of the World Energy,* June 1998.

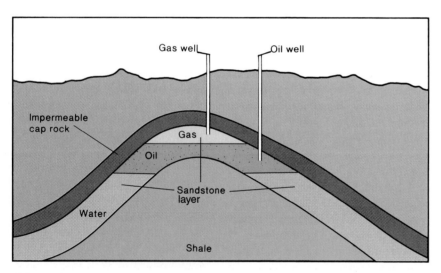

FIGURE 9.5 Crude Oil and Natural Gas Pool Water and gas pressure force oil and gas out of the shale and into sandstone beneath the impermeable rock.

from microscopic marine organisms. When these organisms died and accumulated on the ocean bottom and were buried by sediments, their breakdown released oil droplets. Gradually, the muddy sediment formed rock called shale, which contained dispersed oil droplets. Although shale is common and contains a great deal of oil, extraction from shale is difficult because the oil is not concentrated. If a layer of sandstone formed on the top of the oil-containing rock, and an impermeable layer of rock formed on top of the sandstone, conditions were suitable for oil pools to form. Usually the trapped oil does not exist as a liquid mass but rather as a concentration of oil within sandstone pores, where it accumulates because water and gas pressure force it out of the shale. (See figure 9.5.) These accumulations of oil are more likely to occur if the rock layers were folded by geological forces.

Natural gas, like coal and oil, was formed from fossil remains. In fact, the geological conditions favorable for oil formation are the same as those for natural gas, and the two fuels are often found together. However, in the formation of natural gas, the organic material changed to lighter, more volatile (easily evaporated) hydrocarbons than those found in oil. The most common hydrocarbon in natural gas is the gas methane (CH_4). Water, liquid hydrocarbons, and other gases may be present in natural gas as it is pumped from a well.

The conditions that led to the formation of oil and gas deposits were not evenly distributed throughout the world. Figure 9.6 illustrates the geographic distribution of oil reserves and figure 9.7 illustrates the geographic distribution of natural gas reserves. Some of these deposits are easy to exploit, while others are not.

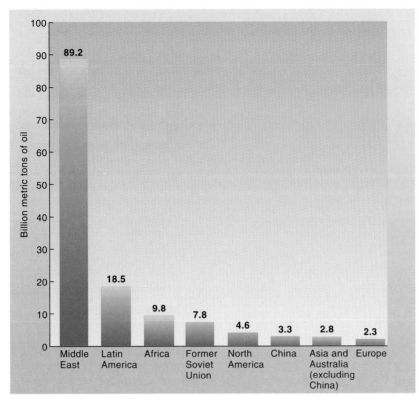

FIGURE 9.6 World's Oil Reserves 1997 The world's supply of oil is not distributed equally. Certain areas of the world enjoy an economic advantage because they control vast amounts of oil. Reserves are given in billions of metric tons of oil.

Source: Data from *BP Statistical Review of World Energy,* June 1998.

ISSUES RELATED TO THE USE OF FOSSIL FUELS

Of the world's commercial energy, over 90 percent is furnished by the three nonrenewable fossil-fuel resources: coal, oil, and natural gas. Coal supplies about 27 percent, oil supplies about 40 percent, and natural gas supplies about 23 percent. Each fuel has advantages and disadvantages and requires special techniques for its production and use.

Coal Use Issues

Coal is the world's most abundant fossil fuel, but it supplies only about 27 percent of the energy used in the world. It varies in quality and is generally classified in three categories: lignite, bituminous, and anthracite. Lignite coal has a high moisture content and is crumbly in nature, which makes it the least desirable form. Bituminous coal is the most widely used because it is the easiest to mine and the most abundant. It supplies about 20 percent of the world's energy requirements. Coal is primarily used for electric power generation and other industrial uses. For most uses, anthracite coal is the most desirable because it furnishes more energy than the other grades and is the cleanest burning. But anthracite is not as common and is usually more expensive because it is found at great depths and is difficult to obtain.

Because coal was formed as a result of plant material being buried under layers of sediment, it must be mined. There are two methods of extracting coal: surface mining and underground mining. **Surface mining** (strip mining) involves removing the material on top of a vein of coal, called **overburden,** to get at the coal beneath. (See figure 9.8.) Coal is usually surface mined when the overburden is less than 100 meters thick. This type of mining operation is efficient because it removes 100 percent of the coal in a vein and can be profitably used for a seam of coal as thin as half a meter. For these reasons, surface mining results in the best utilization of coal reserves. Advances in the methods of surface mining and the development of better equipment have increased the amount of surface mining in the United States from 30 percent of the coal production in 1970 to more than 60 percent today. This trend toward increased surface mining has also occurred in Canada, Australia, and the former Soviet Union.

If the overburden is thick, surface mining becomes too expensive, and the coal is extracted through **underground mining.** The deeply buried coal seam can be reached in two ways: In flat country, where the vein of coal lies buried beneath a thick overburden, the coal is reached by a vertical shaft. (See figure 9.9*a*.) In hilly areas, where the coal seam often comes to the surface along the side of a hill, the coal is reached from a drift-mine opening. (See figure 9.9*b*.)

The mining, transportation, and use of coal as an energy source presents several significant problems. Surface mining disrupts the landscape, as the topsoil and overburden are moved to get at the coal. It is possible to minimize this disturbance by reclaiming the area after mining operations are completed. (See figure 9.10.) However, reclamation rarely, if ever, returns the land to its previous level of productivity. The cost of reclamation is passed on to the consumer in the form of higher coal prices. Underground mining methods do not disrupt the surface environment as much as surface mining does, but subsidence (sinking of the land) occurs if the mine collapses. In addition, large waste heaps are produced around the mine entrance from the debris that must be removed and separated from the coal.

Health and safety are important concerns related to coal mining, which is one of the most dangerous jobs in the world. This is particularly true with underground mining. Many miners suffer from **black lung disease,** a respiratory condition that results from the accumulation of fine coal-dust particles in the miners' lungs. The coal particles inhibit the exchange of gases between the lungs and the blood. The health care costs and death benefits related to black lung disease are an indirect cost of coal mining. Since these costs are partially paid by the federal government, their full

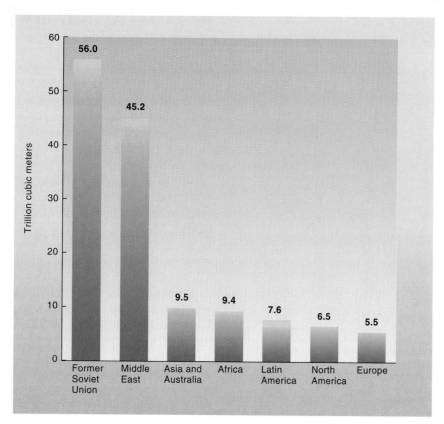

FIGURE 9.7 Natural Gas Reserves 1997 Natural gas reserves, like oil and coal, are concentrated in certain regions of the world. Figures are trillion cubic meters.

Source: Data from *BP Statistical Review of World Energy,* June 1998.

a.

b.

FIGURE 9.9 Underground Mining If the overburden is too thick to allow surface mining, underground mining must be used. (*a*) If the coal vein is not exposed, a vertical shaft is sunk to reach the coal. (*b*) In hilly areas, if the vein is exposed, a drift mine is used in which miners enter from the side of the hill.

FIGURE 9.8 Surface Mining Large power draglines are used to remove the overburden, which is piled to the side. The coal can then be loaded into trucks. When the coal has been removed, the overburden is placed back in the trench.

price is not reflected in the price of coal but is paid by taxpayers in the form of federal taxes.

Because coal is bulky, shipping presents a problem. Generally, the coal can be used most economically near where it is produced. Rail shipment is usually the most economical way of transporting it from the mine. Rail shipment costs include the expense of constructing and maintaining the tracks, as well as the cost of the energy required to move the long strings of railroad cars. In some areas, the coal is transferred from trains to ships.

Coal is dirty and its mining and transport generate a great deal of dust. The large amounts of coal dust released into the atmosphere at the loading and unloading sites can cause local air-pollution problems. If a boat or railroad car is used to transport coal, there is the expense of cleaning it before other types of goods can be shipped. In some cases, the coal can be ground and mixed with water to form a slurry that can be pumped through pipelines. This helps to alleviate some of the air-pollution problems without causing significant water-pollution problems.

Air pollution from coal burning releases millions of metric tons of material into the atmosphere and is responsible for millions of dollars of damage to the

a.

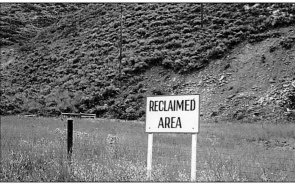

b.

FIGURE 9.10 Surface-Mine Reclamation (*a*) This photograph shows a large area that has been surface mined with little effort to reclaim the land. The windrows created by past mining activity are clearly evident, and little effort has been made to reforest the land. By contrast, (*b*) is an example of proper surface-mining reclamation. The sides of the cut have been graded and planted with trees. The topsoil has been returned, and the level land is now productive farmland.

environment. The burning of coal for electric generation is the prime source of this type of pollution.

Since coal is a fossil fuel formed from plant remains, it contains sulfur, which was present in the proteins of the original plants. Sulfur is associated with **acid mine drainage** and air pollution. Acid mine drainage occurs when the combined action of oxygen, water, and certain bacteria causes the sulfur in coal to form sulfuric acid. Sulfuric acid can seep out of a vein of coal even before the coal is mined. However, the problem becomes most acute when the coal is mined and the overburden is disturbed, allowing rains to wash the sulfuric acid into streams. Streams may become so acidic that they can support only certain species of bacteria and algae. Today, many countries regulate the amount of runoff allowed from

mines, but underground and surface mines abandoned before these regulations were enacted continue to pollute the water.

Currently, a form of acid pollution called acid deposition is becoming serious. Acid deposition occurs when coal is burned and sulfur oxides are released into the atmosphere, causing acid-forming particles to accumulate on a surface. Each year, over 150 million metric tons of sulfur dioxide are released into the atmosphere worldwide. This problem is discussed in greater detail in chapter 17.

The release of carbon dioxide from the burning of coal has become a major issue in recent years. Increasing amounts of carbon dioxide in the atmosphere are said to contribute to global warming. Environmentalists have suggested that the amount of coal used be decreased, since the other fossil fuels (oil and natural gas) produce

less carbon dioxide for an equivalent amount of energy.

Because coal is bulky and dusty and often has a high sulfur content resulting in air pollution, people seek alternative sources of fuel. The most common alternatives for coal are oil and natural gas.

Oil Use Issues

Oil has several characteristics that make it superior to coal as a source of energy. Its extraction causes less environmental damage than does coal mining. It is a more concentrated source of energy than coal, it burns with less pollution, and it can be moved easily through pipes. These characteristics make it an ideal fuel for automobiles. However, it is often difficult to find. Today, geologists use a series of tests to locate underground formations that may contain oil. When a likely area is identified, a test well is drilled to determine if oil is actually present. The many easy-to-reach oil fields have already been tapped. Drilling now focuses on smaller amounts of oil in less accessible sites, which means that the cost of oil from most recent discoveries is higher than that of the large, easy-to-locate sources of the past. As oil deposits on land have become more difficult to find, geologists have widened the search to include the ocean floor. Building an offshore drilling platform can cost millions of dollars. To reduce the cost, as many as 70 wells may be sunk from a single platform. (See figure 9.11.) Total world proven reserves of oil in 1998 were estimated at 1000 billion barrels. Of this, more than three-quarters are in OPEC countries and more than half lie in just four Middle Eastern countries—Saudi Arabia, Iraq, Kuwait, and Iran.

One of the problems of extracting oil is removing it from the ground. If the water or gas pressure associated with a pool of oil is great enough, the oil is forced to the surface when a well is drilled. When the natural pressure is not great enough, the oil must be pumped to the surface. Present technology allows only about one-third of the oil in the ground to be removed. This means that two barrels of oil are left in the ground for every barrel produced. In most oil fields, **secondary recovery** is used to recover more of the

Drilling derrick

Crew quarters

Ocean

Seabed

Platform legs imbedded in seabed

Directionally drilled wells

Impermeable cap rock

Impermeable cap rock

Oil and gas reservoir

Oil and gas reservoir

Reservoir rock

Reservoir rock

a.

FIGURE 9.11 Offshore Drilling Once the drilling platform is secured to the ocean floor, a number of wells can be sunk to obtain the gas or oil.

Source: (b) American Petroleum Institute.

b.

oil. Secondary recovery methods include pumping water or gas into the well to drive the oil out or even starting a fire in the oil-soaked rock to liquefy thick oil. As oil prices increase, more expensive secondary recovery methods will be used. More will be obtained from expensive deep wells and offshore wells.

Processing crude oil to provide useful products generates a variety of problems. Oil as it comes from the ground is not in a form suitable for use. It must be refined. The various components of crude oil can be separated and collected by heating the oil in a distillation tower. (See figure 9.12.) After distillation, the products may be further refined by "cracking." In this process, heat, pressure, and catalysts are used to produce a higher percentage of

volatile chemicals, such as gasoline, from less volatile liquids such as diesel fuel and furnace oils. It is possible, within limits, to obtain many products from one barrel of oil. In addition, petrochemicals from oil serve as raw materials for a variety of synthetic compounds. (See figure 9.13.) All of these processing activities are opportunities for accidental or routine releases that cause air or water pollution. The petrochemical industry is the major contributor to air pollution.

The environmental impacts of producing, transporting, and using oil are somewhat different from those of coal. Oil spills in the oceans have been widely reported by the news media. (See figure 9.14.) However, these accidental spills are responsible only for about a third of the oil

pollution resulting from shipping. Almost 60 percent of the oil pollution in the oceans is the result of normal, routine shipping operations. Oil spills on land can contaminate soil and underground water. The evaporation of oil products and the incomplete burning of oil fuels cause air pollution. These problems are discussed in chapter 17.

Natural Gas Use Issues

Natural gas, the third major source of fossil-fuel energy, supplies 23 percent of the world's energy. The drilling operations to obtain natural gas are similar to those used for oil. In fact, a well may yield both oil and natural gas. As with oil, secondary recovery methods that pump air or water

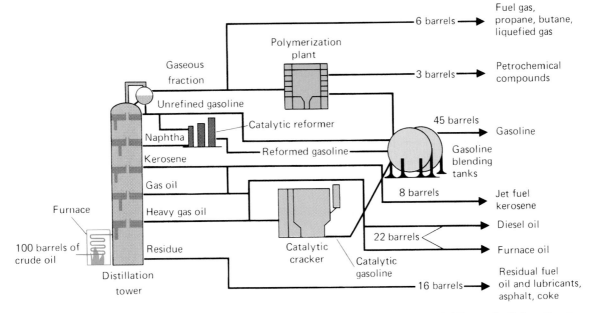

FIGURE 9.12 **Uses of Crude Oil** A great variety of products can be obtained from the distilling and refining of crude oil. A barrel of crude oil produces slightly less than half a barrel of gasoline. This figure shows the many steps in the refining process and the variety of products that can be obtained from crude oil.

From *Man, Energy, Society* by Earl Cook. W.H. Freeman and Company. Copyright ©1976. Reprinted with permission.

FIGURE 9.13 **Oil-Based Synthetic Materials** These common household items are produced from chemicals derived from oil. Although petrochemicals represent only about 3 percent of each barrel of oil, they are extremely profitable for the oil companies.

into a well are used to obtain the maximum amount of natural gas from a deposit. After processing, the gas is piped to the consumer for use.

Transport of natural gas still presents a problem in some parts of the world. In the Middle East, Mexico, Venezuela, and Nigeria, wells are too far from consumers to make pipelines practical, so much of the natural gas is burned as a waste product at the wells. However, new methods of transporting natural gas and converting it into other products are being explored. At −162°C, natural gas becomes a liquid and has only 1/600 of the volume of its gaseous form. Tankers have been designed to transport **liquefied natural gas** from the area of production to the area of demand. In 1997, over 90 billion cubic meters of natural gas were shipped between countries as liquefied natural gas. This is nearly 4.5 percent of the natural gas consumed in the world. Of that amount, Japan imported 58 billion cubic meters.

Many people are concerned about accidents that might cause the tankers to explode. Another, safer process converts natural gas to methanol, a liquid alcohol, and transports it in that form. As the demand for natural gas increases, the amount of it wasted will decrease and new methods of transportation will be employed. Higher prices will make it profitable to transport natural gas greater distances from the wells to the consumers.

Of the three fossil fuels, natural gas is the least disruptive to the environment. A natural gas well does not produce any unsightly waste, although there may be local odor problems. Except for the danger of an explosion or fire, it poses no harm to the environment during transport. Since it is clean burning, it causes almost no air pollution. The products of its combustion are carbon dioxide and water.

Although natural gas is used primarily for heat energy, it does have other uses,

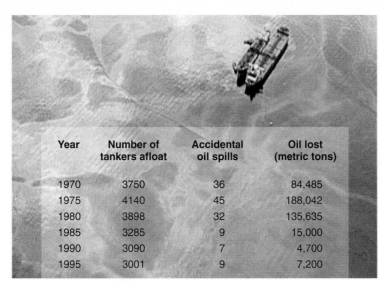

Year	Number of tankers afloat	Accidental oil spills	Oil lost (metric tons)
1970	3750	36	84,485
1975	4140	45	188,042
1980	3898	32	135,635
1985	3285	9	15,000
1990	3090	7	4,700
1995	3001	9	7,200

FIGURE 9.14 Oil Spills Accidents involving oil tankers are sources of water pollution.

Source: Data from Tanker Advisory Center Reference Data for 1973–1995.

FIGURE 9.15 Hydroelectric Power Plant The water impounded in this reservoir is used to produce electricity. In addition, this reservoir serves as a means of flood control and provides an area for recreation.

such as the manufacture of petrochemicals and fertilizer. Methane contains hydrogen atoms that are combined with nitrogen from the air to form ammonia, which can be used as fertilizer.

More than two-thirds of natural gas reserves are located in Russia and the Middle East. In fact, more than a third of total reserves are located in just ten giant fields; six are located in Russia and the remainder are in Qatar, Iran, Algeria, and the Netherlands.

From 1985 to 1997, world production of natural gas rose by more than 25 percent. In Russia, which has the largest natural gas reserves, production increased by 40 percent from 1985 through 1991 but has since fallen as a result of the economic disruption associated with the collapse of the Soviet Union.

Because of oil development from beneath the North Sea, several European countries (the United Kingdom, Norway, and the Netherlands) have sizable reserves

of natural gas. During oil shortages in the 1970s, they increased their use of natural gas to compensate for the increased cost of oil. This trend has continued; natural gas consumption increased by 30 percent from 1985 to 1998.

RENEWABLE SOURCES OF ENERGY CURRENTLY BEING USED

The three nonrenewable fossil-fuel sources of energy—coal, oil, and natural gas—furnish about 90 percent of the world's commercially traded energy. Nuclear power provides over 7 percent of the world's energy. The remainder is supplied by renewable energy sources, including hydroelectric power, tidal power, geothermal power, wind power, solar energy, fuelwood, biomass conversion, and solid waste. These data do not include the use of fuelwood, animal dung, and other locally produced energy sources typically used in the less-developed world.

Hydroelectric Power

People have long used water to power a variety of machines. Some early uses of water power were to mill grain, saw wood, and run machinery for the textile industry. Today, water power is used almost exclusively to generate electricity. As the water flows from higher to lower levels, it supplies the energy to turn a generator and produce electricity. Hydroelectric power plants are commonly located on artificial reservoirs. (See figure 9.15.) The impounded water represents a potential energy source. In some areas of the world where the streams have steep gradients and a constant flow of water, hydroelectricity may be generated without a reservoir. Such sites are usually found in mountainous regions and can support only small power-generating stations. At present, hydroelectricity produces about 2.5 percent of the world's commercially traded energy.

Hydroelectric power potential is distributed among the continents in rough proportion to land area; China alone possesses

Hydroelectric Sites

Slightly over 17 percent of the potential hydroelectric sites of the world have been developed. (See the tables.) The World Energy Conference estimates that the electricity produced by hydropower will increase six times by the year 2020. The less-developed countries, which have developed about 10 percent of their hydropower, will experience most of this growth.

The projected increase will come mainly from the development of plants on large reservoirs. However, construction of "mini-hydro" (less than 10 megawatts) and "micro-hydro" (less than 1 megawatt) plants is also increasing. Such plants can be built in remote places and supply electricity to small areas. China has built over 80,000 such small stations, and the United States has nearly 1500.

About 50 percent of the U.S. hydroelectric capacity has been developed. However, this statistic is somewhat misleading. The Wild and Scenic Rivers Act (1968) prevents the construction of dams on designated streams. Presently, 37 potential hydroelectric sites are on streams protected by this act. The hydroelectric generating potential often quoted for the United States includes these areas, even though, at present, construction of plants at these sites is not possible. In fact, some dams on rivers in states such as Maine and Washington are scheduled to be destroyed in order to improve native fisheries. The U.S. political climate, which now favors the protection of certain rivers, might change if the demand for more energy becomes acute.

DEVELOPED HYDROELECTRIC SITES, 1997

Region	Percent of hydropower developed
Asia	17
South America	18
Africa	10
North America	62
Russia	9
Europe	90
Oceania	34

Source: *Data from Survey of Energy Resources, World Energy Conference.*

WORLD CONSUMPTION OF HYDROELECTRICITY AND OTHER RENEWABLE ENERGY BY REGION (QUADRILLION BTU)

Region/Country	History 1995	2000	Projections 2005	2010
OECD	16.8	19.8	21.5	24.0
U.S.*	6.6	7.8	8.3	9.4
Canada	3.1	4.1	4.7	5.3
Mexico	0.3	0.3	0.4	0.5
Japan	1.0	1.8	1.9	2.0
OECD Europe	5.0	5.3	5.7	6.2
U.K.	0.1	0.2	0.2	0.2
France	0.7	0.7	0.7	0.7
Germany	0.3	0.3	0.4	0.4
Italy	0.6	0.6	0.7	0.7
Netherlands	0.0	0.1	0.1	0.2
Other Eur.	3.3	3.5	3.6	3.9
Other OECD	0.4	0.5	0.5	0.7
EE/FSU	2.7	3.1	3.4	3.8
F.S.U.	2.4	2.6	2.8	2.9
E. Europe	0.5	0.5	0.6	1.0
Non-OECD Asia	3.2	4.8	6.0	7.5
China	1.6	2.6	3.5	4.5
Other Asia	1.8	2.1	2.6	3.0
Middle East	0.1	0.3	0.4	0.4
Africa	0.5	0.7	0.8	0.9
Central and South America	4.0	4.0	4.2	4.4
Total world	26.9	32.6	36.4	41.1

*Includes the 50 states, the District of Columbia, and U.S. territories.

Notes: OECD = Organization for Economic Cooperation and Development. EE/FSU = Eastern Europe/Former Soviet Union.

Sources: History data from Energy Information Administration (EIA), *International Energy Annual*, DOE/EIA-0219(92), Jan. 1994, Washington, D.C.; Projections from EIA, *Annual Energy Outlook*, DOE/EIA-0383(95), Table B1, Jan. 1995, Washington, D.C.; and World Energy Projections 1997.

one-tenth of the world's potential. Mountainous regions and large river valleys are the most promising. Besides the United States, the eastern area of the former Soviet Union, and southern Canada, the regions that have done the most to harness hydroelectric energy are Europe and Japan. Europe has exploited almost 60 percent of its potential. Although it has only one-fourth of Asia's resources, it generates nearly twice as much hydroelectric power. In contrast, Africa has developed only 5 percent of its potential, half of which comes from only three dams: Kariba in East Africa, Aswan on the Nile, and Akosombo in Ghana. In some areas of the world, hydroelectric power is the main source of electricity. More than 35 nations already obtain more than two-thirds of their electricity from falling water. In South America, 73 percent of the electricity used comes from hydroelectric power, compared to 44 percent in the developing world as a whole. Norway gets 99 percent of its electricity and 50 percent of all its energy from falling water.

It is important to recognize that the construction of a reservoir for a hydroelectric plant causes environmental and social problems, including loss of fertile farmland, destruction of the natural aquatic ecosystem, relocation of entire communities (including people and buildings), and a reduction in the amount of nutrient-rich silt deposited on downriver agricultural lands. The building of the Tellico Dam in Tennessee was delayed for several years because it might have caused the extinction of the snail darter, a fish that lived only in streams that would be flooded. The construction of the Aswan Dam in Egypt resulted in the displacement of eighty thousand people. It created an environment that increased schistosomiasis, a waterborne disease caused by flatworm parasites that spend part of their life cycle in snails that live in slowly moving water. The irrigation canals built to distribute the water from the Aswan Dam provide ideal conditions for the snails. Now many of the people who use the canal water for cooking, drinking, bathing, and as a sewer are infected with the flatworm parasites.

Even though hydroelectric projects cause problems, new sites continue to be developed. From 1984 to 1998, the energy furnished by hydroelectricity for world use increased by about 27 percent. The most significant change took place in South America, where hydroelectric use increased about 40 percent over that time.

China is currently constructing a huge hydroelectric dam, known as the Three Gorges Dam, on the Yangtze River. If it is completed, it would be the largest hydroelectric dam in the world. (See Global Perspectives: The Three Gorges Dam.)

Today's large dams rank among humanity's greatest engineering feats. For example, the Itaipu Dam completed in 1982 on the Parana River between Brazil and Paraguay will eventually generate 12,600 megawatts, making it the largest power complex on Earth (until the completion of the Three Gorges Dam). The lakes created by major dams also number among the planet's largest freshwater bodies. Ghana's Lake Volta covers 8500 square kilometers (3280 sq mi), an area the size of Lebanon. Hydroelectric dam projects figure prominently in the economic and investment plans of many developing countries. Egypt electrified virtually all of its villages with power from Aswan. From the 1950s onward, large-scale hydropower development in these countries has occurred primarily because energy-intensive industries need cheap electricity, and because global-lending institutions have been willing to advance multibillion-dollar loans. According to the World Bank, developing countries will need to raise an estimated $100 billion (U.S.) by the year 2000 for hydroelectric plants currently in the planning stage.

Tidal Power

Another source of energy related to local geological conditions is tidal flow. The gravitational pull of the sun and the moon, along with the earth's rotation, causes tides. The tidal movement of water represents a great deal of energy. For years, engineers have suggested that this moving water could be used to produce electricity. The principle is the same as that employed in a hydroelectric plant. As water flows from a higher level to a lower one, it can be used to generate electricity. The greater the difference between high and low tides the more energy can be extracted.

Since tidal changes are greatest near the poles and are accentuated in narrow bays and estuaries, suitable sites of constructing power plants are limited. In the 1930s, the United States explored the possibility of constructing a tidal electrical generating facility at Passamaquoddy, Maine, on the Bay of Fundy. After spending considerable time and money on a feasibility study, it abandoned the idea.

In 1966, France constructed a commercial tidal generating station. (See figure 9.16.) Located on the Rance River Estuary on the Brittany coast, it is the world's only large tidal generating station. It was built to generate 240 megawatts, but because of the tides, it usually produces only 62 megawatts. This satisfies the electric power needs of about 100,000 people (less than 0.2 percent of the population of France). There is also a 16-megawatt plant in Nova Scotia, Canada.

The British government studied the feasibility of constructing a 7200-megawatt tidal station in the Severn Estuary in southwestern England. It would have thirty times the capacity of the French station. The facility could be built for a price comparable to that of a coal-fired plant of similar generating capacity, but it would disrupt the normal estuary flow and would concentrate pollutants in the area. The plant has not been built.

Geothermal Power

The earth's core, with temperatures as high as 4400°C, is a molten mass of material

FIGURE 9.16 Tidal Generating Station
The Rance River Estuary Power Plant in France is the world's largest tidal electrical generating station.

The Three Gorges Dam

A project that began in late 1997 is already being compared to the building of the pyramids and the Great Wall of China. China's Three Gorges Dam across the world's third longest river, the Yangtze, is a project that when completed will be visible from the Moon and will be the largest dam in the world. The Three Gorges Dam will stretch 2 kilometers (1.3 mi) across the Yangtze River, tower 185 meters into the air (610 ft), and create a 600-kilometer (385-mi) reservoir behind it. The dam is expected to cost in excess of $40 billion before completion in 2009. The government of China contends that the dam is necessary for several reasons. The primary reason is the 18,000 megawatts of electric power that the dam will generate. China's rapidly developing economy needs the energy for its expanding industrial centers. It is also argued that the project will transform the Yangtze River, especially the upper regions, into a more navigable and hence, economic, waterway. The final major argument for the dam relates to flood control in the middle and lower reaches of the river. These parts of the river have been prone to frequent and disastrous floods. The economic arguments, however, are not the only issues being debated in relation to the Three Gorges Dam.

Scientists in China and from other nations have warned that the project could threaten migratory fish, concentrate water pollution, and endanger to the point of extinction the Chinese alligator, river dolphins, the Siberian white crane, and the Chinese sturgeon. Environmental concerns relating to the project were instrumental in the decision of the United States to withdraw technical assistance in 1993. The impact on the regional Chinese population is also a critical issue. Upon completion the lake formed by the dam will inundate 153 towns, 4,500 villages, numerous archeological sites, and the scenic canyons of the Three Gorges that have inspired poets and painters for centuries.

- Do you feel that the benefits of this massive project exceed its economic and environmental costs?
- Would you have supported the project in its developmental phase?

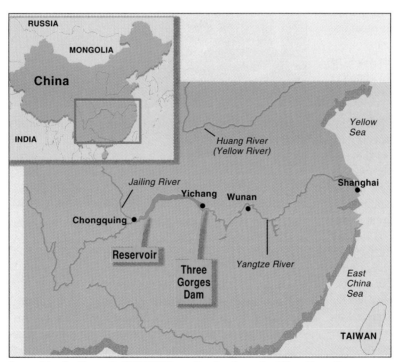

Source: (Magellan Geographix/ABC NEWS.com. Data from International Rivers Network)

possessing vast amounts of energy. In some regions, this material sometimes breaks through the earth's crust and produces volcanoes. In other regions, the hot material is close enough to the surface to heat underground water and form steam. Geysers and hot springs are natural areas where this steam and hot water come to the surface. In areas where the steam is trapped underground, **geothermal energy** is tapped by drilling wells to obtain the steam. The steam is then used to power electrical generators. At present, geothermal energy is practical only in areas where this hot mass is near the surface. (See figure 9.17.)

The United States has about half of the world's geothermal electrical generating capacity, but more than 130 generating plants are operating in twelve other countries. In addition to the Philippines, which have half the generating power that the United States does, Italy, Mexico, Japan, New Zealand, and Iceland produce sizable amounts of electricity by geothermal methods.

California produces 40 percent of the world's geothermal electricity, about 1900 megawatts. One megawatt provides energy for about one thousand households. The Pacific Gas and Electric Company (PG&E) has been producing electricity from geothermal energy since 1960. PG&E's complex of generating units located north of San Francisco is the largest in the world and provides 700 megawatts of power, enough for 700,000 households or about 2.9 million people.

The state of Hawaii has considered building two 25-megawatt geothermal plants on the island of Hawaii. Most of the electricity currently generated for the state comes from oil-fired power plants; geothermal-generated electricity would be less costly and reduce the risk of accidental oil spills. However, the plants would be built in a rainforest area that environmentalists believe would be destroyed by the construction.

In addition to producing electricity, geothermal energy is used directly for space heating. In Iceland, half of the geothermal energy is used to produce electricity and half is used for heating. In the capital, Reykjavík, all of the buildings are heated with geothermal energy at a cost that is less than 25 percent of what it would be if oil were used.

Geothermal energy does present some environmental problems. The steam contains hydrogen sulfide gas, which has the odor of rotten eggs and is an unpleasant form of air pollution. The minerals in the steam corrode pipes and equipment, causing maintenance problems. The minerals are also toxic to fish.

Other ways to use the heat in the earth are less dependent on special geologic features. All objects contain heat energy, which can be extracted from and transferred to other locations. A refrigerator extracts heat from the interior and exports it to the coils on the back of the unit. Similarly, pipes placed within the earth can extract heat and transfer it to a home.

FIGURE 9.17 Geothermal Power Plant In this plant, the steam obtained from geothermal wells is used in the production of electricity.

FIGURE 9.18 Wind Energy Fields of wind-powered generators such as these can produce large amounts of electricity.

It is important to recognize that such an application of geothermal energy requires the expenditure of energy through an electric-powered device, so the amount of heat energy extracted costs something. In some instances, obtaining heat in this manner is less expensive than other traditional sources, such as oil or natural gas, but it is not a "free" source of heat.

Wind Power

As the sun's radiant energy strikes the earth, it is converted into heat, which warms the atmosphere. The earth is unequally heated because various portions receive different amounts of sunlight. Since warm air is less dense and rises, cooler, denser air flows in to take its place. This flow of air is wind. For centuries, wind has been used to move ships, grind grains, pump water, and do other forms of work. In more recent times, wind has been used to generate electricity. (See figure 9.18.)

In 1996, California was using about 14,600 wind turbines to produce about 300 megawatts of electricity at a cost competitive with newly constructed coal or nuclear plants. This is about 40 percent of the world's electricity produced by wind. The other major countries involved in wind power are Germany, Denmark,

Electricity from the Ground Up

The naturally occurring heat beneath the planet's surface powers volcanoes, hot springs, and geysers such as Yellowstone National Park's Old Faithful. In a few places in the world, such as Iceland and parts of California, natural heat and natural water come together in sufficient quantities to provide energy. But in most places where there is abundant subsurface water, useful heat is five kilometers or more below the surface, beyond the reach of economical drilling technology. In places where heat is within reach, there is often no subsurface water.

In the search for cheap, nonpolluting energy, scientists at the Los Alamos National Laboratory in New Mexico have begun a process to "mine heat." Mining heat is actually a new effort to use geothermal energy. In this process, water is pumped at high pressure three kilometers down into the earth through a well. When the water comes back up through a parallel well, it has been heated beyond the boiling point by the earth's natural heat. The hot dry rock project is an attempt to bring heat and water together as a geothermal energy source.

On average, temperatures below the earth's surface increase about 27°C per kilometer. In much of the western United States, however, residual heat from ancient volcanoes increases temperatures by more than 69°C/km. At the Los Alamos site, the temperature at the base of the well is about 240°C.

According to researchers at Los Alamos, a fairly conservative estimate is that there are at least 500,000 quads (quadrillion British thermal units, or Btus) of useful heat in hot dry rock at accessible drilling depths beneath the United States. This is about 6000 times the total amount of energy used in the country in one year.

The fundamental question now being asked is if the energy can be developed economically for commercial use. At this point, the answer is no; however, if new technology is developed, the answer could change. Other questions also must be addressed: Will the heat in a particular well last long enough to justify the cost? Can the system be sealed to minimize water leakage? Can a constant flow of water be sustained?

Even if it proves commercially competitive, hot dry rock mining might encounter obstacles involving site access, hookups to distant power grids, and legal disputes over water rights. If such problems can be overcome, maybe one day the energy used to run your computer will come from water heated kilometers beneath the surface of the Earth.

⑤ Steam from the hot water generates electricity.

① Recirculated water is pumped at high pressure 4 to 6 miles below the earth's surface where the temperature reaches up to 400° F hotter than the earth's surface.

④ Pipes recapture the hot water and carry it back to the earth's surface.

② As the water is forced into the rock, the extreme pressure causes the rocks to break.

③ "Hot rocks" heat the water as it percolates through the rock fissures.

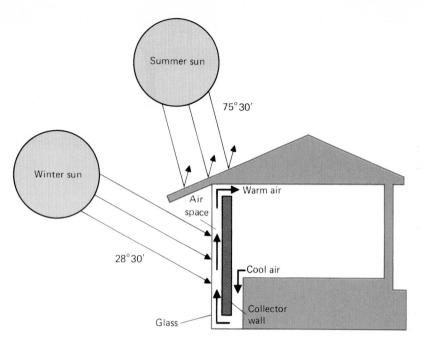

FIGURE 9.19 Passive Solar Heating The length of overhang in this home is designed for solar heating at the latitude of St. Louis, Missouri (38°N). In this design, a wall 30 to 40 centimeters thick is used to collect and store heat. The collector wall is located behind a glass wall and faces south. During a midwinter day, when the sun's angle is 28 degrees, light energy is collected by the wall and stored as heat. At night, the heat stored in the wall is used to warm the house. Natural convection causes the air to circulate past the wall, and the house is heated. During a midsummer day, when the sun's angle is 75 degrees, the overhang shades the collector wall from the sun.

India, and the Netherlands. The California Energy Commission plans to produce 8 percent of the state's electricity by wind by the year 2000, and many other regions of the country plan wind power installations. India, China, Germany, and Spain have plans for major increases in wind-generated electricity through the year 2005. In Inner Mongolia, nomadic herders carry with them two thousand small, portable, wind-driven generators that provide electricity for light, television, and movies in their tents as well as for electric fences to contain their animals.

A steady and dependable source of wind makes the use of wind power more productive in some regions than others. Wide, open areas, such as the Great Plains in North America, are better suited for wind power than are heavily wooded areas. Wind-generated electricity is usually used in conjunction with other sources of electricity that take over when the wind does not blow. Wind generators do have some negative effects. The moving blades are a hazard to birds and produce a noise that some find annoying. In addition, some people consider the sight of a large

number of wind generators, such as are found in California, to be visual pollution.

Solar Energy

The sun is often mentioned as the ultimate answer to the world's energy problems. It provides a continuous supply of energy that far exceeds the world's demands. In fact, the amount of energy received from the sun each day is six hundred times greater than the amount of energy produced each day by all other energy sources combined. The major problem with solar energy is its intermittent nature. It is only available during the day and in cloudy weather is less available than when it is sunny. All systems that use solar energy must store or use supplementary sources of energy when sunlight is not available. Because of differences in the availability of sunlight, some parts of the world are more suited to the use of solar energy than others.

Solar energy is utilized in three ways:

1. In a passive heating system, the sun's energy is converted directly into heat for use at the site where it is collected.

2. In an active heating system, the sun's energy is converted into heat, but the heat must be transferred from the collection area to the place of use.

3. The sun's energy also can be used to generate electricity, which may be used to operate solar batteries or may be transmitted along normal transmission lines.

Passive Solar Systems

Anyone who has walked barefoot on a sidewalk or a blacktopped surface on a sunny day has experienced the effects of passive solar heating. In a **passive solar system,** light energy is transformed to heat energy when it is absorbed by a surface. Some of the earliest uses of passive solar energy were to dry food and clothes and to evaporate seawater to produce salt. In fact, solar energy is still used for these purposes. Homes and buildings can also be designed to use passive solar energy for heating. (See figure 9.19.) Such systems require a large window through which sunlight enters and a large mass that collects and stores the heat. This large mass may be a thick wall or floor that would be a normal part of the room. Since there are no moving parts, a passive solar system is maintenance free. None of the energy is used to transfer heat within the system, and there are no operating costs. However, passive solar design is usually only practical in new construction.

Active Solar Systems

In addition to a solar collector, an **active solar system** requires a pump and a system of pipes to transfer the heat from the site of its production to the area to be heated. (See figure 9.20.) Active solar systems are most easily installed in new buildings, but in some cases can be installed in existing structures. A major consideration in the use of an active solar system is the initial cost of installation. An active system requires a specially designed collector, consisting of a series of liquid-filled tubes; a pump; and pipes to transfer the warm liquid from the collector to the space to be heated. Because an active system has moving parts, it also has operation and maintenance costs like any other heating or cooling system.

FIGURE 9.20 Active Solar System
Collectors provide 65 percent of the hot water in this dental school building in California.

Rock, water, or specially produced products are used to store heat. The hot liquid in the pipes heats the storage medium, which releases its heat when the sun is not shining.

Solar-Generated Electricity

When the first **photovoltaic cell,** a bimetallic unit that allows the direct conversion of sunlight to electricity, was developed by Bell Laboratories in 1954, it was regarded as an expensive novelty. However, as more efficient batteries were developed and production costs were reduced, practical uses were found for photovoltaic cells. By the mid-1980s, more than 60 million solar calculators were being produced annually. These calculators used over 10 percent of the photovoltaic cells manufactured.

Photovoltaic cells appear to be emerging as sources of small amounts of electricity for special uses like calculators and for running equipment in remote regions. The normal system of generating electricity in large, centrally located plants and distributing it by high tension lines is costly, so it is only practical in highly populated areas. Photovoltaic cells are more practical in remote regions of the world. Over ten thousand solar-electric homes have been built in parts of Alaska and the Australian outback. (See figure 9.21.) The French government has subsidized the installation of over two thousand solar electric units on eighteen islands in the Pacific. These units provide electricity for a thousand homes and five hospitals. Many of the developing countries of the world will introduce electricity to villages through the use of photovoltaics rather than the use of generators that require fuel and distribution lines. Two-thirds of the current U.S. production of photovoltaic cells is being exported.

The price of photovoltaic cells has been falling as better technology is developed. Eventually, the cells may become competitive with other energy sources, particularly as the cost of fossil fuels rises.

Solar energy is also being used to generate electricity in a more conventional way. A company named Luz International in California has built solar collector troughs that can heat oil in pipes to 390°C. (See figure 9.22.) This heat can be transferred to water, which is turned into steam that is used to run conventional electricity-generating turbines. As with photovoltaic cells, the cost of producing electricity in this manner is falling and is becoming competitive with conventional sources.

Limitations of Solar Energy

Solar energy provides less than 1 percent of the world's energy for several reasons. The most obvious is that it only works during the day, which means that some type of heat or electrical storage mechanism is needed for night use, which adds to the expense of relying on solar energy. The fact that solar heating is most practical in new construction also limits its use. In colder climates, solar heat is inadequate as the sole source of heat, and some type of a conventional heating system is required for backup. Climate is also a problem, since many areas have extensive cloudy periods, which reduce the amount of energy that can be collected. Although the price of collectors and related equipment has decreased in recent years, many collector systems are still expensive. For example, electricity from photovoltaic cells is still more expensive than conventional electric generating systems. Prices of photovoltaic cells continue to fall, however, and the cost of generating electricity from conventional fuels is rising.

Biomass Conversion

Biomass is any accumulation of organic material produced by living things. The most commonly used biomass sources are fuelwood, agricultural residue from the harvesting of crops, crops grown for their energy content, and animal waste. (Because of its major impact on world energy resources, fuelwood will be discussed

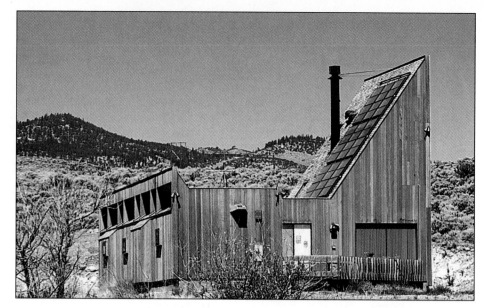

FIGURE 9.21   Solar Energy In some remote areas, solar energy is an economical method of electricity production.

FIGURE 9.22   Solar Generation of Electricity This solar-powered electricity generating plant is capable of generating electricity at a cost that is competitive with other methods of generating electricity.

separately in the next section.) These traditional, often noncommercial sources of fuel provide more than 10 percent of the world's energy, but are not reported in most statistics about global energy. In many developing countries these sources of fuel are a large proportion of the energy available.

Biomass conversion is the process of obtaining energy from the chemical energy stored in biomass. It is not a new idea; burning wood is a form of biomass conversion that has been used for thousands of years. Biomass can be burned directly as a source of heat for cooking, burned to produce electricity, converted to

alcohol, or used to generate methane. (See figure 9.23.) The People's Republic of China has 500,000 small methane digesters in homes and on farms, India has 100,000, and Korea has 50,000. Brazil is the largest producer of alcohol from biomass. The low price of sugar coupled with the high price of oil have prompted Brazil to use its large crop of sugar cane as a source of energy. Alcohol provides 50 percent of Brazil's automobile fuel.

Biomass conversion raises some environmental and economic concerns. Countries that use large amounts of biomass for energy are usually those that have food shortages. Biomass conversion

FIGURE 9.23 **Methane Digester** In the digester unit, anaerobic bacteria convert animal waste into methane gas. This gas is then used as a source of fuel. The sludge from this process serves as a fertilizer. In many less-developed countries, this type of digester has the advantages of providing a source of energy and a supply of fertilizer and managing animal wastes, which helps reduce disease.

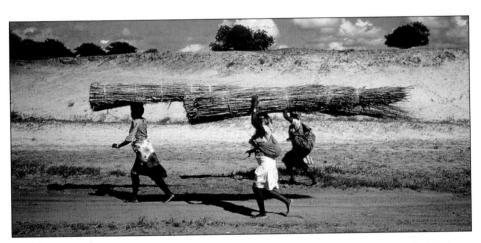

FIGURE 9.24 **Desertification** The demand for fuelwood in many regions has resulted in the destruction of forests. This is a major cause of desertification.

means that fewer nutrients are being returned to the soil, and this compounds the food shortage. If the price of food rises or the price of oil falls, there could be less biomass conversion.

The energy required to produce usable energy stocks from biomass must be taken into account. Growing corn to produce alcohol requires large energy inputs. The amount of energy present in the alcohol produced from the corn is actually less than the amount of energy that went into producing the alcohol. Obviously, this makes no sense from an energy point of view. However, the convenience of a liquid fuel may be worth paying for in economic terms.

Fuelwood

In less-developed countries, fuelwood has been the major source of fuel for centuries. In fact, fuelwood is the primary source of energy for nearly half of the world's population. In these regions, the primary use of fuelwood is for cooking.

The use of fuelwood as a prime energy source, a rapid population increase, and the high cost of other types of fuel have combined to create some serious environmental problems in many areas of the world. It is estimated that 1.3 billion people are not able to obtain enough fuelwood or must harvest fuelwood at a rate that exceeds its growth. This has resulted in the destruction of much forestland in Asia and Africa and has hastened the rate of desertification in these regions. (See figure 9.24.)

Because of its bulk and low level of energy compared to equal amounts of coal or oil, fuelwood is not practical to transport over a long distance, and most of it is used locally. In the United States, Norway, and Sweden, fuelwood furnishes 10 percent of the energy for home heating. Canada obtains 3 percent of its total energy, not just home-heating energy, from fuelwood. Most of this energy is used in forest product industries, such as lumbering and paper mills.

Burning wood is also a source of air pollution. Studies indicate that more than 75 organic compounds are released when wood is burned, 22 of which are hydrocarbons known or suspected to be carcinogens. However, these are usually released in small amounts. No studies have been conducted to determine if wood burning produces them in large enough quantities to be harmful. Often, woodstoves are not operated in the most efficient manner, and high amounts of particulate matter and other products of incomplete combustion, such as carbon monoxide, are released, contributing to ill health and death.

In areas with a high population density, the heavy use of fuelwood releases large amounts of fly ash into the air. In Missoula, Montana, in recent years, 55 percent of the particles in the air during the summer were from burning wood. In the winter, wood was responsible for 75 percent of the particles. A number of steps have been taken to reduce air pollution resulting from burning wood. Some cities, such as London, have a total ban on burning wood. Vail, Colorado, permits only one wood-burning stove per dwelling. Many areas require woodstoves to have special pollution controls that reduce the amount of particulates and other pollutants released.

Solid Waste

Residents of New York City discard in excess of 25,000 metric tons of waste each day, or 1.8 kilograms per person. In fact, New Yorkers lead the world in the production of municipal trash. High-income cities such as New York usually produce more waste than low-income cities. (See table 9.2.) About 80 percent of this waste is combustible and, therefore, represents a potential energy source. (See figure 9.25.)

"Trash power," the use of municipal waste as a source of energy, requires several steps. First, the waste must be sorted so that the burnable organic material is separated from the inorganic material. The sorting is accomplished most economically by the person who produces the waste. That means the producer must separate trash before putting it out for pickup. It must be separated into garbage, burnable materials, glass, and metals, and picked up by compartmentalized collecting trucks. Second, the efficient use of waste as a source of fuel requires a large volume and a dependable supply. If a community constructs a facility to burn 200 metric tons of waste a day, the community must generate and collect 200 metric tons of waste each day.

Communities have been burning trash as a means of reducing the volume of waste for a number of years. The first

TABLE 9.2	AMOUNT OF SOLID WASTE PRODUCED PER CAPITA
High-income cities	Kilograms per day
New York, United States	1.8
Singapore	1.6
Tokyo, Japan	0.94
Rome, Italy	0.72
Low-income cities	
Tunis, Tunisia	0.56
Medellín, Colombia	0.54
Calcutta, India	0.51
Kano, Nigeria	0.46

FIGURE 9.25 Waste to Energy Municipal trash can be burned to produce heat and electricity. This refuse pit is used to feed hoppers of high-temperature furnaces.

consolidated incineration of waste was done in Nottingham, England, in 1874. Burning the municipal waste of Munich, Germany, not only reduces the volume of the waste but supplies the energy for 12 percent of the city's electricity. Rotterdam, the Netherlands, operates a 55-megawatt power plant from its garbage. In the United States, only 3 percent of household waste is burned. This is low in comparison to the 26 percent burned in Japan, 51 percent in Sweden, and 75 percent in Switzerland.

Burning trash is not a way to make a profit. It is a way to decrease the cost of trash disposal because it reduces the need

for landfill sites. Baltimore produces gas from 1000 metric tons of waste per day. This gas generates steam for the Baltimore Gas and Electric Company. The daily waste also yields 80 metric tons of a charcoal-like material (char), 70 tons of ferrous metals, and 170 metric tons of glass. These products all have potential uses. The char can be burned as an additional source of energy. The ferrous metals can be sold, reducing the cost of operating the plant, as well as conserving mineral resources. The glass can be sold for recycling, further reducing the economic costs of the plant.

FIGURE 9.26 **Energy Conservation** The use of fluorescent light bulbs, energy-efficient appliances, and low-emissive glass could reduce energy consumption significantly.

Although the burning of trash reduces the trash volume and furnishes energy, it poses environmental concerns, one of which is air pollution. Many of the older incinerating plants do not comply with today's air-quality standards. Also, much of the waste material, such as bleached paper and plastics, contains-chlorine-containing organic compounds. When burned, these compounds can form dioxins, which are highly toxic and suspected carcinogens. Another problem associated with waste to energy systems is the popularity of recycling. Many of the items that are now recycled such as plastics and wood, have high heat content. The reduction in the amount of these items in the waste stream reduces its value as an energy source.

ENERGY CONSERVATION

Conservation is not a way of generating energy, but it is a way of reducing the need

for additional energy consumption. Some conservation technologies are sophisticated, while others are quite simple. For example, if a small, inexpensive wood-burning stove were developed and used to replace open fires in the less-developed world, energy consumption in these regions could be reduced by 50 percent.

Several technologies that reduce energy consumption are now available. (See figure 9.26.) Highly efficient fluorescent light bulbs that can be used in regular incandescent fixtures give the same amount of light for 25 percent of the energy, and they produce less heat. Since lighting and air conditioning (which removes the heat from inefficient incandescent lighting) account for 25 percent of U.S. electricity consumption, widespread use of these lights could significantly reduce energy consumption. Low-emissive glass for windows can reduce the amount of heat entering a building while allowing light to enter. The use of this glass in new construction and replacement windows could have a major impact on the energy picture. Many other

technologies, such as automatic dimming devices or automatic light-shutoff devices, are being used in new construction.

The shift to more efficient use of energy needs encouragement. Often, poorly designed, energy-inefficient buildings and machines can be produced inexpensively. The short-term cost is low, but the long-term cost is high. The public needs to be educated to look at the long-term economic and energy costs of purchasing poorly designed buildings and appliances.

Electric utilities have recently become part of the energy conservation picture. In some states, they have been allowed to make money on conservation efforts; previously they could only make money by building more power plants. This encourages them to become involved in energy conservation education, because teaching their customers how to use energy more efficiently allows them to serve more people without building new power plants.

The Arctic National Wildlife Refuge has been a source of controversy for many years. The major players are environmentalists who seek to preserve this region as wilderness; the state of Alaska, which funds a major portion of its activities with dividends from oil production; Alaska residents, who receive a dividend payment from oil revenues; oil companies that want to drill in the refuge; and members of Congress who see the oil reserves in the region as important economic and political issues.

In 1960, 8.9 million acres were set aside as the Arctic National Wildlife Range. Passage of the Alaskan National Interest Lands Conservation Act in 1980 expanded the range to 19.8 million acres and established 8.6 million acres as wilderness. The act also renamed the area the Arctic National Wildlife Refuge. There are international implications to this act. The refuge borders the Northern Yukon National Park. Many animals, particularly members of the Porcupine caribou herd, travel across the border on a regular yearly migration. The United States is obligated by treaty to protect these migration routes.

The act requires specific authorization from Congress before oil drilling or other development activities can take place on the coastal plain in the refuge. The coastal plain has the greatest concentration of wildlife, is the calving ground for the Porcupine caribou, and has the greatest potential for oil production. In each case the potential authorization caused a collision of three forces: environmental protection, economic development, and political benefit. Furthermore, there are great differences of opinion within each of the competing interest groups. Some Alaskan citizens support drilling; others oppose it. Members of Congress are similarly split. Even members of the Department of the Interior have provided conflicting testimony about the risks and benefits of drilling for oil in the refuge.

In 1998 the Secretary of the Interior acting under the recommendation of President Clinton cleared the way for oil development on Alaska's North Slope. Under the plan about a third of a 4.6-million-acre study area in the northeastern corner of the federal reserve would be off limits to drilling. Oil leases would be sold on 4 million acres of the government's National Petroleum Reserve west of the Prudhoe Bay oil fields. Some of the leases would allow only slant drilling because the surface is to be protected. Oil companies have repeatedly stated that the oil—an estimated 400 million to 1 billion barrels—can be recovered without endangering wildlife or the fragile arctic ecosystem. Conservationists have argued that none of the reserve should be developed when there is plenty of oil worldwide and prices are cheap. They argue that drilling in the reserve will harm the habitat of millions of migratory birds, caribou, and polar bears. Only time will tell which side, if either, is correct.

- What do you think of the decision to drill?
- Was the decision a compromise between two opposite sides or a politically motivated answer to a controversial issue?
- What would your decision have been?

Migrating caribou in the Arctic National Wildlife Refuge

Summary

A resource is a naturally occurring substance of use to humans that can potentially be extracted using current technology. Reserves are known deposits from which materials can be extracted profitably with existing technology under present economic conditions.

Coal is the world's most abundant fossil fuel. Coal is obtained by either surface mining or underground mining. Problems associated with coal extractions are disruption of the landscape due to surface mining and subsidence due to underground mining. Black lung disease, waste heaps, water and air pollution, and acid mine drainage are additional problems. Oil was originally chosen as an alternative to coal because it was more convenient and less expensive. However, the supply of oil is limited. As oil becomes less readily available, multiple offshore wells, secondary recovery methods, and increased oil exploration will become more common. Natural gas is another major source of fossil-fuel energy.

The primary problem associated with natural gas is transport of the gas to consumers.

Fossil fuels are nonrenewable: The amounts of these fuels are finite. When the fossil fuels are exhausted, they will have to be replaced with other forms of energy, probably renewable forms. Hydroelectric power can be increased significantly, but its development must flood areas and in so doing may require the displacement of people. The use of geothermal and tidal energy is limited by geographic locations. Wind power may be used to generate electricity but may require wide, open areas and a large number of wind generators. Solar energy can be collected and used in either passive or active systems and can also be used to generate electricity. Lack of a constant supply of sunlight is solar energy's primary limitation. Fuelwood is a minor source of energy in industrialized countries but is the major source of fuel in many less-developed nations. Biomass can be burned to provide heat for cooking or to produce electricity, or it can be converted to alcohol or used to generate methane. In some communities, solid waste is burned to reduce the volume of the waste and also to supply energy.

Energy conservation can reduce energy demands without noticeably changing standards of living.

Review Questions

1. Why are fossil fuels important?
2. Distinguish between reserves and resources.
3. What are the advantages of surface mining of coal compared to underground mining? What are the disadvantages of surface mining?
4. Compare the environmental impacts of the use of coal and the use of oil.
5. What are some limiting factors in the development of new hydroelectric generating sites?
6. What factors limit the development of tidal power as a source of electricity?
7. In what parts of the world is geothermal energy available?
8. Why can wind be considered a form of solar energy?
9. Compare a passive solar heating system with an active solar heating system.
10. What problems are associated with the use of solid waste as a source of energy?
11. List three energy conservation techniques.

Critical Thinking Questions

1. Given what you know about the economic and environmental costs of different energy sources, would you recommend that your local utility company use hydroelectricity or coal to supplement electric production? What criteria would you use to make your recommendation?
2. Coal burning electrical power plants in the Midwest have contributed to acid rain in the eastern United States. Other energy sources would most likely be costlier than coal, thereby raising electricity rates. Should citizens of another state be able to pressure these utility companies to change the method of generating electricity? What mechanisms might be available to make these changes? How effective are these mechanisms?
3. Imagine you are an official with the Department of Energy and are in the budgeting process for alternative energy research. Where would you put the money? Why?
4. Given your choices from question 14, what do you think the political repercussions of your decision would be? Why?
5. Do you believe that large dam projects like the Three Gorges Dam project in China or the James Bay Power project in Canada are, on the whole, beneficial or not? What other alternatives would you recommend? Why?
6. Energy conservation is one way to decrease dependence on fossil fuels. What are some things you can do at home, work, or school that would reduce fossil fuel use?
7. What alternative energy resources that the text has outlined are most useful in your area? How might these be implemented?

Key Terms

acid mine drainage *147*
active solar system *156*
biomass *158*
black lung disease *145*
geothermal energy *153*
liquefied natural gas *149*
nonrenewable energy sources *141*
overburden *145*
passive solar system *156*
photovoltaic cell *157*
renewable energy sources *141*
reserves *142*
resources *142*
secondary recovery *147*
surface mining *145*
underground mining *145*

Related Internet Sites

Visit our website at *http://www.mhhe.com/environmentalscience/* for the specific information on each of the topics below.

Earth science, fuel sciences, and mineral economics

Working solar house in Maine

Solar energy uses

Wind energy

The technology and implementation of renewable energy

Wind as a renewable energy source

Geothermal fact sheet

Worldwide renewable energy businesses and organizations

Wind energy technology projects

Alternative energy projects in the northeastern U.S.

10
chapter

Nuclear Energy: Benefits and Risks

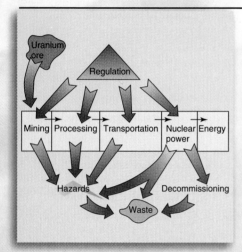

OBJECTIVES

After reading this chapter, you should be able to:

- Explain how nuclear fission has the potential to provide large amounts of energy.
- Describe how a nuclear reactor produces electricity.
- Describe the basic types of nuclear reactors.
- Explain the steps involved in the nuclear fuel cycle.
- List concerns regarding the use of nuclear power.
- Explain the problem of decommissioning a nuclear plant.
- Describe how high-level radiation waste is stored.
- Describe the accident at Chernobyl.
- Explain how a breeder reactor differs from other nuclear reactors.
- List the technical problems associated with the design and operation of a liquid metal fast-breeder reactor.
- Explain the process of fusion.

CHAPTER OUTLINE

The Nature of Nuclear Energy

The History of Nuclear Energy Development

Nuclear Reactors

Breeder Reactors

Nuclear Fusion

The Nuclear Fuel Cycle

Nuclear Material and Weapons Production

Nuclear Power Concerns

   Reactor Safety: The Effects of Three Mile Island and Chernobyl
   Exposure to Radiation
   Thermal Pollution
   Decommissioning Costs
   Radioactive Waste Disposal

Global Perspective: *The Nuclear Legacy of the Soviet Union*

THE NATURE OF NUCLEAR ENERGY

Energy from disintegrating atomic nuclei has a tremendous potential to do good for the people of the world. We routinely use X rays to examine bones for fractures, treat cancer with radiation, and diagnose disease with the use of radioactive isotopes. About 17 percent of the electrical energy generated in the world comes from nuclear power plants. Engineers in the former Soviet Union used nuclear explosions to move large amounts of earth and rock to construct dams, canals, and underground storage facilities. The U.S. government briefly considered the possibility of using nuclear explosions to build a new Panama Canal. (See figure 10.1.)

On the other hand, nuclear energy has the potential to do great harm. The Japanese cities of Hiroshima and Nagasaki were destroyed by nuclear bombs. Military uses of nuclear energy have left a legacy of radioactive wastes. In many cases these wastes have been mismanaged or carelessly disposed of.

In order to understand where nuclear energy comes from, it is necessary to review some of the aspects of atomic structure presented in chapter 3. All atoms are composed of a central region called the nucleus, which contains positively charged protons, and neutrons that have no charge. Moving around the nucleus are smaller, negatively charged electrons. Since the positively charged particles in the nucleus repel one another, energy is needed to hold the protons and neutrons together. However, some isotopes of atoms are **radioactive;** that is, the nuclei of these atoms are unstable and spontaneously decompose. Neutrons, electrons, protons, and other larger particles are released during nuclear disintegration, and a great deal of energy is released as well. The rate of decomposition is consistent for any given isotope. It is measured and expressed as **radioactive half-life,** which is the time it takes for one-half of the radioactive material to spontaneously decompose. Table 10.1 lists the half-lives of several radioactive isotopes.

Nuclear disintegration releases energy from the nucleus as **radiation,** of which there are three major types: **Alpha radiation** consists of a moving particle composed of two neutrons and two protons. Alpha radiation usually travels through air for less than a meter and can be stopped by a sheet of paper or the outer layer of the skin. **Beta radiation** consists of electrons released from nuclei. Beta particles travel more rapidly than alpha particles and will travel through air for a couple of meters. They are stopped by a layer of clothing, glass, or aluminum. **Gamma radiation** is a type of electromagnetic radiation, like X rays, light, and radio waves. It can pass through several centimeters of concrete. If the radiation reaches living tissue, equivalent doses of beta and gamma radiation cause equal amounts of biological damage. Alpha radiation can cause up to twenty times more damage than beta or gamma radiation because it is able to cause more changes in tissue as it passes through. The ingestion or inhalation of radioactive materials places the source of the radiation in direct contact with cells and therefore this kind of exposure is much more dangerous than exposure from distant sources.

The release of neutrons is particularly important in obtaining energy from nuclear disintegration. In addition to releasing alpha, beta, and gamma radiation when they disintegrate, the nuclei of atoms release neutrons. When moving neutrons hit the nuclei of certain other atoms, they can cause those nuclei to split as well. This process is known as **nuclear fission.** If these splitting nuclei also release neutrons, they can strike the nuclei of other atoms, which also disintegrate,

FIGURE 10.1   Uses of Nuclear Energy
Each of these photographs illustrates the uses of the energy available from the splitting of atoms.

TABLE 10.1   THE HALF-LIFE OF SOME RADIOACTIVE ISOTOPES

Radioactive Isotope	Half-life
Iodine 132	2.4 hours
Technetium 99	6.0 hours
Rhodium 105	36.0 hours
Xenon 133	5.3 days
Barium 140	12.8 days
Cerium 144	284.0 days
Cesium 137	30 years
Carbon 14	5730 years
Uranium 234	250,000 years
Chlorine 36	300,000 years
Beryllium 10	4.5 million years
Potassium 40	1.3 billion years
Helium 4	12.5 billion years

resulting in a continuous process called a **nuclear chain reaction.** Only certain kinds of atoms are suitable for the development of a nuclear chain reaction. The two materials commonly used in nuclear reactions are uranium-235 and plutonium-239. In addition, there must be a certain quantity of nuclear fuel (a critical mass) in order for a nuclear chain reaction to occur. It is this process that results in the large amounts of energy released from bombs or nuclear reactors.

THE HISTORY OF NUCLEAR ENERGY DEVELOPMENT

The first controlled fission of an atom occurred in Germany in 1938, but the United States was the first country to develop an atomic bomb. In 1945, the U.S. military dropped atomic bombs on the Japanese cities of Hiroshima and Nagasaki. The incredible devastation of these two cities demonstrated the potential of nuclear energy for destruction and for peaceful uses. For many years, most atomic research involved military applications of nuclear energy as bombs and as power sources for ships. During the 50 years following World War II, the two major military powers of the world—the United States and the former Soviet Union—conducted secret nuclear research projects related to the building and testing of bombs. This continued to be a primary focus of nuclear research until the recent changes in the former Soviet Union, which led to a world in which nuclear war is much less of a concern—although the explosion of nuclear devices in June of 1998 by India and Pakistan has heightened concern about nuclear war somewhat. A legacy of this military research is a great deal of soil, water, and air contaminated with radioactive material. Many of these contaminated sites have come to light recently and require major cleanup efforts. The U.S. Department of Energy has begun to clean up the pollution created by its weapons production activities and will begin shipping transuranic waste to a storage site in New Mexico in the near future.

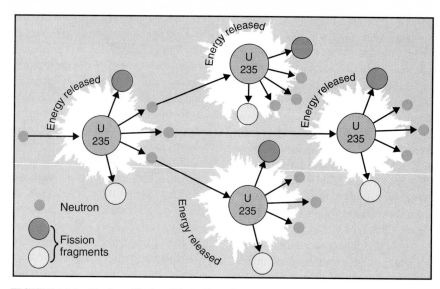

FIGURE 10.2   Nuclear Fission Chain Reaction When a neutron strikes a nucleus of U-235, energy is released, and several fission fragments and neutrons are produced. These new neutrons may strike other atoms of U-235, causing their nuclei to split. This series of events is called a chain reaction.

After World War II, people began to see the potential for using nuclear energy for peaceful purposes rather than as weapons. The world's first electricity-generating reactor was constructed in the United States in 1951, and the Soviet Union built its first reactor in 1954. In December 1953, President Dwight D. Eisenhower, in his "Atoms for Peace" speech, made the following prediction:

"Nuclear reactors will produce electricity so cheaply that it will not be necessary to meter it. The users will pay an annual fee and use as much electricity as they want. Atoms will provide a safe, clean, and dependable source of electricity."

More than 45 years have passed since Eisenhower's predictions. Although nuclear power currently is being used throughout the world as a reliable source of electricity, it has not fulfilled such overly optimistic promises. Several serious accidents have caused worldwide concern about safety, and construction of most new nuclear power projects has stopped. At the same time, many energy experts predict a rebirth of the nuclear power industry as energy demands increase and a new generation of safer nuclear power plants is designed. These experts believe the public will favor nuclear power plants because they do not generate carbon dioxide, which is thought to contribute to global warming.

NUCLEAR REACTORS

A **nuclear reactor** is a device that permits a controlled fission chain reaction. In the reactor, neutrons are used to cause a controlled fission of heavy atoms, such as uranium. **Uranium-235 (U-235)** is a uranium isotope used to fuel nuclear fission reactors. When the nucleus of a U-235 atom is struck by a slowly moving neutron from another atom, the nucleus of the atom of U-235 is split into several smaller particles. Because the nucleus will split it is said to be **fissionable.** When the nucleus is split, two to three rapidly moving neutrons are released, along with large amounts of energy. This energy is an important product of nuclear fission reactions. The neutrons released strike the nuclei of other atoms of U-235 and also cause them to undergo fission, which, in turn, releases more energy and more neutrons, thus resulting in a chain reaction. (See figure 10.2.) Once begun, this chain reaction continues to release energy until the fuel is spent or the neutrons are prevented from striking other nuclei.

In addition to fuel rods containing uranium, reactors contain control rods of cadmium, boron, graphite, or other nonfissionable materials used to control the rate of fission by absorbing neutrons. When control rods are lowered into a reactor, they absorb the neutrons produced by fissioning uranium. There are fewer neutrons to continue the chain reaction, and the rate of fission decreases. If the control rods are withdrawn, more fission occurs, and more particles, radiation, and heat are produced.

The fuel rods housed in a reactor are surrounded by water or some other type of moderator. A **moderator** absorbs energy, which slows neutrons, enabling them to split the nuclei of other atoms more effectively. Fast-moving neutrons are less effective at splitting atoms than slow-moving neutrons. As U-235 undergoes fission, the energy of the fast-moving neutrons is transferred to water; the neutrons slow down, and the water is heated.

In the production of electricity, a nuclear-powered reactor serves the same function as any fossil-fueled boiler. It produces heat, which converts water to steam to operate a turbine that generates electricity. After passing through the turbine, the steam must be cooled, and the water is returned to the reactor to be heated again. Various types of reactors have been constructed to furnish heat for the production of steam. They differ in the moderator used, in how the reactor core is cooled, and in how the heat from the core is used to generate steam. Water is the most commonly used reactor-core coolant and also serves as a neutron moderator. **Light-water reactors (LWR),** which make up 90 percent of reactors operating today, use ordinary water, which contains the lightest, most common isotope of hydrogen, having an atomic mass of one. The two types of LWRs are **boiling-water reactors (BWR)** and **pressurized-water reactors (PWR).**

In a BWR (the type of construction used in about 20 percent of the nuclear reactors in the world), the water functions as both a moderator and reactor-core coolant. (See figure 10.3.) Steam is formed within the reactor and transferred directly to the turbine, which generates electricity. A disadvantage of the BWR is that the steam passing to the turbine must be treated to remove any radiation. Even then, some radioactive material is left in the steam;

FIGURE 10.3   Boiling-Water Reactor (BWR) A boiling-water reactor is a type of light-water reactor that produces steam to directly power the turbine and produce electricity. Water is used as a moderator and as a reactor-core coolant.

FIGURE 10.4   Pressurized-Water Reactor (PWR) A pressurized-water reactor is a type of light-water reactor that uses a steam generator to form steam and a secondary loop to transfer this steam to the turbine.

therefore, the generating building must be shielded.

In a PWR (the type of construction used in over 70 percent of the nuclear reactors in the world), the water is kept under high pressure so that steam is not allowed to form in the reactor. (See figure 10.4.) A secondary loop transfers the heat

FIGURE 10.5 Advanced Gas-Cooled Reactor This type of reactor uses graphite as a moderator and the gas carbon dioxide as the reactor-core coolant. A steam generator forms steam and a secondary loop is used to transfer steam to the turbine.

TABLE 10.2 NUMBER OF OPERATING NUCLEAR POWER PLANTS (1997)

Country	Number of nuclear power reactors
Argentina	2
Armenia	1
Belgium	7
Brazil	1
Bulgaria	6
Canada	21
China	3
Czech Republic	4
Finland	4
France	58
Germany	20
Hungary	4
India	10
Japan	54
Kazakhstan	1
South Korea	12
Lithuania	2
Mexico	2
Netherlands	1
Pakistan	1
Romania	1
Russia	29
Slovakia	4
Slovenia/Croatia	1
South Africa	2
Spain	9
Sweden	12
Switzerland	5
Taiwan	6
Ukraine	15
United Kingdom	35
United States	109
World total	**442**

from the pressurized water in the reactor to a steam generator. The steam is used to turn the turbine and generate electricity. Such an arrangement reduces the risk of radiation in the steam but adds to the cost of construction by requiring a secondary loop for the steam generator.

A third type of reactor that uses water as a coolant is a **heavy-water reactor (HWR)**, developed by Canadians. It uses water that contains the hydrogen isotope deuterium in its molecular structure as the reactor-core coolant and moderator. Since the deuterium atom is twice as heavy as the more common hydrogen isotope, the water that contains deuterium weighs slightly more than ordinary water. An HWR also uses a steam generator to convert regular water to steam in a secondary loop. Thus, it is similar in structure to a PWR. The major advantage of an HWR is that naturally occurring uranium isotopic mixtures serve as a suitable fuel. This is possible because heavy water is a better neutron moderator than is regular water, while other reactors require that the amount of U-235 be enriched to get a suitable fuel. Since it does not require enriched fuel, the operating costs of an HWR are less than that of an LWR.

The **gas-cooled reactor (GCR)** was developed by atomic scientists in the United Kingdom. Carbon dioxide serves as a coolant for a graphite-moderated core. As in the HWR, natural isotopic mixtures of uranium are used as a fuel. (See figure 10.5.)

The various types of reactors represent differing approaches to building safe, economical plants. Each method has its advantages and disadvantages, and its supporters and critics. Modifications will continue to be made to these basic types. Figure 10.6 shows the distribution of nuclear power plants in North America and table 10.2 shows numbers of reactors present in each country.

BREEDER REACTORS

During the early stages of the development of nuclear power plants, breeder reactor construction was seen as the logical step after nuclear fission development. A regular fission reactor produces heat to generate electricity but does not form radioactive products useful as a fuel. A **nuclear breeder reactor** is a nuclear fission reactor that produces heat to be converted to steam to generate electricity and also forms a new supply of radioactive isotopes. If a fast-moving neutron hits a uranium-238 (U-238) nucleus and is absorbed, an atom of fissionable **plutonium-239 (Pu-239)** is produced. (See figure 10.7.) In a breeder reactor, water is not used as a moderator because water slows the neutrons too much and Pu-239 is not produced. Breeder reactors need a moderator that allows the neutrons to move more rapidly and that also has good heat transfer properties.

The **liquid metal fast-breeder reactor (LMFBR)** appears to be the most promising model. In this type of reactor, the fuel rods in the core are surrounded by rods of U-238 and liquid sodium. The energy of the neutrons released from U-235 in the fuel rods heats the sodium to a

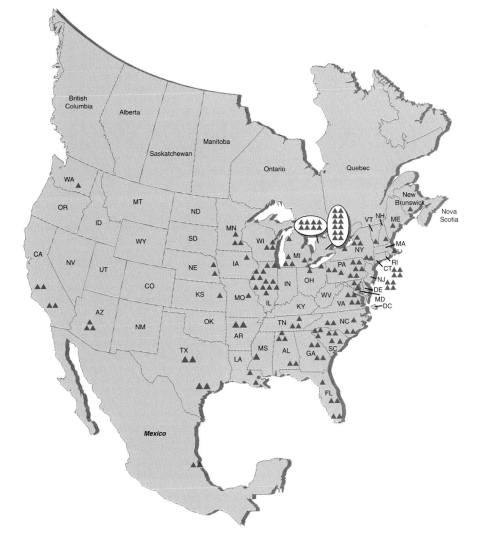

FIGURE 10.6 Distribution of Nuclear Power Plants in North America The United States has the largest number of nuclear power plants of any country in the world—over 100. Canada has about 20 nuclear power plants and Mexico has two.

92 protons   92 protons   93 protons   94 protons
146 neutrons 147 neutrons 146 neutrons 145 neutrons

FIGURE 10.7 Formation of Pu-239 in a Breeder Reactor When a fast-moving neutron (N) is absorbed by the nucleus of a U-238 atom, a series of reactions results in the formation of Pu-239 from U-238. Two intermediate atoms are U-239 and Np-239, which release beta particles (electrons) from their nuclei. (A neutron in the nucleus can release a beta particle and become a proton.) This is an important reaction because, while the U-238 does not disintegrate readily and therefore is not a nuclear fuel, the Pu-239 is fissionable and can serve as a nuclear fuel.

temperature of 620°C. In addition to furnishing heat, these fast-moving neutrons are absorbed by the rods containing U-238, and some of these atoms are converted into Pu-239. After approximately 10 years of operation, during which electricity is produced, the LMFBR will have also produced enough radioactive material to operate a second reactor.

There are some serious drawbacks to the LMFBR. Sodium reacts violently if it comes into contact with water or air. Therefore, costly, highly specialized equipment is required to contain and pump the sodium. If these systems fail, the sodium boils, which allows the chain reaction to proceed at a faster rate and could damage the reactor, leading to a nuclear accident.

There are also problems in the startup of an LMFBR. When a breeder reactor is starting up or going on-line after a shutdown, the solid sodium cannot be moved by the pumps until it becomes a liquid. This presents a technical problem in developing LMFBRs.

Another problem is that reaction rates are extremely rapid and very difficult to regulate. The instrumentation needed to monitor an LMFBR must be of extremely high quality because control of the reaction involves precise adjustments over very short periods of time.

Finally, the product of a breeder reactor, plutonium-239, is extremely hazardous to humans who come in contact with it. And, because plutonium-239 can be made into nuclear weapons, it must be transported, processed, or produced under very close security. The more breeder reactors in use, the more difficult the security problems and the more likely the chance that the small amount of plutonium needed to manufacture a bomb could be stolen.

Because of these problems, no breeder reactors are scheduled for commercial use in the United States. In fact, the only experimental U.S. breeder reactor is scheduled to be shut down. The Clinch River Fast-Breeder Reactor Plant near Knoxville, Tennessee, which was to be operational in the 1970s, never was completed.

In Europe, as in the United States, the development of breeder reactors has slowed. In 1981, after running smoothly for eight years, a 250-megawatt French prototype plant developed a leak in the cooling system that resulted in a sodium fire. This raised some serious questions about the safety of breeder reactors. In 1982, the French government announced that it was scaling down the planned construction of breeder reactors from five to one. Today, it is doubtful that even that one will be built. There are a total of five LMFBRs in operation in the world today, in France, Japan, Kazakhstan, and Russia.

NUCLEAR FUSION

Another aspect of nuclear power that may have promise for the future involves the even more advanced technology of nuclear fusion. When two lightweight atomic nuclei combine to form a heavier nucleus, a large amount of energy is released. This process is known as **nuclear fusion.** The energy produced by the sun is the result of fusion. Most studies of fusion have involved small atoms like hydrogen. Most hydrogen atoms have one proton and no neutrons in the nucleus. The hydrogen isotope deuterium (H^2) has a neutron and a proton. Tritium (H^3) has a proton and two neutrons. When deuterium and tritium isotopes combine to form heavier atoms, large amounts of energy are released. (See figure 10.8.) The energy that would be released by combining the deuterium in 1 cubic kilometer of ocean water would be greater than that contained in the world's entire supply of fossil fuels.

Although fusion could solve the world's energy problems, technology must answer several questions before we can actually use fusion power. Three conditions must be met simultaneously if fusion is to occur: high temperature, adequate density, and confinement. If heat is used to provide the energy necessary for fusion, the temperature must approach that of the center of the sun. At the same time, the walls of the vessel confining the atoms must be protected from the heat, or they will vaporize. However, the main problem is containment of the nuclei. Because they have a positive electric charge, the nuclei repel one another.

Even though, in theory, fusion promises to furnish a large amount of energy, technical difficulties appear to

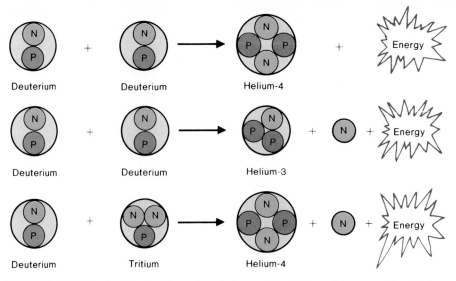

FIGURE 10.8 **Nuclear Fusion** In nuclear fusion, small atomic nuclei are combined to form heavier nuclei. Large amounts of energy are released when this occurs. Different isotopes of hydrogen can be used in the process of fusion. There are three possible types of fusion: (1) two deuterium isotopes can combine to form helium-4 and energy; (2) two deuterium isotopes can combine to form helium-3, a free neutron, and energy; and (3) a deuterium and a tritium isotope can combine to form helium-4, a free neutron, and energy.

prevent its commercial use in the near future. Even the governments of nuclear nations are budgeting only modest amounts of money for fusion research. And, as with nuclear fission and the breeder reactor, economic costs and fear of accidents may continue to delay the development of fusion reactors.

THE NUCLEAR FUEL CYCLE

In order to appreciate the consequences of using nuclear fuels to generate energy, it is important to understand how the fuel is processed. The nuclear fuel cycle begins with the mining operation. (See figure 10.9.) Low-grade uranium ore is obtained by underground or surface mining. The ore contains about 0.2 percent uranium by weight. After it is mined, the ore goes through a milling process. It is crushed and treated with a solvent to concentrate the uranium. Milling produces yellowcake, a material containing 70 to 90 percent uranium oxide.

Naturally occurring uranium contains about 99.3 percent nonfissionable U-238 and only 0.7 percent fissionable U-235.

This concentration of U-235 is not high enough for most types of reactors, so the amount of U-235 must be increased by enrichment. Since the masses of the isotopes U-235 and U-238 vary only slightly, and there is no chemical difference, enrichment is a difficult and expensive process. However, it increases the U-235 content from 0.7 percent to 3 percent.

Fuel fabrication converts the enriched material into a powder, which is then compacted into pellets about the size of a pencil eraser. These pellets are sealed in metal fuel rods about 4 meters in length, which are then loaded into the reactor.

As fission occurs, the concentration of U-235 atoms decreases. After about three years, a fuel rod does not have enough radioactive material to sustain a chain reaction, and the spent fuel rods must be replaced by new ones. The spent rods are still very radioactive, containing about 1 percent U-235 and 1 percent plutonium. These rods are a major source of radioactive waste material produced by a nuclear reactor.

When nuclear reactors were first being built, scientists proposed that spent fuel rods could be reprocessed. The remaining U-235 could be enriched and used to manufacture new fuel rods. Since plutonium is

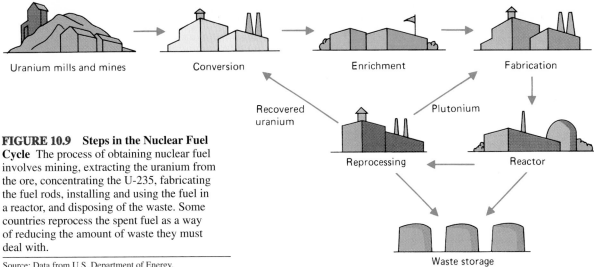

FIGURE 10.9 Steps in the Nuclear Fuel Cycle The process of obtaining nuclear fuel involves mining, extracting the uranium from the ore, concentrating the U-235, fabricating the fuel rods, installing and using the fuel in a reactor, and disposing of the waste. Some countries reprocess the spent fuel as a way of reducing the amount of waste they must deal with.

Source: Data from U.S. Department of Energy.

fissionable, it could also be fabricated into fuel rods. Besides providing new fuel, reprocessing would reduce the amount of nuclear waste. However, the cost of producing fuel rods by reprocessing was found to be greater than the cost of producing fuel rods from ore, and the United States closed its reprocessing facilities. At present, India, Japan, Russia, France, and the United Kingdom operate reprocessing plants that reprocess spent fuel rods as an alternative to storing them as a nuclear waste.

Each step in the nuclear fuel cycle involves the transport of radioactive materials. The uranium mines are some distance from the processing plants. The fuel rods must be transported to the power plants, and the spent rods must be moved to a reprocessing plant or storage area. Each of these links in the fuel cycle presents the possibility of an accident or mishandling that could release radioactive material. Therefore, the methods of transport are extremely carefully designed and tested before they are used. Many people are convinced that the transport of radioactive materials is hazardous, while others are satisfied that the utmost care is being taken and that the risks are extremely small.

Ultimately, both high-level and low-level radioactive wastes must be stored. Thus, each step in the nuclear fuel cycle, from the mining of uranium to the storage of nuclear waste, poses health and environmental concerns.

NUCLEAR MATERIAL AND WEAPONS PRODUCTION

Producing nuclear materials for weapons and other military uses involves many of the same steps used to produce nuclear fuel for power reactors. In fact, the nuclear power industry is an outgrowth of the weapons industry. In the United States, the Department of Energy currently is responsible for nuclear research for both weapons and peaceful uses and stewardship of the facilities used for research and weapons production. Some facilities are used for both processing nuclear fuel and providing materials for weapons. In both cases, uranium must be mined, concentrated, and transported to sites of use. Furthermore, military uses involve the production and concentration of plutonium. The production and storage facilities invariably become contaminated, as does the surrounding land.

Research and production facilities have typically dealt with hazardous chemicals and low-level radioactive wastes by burying them, pumping them into the ground, storing them in ponds, or releasing them into rivers. Despite environmental regulations that prevented such activities, the Department of Energy (formerly the Atomic Energy Commission) maintained that it was exempt from such federal environmental legislation. As a result, the DOE has become the steward of a large number of sites that are contaminated with both hazardous chemicals and radioactive materials. The magnitude of the problem is huge. There are 3365 square miles of DOE properties that are or have been involved in weapons development or production. These include:

- 3700 contaminated sites,
- 330 underground storage tanks with high-level radioactive waste,
- more than a million 55-gallon drums of radioactive, hazardous, or mixed waste in storage,
- 5700 sites where wastes are moving through the soil, and
- millions of cubic meters of low-level and high-level radioactive wastes.

The Department of Energy has pledged to clean up these sites by 2019. Environmental cleanup is now the largest single item in its budget. Several U.S. sites are currently being cleaned up, but things are not going smoothly. Local residents and the states that are hosts to these facilities distrust the DOE and are insisting that the sites be cleaned completely. This may not be technologically or economically possible. Furthermore, environmental cleanup is a new mission for the DOE and the department is having difficulty adjusting. The cleanup process will take many years

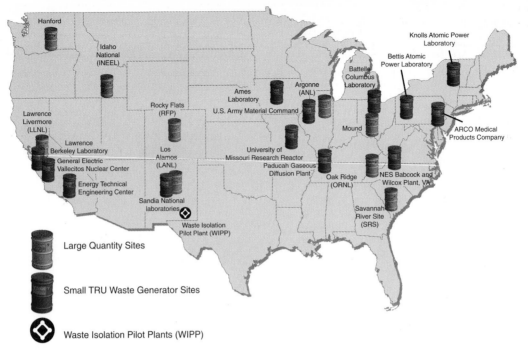

FIGURE 10.10 The U.S. Department of Energy Waste Sites As a result of many activities related to the research and the production of nuclear weapons, the U.S. Department of Energy has responsibility for cleaning up many contaminated sites.

and require the expenditure of tens of billions of dollars. Figure 10.10 shows the location of waste sites the U.S. Department of Energy has responsibility for cleaning up.

An additional problem has arisen as a result of the reduced importance of nuclear weapons. The political disintegration of the Soviet Union and Eastern Europe has made large numbers of nuclear weapons, both in the East and West, unnecessary. This has probably made the world a safer place, but how are nations of the world to dispose of their nuclear weapons? Some nuclear material can be diverted to fuel use in nuclear reactors, and some reactors can be modified to accept enriched uranium or plutonium. The security of these materials, particularly in the former Soviet Union, is a concern. Investigators have uncovered several incidents in which plutonium has been sold to other countries.

NUCLEAR POWER CONCERNS

Nuclear power has provided a significant amount of electricity for the people of the world. Currently, over 7 percent of the energy consumed worldwide and 17 percent of the electricity consumed worldwide comes from nuclear power. However, several accidents have raised questions about safety, radiation has been released into the air and water, the production and transportation of fuel have caused contamination, and the disposal of wastes is a continuing problem.

Reactor Safety: The Effects of Three Mile Island and Chernobyl

Although there have been accidents at nuclear power plants since they were first used, two relatively recent accidents have had the greatest effect on people's attitudes toward nuclear power plant safety: Three Mile Island in 1979 and Chernobyl in 1986.

On March 14, 1979, the valves to three auxiliary pumps at one of the reactors at the Three Mile Island nuclear plant closed. On March 28, 1979, the main pump to the reactor broke down. The auxiliary pumps failed to operate because of the closed valves, and the electrical-generating turbine stopped. At this point, an emergency coolant should have flooded the reactor and stabilized the temperature. The coolant did start to flow, but a faulty gauge indicated that the reactor was already flooded. Relying on this faulty reading, an operator overrode the automatic emergency cooling system and stopped it. Without the emergency coolant, the reactor temperature rose rapidly. The control rods eventually stopped fission, but a partial core meltdown had occurred. Later that day radioactive steam was vented into the atmosphere.

The crippled reactor was eventually defueled in 1990 at a cost of about $1 billion. It has been placed in monitored storage until the companion reactor, which is still operating, reaches the end of its useful life. At that time both reactors will be decommissioned.

Chernobyl is a small city in Ukraine near the border with Belarus, north of Kiev. As is true of many small cities in the world, most people had never heard of it. (See figure 10.11.) However, in the spring of 1986, the world's largest nuclear accident catapulted Chernobyl into the news.

At 1:00 a.m. on April 25, 1986, at Chernobyl Nuclear Power Station-4, a test was begun to measure the amount of electricity that the still-spinning turbine would produce if the steam were shut off. This was important information since the emergency core cooling system required energy for its operation and the coasting turbine could provide some of that energy until another source became available. The amount of steam being produced was reduced by lowering the control rods into the reactor. But the test was delayed because of a demand for electricity and a new shift of workers came on duty. The operators failed to program the computer to maintain power at 700 megawatts, and output dropped to 30 megawatts. This presented an immediate need to rapidly increase the power, and many of the control rods were withdrawn. Meanwhile, an inert gas (xenon) had accumulated on the fuel rods. The gas absorbed the neutrons and slowed the rate of power increase. In an attempt to obtain more power, operators withdrew all the control rods. This was a second serious safety violation.

At 1:00 a.m. on April 26, the operators shut off most emergency warning signals and turned on all eight pumps to provide adequate cooling for the reactor following the completion of the test. Just as final stages of the test were beginning, a signal indicated excessive reaction in the reactor. In spite of the warning, the operators blocked the automatic reactor shutdown and began the test.

As the test continued, the power output of the reactor rose beyond its normal level and continued to rise. The operators activated the emergency system designed to put the control rods back into the reactor and stop the fission. But it was too late. The core had already been deformed, and the rods would not fit properly; the reaction could not be stopped. In 4.5 seconds, the energy level of the reactor increased two thousand times. The fuel rods ruptured, the cooling water turned into steam, and a steam explosion occurred. The lack of cooling water allowed the reactor to explode. The explosion blew the 1000 metric ton concrete roof from the reactor and the reactor caught fire.

In less than 10 seconds, Chernobyl became the scene of the world's worst nuclear accident. A core meltdown had occurred. (See figure 10.12.) It took 10 days

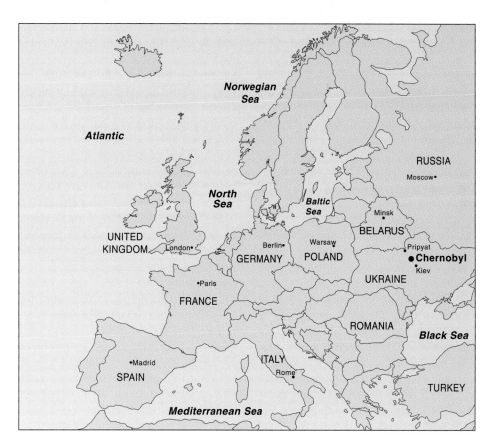

FIGURE 10.11 **Chernobyl** The town of Chernobyl became infamous as the location of the world's worst nuclear power plant accident.

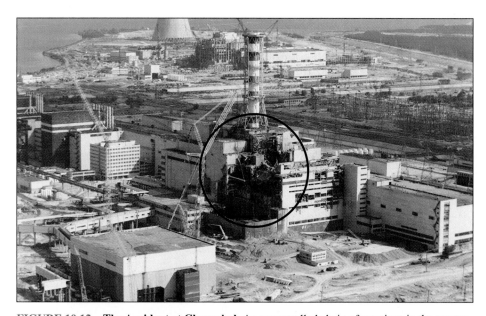

FIGURE 10.12 **The Accident at Chernobyl** An uncontrolled chain of reactions in the reactor of unit four resulted in a series of explosions and fires. See the circled area in the photograph.

to bring the runaway reaction under control. By November, the damaged reactor was entombed in a concrete covering, but the hastily built structure, known as the sarcophagus, may have structural flaws. The Ukrainian government is planning to have a second containment structure built around the current structure. The immediate consequences were 31 fatalities; 500 persons hospitalized, including 237 with acute radiation sickness; and 116,000 people evacuated. Of the evacuees, 24,000 received high doses of radiation. The delayed effects are more difficult to assess. Many people suffer from illnesses they feel are related to their exposure to the fallout from Chernobyl. In 1996 it became clear that at least one delayed effect was strongly correlated with exposure. Children or fetuses exposed to fallout are showing increased frequency of thyroid cancer. The thyroid gland accumulates iodine, and radioactive iodine 131 was released from Chernobyl. It is still too early to tell if there are other delayed health effects.

More than a year after the disaster at Chernobyl, the decontamination of 27 cities and villages within 40 kilometers of Chernobyl was considered finished. This does not mean that the area was cleaned up, only that all practical measures were completed. Some areas were simply abandoned. The largest city to be affected was Pripyat, which had a population of 50,000 and was only 4 kilometers from the reactor. A new town was built to accommodate those displaced by the accident, and Pripyat remains a ghost town. Sixteen other communities within the 40-kilometer radius have been cleaned up, and the original inhabitants have returned, although there is still controversy about the safety of these towns.

One important impact of Chernobyl is that it deepened public concern about the safety of nuclear reactors. Even before Chernobyl, between 1980 and 1986, the governments of Australia, Denmark, Greece, Luxembourg, and New Zealand had officially adopted a "no nuclear" policy. Since 1980, 10 countries have canceled nuclear plant orders or mothballed plants under construction. Argentina canceled 4 plants; Brazil, 8; China, 8; Mexico, 18; and the United States, 54. There have been no orders for new plants in the United States since 1974. Sweden, Austria, and the Philippines have decided to phase out and dismantle their nuclear power plants.

Before Chernobyl, 65 percent of the people in the United Kingdom were opposed to nuclear power plants; after Chernobyl, 83 percent were against them. In Germany, opposition increased from 46 percent to 83 percent. Opposition in the United States rose from 67 to 78 percent. (See figure 10.13.) Even the French, who have the greatest commitment to nuclear power, were against it by 52 percent. The number of new nuclear plants being constructed has been reduced. In 1996 there were 34 nuclear plants under construction—before Chernobyl there were 160. Many of these plants were canceled. In addition, many plants have been shut down prematurely, resulting in the prediction that the amount of energy furnished by nuclear fission reactors will actually decline in the future.

At the same time some people are predicting the death of nuclear power as a viable energy source, others are examining ways to make nuclear power facilities safer. They expect nuclear power to be an important energy source in the future. At both Three Mile Island and Chernobyl, operator error caused or contributed to the accidents. Operators manually stopped normal safety actions from taking place. However, a contributing factor was the design: active mechanical processes had to work properly to shut the reactors down. Many of the new designs for reactors include passive mechanisms that will shut down the reactor, special catching basins for the reactor core should it melt down, and better containment buildings.

Nuclear power will continue to be a part of the energy mix, particularly in countries that lack fossil fuel reserves. Furthermore, as fossil fuels are used up, the pressure for energy will probably create a market for nuclear power technology.

Exposure to Radiation

Although nuclear accidents are spectacular, frightening, and can cause immediate death because of the incredible amount of energy involved, the long-term problems resulting from exposure to radiation are even more worrisome. Radiation is converted to other forms of energy when it is absorbed by matter. When organisms are irradiated, this energy conversion causes damage at a cellular, tissue, organ, or organism level. The degree and kind of damage vary with the kind of radiation, the amount of radiation, the duration of the exposure, and the types of cells irradiated.

Radiation can also cause mutations, which are changes in the genetic messages within cells. Mutations can cause two quite different kinds of problems. Mutations that occur in the ovaries or testes can form mutated eggs or sperm, which can lead to abnormal offspring. Care is usually taken to shield these organs from unnecessary radiation. Mutations that occur in other tissues of the

FIGURE 10.13 Anti-nuclear Demonstration This anti-nuclear demonstration was held in 1998 in Germany as members of the German public protested the transport of nuclear waste across the country to a storage site.

body may manifest themselves as abnormal tissue growths known as cancer. Two common cancers that are strongly linked to increased radiation exposure are leukemia and breast cancer. Because mutations are essentially permanent, they may accumulate over time. Therefore, the accumulated effects of radiation over many years may result in the development of cancer later in life.

Human exposure to radiation is usually measured in **rems** (*r*oentgen *e*quivalent *man*), a measure of the biological damage to tissue. The effects of large doses (1000 to 1,000,000 rems) are easily seen and can be quantified, because there is a high incidence of death at these levels, but demonstrating known harmful biological effects from smaller doses is much more difficult. (See table 10.3.) Moderate doses (10 to 1000 rems) are known to increase the likelihood of cancer and birth defects. The higher the dose, the higher the incidence of abnormality. Lower doses may cause temporary cellular changes, but it is difficult to demonstrate long-term effects. Thus, the effects of low-level, chronic radiation generate much controversy. Some people feel that all radiation is harmful, that there is no safe level, and that special care must be taken to prevent exposure. (See figure 10.14.) Others feel that the increased risk of low-level radiation is extremely small and that current radiation standards are adequate to protect the public, especially in light of the benefits of radiation, such as medical diagnoses and electrical energy. Current research is trying to assess the risks associated with repeated exposure to low-level radiation.

Each step in the nuclear fuel cycle poses a radiation exposure problem, beginning with the mining of uranium. Although the radiation level in the ore is low, miners' prolonged exposure to low-level radiation increases their rates of certain cancers, such as lung cancer. Uranium miners who smoke have an even higher lung cancer rate. After mining, the ore must be crushed in a milling process. This releases radioactive dust into the atmosphere, so workers are chronically exposed to low levels of radioactivity. The crushed rock is left on the surface as mine

TABLE 10.3 RADIATION EFFECTS

Source	Dose	Biological effect
Nuclear bomb blast or	100,000 rems/incident	Immediate death
exposure in a nuclear facility	10,000 rems/incident	Coma, death within one to two days
X rays for cancer patients	1000 rems/incident	Nausea, lining of intestine damaged, death in one to two weeks
	100 rems/incident	Increased probability of leukemia
	10 rems/ incident	Early embryos may show abnormalities
Upper limit for occupationally exposed people	5 rems/year	Effects difficult to demonstrate
X ray of the intestine	1 rem/procedure	Effects difficult to demonstrate
Upper limit for release from nuclear installations (except nuclear power plants)	0.5 rem/year	Effects difficult to demonstrate
Natural background radiation	0.2–0.3 rem/year	Effects difficult to demonstrate
Upper limit for release by nuclear power plants	0.005 rem/year	Effects difficult to demonstrate

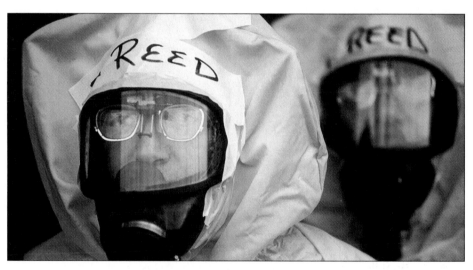

FIGURE 10.14 Protective Equipment Persons working in an area subjected to radiation must take steps to protect themselves. These workers are wearing protective clothing and filtering the air they breathe.

tailings. There are over 150 million metric tons of low-level radioactive mine tailings in the United States, and at least as much in the rest of the world. These tailings constitute a hazard because they can be dispersed into the environment. (See figure 10.15.)

In the enrichment and fabrication processes, the main dangers are exposure to radiation and accidental release of radioactive material into the environment. Transport also involves exposure risks. Most radioactive material is transported by highways or railroads. If a transporting vehicle were involved in an accident, radioactive material could be released into the environment. Even after fuel rods have been loaded into the reactor, people who work in the area of the reactor risk radiation exposure. Countries that reprocess spent fuel rods must be concerned about the exposure of workers and the possible theft of rods or reprocessed material by terrorist organizations that would use the radioactive material to construct an atomic bomb.

③ Radon diffusion

④ Windblown particles

⑤ Windblown deposition

② Water erosion

⑥ Direct gamma

① Leaching

⑦ Potable Aquifer

FIGURE 10.15 Uranium Mine Tailings Even though the amount of radiation in the tailings is low, the radiation still represents a threat to human health. Radioactivity may be dispersed throughout the environment and come in contact with humans in several ways: (1) radioactive materials may leach into groundwater; (2) radioactive materials may enter surface water through erosion; (3) radon gas may diffuse from the tailings and enter the air; (4) persons living near the tailings may receive particles in the air, (5) come in contact with objects that are coated with particulate, or (6) receive direct gamma radiation.

Source: From U.S. Environmental Protection Agency.

Thermal Pollution

Thermal pollution is the addition of waste heat to the environment. This is a problem particularly in aquatic environments, since many aquatic organisms are very sensitive to changes in temperature. All industrial processes release waste heat, so the problem of thermal pollution is not unique to nuclear power plants. In both fossil-fuel and nuclear plants, generating steam to produce electricity results in a great deal of waste heat. In a fossil-fuel plant, half of the heat energy produces electricity, and half is lost as waste heat. In a nuclear power plant, only one-third of the heat generates electricity, and two-thirds is waste heat. Therefore, the less efficient nuclear power plant increases the amount of thermal pollution more than does an equivalent fossil-fuel power plant. To reduce the effects of this waste heat, utilities build costly cooling facilities. Cooling processes usually involve water, so nuclear power plants often are constructed next to a water source. In some cases, water is drawn directly from lakes, rivers, or oceans and returned. In other cases, it is supplied by giant cooling towers. (See figure 10.16.)

Decommissioning Costs

All industrial facilities have a life expectancy, that is, the number of years they can be profitably operated. The life expectancy for an electrical generating plant, whether fossil-fuel or nuclear, is about 30 to 40 years, after which time the plant is demolished. With a fossil-fuel plant, the demolition is relatively simple and quick. A wrecking ball and bulldozers reduce the plant to rubble, which is trucked off to a landfill. The only harm to the environment is usually the dust raised by the demolition.

Demolition of a nuclear plant is not so simple. In fact, nuclear plants are not demolished, they are decommissioned. **Decommissioning** involves removing the fuel, cleaning surfaces, and permanently preventing people from coming into contact with the contaminated buildings or

FIGURE 10.16 Cooling Towers Cooling towers draw air over wet surfaces. The evaporation of the water cools the surface, removing heat from the process and releasing the heat into the air.

equipment. Today, over 70 nuclear plants in the world are shut down and waiting to be decommissioned, and the number of plants scheduled for decommissioning will grow. By 2000, 62 of the 110 nuclear plants in the United States will be 20 years old or older. The Nuclear Regulatory Commission authorizes their operation for 40 years but is considering extending authorization for an additional 20 years. This would put off the need for decommissioning for a few years, but eventually it will be necessary. In the United States, many small experimental facilities have been decommissioned and fourteen U.S. plants have been removed from service and are already awaiting decommissioning. Canada has decommissioned three plants and has one in mothballs.

The decommissioning of a plant is a two-step process. In stage 1 the plant is shut down and all the fuel rods are removed; all of the water used as a reactor-core coolant, as a moderator, or to produce steam is drained; and the reactor and generator pipes are cleaned and flushed. The spent fuel rods, the drained water, and the material used to clean the pipes are all radioactive and must be safely stored or disposed of. This removes 99 percent of the radioactivity. Stage 2 consists of dismantling all the parts of the plant except the reactor and safely containing the reactor.

Utilities have three decommissioning options: (1) decontaminate and dismantle the plant as soon as it is shut down; (2) put the plant in mothballs for 20 to 100 years to allow radioactive materials that have a short half-life to disintegrate, and then dismantle the plant; (3) entomb the plant by covering the reactor with reinforced concrete and placing a barrier around the plant.

Originally, entombment was thought to be the best method. But because of the long half-life of some of the radioactive material and the danger of groundwater contamination, this method now appears to be the least favorable. Most U.S. utilities plan to allow about 60 years to elapse between stage 1 and stage 2. Japan, which has greater needs for land, plans to mothball its plants for 5 to 10 years before dismantling them. The costs associated with decommissioning include dismantling the plant and then packaging, transporting, and burying the wastes. Because of contamination and activated material, ordinary dismantling methods cannot be used to demolish the buildings. Special remote-control machinery must be developed. Techniques that do not release dust into the atmosphere must be used, and workers must wear protective

clothing. Waste disposal is estimated to be 40 percent of the decommissioning costs. Table 10.4 shows the volume of material that would be handled from a 1100-megawatt power plant.

Recent experience with the cost of decommissioning indicates that the cost for decommissioning a large plant will be between $200–400 million, about 5 percent of the cost of generating electricity. Although the mechanisms vary with countries, the money for decommissioning is generally collected over the useful life of the plant.

Radioactive Waste Disposal

When the world entered the atomic age, the problem of the disposal of nuclear waste was not fully appreciated. Low-level radioactive waste is generated by nuclear power plants, military facilities, hospitals, and research institutions. High-level radioactive waste results from spent fuel rods, obsolete nuclear weapons, and the wastes generated by their manufacture.

High-Level Radioactive Waste

In the United States, 380,000 cubic meters of highly radioactive military waste are temporarily stored at several U.S. sites. (See figure 10.17.) A high-level waste site has been constructed near Carlsbad, New Mexico. Known as the Waste Isolation Pilot Plant, it was built to house the high-level radioactive waste from the U.S. weapons program. The wastes are commonly

referred to as **transuranic wastes** which consist primarily of various isotopes of plutonium. This facility began accepting waste in March 1999. In addition to the high-level waste from weapons programs, two million cubic meters of low-level radioactive military and commercial waste are buried at various sites. In addition, about 30,000 metric tons of high-level radioactive waste from spent fuel rods are being stored in special storage ponds at nuclear reactor sites. Many plants are running out of storage space and have been authorized by the Nuclear Regulatory Commission to store radioactive waste in aboveground casks, because the storage ponds will not accommodate additional waste.

In other countries, spent fuel rods are either stored in special ponds or sent to reprocessing plants. Even though reprocessing is more expensive than manufacturing fuel rods from ore, some countries reprocess as an alternative to waste storage. The reprocessing plants in France and the United Kingdom accept domestic and foreign fuel rods, and any fuel rods fabricated by the former Soviet Union can be returned to reprocessing plants in Russia.

At this point in time, no country has a permanent storage solution for the disposal of high-level radioactive waste. Most experts feel the best solution is to bury it in a stable geologic formation. Several countries are investigating storage in salt deposits. The waste would be placed in borosilicate glass containers. Each container would then be sealed in a

Material	Volume (cubic meters)
Radioactive	618
Activated	
Metal	484
Concrete	707
Contaminated	
Metal	5,465
Concrete	10,613
Total	17,887

TABLE 10.4 LOW-LEVEL RADIOACTIVE CONTAMINANTS FROM DECOMMISSIONING A 1100-MEGAWATT PRESSURIZED-WATER REACTOR

Source: Data from *Worldwatch Paper 69*, "Decommissioning: Nuclear Power's Missing Link," and U.S. Nuclear Regulatory Commission.

The Nuclear Legacy of the Soviet Union

The tremendous political changes that have occurred in the former Soviet Union and Eastern Europe have brought to light the magnitude of nuclear contamination caused by those nations' unwise use of nuclear energy. The Soviet Union used nuclear energy for several purposes: generating power for electricity and nuclear-powered ships, testing and producing weapons, and blasting to move earth. The accident at Chernobyl was only the most recent and most public of many problems associated with the use of nuclear energy in the Soviet Union. About 13 nuclear reactors of the Chernobyl type are still operating in the former Soviet Union. These are considered unsafe by most experts, including former Soviet scientists.

The production of nuclear fuels and weapons and the reprocessing of nuclear fuels, all of which produced nuclear wastes, occurred at three major sites: Chelyabinsk, Tomsk, and Krasnoyarsk. These facilities were secret until the recent breakup of the Soviet Union. At Chelyabinsk, radioactive waste was dumped into the Techa River and Lake Karachai. Other nuclear wastes were secretly dumped into the ocean or were buried in shallow dumps. It is estimated that there are 600 secret nuclear dump sites in and around Moscow alone.

More than 100 nuclear bombs were exploded to move earth for mining purposes, to create underground storage caverns, or to "dig" canals. In addition, 467 nuclear blasts were conducted to test weapons at a site near Semipalatinsk in Kazakhstan. As a result, nuclear fallout contaminated farmland to the northeast.

The Barents Sea and the Kara Sea have been polluted by nuclear tests and the dumping of nuclear waste. At least 15 nuclear reactors from obsolete ships were dumped into the Kara Sea. Seals that live in the area have high rates of cancer.

None of these problems is unique to the former Soviet Union. Similar secrecy surrounded the early development of nuclear power in the United States and the rest of the world. What makes the situation in the former Soviet Union different is its magnitude, caused by decades of secrecy and a commitment to nuclear weapons as the primary deterrent against enemies. It can be argued that the threat posed to the Soviet Union by its former enemies contributed to the reckless development of nuclear power.

---

thick-walled, stainless steel canister. Due to the high temperature of radioactive waste, the containers would be stored aboveground for 10 years. At the end of this time, the temperature of the container would be reduced, and the material could be buried in a salt deposit 600 meters below the surface.

Sweden intends to store its waste 500 meters underground in granite. Site construction is not planned until 2010, when Sweden's 12 reactors are scheduled to be decommissioned. (Sweden plans to discontinue nuclear power generation of electricity, but some now doubt that this will occur.) France plans to use its reprocessing plants and "temporary" storage indefinitely.

The politics concerning the disposal of high-level radioactive waste are probably as critical as developing a suitable method. No communities want a radioactive disposal site in their area. In December 1982, the U.S. Congress passed legislation calling for a high-level radioactive disposal site for the storage of spent fuel rods from nuclear power plants to be selected by March 1987 and to be completed by 1998. In 1984, the Department of Energy stated that it was already three years behind schedule. Final site selection occurred in 1989, nearly three years later than the date set by legislation. The location is Yucca Mountain, Nevada. The site was chosen for several reasons. It is in an unpopulated area near the Nevada Test Site where several nuclear devices were exploded. It is considered to be stable geologically. It is very unlikely to experience earthquakes or volcanoes. It is a very dry area and the water table is about 600 meters below the mountain, so

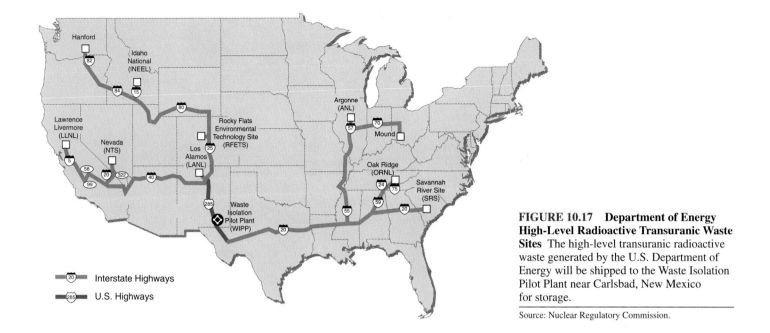

FIGURE 10.17 Department of Energy High-Level Radioactive Transuranic Waste Sites The high-level transuranic radioactive waste generated by the U.S. Department of Energy will be shipped to the Waste Isolation Pilot Plant near Carlsbad, New Mexico for storage.

Source: Nuclear Regulatory Commission.

FIGURE 10.18 High-Level Nuclear Waste Disposal Current plans for the disposal of high-level nuclear waste involve placing materials in tunnels underground at Yucca Mountain, Nevada. The material could be stored about 300 meters below the surface and about 300 meters above the water table. Because of the dry climate it is considered unlikely that water infiltrating the soil would move nuclear material downward to the water table.

groundwater will not likely be contaminated. These are important considerations because the site must be stable enough to last thousands of years.

Work has begun on the series of tunnels that would serve as storage places for the waste. (See figure 10.18.) However, current work is primarily exploratory and is seeking to characterize the likelihood of earthquake damage and the movement of water through sediments. If completed, the facility would hold about 70,000 metric tons of spent fuel rods and other highly radioactive material. It will not be completed before 2015, and by that time the total amount of waste produced by nuclear power plants will exceed the storage capacity of the site. Furthermore, the amount of waste from nuclear weapons production greatly exceeds that produced by nuclear power plants. In 1998 a geologic study suggested that the site may not be as stable as first thought and this is likely to further delay the completion of the project. Local protests against the site are continuing. Nevada can refuse the site (its refusal could be overridden by Congress), and Congress has not appropriated adequate funds to proceed rapidly. Therefore, it is still uncertain if or when it will be open to receive waste.

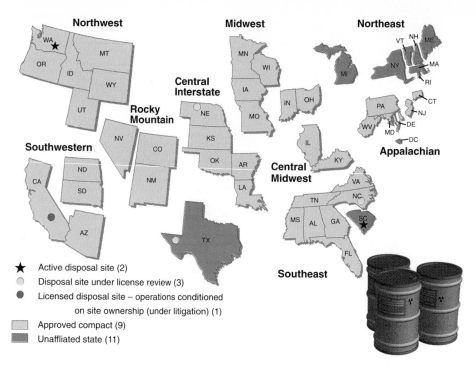

FIGURE 10.19 Low-Level Radioactive Waste Sites Each state is responsible for the disposal of its low-level radioactive waste. Many states have formed compacts and selected a state to host the disposal site. None of the proposed sites is currently accepting waste. Many other states do not have an acceptable site and are temporarily relying on a site in South Carolina to accept their waste. Eventually they will need to find other acceptable sites. The map shows the current compacts and proposed disposal sites.

Source: Nuclear Regulatory Commission.

Low-Level Radioactive Waste

Low-level radioactive waste includes the cooling water from nuclear reactors, material from decommissioned reactors, radioactive materials used in the medical field, protective clothing worn by persons working with radioactive materials, and materials from many other modern uses of radioactive isotopes. Disposal of this type of radioactive waste is very difficult to control. Estimates indicate that much of it is not disposed of properly.

In 1970, a U.S. moratorium halted the dumping of radioactive waste in the oceans. Prior to this, the United States placed some 90,000 barrels of radioactive waste on the ocean floor. European countries also have dumped both high-level and low-level radioactive material into the Atlantic Ocean. Before 1983, when ocean

dumping was halted, these countries disposed of 90,000 metric tons of radioactive waste in the ocean.

The United States produces about 800,000 cubic meters of low-level radioactive waste per year. This is presently being buried in disposal sites in Nevada, South Carolina, and Washington. However, these states balked at accepting all the country's low-level waste. In 1980, Congress set a deadline of 1986 (later extended to 1993) for each state to provide for its own low-level radioactive waste storage site. Later, a change allowed several states to cooperate and form regional coalitions, called compacts. Under this arrangement, one state would provide a disposal site for the entire compact. Several states did not join compacts and several others have since withdrawn from the

compacts they were originally a part of. Figure 10.19 shows the current compacts and the status of their disposal sites. Several of the proposed sites are under severe pressure from activists who do not want such sites in their locality. As a result most states still do not have a permanent place to deposit low-level radioactive wastes and are relying on temporary storage facilities in Richland, Washington or Barnwell, South Carolina. As a further example of the creative politics created by the need to have low-level radioactive waste disposal sites and the unwillingness to host such a site, the states of Texas, Maine, and Vermont have applied to be a compact with Texas serving as the host for the site.

Summary

Nuclear fission is the splitting of the nucleus of the atom. The resulting energy can be used for a variety of purposes. The splitting of U-235 in a nuclear reactor can be used to heat water to produce steam that generates electricity. Various kinds of nuclear reactors have been constructed, including boiling-water reactors, pressurized-water reactors, heavy-water reactors, gas-cooled reactors, and experimental breeder reactors. Scientists are also conducting research on the possibilities of using fusion to generate electricity. All reactors contain a core with fuel, a moderator to control the rate of the reaction, and a cooling mechanism to prevent the reactor from overheating.

The nuclear fuel cycle involves mining and enriching the original uranium ore, fabricating it into fuel rods, using the fuel in reactors, and reprocessing or storing the spent fuel rods. The fuel and wastes must also be transported. At each step in the cycle, there is danger of exposure. During the entire cycle, great care must be taken to prevent accidental releases of nuclear material.

The use of nuclear materials for military purposes has resulted in the same kinds of problems caused by nuclear power generation. The reduced risk of nuclear warfare has resulted in a need to destroy nuclear weapons and clean up the sites contaminated by their construction.

Public acceptance of nuclear power plants has been declining. Initial promises of cheap electricity have not been fulfilled because of expensive construction, cleanup, and decommissioning costs. The accidents at Three Mile Island in the United States and Chernobyl in Ukraine have accelerated concerns about the safety of nuclear power plants.

Other concerns are the unknown dangers associated with exposure to low-level radiation, the difficulty of agreeing to proper long-term storage of high-level and low-level radioactive waste, and thermal pollution.

Review Questions

1. How does a nuclear power plant generate electricity?
2. Name the steps in the nuclear fuel cycle.
3. What is a rem?
4. What is a nuclear chain reaction?
5. How will nuclear fuel supplies, the cost of decommissioning facilities, and storage and ultimate disposal of nuclear wastes influence the nuclear power industry?
6. What happened at Chernobyl, and why did it happen?
7. Describe a boiling-water reactor, and explain how it works.
8. How is plutonium-239 produced in a breeder reactor?
9. Why is plutonium-239 considered dangerous?
10. Why is fusion not currently being used as a source of energy?
11. What are the major environmental problems associated with the use of nuclear power?
12. List three environmental problems associated with the construction and subsequent de-emphasis of nuclear weapons.

Critical Thinking Questions

1. Recent concerns about global warming have begun to revive the nuclear industry in the United States. Do you think nuclear power should be used instead of coal for generating electricity? Why?
2. Why has nuclear power not become the power source that is "too cheap to meter," as foreseen by President Eisenhower?
3. Should there be international regulations concerning nuclear power plants, their operation, and decommissioning? Who should be responsible for developing these regulations? What kinds of regulations, if any, would you like to see?
4. Disposal of radioactive wastes is a big problem for the nuclear energy industry. What are some of the things that need to be evaluated when considering nuclear waste disposal? What criteria would you use to judge whether a storage proposal is adequate or not?
5. Nuclear weapons testing has released nuclear radiation into the environment. These tests have always been justified as necessary for national security. Do you agree or not? What are the risks? What are the benefits?
6. If your electric utility were to build a nuclear power reactor in your area, what kind would you advise them to build? Why?

Key Terms

alpha radiation *165*	nuclear fission *165*
beta radiation *165*	nuclear fusion *170*
boiling-water reactor (BWR) *167*	nuclear reactor *166*
decommissioning *176*	plutonium-239 (Pu-239) *168*
fissionable *166*	pressurized-water reactor
gamma radiation *165*	(PWR) *167*
gas-cooled reactor (GCR) *168*	radiation *165*
heavy-water reactor (HWR) *168*	radioactive *165*
light-water reactor (LWR) *167*	radioactive half-life *165*
liquid metal fast-breeder reactor	rem *175*
(LMFBR) *168*	thermal pollution *176*
moderator *167*	transuranic waste *177*
nuclear breeder reactor *168*	uranium-235 (U-235) *166*
nuclear chain reaction *166*	

Related Internet Sites

Visit our website at http://www.mhhe.com/environmentalscience/ for specific information on each of the topics below.

Technology to clean up industrial and nuclear waste

Department of Energy—Office of High Energy and Nuclear Physics

Nuclear Energy Institute—electricity for cleaner air

Civil nuclear industry and related information

Potential locations for storing radioactive wastes and spent nuclear fuel

Fusion energy

Links to other nuclear pages

Education about fusion energy for laypeople

Advantages of fusion over other energy sources

Canadian Coalition for Nuclear Responsibility

Links to other nuclear energy sites

PART FOUR

Human Influences on Ecosystems

The success of the human species is a result of our ability to change our surroundings so that we have tools, food, water, and shelter. Our use of natural resources is important to many aspects of modern society. **Chapter 11** discusses how humans exploit mineral resources and use natural ecosystems. Often, this means that existing natural ecosystems are altered significantly by extracting resources, or natural ecosystems are replaced by agricultural ecosystems. In some cases, organisms are driven to extinction by human activities.

Chapter 12 emphasizes the use of the land surface, while **chapter 13** discusses the nature and wise use of soil. Land and soil are not the same. Land is the part of the world that is not covered by water, while soil is a thin layer on the land's surface that supports plant growth. Land-use decisions must take into account the nature of the soil.

Agriculture allows us to support about 6 billion people on the earth. The use of the soil, discussed in chapter 13, and the use of pesticides, discussed in **chapter 14,** are both important to food production. Misuse of the soil or pesticides can lead to the loss of usable farmland or to the poisoning of people.

Water is necessary for human life. It is important as a nutrient, but also for agriculture, industrial processes, transportation, and many other functions. **Chapter 15** deals with a wide range of water-use issues and conservation practices.

chapter 11

Human Impact on Resources and Ecosystems

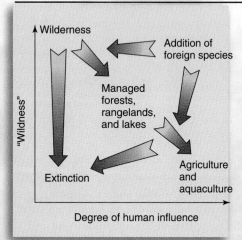

Managed forests, rangelands, and lakes

Wilderness

"Wildness"

Addition of foreign species

Extinction

Agriculture and aquaculture

Degree of human influence

OBJECTIVES

After reading this chapter, you should be able to:

- Recognize that humans have an increasing impact on natural ecosystems.
- Define pollution.
- Differentiate between renewable and nonrenewable resources.
- Recognize that mineral resources are unevenly distributed, which creates international trade in these commodities.
- List three types of costs associated with mineral exploitation.
- Understand that some wilderness areas still have minimal human influence.
- Appreciate the ways humans modify forests.
- Identify causes of desertification.
- Recognize that aquatic systems are modified by terrestrial changes.
- Identify changes that occur to aquatic systems as a result of human activity.
- Recognize that wildlife management focuses on specific species.
- Appreciate that waterfowl management is an international problem.
- Recognize that extinction is a natural process.
- Recognize that humans increase the rate of extinction.
- Identify ways that humans cause extinctions.
- Recognize that many extinctions can be prevented if societies are willing to preserve crucial habitats and prevent the hunting of endangered species.

CHAPTER OUTLINE

The Changing Role of Human Impact

Historical Basis of Pollution

Renewable and Nonrenewable Resources

Costs Associated with Resource Exploitation

Mineral Resources
  Steps in Mineral Exploitation
  Recycling of Mineral Materials

Exploitation and Modification of Terrestrial Ecosystems
  Agricultural Ecosystems
  Forest Resources
  Managing Forest Ecosystems

Environmental Close-Up: *The Northern Spotted Owl*
  Management of Rangeland Ecosystems

Areas with Minimal Human Impact— Wilderness and Remote Areas

Managing Aquatic Ecosystems

Global Perspective: *The History of the Bison*
  Managing Marine Ecosystems
  Managing Freshwater Ecosystems

Environmental Close-Up: *Farming, Fish Kills, and* Pfiesteria piscicida

Managing Ecosystems for Wildlife

Environmental Close-Up: *Native American Fishing Rights*

Natural Selection and Extinction

Human-Accelerated Extinction

Why Worry About Extinction?

What Is Being Done to Prevent Extinction?

Environmental Close-Up: *The California Condor*

Issues and Analysis: *Costa Rican Forests Yield Tourists and Medicines*

THE CHANGING ROLE OF HUMAN IMPACT

At one time, a human was just another consumer somewhere in the food chain. Humans fell prey to predators and died as a result of disease and accident just like other animals. The tools they used were primitive, so these people did not have a long-term effect on their surroundings. They only minimally exploited mineral and energy resources. They mined chert, obsidian, and raw copper for the manufacture of tools, and used certain other mineral materials, such as salt, clay, and ocher, as nutrients, for making pots, or as pigments. Aside from occasional use of surface seams of coal or surface oil seeps, early humans met most of their energy needs by muscle power or the burning of biomass.

As human populations grew, and as their tools and methods of using them became more advanced, the impact that a single human could have on his or her surroundings increased tremendously. The purposeful use of fire was one of the first events that marked the capability of humans to change ecosystems. Although in many ecosystems fires were natural events, the use of fire by humans to capture game and to clear land for gardens could destroy climax communities and return them to earlier successional stages more frequently than normal.

As technology advanced, wood was needed for fuel and building materials, land was cleared for farming, streams were dammed to provide water power, and various mineral resources were exploited to provide energy and build machines. These modifications allowed larger human populations to survive, but always at the expense of previously existing ecosystems.

Today, with about 6 billion people on the earth, nearly all of the surface of the earth has been affected in some way by human activity. Even the polar ice caps show the effects of human activity. Various kinds of organic pollutants and lead residues from the burning of leaded gasoline can be identified in the layers of ice that build up from the continuous accumulation of snow in these areas. (The amount of lead is currently decreasing because some countries have been making the transition from burning leaded fuel to unleaded fuel in automobiles.)

HISTORICAL BASIS OF POLLUTION

Pollution is usually defined as something that people produce in large enough quantities that it interferes with our health or well-being. Two primary factors that affect the amount of damage done by pollution are the size of the population and the development of technology that "invents" new forms of pollution.

When the human population was small and people lived in a simple manner, the wastes produced were biological and so dilute that they usually did not constitute a pollution problem. People used what was naturally available and did not manufacture many products. Humans, like any other animal, fit into their natural ecosystems. Their waste products were **biodegradable** materials. Biodegradable materials are a source of food for decomposers, the organisms that break the material down into simpler chemicals, such as water and carbon dioxide.

Pollution began when human populations became so concentrated that their waste materials could not be broken down as fast as they were produced. As the population increased, people began to congregate and establish cities. The release of large amounts of smoke and other forms of waste into the air caused an unhealthy condition because the pollutants were released faster than they could be absorbed and dispersed by the atmosphere.

Throughout history, humans have made numerous attempts to eliminate the misery caused by hunger and disease. In general, we rely on science and technology to improve our quality of life. However, technological progress often offers short-term solutions that in the process of solving one problem can create new forms of pollution.

The development of the steam engine allowed machines to replace human labor but increased the amount of smoke and other pollutants in the air and increased the need for fuel. The modern chemical industry has produced many extremely valuable synthetic materials (plastics, pesticides, medicines) but has also produced toxic pollutants.

Even identifying pollution is not always easy. To some the smell of a little wood smoke in the air is pleasant; others do not like the odor. A business may consider advertising signs valuable and necessary; others consider them to be visual pollution. (See figure 11.1.) Even the presence of chemicals in drinking water can be difficult to classify as a clear-cut example of pollution. Toxic heavy metals such as arsenic certainly should be considered hazardous, but we do not know how much arsenic contamination is allowable before harm occurs. Are the small quantities of arsenic in groundwater great enough that, over a period of time, they will accumulate and cause damage to individuals? Is the arsenic a normal part of the groundwater, or is it the result of past heavy spraying of arsenic-containing pesticides on apple trees?

Certainly, if the arsenic is the result of human activity and is causing health problems, it is a pollutant. If it is a natural part of the groundwater, it may still be a health hazard but is technically not a pollutant.

RENEWABLE AND NONRENEWABLE RESOURCES

Modern technologies have allowed us to exploit natural resources to a much greater extent than our ancestors were capable of. **Natural resources** are structures and processes that humans can use for their own purposes but cannot create. If the supply of a resource is very large and the demand for it is low, the resource may be thought of as free. Sunlight, oceans, and air are often not even thought of as natural resources because their supply is so large. If a resource has always been rare or it has been severely reduced by consumption, it is expensive. Pearls and precious metals are expensive because they have always been rare. In the past, land and its covering of soil was considered a limitless natural resource, but as the population grew and the demand for food, lodging, and transportation increased, we began to realize

Fish kill—an indication of water contamination

Smoke from stack—contains particulate material, which could cause lung problems

Feed lot—odor pollution as well as a source of water contamination from surface runoff

Strip—visual pollution, which is an annoyance but not a health hazard

Smog—an indication that thermal inversion has kept the air contamination in the valley

Traffic—fumes (HC, NO_x, PAN, etc.) from internal combustion engines cause eye irritation

Nuclear power plant cooling towers—possible thermal pollution and radiation hazards

Litter—an indication that we need to become more aware of how we dispose of materials

Health hazard

Annoyance

FIGURE 11.1 **Forms of Pollution** Pollution is produced in many forms. Some are major health concerns, whereas others merely annoy.

that land is a finite, nonrenewable resource. The economic value of land is highest in metropolitan areas where open land is unavailable. Unplanned, unwise or inappropriate use can result in severe damage to the land and its soil. (See figure 11.2.) The landscape is also a natural resource, as we see in countries with a combination of mountainous terrain and high rainfall that can be used to generate hydroelectric power. Rivers, forests, scenery, climates, and wildlife populations are additional examples of natural resources.

Natural resources are usually categorized as either renewable or nonrenewable. **Renewable resources** can be formed or regenerated by natural processes. Soil, vegetation, animal life, air, and water are renewable primarily because they naturally undergo repair and a cleansing process. However, just because a resource is renewable does not mean that it is inexhaustible. Table 11.1 shows how an increase in population will result in a decrease in the amount of resource available to each person in the world even if production

increases. Overuse of renewable resources can result in their irreversible degradation. **Nonrenewable resources** are not replaced by natural processes, or the rate of replacement is so slow as to be ineffective. Therefore, when nonrenewable resources are used up, they are gone, and a substitute must be found or we must do without.

COSTS ASSOCIATED WITH RESOURCE EXPLOITATION

Costs are always associated with the exploitation of any natural resource. These costs fall into three categories. First, the **economic costs** are those monetary costs necessary to exploit the resource. Money is needed to lease or buy land, build equipment, pay for labor, and buy energy to run the equipment. A second category is the **energy cost** of exploiting the resource. It takes energy to extract, concentrate, and transport mineral materials to manufacturing sites. Since energy costs money, energy costs are ultimately converted to economic costs. When energy is inexpensive, inefficient processes may be profitable; however, when the cost of energy rises, energy-intensive processes will be eliminated.

A third way to look at costs is in terms of environmental effects. Air pollution, water pollution, animal extinction, and loss of scenic quality are all **environmental costs** of resource exploitation. Environmental costs are often deferred costs. They may not even be recognized as costs at first but become important after several years. Environmental costs may also be lost opportunities or lost values because the resource could not be used for another purpose. Lately, environmental costs are being converted to economic costs as more strict controls on pollution are enforced. It takes money to clean up polluted water and air, or to reclaim land that has been degraded by extracting or using resources.

These three categories of costs (economic, energy, and environmental) apply to the several steps involved in the extraction and use of minerals. These steps are exploration, mining, refining, transportation, and manufacturing.

FIGURE 11.2 Mismanagement of a Renewable Resource Although soil is a renewable resource, extensive use can permanently damage it. Many of the world's deserts were formed or extended by unwise use of farmland. This photograph shows a once-productive farm, now abandoned to the wind and sand because the soil was mistreated and allowed to erode.

| TABLE 11.1 | POPULATION SIZE AND AVAILABILITY OF RENEWABLE RESOURCES, CIRCA 1990, WITH PROJECTIONS FOR 2010 |

	Circa 1990 (million)	2010 (million)	Total change (percent)	Per capita change (percent)
Population	5290	7030	+33	
Fish catch (tons)[1]	85	102	+20	−10
Irrigated land (hectares)	237	277	+17	−12
Cropland (hectares)	1444	1516	+5	−21
Rangeland and pasture (hectares)	3402	3540	+4	−22
Forests (hectares)[2]	3413	3165	−7	−30

[1]Wild catch from fresh and marine waters, excludes aquaculture.

[2]Includes plantations; excludes woodlands and shrublands.

Sources: Population figures from U.S. Bureau of the Census, Department of Commerce, *International Data Base*, unpublished printout, November 2, 1993; 1990 irrigated land, cropland, and rangeland from U.N. Food and Agriculture Organization (FAO), *Production Yearbook 1991*; fish catch from M. Perotti, chief, Statistics Branch, Fisheries Department, FAO, private communication, November 3, 1993; forests from FAO, *Forest Resources Assessment 1990*, *1992 and 1993*. For detailed methodology, see *State of the World 1994*, among other sources.

MINERAL RESOURCES

Mineral resources are one of the major kinds of nonrenewable resources. The distribution of these resources is not uniform throughout the world. The geological and biological processes that formed mineral resources in certain places on the earth occurred many millions of years ago. Since that time, landmasses have been divided into political entities. Some have a greater wealth of mineral resources than others. Because no country has within its territorial limits all of the mineral resources it needs, an international exchange has developed. In particular, the industrially developed countries import many of the minerals they need from countries that have the resources but no economic ability to develop them. Table 11.2 shows several minerals and the countries that supply them. North America lacks a large number of materials and must import them from other countries, often in less-developed areas of the world.

Not only are mineral resources unevenly distributed, but those that are easiest to use and the least costly to extract have been exploited. As we continue to use mineral resources, they will be harder to find and more costly to develop. As with energy, North America is one of the primary consumers of the world's mineral resources. Reasonable estimates are that North America consumes over 30 percent of the minerals produced in the world each year, which is a disproportionate share, given that the combined population of the United States and Canada is about 5.1 percent of the world's population.

Steps in Mineral Exploitation

The costs associated with locating new sources of minerals (exploration) are primarily economic because exploration takes time and new technology. There are also some energy costs and some very small environmental costs. As the better sources of mineral resources are used up, it will be necessary to look for minerals in areas that are more difficult to explore, such as under the oceans. Therefore, both economic and energy costs will increase. Some areas, such as national parks and preserves, have been off-limits to mineral exploration. As current reserves of mineral resources are used up, pressures will build to explore in these protected areas, resulting in increased environmental costs.

Once a mineral resource has been located and the decision made to exploit it, the resource must be taken from the earth. Mining involves large expenditures of money to pay for labor and the construction of machines and equipment. Energy must be purchased for the operation of the equipment.

In addition to the economic and energy costs there are significant environmental costs. Mining affects the environment in several ways. All mining operations involve the separation of the valuable mineral from the surrounding rock. The surrounding rock must then be disposed of in some way. These pieces of rock are usually piled on the surface of the earth, where they are known as mine tailings and present an eyesore. It is also difficult to get vegetation to grow on these deposits. Some mine tailings contain materials (such as asbestos, arsenic, lead, and radioactive materials) that can be harmful to humans and other living things.

Many types of mining operations require vast quantities of water for the extraction process. The quality of this water is degraded, so it is unsuitable for drinking, irrigation, or recreation. Since mining disturbs the natural vegetation in an area, water may carry soil particles into streams and cause erosion and siltation. Some mining operations, such as strip mining, rearrange the top layers of the soil, which lessens or eliminates its productivity for a long time. (See figure 11.3.) Strip mining has disturbed approximately 75,000 square kilometers of U.S. land, an area equivalent to the state of Maine.

After ore is mined, it must be processed to remove the desired materials from the rocks in which they are embedded. Table 11.3 shows the percentage of desired material present in several ores. Ore processing poses economic costs: physical facilities and equipment must be built, people must be employed, and fuel must be purchased for the operation. The energy costs are also significant because refining is basically the process of concentrating the desired material. Natural physical processes tend to disperse materials; therefore, energy must be expended to concentrate them. In many cases, this energy expenditure is the major cost associated with obtaining the mineral. As energy costs increase and ores become more diluted, the economics of producing some materials may change so drastically that seeking substitute materials that require less energy expenditure is the best alternative.

Extracting materials from ores also involves environmental costs in the form of air and water pollution. These environmental costs are being converted to economic costs as regulations on industry require less environmental damage than had been previously tolerated.

Transportation is another component of the overall cost of extracting minerals from the earth. Transportation is involved

TABLE 11.2 DISTRIBUTION OF SOME MINERAL RESOURCES

Mineral	Percent consumed in U.S. that was imported	Major supplier countries (1990–93)
Columbium	100	Brazil, Canada
Mica (sheet)	100	India, Belgium
Strontium	100	Mexico
Manganese	100	South Africa, France, Brazil
Bauxite	99	Jamaica, Guinea, Australia
Platinum group	91	South Africa, Russia, United Kingdom
Tantalum	86	Germany, Australia, Canada
Tin	84	Brazil, Bolivia, China
Cobalt	79	Zaire, Zambia, Canada
Chromium	75	South Africa, Zimbabwe, Turkey
Potash	74	Canada
Nickel	66	Canada, Australia, Norway
Cadmium	50	Canada, Australia, Mexico
Zinc	41	Canada, Mexico, Peru

Source: Data from *Statistical Abstract of the United States, 1997.*

FIGURE 11.3 A Strip-Mining Operation It is easy to see the important impact a mine of this type has on the local environment. Unfortunately, many mining operations are located in areas that are also known for their scenic beauty.

TABLE 11.3 METALS IN ORES

Metal	Percent metal needed in the ore for profitable extraction	Price range per kilogram
Iron	30.0	
Chromium	30.0	Less than a dollar
Aluminum	20.0	
Nickel	1.5	
Tin	1.0	Several dollars
Copper	0.5	
Uranium	0.1	Tens of dollars

TABLE 11.4 COSTS ASSOCIATED WITH MINERAL EXPLOITATION

Steps in exploitation	Economic costs	Energy costs	Environmental costs
Exploration	High because personnel and technology are expensive	Small	Small
Mining	High because of construction of plants and the labor to run them	High because of the need to move large amounts of material to separate valuable ores from other material	High because of changes in the landscape and large amounts of wastes produced. Many mining operations use large amounts of water.
Refining	High because of the labor, equipment, and energy needed to concentrate ores	High because of the need to concentrate ores	Air and water pollution from refining facilities can be a problem.
Transportation	Needed at all stages of the process to move products from the extraction site to the manufacturing site	Energy is involved in all aspects of transportation.	Air pollution can be a problem. The construction of new transportation routes may influence scenic and other ecosystem values.
Manufacturing	Manufacturing plants and labor are needed.	Large amounts of energy are used in all manufacturing processes.	Air pollution and water pollution can be a problem, but most manufacturing sites are strictly regulated.

in the actual mining process, in getting the ore to the refinery, and in moving the concentrated mineral to the site where it will be made into a finished product. Transportation costs are primarily in the form of money and energy.

Finally, there is the cost of making the finished product from the concentrated ore. Table 11.4 reviews these steps and the kinds of costs associated with each step.

Recycling of Mineral Materials

To fully appreciate mineral resource use we should consider the concept of recycling. Many minerals extracted from the earth are not actually consumed or used up; they are just temporarily within a structure or a process. When the material is reclaimed and used again for another structure or process, it is recycled. (See chapter 18.) In order for recycling to be economic and effective, several conditions must be met. The material must be able to be collected relatively easily and the material must have a high economic value.

Empty aluminum beverage cans and waste oil are no longer useful to the consumer. However, the aluminum atoms can be reprocessed into new cans or other aluminum products and the oil can be reprocessed or burned as fuel. For recycling to be economically practical, the obsolete items must be recaptured before they have dispersed into the environment. For example, waste oil is relatively easy to recapture at the gasoline station where engine oil is replaced, but difficult to recapture if the oil is dumped on the ground or into a sewer.

In many industries, however, the value of the used material is very low and new materials can be produced for less money than it would cost to recycle the material. For example, disassembling a building and reusing its parts is more costly than using new construction materials.

One reason recycling is not more common is that, historically, the monetary costs for energy have been extremely low. As the cost of energy rises, more attention will be given to recycling minerals rather than mining new ones. Manufacturing products in an environmentally harmonious way will become economically advantageous in the future.

EXPLOITATION AND MODIFICATION OF TERRESTRIAL ECOSYSTEMS

Agricultural Ecosystems

Throughout the world forests and grasslands have been converted to agricultural land and little of the original vegetation remains, since nearly all agricultural practices involve the removal of the original vegetation and substitution of exotic domesticated crops and animals for the original inhabitants. About 40 percent of the world's land surface has been converted to cropland and permanent pasture. Typically, the most productive natural ecosystems (forests and grasslands) are the first to be modified by human use and the most intensely managed. As the human population grows, it needs more space to grow food. Historically, the pressures to modify the environment were greatest in areas that had dense populations. Many of these early centers of agricultural development have been altered so that the land is no longer capable of growing the food they need. Much of the Mediterranean and Middle East once supported extensive forests and agriculture but now consist of dry scrubland. Today agricultural land is being pushed to feed more people and its wise use is essential to the health and welfare of the people of the world. Chapters 13 and 14 will take a close look at patterns of use of agricultural land.

Forest Resources

Most forest ecosystems have been extensively modified by human activity. Originally, almost half of the United States, three-fourths of Canada, almost all of Europe, and significant portions of the rest of the world were forested. The forests were removed for fuel, building materials, to clear land for farming, and just because they were in the way. This activity returned the forests to an earlier successional stage, which resulted in the loss of certain animal and plant species that required mature forests as part of their niche.

Continued modification of forests is having negative effects on plants and animals that require mature forests as their habitat. Today in Australia, a small squirrel-sized marsupial known as a numbat is totally dependent on old forests that have termite- and ant-infested trees. The insects are numbats' major source of food, and the hollow trees and limbs caused by the insects' activities provide numbats with places to hide. Clearly, tree harvesting will have a negative impact on this species, which is already in danger of extinction. (See figure 11.4.) In addition to serving as habitats for many species of plants and animals, forests provide many other services. Forested areas provide firewood, lumber, climate modification, and are essential as protectors of watersheds. If the trees are removed, the climate often becomes less humid and since there is less vegetation to hold the water, flooding is more common, and much valuable water is lost as runoff.

In general, all continents contain significant amounts of forested land. Table 11.5 shows that temperate parts of the world (particularly North America and Europe which have large forested areas) are maintaining their forests while tropical forests are being lost. In a ten-year period nearly 4 percent of the tropical forests were lost. Tropical forests have a diverse mixture of tree species that require different kinds of harvesting techniques than have traditionally been used in northern temperate forests. Also, because of soil characteristics, tropical forests are not as likely to regenerate after logging as are many of the temperate forests. If they do not regenerate, they must be considered nonrenewable resources. If these forests are to be used, a new set of forestry principles will be needed to establish a renewable tropical forest industry. Currently, few of these forests are being managed for long-term productivity; they are being harvested on a short-term economic basis only, as if they were nonrenewable resources. Tropical forests in Africa and Latin America are being deforested at a rate of 0.6 percent per year and those in Asia are being deforested at about 1 percent per year.

The deforestation of large tracts of land significantly impacts climate. Forested land very effectively traps rainfall and presents its rapid runoff. Furthermore,

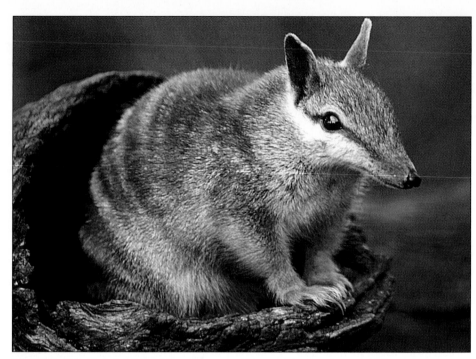

FIGURE 11.4   A Specialized Marsupial—the Numbat This small marsupial mammal requires termite- and ant-infested trees for its survival. The insects serve as food and the hollow limbs and logs provide hiding places. Loss of old-growth forests with diseased trees will lead to the numbat's extinction.

TABLE 11.5   CHANGES IN THE EXTENT OF FORESTED OR WOODED LAND 1981–1990

Region	Million hectares of forested or wooded land (1990)		Percent change 1981–1990	
All tropical countries	2728		−3.8	
Tropical Africa		1084		−2.2
Tropical Asia and Oceana		453		−4.3
Tropical Latin America and Caribbean		1191		−4.5
All temperate countries	2392		+0.1	
Temperate Africa		53		−7.2
Temperate Asia and Oceana		207		+5.3
Temperate Latin America		68		−5.3
Temperate North America and Europe		2064		0.0
World	5120		−1.9	

Source: Food and Agriculture Organization of the United Nations.

the large amount of water transpired from the leaves of trees tends to increase the humidity of the air in forested areas. The shade provided by trees and the cooling provided by the evaporation of water also tend to moderate temperature extremes. Destruction of huge areas of forest can result in regional climate change, generally resulting in a hotter, more arid climate. More recently, people have become concerned about preserving the potential of forests to trap carbon dioxide. Trees trap large amounts of carbon dioxide as a result of photosynthesis and may help to prevent increased carbon dioxide levels that contribute to global warming. (See chapter 17 for a discussion of global warming.)

Another complicating factor is that human population pressures are the greatest in tropical regions of the world. More people need more food, which means that forestland will be converted to agriculture and the value of forest for timber, fuel, watershed protection, wildlife habitat, and carbon dioxide storage will be lost.

Managing Forest Ecosystems

Whenever a resource is exploited two major interests come into conflict: the economy and the environment. Economic factors are easy to measure. The cost of exploitation and the financial return are the primary issues. The environmental value of a resource is often difficult to put into monetary terms and environmentalists must rely on ethical or biological arguments to make their points. Modern forest management practices in many parts of the world involve a compromise that allows economic exploitation while maintaining some of the environmental value of the forest.

The forests of the world are known quantities. The economic worth of the standing timber can be assessed, and the importance of forests for wildlife and watershed protection can be given a value. However, the very act of harvesting trees must alter the original "wilderness" character of the forest. Logging removes the trees and, therefore, the habitat for many kinds of animals that require mature stands of timber. The pine martin, grizzly bear, and cougar all require forested habitat that is relatively untouched by human activity. An area that has recently been logged is not very scenic, and the roads and other changes often irreversibly alter the area's wilderness character. Obviously, wilderness and logging cannot coexist. Therefore, it becomes necessary to designate specific forests as wilderness or harvestable resources.

Several concerns must be addressed when it is decided that a forested area will be logged. Removing trees exposes the soil to increased erosion. Because the soil's water-holding ability is related to the amount of organic material and roots it has, denuded land allows water to run off rather than sink into it. Soil particles can wash into streams, where they cause siltation. The loss of soil particles reduces the soil's fertility. The particles that enter streams may cover spawning sites and eliminate fish populations. If the trees along the stream are removed, the water will be warmed by the

increased sunlight. This may also have a negative effect on the fish population. The process of cutting the trees and transporting the log also disturbs the wildlife in the area.

The roads necessary for moving equipment and removing logs are another problem. Constant travel over these roads removes vegetation and exposes the bare soil to more rapid erosion. When the roads are not properly located and constructed, they eventually become gullies and serve as channels for the flow of water. In many parts of the world the construction of logging roads increases access to the forest and results in colonization by peasant "squatters" who seek to clear the forest for agriculture. Most of these environmental problems can be minimized by using properly engineered roads and appropriate harvesting methods.

One of the most controversial logging practices is **clear-cutting.** (See figure 11.5.) As the name implies, all of the trees in a large area are removed. This is a very economical method of harvesting but it exposes the soil to significant erosive forces. If large blocks of land are cut at one time it may slow reestablishment of forest and have significant effects on wildlife. On some sites, clear-cutting is a reasonable method of harvesting trees. Sites with gentle slopes, where regrowth is rapid, experience little erosion and environmental damage is limited. This is especially true if a border of undisturbed forest is left along the banks of any streams in the area. The roots of the trees help to retain the stream banks and retard siltation. The shade provided by the trees also helps to prevent warming of the water, which might be detrimental to some fish species.

Clear-cutting can be very destructive on sites with steep slopes or where regrowth is slow. Under these circumstances, it may be possible to use **patchwork clear-cutting.** With this method, smaller areas are clear-cut among patches of untouched forest. This reduces many of the problems associated with clear-cutting and can also improve conditions for species of game animals that flourish in successional forests but not in mature forests. For example, deer, grouse, and rabbits benefit from a mixture of mature forest and early-stage successional forest.

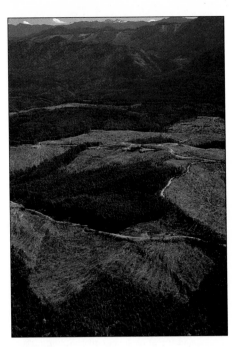

FIGURE 11.5 A close clustering of large, clear-cut plots can lead to a loss of species and accelerated soil erosion.

Clear-cut sites where natural reseeding or regrowth is slow may need to be replanted with trees, a process called **reforestation.** Reforestation is especially important for many of the conifer species, which often require bare soil to become established. Many of the deciduous trees will resprout from stumps or grow quickly from the seeds that litter the forest floor, so reforestation is not as important in deciduous forests.

Selective harvesting of some species of trees is also possible but is not as efficient or as economical as other methods, from the point of view of the harvesters. It allows them, however, to take individual, mature, high-value trees without causing much change to the forest ecosystem. In many tropical forests, high-value trees, like mahogany, are often harvested selectively. However, there may still be extensive damage to the forest by the construction of roads and the damage to noncommercial trees by the felling of the selected species.

Proper planning and effort can integrate several uses of the forest. The forest products industry would prefer even-aged stands of a single species of tree for the most efficient harvest. Wildlife managers would prefer to see a variety of tree

ENVIRONMENTAL
CLOSE-UP

The Northern Spotted Owl

In June 1990, the northern spotted owl was listed as a threatened species. Threatened species are those that are likely to become endangered if current conditions do not change. The northern spotted owl lives in old-growth coniferous forests in the Pacific Northwest. As with most carnivores at the top of the food chain, the northern spotted owl requires large areas of forest for hunting. Listing the owl as threatened requires that the U.S. Forest Service develop and implement a plan to protect it. Since the owl is found only in mature old-growth forests, and since it requires large areas for hunting, the plan must set aside large tracts of forested land and protect them from logging.

Logging in the Pacific Northwest has already used up most of the trees on private land. Most of the remaining old-growth forest, which is about 10 percent of its original area, is on land managed by the U.S. Forest Service. Many of the trees in these forests are several centuries old, and forest replacement following logging takes an extremely long time. This timescale would not allow for the continued existence of the

northern spotted owl if major sections of the remaining old-growth forest were logged.

The timber industry is an important element in the economy of the Pacific Northwest. The U.S. Forest Service had been selling about 300 square kilometers of forested land per year. If the old-growth forest is not available for harvest, forestry workers will lose their jobs. Those who make their living from the forest argue that there are already adequate wilderness areas and areas unfit for logging to supply the northern spotted owl with all the space it needs. They also argue that their livelihood is endangered in a way similar to the owl, and ask who is more important.

---

species at different stages of maturity to encourage many species of wildlife. A single-species, even-aged forest does not support as wide a variety of wildlife as does a mixed-age forest. However, with proper planning trees can be harvested in small enough patches to encourage wildlife but in large enough sections to be economical. Some older trees, or some patches of old-stand timber, can be left as refuges for species that require them as a part of their niche. For example, in the southern pine forests of the United States, the red-cockaded woodpecker requires old, diseased pine trees in which to build its nests. (See figure 11.6.) A well-managed forest plantation does not provide these sites but recognition of the habitat requirements of this species and a willingness to retain some old, diseased trees can allow harvesting of the timber while protecting the habitat of the woodpecker.

Many lumber companies maintain forest plantations as crops and manage them in the same way farmers manage crops. They plant single species, even-aged forests of fast-growing hybrid trees that have been developed in the same way as high-yielding agricultural crops. Competing species are controlled by the use

of fire in some forests, and insects are controlled by aerial spraying. In these intensively managed forests, some single-species plantations mature to harvestable size in 20 years, rather than in the approximately 100 years typical for naturally reproducing mixed forests. However, the quality of the lumber products is reduced. Such forests have low species diversity and are not as valuable for wildlife and other uses as are more natural mixed-species forests. Furthermore, the trees planted in many managed forests may be exotic species. *Eucalyptus* trees from Australia have been planted in South America, Africa, and other parts of the world and most of the forests in northern England and Scotland have a mixture of native pines and imported species from the European mainland and North America.

Management of Rangeland Ecosystems

Rangelands consist of the many arid and semiarid lands of the world that support grasses or a mixture of grasses and drought-resistant shrubs. These lands are too dry to support crops but are often used to raise low-density populations of domesticated or semidomesticated

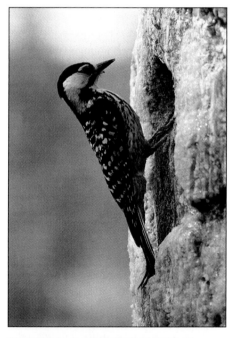

FIGURE 11.6 Red-Cockaded Woodpecker The red-cockaded woodpecker requires old, living pines that have a disease known as red heart in which to build its nest. Diseased, old trees are rare in intensely managed forests. Therefore, the birds are endangered.

animals in permanent open ranges or nomadic herds. Usually the animals are introduced species not native to the region.

◼ Nomadic herding   ▢ Stock raising on ranges

FIGURE 11.7 Use of Rangelands The arid and semiarid regions of the world will not support farming without irrigation. In many of these areas livestock can be raised. Permanent ranges occur where rainfall is low but regular. Nomadic herders can utilize areas that have irregular, sparse rainfall.

Consequently these areas have been modified by the animals' selective eating habits, which tend to reduce certain species of plants and encourage others. Sheep, cattle, and goats that are native to Europe and Asia have been introduced into the Americas, Australia, New Zealand, and many areas of Africa. The introduction of exotic species has reduced the numbers of certain native species. For example, sheep and cattle have replaced the bison in North America, and many native antelope species in Africa have been negatively affected by cattle. In an effort to increase the productivity of rangelands, management techniques may specifically eliminate certain species of plants not useful as food for the grazing animals, or specific grasses may be planted that are not native to the area. Because rainfall is low and often unpredictable it is important to regulate the number of livestock on the range. In areas where the animals are on permanent pastures the numbers of animals can be adjusted to meet the capacity of the range to provide forage. In many parts of the world nomadic herders simply move their animals from areas where forage is poor to areas that have better forage. This is often a seasonal activity that involves movement of animals to higher elevations in the summer or to areas where rain has fallen recently. (See figure 11.7.)

In many parts of the world, where human population pressures are great, overgrazing is a severe problem. As populations increase, desperate people attempt to graze too many animals on the land. They also cut down the trees for firewood. If overgrazed, many plants die, and the loss of plant cover subjects the soil to wind erosion, resulting in a loss of fertility, which further reduces the land's ability to support vegetation. The cutting of trees for firewood has a similar effect, but it is especially damaging because many of these trees are legumes, which are important in nitrogen fixation. Their removal further reduces soil fertility. This severe overuse of the land results in conversion of the land to a more desertlike ecosystem. This process of converting arid and semiarid land to desert because of improper use by humans is called **desertification.** Desertification can be found throughout the world but is particularly prevalent in northern Africa and parts of Asia, where rainfall is irregular and unpredictable, and where many people are subsistence

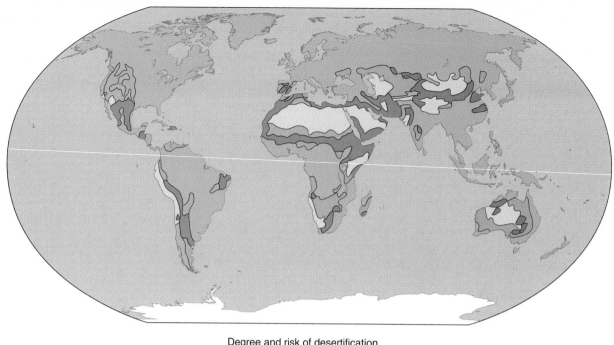

Degree and risk of desertification

High   Moderate   Existing desert

FIGURE 11.8 **Desertification** Arid and semiarid areas can be converted to deserts by overgrazing or unsuccessful farming practices. The loss of vegetation increases erosion by wind and water, increases the evaporation rate, and reduces the amount of water that infiltrates the soil. All of these conditions encourage the development of desertlike areas.

(Top) Sources: *United Nations Map of World Desertification,* United Nations Food and Agriculture Organization, United Nations Educational, Scientific, and Cultural Organization, and the World Meteorological Organization for the United Nations Conference on Desertification, 1977, Nairobi, Kenya.

farmers or nomadic herders who are under considerable pressure to provide food for their families. (See figure 11.8.)

AREAS WITH MINIMAL HUMAN IMPACT— WILDERNESS AND REMOTE AREAS

There are still many areas of the world that have had minimal human impact. Some of these are remote areas with harsh environmental conditions, such as the continent of Antarctica, northern arctic areas, the tops of some tall mountains, or extremely arid areas, such as desert areas of Africa, central Australia, and central Asia. Most of these have been explored and found to be too harsh to sustain agriculture and therefore dense human habitation is impossible.

Many other areas of the world, such as tropical rainforests, currently support ecosystems that have small human populations but are under threat because these ecosystems are capable of supporting agriculture or other human uses. The primary factor that will determine the survival of these natural areas is population pressure.

Many countries have established parks and other special designations of land use to protect areas of natural beauty or communities of organisms thought worthy of protection. This has been most noticeable in Africa, Central and South America, North America, and Australia. Until recently, these continents had large amounts of land that were relatively untouched. (See figure 11.9.) Parks and protected lands allow a variety of uses, depending on the country and its laws. Some are used primarily as tourist attractions, which generate funds for the country. Others are set up primarily to protect a species or community of organisms, while others simply seek to restrict use so that certain environmental standards are met. For example, harvesting in a forest might be restricted to a certain number of trees each year, and the number of people who visit a particular scenic site might be limited. A particularly sensitive issue is the designation of certain areas as wilderness. Although definitions of wilderness vary from country to country, the U.S. Congress, in the Wilderness Act of 1964, defined **wilderness** as "an area where the earth and its community of life are untrampled by man, where man himself is a visitor who does not remain."

Recently, there has been intense pressure on members of Congress from oil-producing states such as Alaska to allow oil exploration in areas currently designated as wilderness, such as the Arctic National Wildlife Refuge. As population increases and resources become more scarce, this pressure will become greater throughout the world.

The oceans of the world represent a major area that has been modified very little by human activities. Some areas, particularly those near the shore, receive intense use, but many areas of the open ocean are poorly understood and little affected by humans.

MANAGING AQUATIC ECOSYSTEMS

Aquatic ecosystems are divided into two categories: marine and freshwater. The primary renewable resources obtained

Growth of global network of protected areas, 1890-1995

FIGURE 11.9 **Growth in Major Park and Wilderness Areas of the World** Although there has been a steady increase in the number of protected areas and the number of hectares protected, many of the areas are in more-developed countries that can afford to set aside land for nonproductive uses.

Source: Data from International Union for Conservation of Nature and Natural Resources (IUCN), the United Nations List of National Parks and Protected Areas, 1985, Gland, Switzerland, and more recent 1995 data.

from these ecosystems are fish, shellfish, and crustaceans (lobsters, shrimp). Both marine and freshwater ecosystems present a distinct set of management issues. These are typically related to ownership issues, harvesting methods, and allowable harvest quotas. This section primarily will deal with the food production ability of aquatic ecosystems. Other issues related to water resources will be covered in chapter 15.

Managing Marine Ecosystems

The United Nations estimates that 70 percent of the world's marine fisheries are being overexploited or are in danger

of overexploitation as the number of fishers increases. In order to preserve the fishery total fishing capacity should be reduced by about 30 percent. The total landings of fish peaked at just over 85 million tonnes in 1989 and has been steady since then. However, the nature of the catch has changed as many of the more desirable species have declined. Figure 11.10a,b show the change in species composition for both bottom-dwelling species and surface-dwelling open-ocean fish. The commercial fishing industry has been attempting to market fish species that previously were regarded as unacceptable to the consumer. These activities are the result of reduced catches of desired species.

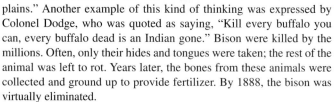

The History of the Bison

The original ecosystem in the central portions of North America was a prairie dominated by a few species of grasses. The eastern prairie, where moisture was greater, had grasses up to 2 meters tall, while the dryer western grasslands were populated with shorter grasses. Many kinds of animals lived in this area, including prairie dogs, grasshoppers, many kinds of birds, and bison. The bison was the dominant organism. Millions of these animals roamed the prairies of North America with few predators other than the Native Americans, who used the bison for food, their hides for shelter, and their horns for tools and ornaments. The relationship between the bison and Native Americans was a predator-prey relationship in which the humans did not significantly reduce bison numbers.

When Europeans came to North America, they changed this relationship drastically. European-born Americans sought to convert the prairie to agriculture and ranching. However, two things stood in their way: the Native Americans, who resented the intrusion of the "white man" into their territory, and the bison. Since many of the Native American tribes had horses and a history of warlike encounters with other tribes, they attempted to protect their land from this intrusion. The bison was a competitor in the eyes of the settlers, since it was impossible to use the land for agriculture or ranching with the millions of bison occupying so much area. The U.S. government established a policy of controlling the bison and the Native Americans: Since bison were the primary food source of Native Americans in many areas, eliminating them would result in the starvation of many Native Americans, which would eliminate them as a problem for the frontier settler.

In 1874, the Secretary of the Interior stated that "the civilization of the Indian was impossible while the buffalo remained on the plains." Another example of this kind of thinking was expressed by Colonel Dodge, who was quoted as saying, "Kill every buffalo you can, every buffalo dead is an Indian gone." Bison were killed by the millions. Often, only their hides and tongues were taken; the rest of the animal was left to rot. Years later, the bones from these animals were collected and ground up to provide fertilizer. By 1888, the bison was virtually eliminated.

A few bison were left in the Canadian wild and in remote mountain areas of the United States, while others survived in small captive herds. Eventually, in the early 1900s, the U.S. government established the national Bison Range near Missoula, Montana. The Canadian government established a bison reserve in Alberta. These animals have proven to be useful since many people now desire meat with a lower fat content, which bison have. Cross-breeding cattle with bison has led to a new breed of cattle with reduced fat content known as a beefalo. The animal that was barely saved from extinction by a few thoughtful individuals may contribute to better health for all.

Examples of "newly discovered" fish in this category are monkfish and orange roughy.

One of the major problems associated with the management of marine fisheries resources is the difficulty in achieving agreement on limits to the harvest. The oceans do not belong to any country and therefore every country feels they have the right to exploit the resource wherever they want. Therefore, international agreements are required to manage the resource. This is very difficult to achieve and each country seeks to exploit the resource for its advantage without regard for the sustainable use of the resource. (See Global Perspective: The Tragedy of the Commons, page 327.) Figure 11.11 shows the major countries involved in harvesting marine fishes and their approximate catch. Many other countries maintain small, low technology fishing operations that primarily supply fish to the local population.

For several reasons, the most productive areas of the ocean are those close to land. In shallow water, the entire depth (water column) is exposed to sunlight. Plants and algae can carry on photosynthesis, and biological productivity is high. The nutrients washed from the land also tend to make these waters more fertile. Furthermore, land masses modify currents that bring nutrients up from the ocean bottom. Many of the commercially important fish and other seafood species are bottom-dwellers, but fishing for them at great depths is not practical. Therefore, fishing pressure is concentrated on areas where the water is shallow and relatively nutrient-rich. Countries have established control over the waters near them by establishing a 200-nautical-mile limit within which they control the fisheries. This has not solved conflicts however as neighboring countries dispute the fishing practices of waters they both claim.

Canada and the United States have had continuing arguments over the management of the Pacific salmon fishery and the North Atlantic cod fishery until its collapse in the mid-1990s. Another problem associated with these shallow-water, near-shore fisheries is the method used to harvest the fish. The bottom-dwelling fish are generally harvested with trawls—nets that are dragged along the bottom. These nets capture many different species, many of which are not commercially valuable. Typically 25 percent of the catch consists of species that have no commercial value. These are discarded by being thrown overboard. However, they are usually dead and their removal further alters the ecological nature of the seafloor.

Open-ocean species of fish are also subject to overexploitation. During the 1960s anchovy fishing off the coast of Peru was a major industry. From 1971 to

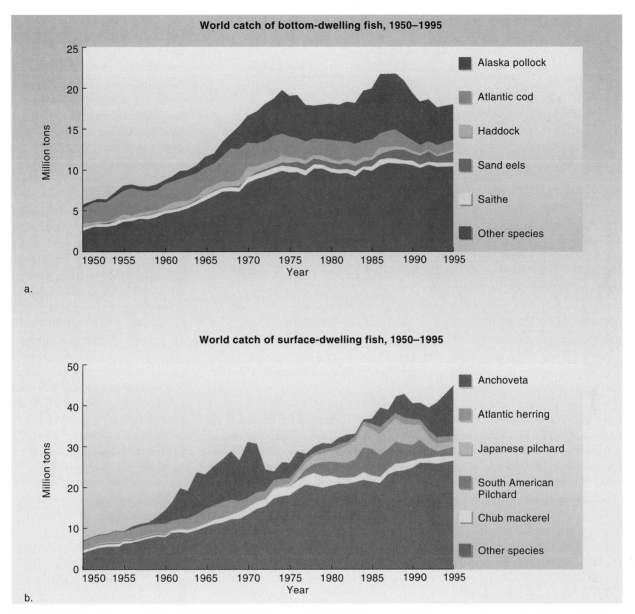

FIGURE 11.10   Changes in World Marine Fish Harvests 1950–1995 Increased fishing pressure has resulted in the significant reduction in many of the traditional fishing stocks. As traditional species have been overfished other, less desirable species have been replacing them. This is true for both the bottom-dwelling species (*a*) and those that live in the open ocean (*b*).

Source: Food and Agriculture Organization of the United Nations.

1972, the catch dropped dramatically. Overfishing was believed to be one of the major contributing factors. This was aggravated by an increase in the area's water temperature, which prevented nutrient-rich layers from rising to the euphotic zone. Thus, productivity at all trophic levels decreased. Currently, the Peruvian anchovy fishery has rebounded but is considered to be overexploited and not sustainable. (See figure 11.10*b*.)

Fish farming (aquaculture) is becoming increasingly important as a source of marine fish production. Salmon farming has been particularly successful. This involves raising the salmon in "pens" in the ocean which allows the introduction of food and other management techniques to achieve rapid growth of the fish. In 1995 British Columbia produced salmon worth $167 million from 80 farms. This compared to $75 million from wild-caught fish. Many other countries, including Norway, Scotland, and the United States, also farm salmon. However, even this method of producing fish is not without its

environmental effects, since large numbers of fish release waste into the water and this fertilizer can increase the amount of algae in the water.

Managing Freshwater Ecosystems

Because freshwater ecosystems are small and more intimately associated with human activities, few have not been considerably altered. Changes in water quality and the introduction of exotic species are

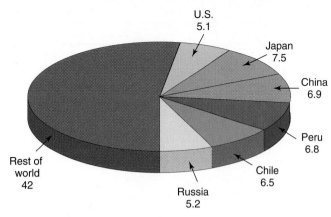

Million metric tons of marine fish harvested

U.S.
5.1

Japan
7.5

China
6.9

Peru
6.8

Chile
6.5

Russia
5.2

Rest of
world
42

FIGURE 11.11   Major Harvesters of Fish   Six countries account for about half of the approximately 80 million metric tons of marine fish harvested each year.

Source: Food and Agriculture Organization of the United Nations.

two primary alterations. The warming of water due to thermal pollution and the addition of nutrients have made it impossible for some native species to survive. The effects of water pollution are discussed in chapter 15.

Although many places in the world harvest freshwater fish for food, the quantities are relatively small compared to those produced by marine fisheries. Furthermore, the management of the resource is much more intense since the bodies of water are smaller and human populations have greater access to the resource. Fish management in much of the world includes managing for both recreational and commercial food purposes. The fisheries resource manager must try to satisfy two interest groups. Both sports fishers and those who harvest for commercial purposes must adhere to regulations. However, the regulations usually allow the commercial fisher to use different harvesting methods, such as nets. In much of North America freshwater fisheries are primarily managed for sport fishing. The management of freshwater fish populations is similar to that of other wild animals. Fish require cover, such as logs, stumps, rocks, and weed beds, so they can escape from predators. They also need special areas for spawning and raising young. These might be a gravel bed in a stream for salmon, a sandy area in a lake for bluegills or bass, or a marshy area for pike. A freshwater fisheries biologist tries to manipulate some of the features of the habitat to

enhance them for the desired species of game fish. This might take the form of providing artificial spawning areas or cover. Regulation of the fishing season so that the fish have an opportunity to breed is also important.

In addition to these basic concerns, the fisheries biologist pays special attention to water quality. Whenever people use water or disturb land near the water, water quality is affected. For example, toxic substances kill fish directly, and organic matter in the water may reduce the oxygen. The use of water by industry or the removal of trees lining a stream warms the water and makes it unsuitable for certain species. Poor watershed management results in siltation, which covers spawning areas, clogs the gills of young fish, and changes the bottom so food organisms cannot live there. The fisheries biologist is probably as concerned about what happens outside the lake or stream as what happens within it.

The introduction of exotic fish species has greatly affected naturally-occurring freshwater ecosystems. The Great Lakes, for example, have been altered considerably by the accidental and purposeful introduction of fish species. The sea lamprey, smelt, carp, alewife, brown trout, and several species of salmon are all new to this ecosystem. (See figure 11.12.) The sea lamprey is parasitic on lake trout and other species and nearly eliminated the native lake trout population. Controlling the lamprey problem requires use of a very specific larvicide that kills the immature

lamprey in the streams. This technique works because mature lamprey migrate upstream to spawn, and the larvae spend a year or two in the stream before migrating downstream into the lake. With the partial control of lamprey, lake trout populations have been increasing. (See figure 11.13.) Recovery of the lake trout is particularly desirable, since at one time it was an important commercial species.

Another accidental introduction to the Great Lakes, the alewife, a small fish of little commercial or sport value, became a problem during the 1960s, when alewife populations were so great that they died in large numbers and littered beaches. Various species of salmon were introduced about this time in an attempt to control the alewife and replace the lake trout population, which had been depressed by the lamprey. While this salmon introduction has been an economic success and has generated millions of dollars for the sport fishing industry, it may have had a negative impact on some native fish like the lake trout, with which the salmon probably compete. Salmon also migrate upstream, where they disrupt the spawning of native fish. Most species of salmon die after spawning, which causes a local odor problem.

Aquaculture in freshwater systems typically involves the construction of ponds which allow the close management of the fish. Very little raising of fish occurs in open waters although the immature stages of many species may be raised in special nursery facilities (fish hatcheries) prior to their release into the wild.

Streams and rivers are modified for navigation, irrigation, flood control, or power production purposes, all of which may alter the natural ecosystem and change the numbers or kinds of fish present. These topics are discussed in greater detail in chapter 15.

MANAGING ECOSYSTEMS FOR WILDLIFE

Many kinds of terrestrial wild animals are desired as game animals or for other purposes, and efforts are made to improve

Lake trout

Brown trout

Native	Introduced
Brook lamprey	Sea lamprey
Lake trout Brook trout Whitefish Herring Ciscoes	Brown trout Rainbow trout Pink salmon Coho salmon Chinook salmon Atlantic salmon
Suckers	
Chubs Shiners	Carp
Catfish Bullheads	
	Smelt Alewife
White bass	White perch
Smallmouth bass Rock bass	Crappies Sunfish
Yellow perch Walleye	

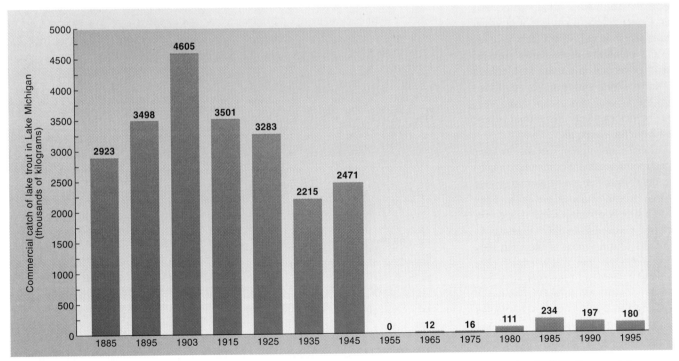

Yellow perch

Smelt

FIGURE 11.12 Native and Introduced Fish Species in the Great Lakes The Great Lakes have been altered considerably by the introduction of many non-native fish species. Some were introduced accidentally (lamprey, alewife, and carp), and others were introduced on purpose (salmon, brown trout, rainbow trout).

FIGURE 11.13 The Impact of the Lamprey on Commercial Fishing The lamprey entered the Great Lakes in 1932. Because it is an external parasite on lake trout, it had a drastic effect on the population of lake trout in the Great Lakes. As a result of programs to prevent the lamprey from reproducing, the number of lamprey has been reduced somewhat, and the lake trout population is recovering with the aid of stocking programs.

Farming, Fish Kills, and *Pfiesteria piscicida*

During the summer of 1997 several instances of fish kills by *Pfiesteria piscicida* were suspected along the East Coast of the United States. The *Pfiesteria piscicida* is a dinoflagellate that has over 20 stages in its life cycle. Four of the stages release a toxin that causes fish to become lethargic, develop open sores, and often die. *Pfiesteria piscicida* appear to be triggered to transform into the toxic forms when they detect the presence of large numbers of fish. Fish are often caused to concentrate when nutrient-rich waters become depleted of oxygen. The fish move from oxygen-depleted water to the areas that have the best oxygen

concentrations. It appears that *Pfiesteria piscicida* detect some component of the excrement of these concentrated populations of fish and develop into its toxic forms. The *Pfiesteria piscicida* organism then feeds on the tissues of the sick and dying fish. Humans working with *Pfiesteria* have been affected by the organism as well. Symptoms include sores, lightheadedness, and asthmalike symptoms, among others.

It appears that the problem occurs in areas of the ocean that are shallow and nutrient-rich. Human and swine effluent overflows have been implicated in some cases, as has fertilizer runoff from farming operations.

conditions for these species. Several techniques, some of which result in ecosystem modifications, are used to enhance certain wildlife populations. These techniques include habitat analysis, population census methods, stocking areas with game species, predator control, establishing refuges, and habitat management.

Managing a particular species requires understanding the habitat needs of that species. An animal's habitat must provide the following: food, water, and cover. **Cover** refers to any set of physical features that conceals or protects animals from the elements or enemies. Several kinds of cover are important and include safe places to rest and raise young, places where the animal can escape from enemies, and places where the animal is protected from the elements.

Animals are highly specific in their habitat requirements. For example, bobwhite quail must have a winter supply of food that protrudes above the snow. They require a supply of small rocks called grit, which the birds use in their gizzards to grind food. During most of the year, they need a field of tall grass and weeds as cover from natural predators. However, such protection is not available during the winter, so a thicket of brush is necessary. The thicket serves as cover from predators and shelter from the cold winter weather. Most areas provide suitable cover for resting, but for sleeping, quail prefer an open, elevated location, which allows the birds to quickly

take flight if attacked at night. When raising young, they require grassy areas with some patches of bare ground where the young can sun themselves and dry out if they get wet. All of these requirements must be available within a radius of 400 meters, because this is the extent of a quail's normal daily travels. (See figure 11.14.)

Once habitat requirements are known, the desirable population size must be ascertained. The population must be checked regularly to see if it is within acceptable limits. Population censuses are used to provide these data. If the population is below the desired number, organisms may be artificially introduced. This is

FIGURE 11.14 Cover Requirements for Quail Quail need several kinds of cover to be successful in an area. When raising young, they need areas of tall grass and weeds to provide protection from predators. In the winter, they need thickets to provide cover from weather and predators.

Native American Fishing Rights

Throughout many parts of the United States, particularly in the Pacific Northwest and the Great Lakes states, there is a continuing controversy over fishing rights of Native Americans. This conflict is unique because it involves treaties that were made 100 to 150 years ago between the U.S. government and Native American nations. It has become a major political, economic, social, and legal issue in some states, such as Washington, Michigan, Wisconsin, and Minnesota, and has involved the entire court system, from local courts to the U.S. Supreme Court. The controversy revolves around the interpretation of treaty language. It has, on several occasions, turned to violence and has divided many communities.

According to the wording of many treaties entered into in the 1800s, the rights of Native Americans to fish would not be infringed upon by the states. Native Americans claim the treaties give them the legal authority to engage in commercial fishing enterprises even when such fishing may be restricted or banned altogether for the general public.

On the other side of the argument are many state officials and sport fishers who believe that Native Americans are seriously endangering populations of such fish as salmon and trout by their uncontrolled harvesting for commercial purposes. They further argue that many species being taken by Native Americans belong to the entire state and not only to a certain group, because the fish are stocked or planted by the state. Another concern is the fishing techniques used by Native Americans. In the 1800s, when the treaties were signed, commercial fishing technology was limited. Today, however, Native American commercial fishers use nylon nets, power boats, depth finders, and other technological aids that enable them to catch much larger quantities of fish than they could with traditional fishing practices.

In the early 1970s, when sport fishers complained that stocks of fish in Lake Michigan were being depleted because of Native Americans' gill-net fishing, the state of Michigan tried to regulate Native American fishing. The issue ended up in court, and in 1978, a U.S. district court judge in Grand Rapids upheld fishing rights granted under 1836 and 1855 treaties between two Chippewa tribes and the U.S. government. The federal judge ruled that the state had no authority in the matter because the issue in question involved a federal treaty, and the state did not have the right to make regulations contrary to the treaty. Only Congress had such power. Subsequent to the decision, the tribes and the state negotiated changes in the areas in which the tribes could fish in an effort to reduce the potential for conflict between Native American fishers and sport or other commercial fishers.

A similar case was decided by a U.S. district judge in Tacoma, Washington, who interpreted treaties signed in 1854 and 1855 and ruled that Native Americans could catch up to 50 percent of the salmon that passed through their tribal land on the way to other parts of the state. In the face of protests by the non-native commercial fishing industry, the federal judge took over the regulation of salmon fishing in the state. Commercial fishers, who outnumbered Native American fishers by a ratio of approximately eight to one, feared for their livelihood. The case eventually went to the U.S. Supreme Court. In 1979, the Court ruled, in a six-to-three decision, to uphold the federal treaties that entitled Native Americans to half the salmon caught in the area of Puget Sound in Washington state. It further held that the 50 percent figure had to include the salmon that Native Americans caught on their lands for home consumption and religious ceremonies as well as the fish they caught for commercial purposes. The high court also voted to uphold the right of the district court to continue supervising the fishing industry because of the state's resistance to the interpretation of the treaties.

Clearly, the issues surrounding Native American fishing rights are complex and broad in scope. There are purely biological questions involving a resource and its wise use, economic issues involving families' livelihoods, cultural and religious concerns pertaining to Native Americans' use of a resource, legal questions involving federal and state conflicts, and moral questions relating to the unfair treatment of Native Americans in the past. In such conflicts, there is seldom one right answer. While the courts have ruled on the cases in Michigan and Washington, and the states and Native Americans are trying to use the fishing resource wisely, the problem still exists and is likely to continue for years to come.

---

referred to as stocking. Many game species are introduced exotics, and areas are stocked with these introduced species. The ring-necked pheasant was originally from Asia but has been introduced into Europe, Great Britain, and North America. All of the large game animals of New Zealand are introduced species, since there were none in the original biota of these islands. Several species of deer have been introduced into Europe from Asia. Many of these are raised in deer parks, where the animals are similar to free-ranging domestic cattle. They may be hunted for sport or slaughtered to provide food.

Since wildlife management often involves the harvesting of animals by hunting for sport and for meat, seasons are usually regulated to assure adequate reproduction and provide the largest possible healthy population during the hunting season. Hunting seasons usually occur in the fall so that surplus animals are taken before the challenges of winter. Winter taxes the animal's ability to stay warm and is also a time of low food supplies in most temperate regions. A well-managed wildlife resource allows for a large number of animals to be harvested in the fall and still leaves a healthy population to survive the winter and reproduce during the following spring (see figure 11.15).

Often, predator or competitor populations have been reduced to allow optimal survival of game species. In intensely managed European systems, gamekeepers are often charged with the responsibility of killing predators or

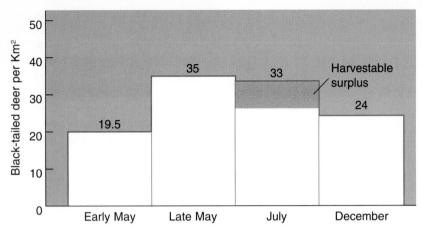

FIGURE 11.15 **Managing a Wildlife Population** The seasonal changes in this population of black-tailed deer are typical of many game species. The hunting season is usually timed to occur in the fall so that surplus animals will be harvested before winter, when the carrying capacity is lower.

Source: Data from R.D. Taber and R.F. Dasmann, "The Dynamics of Three Natural Populations of the Deer *Odocoileus hemionus columbianus*," in *Ecology* 38(2):233–46, 1957.

FIGURE 11.16 **Habitat Protection** The area on the left side of the fence has been protected from cattle grazing. This area provides a haven for many native species of plants and animals that cannot survive in heavily grazed areas.

unwanted competitors. With more wild and less intensely managed species, as is typical in North America, very little predator or competitor control is exercised. At one time, it was thought that populations of game species could be increased substantially if predators were controlled. In Alaska, for example, the salmon-canning industry claimed that bald eagles were reducing the salmon population, and a bounty was placed on the bald eagle. From 1917 to 1952, 128,000 eagles were killed for bounty money in Alaska. This theory of predator control to increase populations of game species has not proven to be valid in most cases, however, since the predators do not normally take the prime animals anyway. They are more likely to capture sick or injured individuals not suitable for game hunting.

For some species, such as ground nesting birds, it may make some sense to control predators that eat eggs or capture the young, but in most cases, humans have a greater impact through habitat modification and hunting than do the natural predators. Although at one time they were thought to be helpful, bounties and other forms of predator management have largely been eliminated in North America. In fact, the pendulum has swung the other way and predator control is not considered to be cost effective in most cases. One exception to this general trend is the hunting

and trapping of wolves in Alaska and Canada. One rationale for the taking of wolves is that they kill moose and caribou. Since many of the people who live in these areas rely on wild game as a significant part of their food source, controlling wolf populations is politically popular. However, the number of wolves taken is regulated. Since in many parts of the world the major predators of game species are humans, regulation of hunting is a form of predator control.

Many species may require refuges where they are protected or where the habitat is manipulated to assure their reproduction. Many native prairie species benefit from the exclusion of introduced grazing animals like cattle and sheep. (See figure 11.16.) The Kirtland warbler is an endangered species that nests in special managed areas in the central part of Michigan, although evidence now indicates that some of these birds may be mating in the neighboring states of Wisconsin and Minnesota. The nesting cover required for this species is young, even-aged stands of jack pine. The areas where these birds nest have been protected, and, in addition, the habitat has been modified to provide for optimal nesting. Portions of their nesting area are periodically burned to assure future nesting sites. Fire is a very important part of the ecological niche of the Kirtland warbler because many jack

pine cones will only open if subjected to fire and the seedings require unshaded bare soil to become established.

Burning old stands of jack pine to allow for regrowth is an example of **habitat management.** Once the food habits, predators, and cover requirements of a game species are well understood, the habitat can be altered to optimize it for the desired species. Habitat management may take the form of encouraging some species of plants that are the preferred food of the game species. For example, habitat management for deer may involve encouraging the growth of many young trees, saplings, and low-growing shrubs by cutting the timber in the area and allowing the natural regrowth to supply the food and cover the deer need. Both forest management and deer management may have to be integrated in this case because some other species of animals, like squirrels, will be excluded if mature trees are cut, since they rely on the seeds of trees as a major food source.

Given suitable habitats and protection, most wild animals can maintain a sizable population. In general, organisms produce more offspring than can survive. Figure 11.17 illustrates the reproductive potential of both quail and white-tailed deer. High reproductive potential, protection from hunting, and the management and restoration of suitable habitats have reversed the drastic population declines of deer in some parts of the world. In Pennsylvania, where deer were once extinct, the number is now

Quail

	Adults	Young	Total
1st year	2	14	16
2nd year	16	112	128
3rd year	128	896	1024

Deer

	Adults	Yearlings	Fawns	Total
1st year	2	0	2	4
2nd year	2	2	2	6
3rd year	4	2	4	10
4th year	6	4	6	16
5th year	10	6	10	26

FIGURE 11.17 **Reproductive Potential** If we assume no mortality, animals have a reproductive capacity far above what is required to just keep the population stable. In the real world, mortality generally keeps populations from growing beyond the ability of their habitat to sustain them.

From Aldo Leopold, *Game Management.* Copyright © 1933 by Allyn and Bacon. Reprinted by permission.

in excess of 900,000. Even in predominately agricultural states, deer have made a comeback. In Indiana, a herd of over 280,000 deer has grown from an initial stocking of 35 individuals.

Other animals have shown comparable changes in population size. In Zimbabwe, unlike in most African countries, the number of elephants increased from about 200 in 1900 to 40,000 in 1987. This is probably not a sustainable population, and the government would like to reduce the number to about 30,000. The wild turkey population in the United States has increased from about 20,000 birds in 1890 to 4 million birds today.

Waterfowl (ducks, geese, swans, rails, etc.) present some special management problems because they are migratory. **Migratory birds** can fly thousands of kilometers and, therefore, can travel north in the spring to reproduce during the summer months and return to the south when cold weather freezes the ponds, lakes, and streams that serve as their summer homes. (See figure 11.18.) Because many waterfowl nest in Canada and winter in parts of the United States and Central America, an international agreement between Canada, the United States, and Mexico is necessary

to manage and prevent the destruction of this wildlife resource. Habitat management has taken several forms. In Canada, where much of the breeding occurs, government and private organizations such as Ducks Unlimited have worked to prevent the draining of small ponds and lakes that provide nesting areas for the birds. In addition, new impoundments have been created where it is practical. Because birds migrate southward during the fall hunting season, a series of wildlife refuges provide havens from hunters. In addition, these refuges may be used to raise local populations of birds. During the winter, many of these birds congregate in the southern United States. Refuges in these areas are important overwintering areas where the waterfowl can find food and shelter.

NATURAL SELECTION AND EXTINCTION

Extinction is the death of a species, the elimination of all the individuals of a particular kind. It is a natural and common event in the long history of biological evolution. Of the estimated 500 million

species of organisms that are believed to have ever existed on earth since life began, perhaps 5 million to 10 million are currently active. This represents an extinction rate of 98 to 99 percent. Obviously, these numbers are estimates, but the fact remains that extinction has been the fate of most species of organisms. However, the process of **speciation** produces new species. Thus, new species are continually added while at the same time those that are not able to adapt to a changing environment are eliminated.

Recall that a species is a population of organisms. Every individual is a member of a species and shares many characteristics that are typical of the species. For example, all dogs have four legs, two eyes, hair, and a tail of some kind. But there is also variation among individuals within a species. Some are taller, can run faster, have quicker reflexes, or are better able to hear, smell or see. This individual variation allows some individual organisms to cope better with their surroundings. Most of these characteristics are at least partly the result of the genetic material, that is, genes and DNA, that the individuals inherited from their parents.

Individuals that have a combination of characteristics that allow them to be more successful are more likely to live long enough to reproduce, and are also likely to have larger numbers of offspring. Their offspring are likely to have many of the characteristics that made their parents successful and will outnumber the offspring of parents that were not well adapted. This process is known as natural selection. Thus the characteristics typical of the species change slightly over time.

If two populations of the same species are isolated from one another for a long time, and the environments of the two locations are very different, the process of natural selection can result in the two populations becoming so different from one another that they do not interbreed and so are considered different species. (See figure 11.19.) Keep in mind that environments change over time, so a set of characteristics will not always be suitable. For example, at one time, much of Canada and the northern parts of the United States were covered with glaciers. The climate was quite different from what

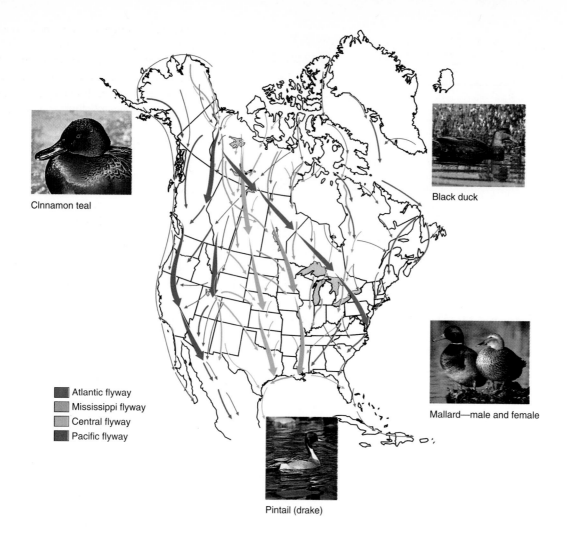

Cinnamon teal

Black duck

Mallard—male and female

Atlantic flyway
Mississippi flyway
Central flyway
Pacific flyway

Pintail (drake)

FIGURE 11.18 **Migration Routes for North American Waterfowl** Migratory waterfowl follow traditional routes when they migrate. These have become known as the Atlantic, Mississippi, Central, and Pacific flyways. Many of these waterfowl are hatched in Canada, migrate through the United States, and winter in the southern United States or Mexico.

exists today and the plants and animals that existed were also different.

Because the environment is continually changing, a species must adapt or become extinct. If this adaptation occurs over a long period, the original species may evolve into a new species. In this situation, the original species eventually becomes extinct, but before it does, many of its genes (particularly the favorable ones) are passed on to the newly evolving species. An example of this is the extinction of the dinosaurs. As earth's environment changed, it became less favorable

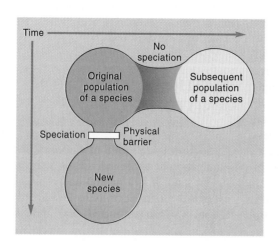

FIGURE 11.19 **Speciation**
Populations of a species may become isolated from one another for a long time as a result of a barrier. As they encounter new environmental conditions, selection occurs, and the isolated population becomes adapted to local environmental conditions. If isolation is total and lasts for a long time, a new species may develop. If no barrier prevents gene exchange between populations, a new species is not likely to develop, but the population will still adapt to new environmental conditions.

TABLE 11.6 PROBABILITY OF BECOMING EXTINCT

Most likely to become extinct	Least likely to become extinct
Low population density	High population density
Found in small area	Found over large area
Specialized niche	Generalized niche
Low reproductive rates	High reproductive rates

for dinosaurs. However, before their extinction, many of the favorable genes in this reptile line formed the basis for the newly evolving mammals and birds. This form of gradual extinction and replacement by other, better-adapted groups is common in evolutionary history. The fossil record shows many groups of organisms that flourished for millions of years before more advanced groups replaced them.

Studies of modern local extinctions suggest that certain kinds of species are more likely than others to become extinct. (See table 11.6.) Species that have small populations of dispersed individuals are more prone to extinction because successful breeding is more difficult than in species that have large populations of relatively high density. Some kinds of organisms, such as carnivores at higher trophic levels in food chains, typically have low populations but also have low rates of reproduction compared to their prey species. Organisms in small, restricted areas are also prone to extinction because an environmental change in their locale can eliminate the entire species at once. Organisms scattered over large areas are much less likely to be negatively affected by one event. Specialized organisms are also more likely to become extinct than are generalized ones. Since specialized organisms rely on a few key factors in the environment, anything that negatively affects these factors could result in their extinction, whereas generalists can use alternate resources.

Rabbits and rats are good examples of animals that are not likely to become extinct soon. They have high population density and a wide geographic distribution. In addition, they have high reproductive rates and are generalists that can live under a variety of conditions and use a variety of items as food. The cheetah is much more likely to become extinct because it has a low population density, is restricted to certain parts of Africa, has low reproductive rates, and has very specialized food habits. It must run down small antelope, in the open, during daylight, by itself. Similarly, the entire wild whooping crane species consists of about 130 individuals that are restricted to small winter and summer ranges that must have isolated marshes. (Captive and experimental populations bring the total number to about 200 individuals.) In addition, their rate of reproduction is low.

HUMAN-ACCELERATED EXTINCTION

Humans are among the most successful organisms on the face of the earth. We are adaptable, intelligent animals with few enemies. As our population increases, we displace other kinds of organisms. This has resulted in an accelerated rate of extinction. Wherever humans become the dominant organism, extinctions occur. Sometimes, we use other animals directly as food. In doing so, we reduce the population of our prey species. Since our population is so large and because we have an advanced technology, catching or killing other animals for food is relatively easy. In some cases, this has led to extinctions. The passenger pigeon in North America, the moas (giant birds) of New Zealand, and the bison and wild cattle of Europe were certainly helped on their way to extinction by people who hunted them for food.

We use organisms for a variety of purposes in addition to food. Many plants and animals are used as ornaments. Flowers are picked, animals skins are worn, and animal parts are used for their purported aphrodisiac qualities. In the United States, many species of cactus are being severely reduced because people like to have them in their front yards. In other parts of the world, rhinoceros horn is used to make dagger handles or is powdered and sold as an aphrodisiac. Because some people are willing to pay huge amounts of money for these products, unscrupulous people are willing to take the chance of poaching these animals for the quick profit they can realize. Table 11.7 shows some of the organisms, or their parts, that are highly prized by buyers. The World Wildlife Fund estimates that illegal trade in wild animals globally produces $2 billion to $30.5 billion per year. These activities have already resulted in local extinctions of some plants and animals and may be a contributing factor to the future extinction of some species.

Some organisms are extinct because they were regarded as pests. Many large predators have been locally exterminated because they preyed on the domestic animals that humans use for food. Mountain lions and grizzly bears in North America have been reduced to small, isolated populations, in part because they were hunted to reduce livestock loss. Tigers in Asia and the lion and wolf in Europe were reduced

TABLE 11.7 TYPICAL SPECIES PRICES (1990 RATES)

International species	Price ($U.S.)	North American species	Price ($U.S.)
Olive python	1500	Bald eagle	2500
Rhinoceros horn	12,500/pound	Golden eagle	200
Tiger skin (Siberian)	3500	Gila monster	200
Tiger meat	130/pound	Peregrine falcon	10,000
Cockatoo	2000	Grizzly bear	5000
Leopard	8500	Grizzly bear claw necklace	2500
Snow leopard	14,000	Polar bear	6000
Elephant tusk	250/pound	Black bear paw pad	150
Walrus tusk	50/pound	Reindeer antlers	35/pound
Mountain gorilla	150,000	Mountain lion	500
Giant panda	3700	Mountain goat	3500
Ocelot	40,000/coat	Saguaro cactus	15,000
Imperial Amazon macaw	30,000		

Source: Data from U.S. Fish and Wildlife Service.

a.

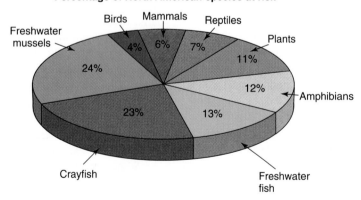

b.

FIGURE 11.20 Causes of Extinction Many organisms have become extinct as a result of human activities. The most important of these is the indirect destruction of habitat by humans modifying the environment for their own purposes. Part (*b*) shows the species at risk in North America (percentage by group).

(*b*) Source: Data from the Nature Conservancy and the Network of Natural Heritage Programs and Conservation Data Centers 1993.

or eliminated for similar reasons. Even though commercial hunters killed thousands of passenger pigeons, their ultimate extinction was caused primarily by the increased agricultural use of the forests. Passenger pigeons ate the acorns of oaks and the beechnuts from beech, and also relied on the forests for communal nesting sites. When the forests were cleared, the pigeons became pests to farmers, who shot them to protect their crops from being eaten by the birds. Some extinctions of pest species are considered desirable. Most people would not be disturbed by the extinction of black widow spiders, mosquitoes, rats, or fleas. In fact, people work hard to drive some species to extinction. For example, in the *Morbidity and Mortality Weekly Report* (October 26, 1979), the U.S. Centers for Disease Control triumphantly announced that the virus that causes smallpox was extinct in the human population after many years of continuous effort to eliminate it.

The most important cause of extinctions related to human activity is habitat alteration. (See figure 11.20.) Whenever humans populate an area, they change it by converting the original ecosystem into something that supports themselves. Forests and grasslands have been converted to agricultural and grazing lands. In addition, humans have introduced new species to grow or graze on these lands. These new plants and animals compete with the native organisms for nutrients and living space, and often, the native organisms lose. (See Global Perspective: The History of the Bison.) The African elephant and various species of rhinoceros are in danger because they are hunted for ivory or horn, but it was the original fragmentation of their habitat for agricultural purposes that first placed the animals in conflict with humans.

In much of the tropical world today, rainforests are being cleared to provide grazing land or agricultural land for an expanding human population. This activity destroys existing forests and fragments them into small islands. Scientists studying the effects of this activity have noticed that, as a forest is reduced to small patches, many species of birds disappear from the area. The same kinds of activities certainly happened in Europe, where little of the original forest is left. In North America, the eastern deciduous forests were reduced, which probably resulted in the extinctions of some animals and plants. Almost all of the original prairie in the United States has been replaced with agricultural land, resulting in the loss of some species.

Humans also cause extinction in less direct ways. The building of dams changes the character of rivers, making them less suitable for some species. Air and water pollution may kill all life in an area. For example, acid precipitation has lowered the pH of some lakes so much that all life has been eliminated. In other cases, indirect human activity may selectively eliminate species that are less tolerant to the pollutants being released, or the introduction of exotic species may eliminate existing species. The accidental introduction of the chestnut blight fungus into the United States resulted in the loss of the chestnut tree over much of the Appalachian region. Some species that were dependent on these trees probably were also eliminated.

WHY WORRY ABOUT EXTINCTION?

If extinction is a natural event, and if the species that become extinct are those that cannot effectively cope with the activities of humans, why should we worry about them? There are several answers to that question. First of all, strictly from a selfish point of view, many species we know little about may be useful to us. Many plants have chemicals in them that can be used as medicines. If we drive them into extinction, we may be eliminating a potentially useful product. Also, as the human population increases, the need for different kinds of food plants and animals will increase. Once they are eliminated, we have lost the opportunity to use them for our own ends. Most of the wild ancestors of our most important food grains, like maize (corn), wheat, and rice, are thought to be extinct. What would our world be like if these plants had not been domesticated

before they went extinct as wild populations? Because of this concern, many agricultural institutes and universities maintain "gene banks" of wild and primitive stocks of crop plants.

Another interesting answer to the question is that certain organisms in some ecosystems appear to play pivotal roles. Accidental extinction of one of these species could be devastating to the ecosystem and the humans that use it. For example, the sardine fishery off the coast of southern California and the anchoveta fishery off the coast of Peru appear to have been fundamentally altered by overfishing. Although the details are not known, it is thought that, once the population was significantly reduced, other organisms filled their niche, making it difficult for their population to return to its original size. The people who depended on these fisheries now must find other ways to make a living.

A third answer, according to many people, is that all species have a fundamental right to exist without being needlessly eliminated by unthinking activity of the human species. This philosophical statement has nothing to do with economic or social value; it is an ethical position.

WHAT IS BEING DONE TO PREVENT EXTINCTION?

Efforts to prevent human-caused extinctions are difficult to assess. Some countries have enacted legislation to protect species that are in danger of becoming extinct. Species that receive special protection usually are given some sort of designation, such as endangered or threatened. **Endangered species** are those that have such small numbers that they are in immediate jeopardy of becoming extinct. **Threatened species** could become extinct if a critical factor in their environment were changed.

Most of the interest in preventing human-caused extinctions comes from developed countries. There, however, the problem is less acute because vulnerable

species have already been eliminated. Extinction is a greater potential problem in tropical, less-developed countries. Many biologists estimate that there may be as many species in the tropical rainforests of the world as in the rest of the world combined. Unfortunately, extinction prevention is not a major issue in many less-developed countries. This difference in level of interest is understandable since the developed world has surplus food, higher disposable income, and higher education levels, while people in many of the less-developed countries, where population growth is high, are most concerned with immediate needs for food and shelter, not with long-term issues like extinction.

Nevertheless, many of the governments of less-developed countries have responded to pressures and suggestions from outside sources and have established preserves and parks to protect species in danger of extinction. This does not solve the problem, however, since the areas must be protected from poachers and unauthorized agricultural activity of people who sneak into them. These people are responding to basic biological and economic pressures to provide food for their families. Hiring security forces to patrol such areas is expensive for many of these countries.

Even in countries where interest in extinction prevention is relatively high, there is a cultural bias in favor of protecting certain kinds of organisms. Most endangered or threatened species are birds, mammals, some insects (particularly butterflies), a few mollusks and fish, and certain categories of plants. Bacteria, fungi, reptiles, most insects, and many other inconspicuous organisms rarely show up on endangered species lists, even though they play vital roles in the nitrogen cycle, in the carbon cycle, and as decomposers.

Several international organizations work to prevent the extinction of organisms. The World Conservation Union (IUCN) estimates that, by the year 2000, at least 500,000 species of plants and animals may be exterminated. The IUCN classifies species in danger of extinction into four categories: endangered, vulnerable, rare, and indeterminate. Endangered species are those whose survival is

ENVIRONMENTAL
CLOSE-UP

The California Condor

The California condor (*Gymnogyps californianus*) is thought to have been adapted to feed on the carcasses of large mammals found in North America during the Ice Age. With the extinction of the large, Ice Age mammals, their major food source disappeared. By the 1940s, their range had shrunk to a small area near Los Angeles, California. Further fragmentation of their habitat caused by human activity, death by shooting, and death by eating animals containing lead shot reduced the total population to about 17 animals by 1986. A low reproductive potential made it difficult for the species to increase in numbers. They do not become sexually mature until six years of age, and females typically lay one egg every two years.

In 1987, all the remaining wild condors were captured to serve as breeding populations. The plan was to raise young condors in captivity, leading to a large population that would ultimately allow for the release of animals back into the wild. Captive condors bred successfully, and females could be induced to lay a second egg if the first egg was removed from the nest. This increased the number of offspring produced per female and resulted in a captive population of 103 individuals by 1996. In 1991, two condors were released into the wild, and six more were released in 1992. In 1996, a second population of condors was introduced in Arizona near the Utah border. There has been unanticipated mortality from contact with power lines and other factors. However, the behavior of the animals seems natural and there is hope that they may eventually reproduce in the wild.

unlikely if the conditions threatening their extinction continue. These organisms need action by people to preserve them, or they will become extinct. Vulnerable species are those that have decreasing populations and will become endangered unless causal factors, such as habitat destruction, are stopped. Rare species are primarily those that have small worldwide populations and that could be at risk in the future. Indeterminate species are those that are thought to be extinct, vulnerable, or rare, but so little is known about them that they are impossible to classify.

Although the IUCN is a highly visible international conservation organization, it has very little power to effect change. It generally seeks to protect species in danger by encouraging countries to complete inventories of plants and animals within their borders. It also encourages the training of plant and animal biologists within the countries involved. (There is currently a critical shortage of plant and animal biologists who are familiar with the organisms of the tropics.) The IUCN also encourages the establishment of preserves to protect species in danger of extinction.

The U.S. Endangered Species Act was passed in 1973. This legislation gave the federal government jurisdiction over any species that were designated as endangered. About 300 U.S. species and subspecies have been so designated by the Office of Endangered Species of the Department of the Interior. (See figure 11.21.) The Endangered Species Act directs that no activity by a governmental agency should lead to the extinction of an endangered species and that all governmental agencies must use whatever measures are necessary to preserve these species.

Blackfooted ferret

Whooping crane

Mission blue butterfly

Galápagos tortoise

Giant panda

Persistent trillium

FIGURE 11.21 **Endangered Species** Many species have been placed on the endangered species list. These are plants and animals that are present in such low numbers that they are in immediate danger of becoming extinct.

The key to preventing extinctions is preservation of the habitat required by the endangered species. Consequently, many U.S. governmental agencies and private organizations have purchased sensitive habitats or have managed areas to preserve suitable habitats for endangered species. Setting aside certain land areas or bodies of water forces government and private enterprise to confront the issue of endangered species. The question eventually becomes one of assigning a value to the endangered species. This is not an easy task; often, politics become involved, and the endangered species does not always win.

A case in point is the controversy that surrounded Tennessee's Tellico Dam project in 1978. The U.S. Supreme Court declared that completion of the $116 million federal project would result in a violation of the Endangered Species Act because the dam would threaten the survival of an endangered species called the snail darter, a tiny fish about 8 centimeters long that lived in the stream that would be altered by the dam's construction. Developers, however, were not deterred. They lobbied in Congress to have all federally funded projects exempted from the act. Conservationists lobbied for the preservation of the act as it was originally written. Eventually, in 1978, Congress amended the Endangered Species Act so that exemptions to the act could be granted for federally declared major disaster areas or for national defense, or by a seven-member Endangered Species Review Committee. Because this group has the power to sanction the extinction of an organism, it has been nicknamed the "God Squad." If the committee found that the economic benefits of a project outweighed the harmful ecological effects, it would exempt a project from the Endangered Species Act. At their first meeting, the review board denied the request to exempt the Tellico Dam project on the grounds that the project was economically unsound. This should have stopped the Tellico Dam project. However, nine months later, as a result of several political maneuvers, Congress appropriated money to complete the dam. It is now complete and full of water. The snail darters that once dwelled on the site were transplanted to nearby rivers and by 1981 appeared to be reproducing.

The amendments to the Endangered Species Act also weakened the ability of the U.S. government to add new species to the endangered and threatened lists. Before a species can be listed, it is now necessary to determine the boundaries of its critical habitat, prepare an economic impact study, and hold public hearings—all within two years of the proposal of the listing. The political aspects of endangered species continues to be important in the U.S. Congress. Intense lobbying by environmental and business interests seeks to shape the reauthorization of the act to suit their interests.

The Central American republic of Costa Rica has been very successful in protecting a significant amount of its remaining forests. It has a rich variety of different kinds of tropical forests and other natural resources. Mangrove swamps, cloud forests, rainforests, dry tropical forests, volcanoes, and beaches on both the Pacific Ocean and the Caribbean Sea are all part of the mix found in an area about the size of West Virginia.

Nearly 20 percent of the land in Costa Rica is protected as parks, reserves, and refuges. This is the result of several factors. The government is committed to preserving natural areas. A major part of the parks program involves educating the local people about the values of the parks, including the biological value of the large number of species of plants and animals and the economic value of the parks as a tourist attraction. The job of educating the people is made easier by the fact that 93 percent of the people are literate.

Many jobs in Costa Rica are in the developing ecotourism market. People who wish to visit natural areas require guides, transportation, food, and lodging. The jobs created in these industries encourage local people to preserve their natural resources because their livelihood depends on it. Furthermore, when people are employed, they are less likely to try to convert forested land into farmland.

In addition to using the natural resources of forests for tourists, the government of Costa Rica has an agreement between its Instituto Nacional de Biodiversidad (INBio) and Merck and Company, Inc., to prospect for possible drugs from many tropical plants and animals found in the forests. In return, Merck will pay $1 million to the Costa Rican government. Specially trained local parataxonomists (technicians trained in the identification of plants and animals) are involved in this search, resulting in additional jobs.

- What conditions have contributed to the apparent success of Costa Rica in protecting its forests?
- What are possible sources of failure in other countries?
- Who benefits from ecotourism?

MAP OF COSTA RICA
Not to Scale

National Parks and Wildlife Refuges
1. Isla Bolaños Wildlife Refuge
2. Santa Rosa National Park
3. Lomas Barbudal Biological Reserve
4. Palo Verde National Park and the Dr. Rafael Lucas Rodriguez Caballero Wildlife Refuge
5. Barra Honda National Park
6. Peñas Blancas Forest Refuge
7. Guayabo, Negritos, and Pájaros Islands Biological Reserves
8. Ostional National Wildlife Refuge
9. Carara Biological Refuge
10. Curú National Wildlife Reserve
11. Cabo Blanco-Absolute Nature Reserve
12. Manuel Antonio National Park
13. Caño Island Biological Reserve
14. Golfito National Wildlife Reserve
15. Corcovado National Park
16. Isla del Coco National Park
17. Rincón de la Vieja National Park
18. Poás Volcano National Park
19. Braulio Carrillo National Park
20. Irazú Volcano National Park
21. Guayabo National Monument
22. Tapantí National Wildlife Reserve
23. Chirripó National Park and Costa Rica-Panamá International Friendship Park
24. Hitoy-Cerere Biological Reserve
25. Caño Negro National Wildlife Refuge
26. Barra del Colorado National Wildlife Refuge
27. Tortuguero National Park
28. Cahuita National Park
29. Gandoca-Manzanillo National Wildlife Refuge
30. Guanacaste National Park

Summary

Pollution is the result of technological advancements and increased population density. It is defined as anything produced by humans in a quantity that interferes with the health or well-being of an organism.

Natural resources are structures and processes that can be used by people but cannot be created by them. Renewable resources can be regenerated or repaired; nonrenewable resources are consumed.

Mineral resources must be extracted from ores, a process that requires energy and money, and also changes the environment. Mineral resources are not evenly distributed around the world, so most countries must purchase some of their minerals from foreign sources. The major steps of mineral exploitation are exploration, mining, processing the ore, transportation, and manufacturing the finished product. Ultimately, all costs of mineral exploitation are reduced to monetary costs. Recycling reduces the demand for new sources of mineral deposits but does not necessarily save money.

As the human population increases, we alter more and more ecosystems as we attempt to feed and house ourselves. However, some remote areas have been changed very little by humans. Tropical rainforests represent some of the last large wilderness areas, but they are being rapidly converted to other uses by the constant pressure of growing populations. Many of these wilderness areas should be protected because of the rich diversity of species they contain.

We change natural ecosystems by replacing them with agricultural ecosystems, we alter species mixtures by introducing plants and animals, and we reduce populations by harvesting trees and animals for our use. Forests must be cut in a manner that allows for regrowth so that the soil is not exposed to the erosional effects of wind and water. Cutting small areas and reforestation help to prevent these problems.

Grazing of arid and semiarid lands can be a valuable way to provide food for people. However, the land is often overgrazed and then may be degraded to a desert that is not capable of supporting the animals needed to feed growing populations.

Aquatic ecosystems are modified by pollution and activities that occur on the land adjacent to the water. Land that is devoid of vegetation erodes and fills streams and lakes with sediment. Warming of the water is also likely to occur if trees are removed from the stream side. Many exotic species of fish have been introduced into the freshwater ecosystems of the world. This alters normal food chains and reduces the populations of native species.

Often, areas are managed for certain species of wildlife. This involves careful planning and habitat manipulation to provide the best possible population for hunting. Some areas are intensely managed, as in many European game parks, while in other parts of the world, the game animals lead a more normal wild life. Waterfowl present a unique problem because they migrate across international boundaries.

Extinction is a normal consequence of not being able to adapt to changes in the environment. However, since humans have such a great influence on nearly every ecosystem in the world, they have been the cause of increased rates of extinction. Many people recognize the value of species, both as possible helpers of humans and for their own intrinsic worth, and are trying to preserve sensitive habitats so that species will not be driven to extinction because of the appropriation of their habitats for other uses.

Review Questions

1. Name three ways humans directly alter ecosystems.
2. Why is the impact of humans greater today than at any time in the past?
3. Define pollution. Has the definition changed with time?
4. What are three kinds of costs associated with resource exploitation?
5. Why is recycling usually more energy efficient than mining new raw materials?
6. List three problems associated with forest exploitation.
7. What is desertification? What causes it?
8. What effects do increased temperature and increased organic matter have on aquatic ecosystems?
9. List six techniques utilized by wildlife managers.
10. What special problems are associated with waterfowl management?
11. What is extinction, and why does it occur?
12. Why should humans worry about extinction? Do you?
13. List three actions that can be taken to prevent extinctions.

Critical Thinking Questions

1. Brazil, and other developing countries, need to get foreign currencies to fund economic growth. Should industrialized countries have some say over the way Brazil harvests its forests and extracts its mineral resources? What role should the industrialized world have in helping Brazil preserve its natural resources?
2. Imagine you are a mining company executive trying to determine the costs of your mining operation. Create a list of your costs. How important are environmental costs to your company? Why?
3. Logging, mineral extraction, and grazing on federal lands have been subsidized in order to stimulate economic growth and support rural communities. What place should ecological value have in determining the way land is used? Should subsidies be continued? Justify your stand on this issue.
4. Perhaps 98–99 percent of all species have beome extinct, many long before human arrived on the scene. Why should we be concerned about extinction of organisms today?

5. Would you support clearing of forests and plowing of grasslands that have significant ecological importance in order to support agriculture in countries that have significant hunger? Where do you draw the line between preserving ecosystems and human interest?

6. A Pacific Northwest Native American tribe has been permitted to hunt an endangered whale species to preserve its traditional culture and to uphold its treaty rights Now there is fear that other countries will hunt the whales, too. Do you think the tribe should be denied its rights? Why, or why not?

7. White-tailed deer are a growing problem in much of the Midwest because urban sprawl has created good deer habitat with lots of food and few predators. Many mid-sized cities are considering sponsoring hunts to lower the deer population. These proposals have been met with a mixed reaction. What do you think about this? What values, perspectives, and beliefs do you hold reguarding this issue?

8. Pharmaceutical companies are helping some developing countries preserve their rainforests so these companies can look for organisms with possible pharmacological value. How do you feel about these arrangements? What limits would you place on the pharmaceutical companies, if any? Why?

Key Terms

biodegradable *187*
clear-cutting *193*
cover *202*
desertification *195*
economic costs *188*
endangered species *209*
energy costs *188*
environmental costs *188*
extinction *205*
habitat management *204*
migratory birds *205*

natural resources *187*
nonrenewable resources *188*
patchwork clear-cutting *193*
pollution *187*
reforestation *193*
renewable resources *188*
selective harvesting *193*
speciation *205*
threatened species *209*
wilderness *197*

Related Internet Sites

Visit our website at http://www.mhhe.com/environmentalscience/ for specific information on each of the topics below.

Ecosystem management in Oregon

Ecotourism as a tool for conservation and sustainable development

Sustainable tourism initiatives

Impact of ecotourism

Preservation of endangered species, biodiversity, and habitats

The Nature Conservancy, a searchable index

Australia's conservation of endangered species

Whale and marine mammal research

Wildlife and endangered species

Texas Department of Range, Wildlife, and Fisheries Management

Marine mammal conservation

Facts and information on endangered species

Renewable and nonrenewable energy sources

Travel programs that focus on environmental awareness, wilderness programs, and local cultures

Refuge management and species information

12

chapter

Land-Use Planning

Cities   Suburbs

Using land surfaces

Conflicting desires
• Industry
• Housing
• Transportation   } Planning resolves conflicts
• Commercial development
• Recreation

OBJECTIVES

After reading this chapter, you should be able to:

• List the relative amounts of land used for crops, livestock, urban development, and other uses in the United States.
• Explain the impact that water has on the location and development of cities.
• Explain why farmland surrounding cities was used for housing.
• Explain how taxation may influence land use.
• Explain why floodplains and wetlands are often mismanaged.
• List several exclusionary land uses and several multiple land uses.
• Describe the economic impact of recreation.
• Recognize that people desire outdoor recreation.
• Explain why recreational areas are needed in urban locations.
• Explain why some land must be designated for particular recreational uses, such as wilderness areas, and why that decision sometimes invites disagreement from those who do not desire to use the land in the designated way.
• List the steps in the development and implementation of a land-use plan.
• Describe methods of enforcing compliance with land-use plans.
• List the problems associated with local land-use planning.
• Explain the advantage of regional planning and the problems associated with it.

CHAPTER OUTLINE

The Need for Planning

Historical Forces That Shaped Land Use
    The Importance of Waterways
    The Rural to Urban Shift

The Death of the Central City and the Rise of Suburbia

Global Perspective: *Urbanization in the Developing World*

Some Problems Associated with Unplanned Urban Growth

    Transportation Problems
    Loss of Open Space
    Loss of Farmland
    Floodplain Problems

Environmental Close-Up: *Wetlands Loss in Louisiana*

    Wetlands Misuse
    Other Land-Use Considerations

Land-Use Planning: Principles and Challenges

    Urban Transportation Planning
    Urban Recreation Planning

National and Regional Planning
    Multiple Land Use
    The Use of Public Land for Outdoor
       Recreation

Issues and Analysis: *Decision Making in Land-Use Planning: The Malling of America*

THE NEED FOR PLANNING

A large proportion of the land surface of the world (about 1/3 to 1/2) has been changed by human activity. Most of this change has occurred without evaluating the land and determining the most logical use for the land. Currently most land-use decisions are made based on economic considerations or the short-term needs of a growing population. Land should be considered a nonrenewable resource because it cannot be newly formed by natural processes, and once land has been converted to a particular use it is generally unavailable for other purposes. As the population of the world grows there will be increased competition for the use of the land and systematic land-use planning will become more important. Furthermore, as the population of the world becomes more urbanized and cities grow, urban planning becomes critical.

HISTORICAL FORCES THAT SHAPED LAND USE

Currently in the United States, about 47 percent of the land is used for crops and livestock, about 45 percent is forests and natural areas, and nearly 5 percent is used intensively by people in urban centers and as transportation corridors. Canada is 54 percent forested and wooded and uses only 8 percent of its land for crops, and livestock. Less than 1 percent of the land is in urban centers and transportation corridors. A large percentage of its remaining land is wilderness in the north.

This pattern of land use differs greatly from the original conditions experienced by the early European colonists who immigrated to the New World. The promise of open space and available land was a primary incentive for those who left their homes. The first colonists converted portions of the original landscape to farming and as the North Ameri-

can population increased, villages and cities developed.

The Importance of Waterways

Waterways provided the primary method of transportation which allowed for exploration and the development of commerce in the early North America. Those towns and villages with good transportation could easily receive raw materials and distribute manufactured goods. Some of these grew into major industrial or trade centers. Thus, early towns were usually built near rivers, lakes, and oceans. Typically, cities developed as far inland as rivers were navigable. Abrupt changes in elevation required transshipment of goods, and cities often grew at these points (Buffalo, New York; Sault Sainte Marie, Ontario). In addition to transportation, bodies of water provided drinking water, power, and waste disposal for growing villages and towns. Without access to water, St. Louis, Montreal, Chicago, Detroit, Vancouver, and other cities would not have developed. (See figure 12.1.) The

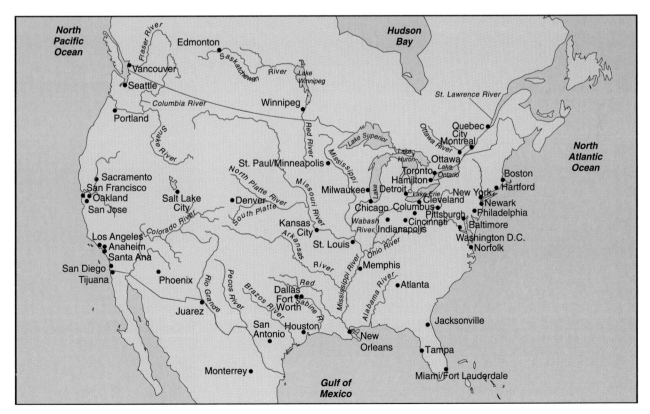

FIGURE 12.1 Water and Urban Centers Note that most of the large urban centers are located on water. Water is an important means of transportation and was a major determining factor in the growth of cities. The cities shown have populations of 1,000,000 or more (Except Edmonton, Winnipeg, and Quebec City).

availability of other natural resources, such as mineral resources, good farm land, or forest resources were also important in determining where people congregated to form villages and towns. Industrial development began on the waterfront since water supplied transportation, waste disposal, and power. Large factories replaced small gristmills, sawmills, and blacksmith shops. Industry eventually took over most of the waterfront.

The Rural to Urban Shift

North America remained essentially rural until industrial growth began in the last third of the 1800s and the population began a trend toward greater urbanization. (See figure 12.2.) There were several forces that led to this rural to urban transformation. First, the Industrial Revolution led to improvements in agriculture that required less farm labor at the same time industrial jobs became available in the city. Thus people migrated from the farm to the city. The average person was no longer a farmer but, rather, a factory worker, shopkeeper, or clerk living in a tenement or tiny apartment near where they worked. This pattern of rural to urban migration

occurred throughout North America, and it is still occurring in developing countries today. A second factor that affected the growth of cities was the influx of immigrants from Europe. Although some became farmers many of these new citizens tended to settle in towns and cities, where jobs were available. Many immigrants congregated in certain parts of cities. As a result, today most large cities have ethnic sections, with populations of Irish, Italian, Chinese, German, and Greek descent, that give North American cities a unique character. A third reason for the growth of cities was that they offered a greater variety of cultural, social, and artistic opportunities than did rural communities. Thus, cities were attractive for cultural as well as economic reasons.

THE DEATH OF THE CENTRAL CITY AND THE RISE OF SUBURBIA

During the early stages of industrial development there was little control of industry activities, so the waterfront became a polluted, unhealthy, undesir-

able place to live. As roads and rail transport became available anyone who could afford to do so moved away from the original, industrial city center. The more affluent moved to the outskirts of the city and the development of suburban metropolitan regions began. Thus the agricultural land surrounding the towns was converted to housing. Until land transportation systems became well developed, farms needed to be close to the city so that farmers could transport their produce to the markets in the city. This rich farmland adjacent to the city was ideal for the expansion of the city. As the population of the city grew, demand for land increased. As the price of land in the city rose, people and businesses began to look for cheaper land farther away from the city. Developers and real estate agents were quick to respond and to help people acquire and convert agricultural land to residential or commercial uses. Land was viewed as a commodity to be bought and sold for a profit, rather than as a nonrenewable resource to be managed. As long as money could be made by converting agricultural land to other purposes, it was impossible to prevent such conversion.

FIGURE 12.2 Rural to Urban Population Shift In 1800, the United States was essentially a rural country. Industrialization in the late 1800s began the shift to an increasing urban population. In 1995, 75 percent of the U.S. population was urban.

(Graph) Sources: Data from the *Statistical Abstract of the United States,* 1993 and 1995 data from Population Reference Bureau, Inc., Washington, D.C.

Urbanization in the Developing World

Traditionally most of the population of the developing world has been rural. However, in recent years there has been a rapid growth in the number of people migrating to the cities. By 2025 about 5 billion people are expected to be living in urban centers. Most of the increase in urban areas will be in the developing world. Many migrate to the cities because they feel they will have greater access to social services and other cultural benefits than are available in rural areas. Many also feel there are more employment opportunities. However, the increase in the urban population is occurring so rapidly that it is very difficult to provide the services needed by the population and jobs are not being created as fast as the urban population is growing. Thus many of the people live in poverty on the fringes of the city in shantytowns that lack water, sewer, and other services. Often these shantytowns are constructed without permission only a short distance from affluent urban dwellers. Because the poor lack safe drinking water and sewer services, they pollute the local water sources and disease is common. Because they burn wood and other poor quality fuels in inefficient stoves, air pollution is common. The additional people also create traffic problems of staggering proportions. The following chart shows the ten cities with the largest rate of population increase. All are in the developing world.

Name of city	Annual rate of population growth (1990–1995)	Size of population– 1995 (millions)
Dhaka, Bangladesh	5.74	7.8
Lagos, Nigeria	5.68	10.3
Jakarta, Indonesia	4.35	11.5
Karachi, Pakistan	4.27	9.9
Bombay, India	4.22	15.1
Delhi, India	3.80	9.9
Istanbul, Turkey	3.67	7.8
Manila, Philippines	3.05	9.3
Tianjin, China	2.88	10.7
Lima, Peru	2.81	7.5

Source: Data from *World Resources 1996–1997.*

The development of land around cities in North America has destroyed many natural areas that people had long enjoyed. The Sunday drive from the city to the countryside has become more difficult as people must drive farther to escape the ever growing suburbs. The unique character of neighborhoods and communities has been changed by the erection of shopping malls, apartment complexes, and expressways. Most of these alterations have occurred without considering how they will affect the lives of the people who live in the area and the nature of their community.

As cities continued to grow, certain sections within each city began to deteriorate. Industrial activity continued to be concentrated near water in the city's center. Industrial pollution and urban crowding turned the core of many cities into undesirable living areas. In the early 1900s, people who could afford to leave began to move to the outskirts. This trend continued after World War II, in the 1940s, 1950s, and 1960s, as a strong economy allowed more people to buy homes. Most of these new single family homes were in attractive suburbs, away from the population and congestion of the central city. These collections of houses were built on large lots that provided outdoor space for family activities. The blocks of homes were added to the periphery of the city in a decentralized pattern, which decreased energy efficiency, increased the cost of supplying utility services, and made it very difficult to establish efficient public transportation networks. However, public transport was not considered important because rising automobile ownership and improved highway systems would accommodate the transportation needs of the suburban population. The convenience of a personal automobile escalated decentralized housing patterns, which, in turn, required better highways, which led to further decentralization. (See figure 12.3.)

By 1960, unplanned suburban growth had become known as **urban sprawl.** Large housing tracts surrounded cities, which made it difficult for people to find open space. A city dweller could no longer take a bus to the city limits and enjoy the open space of the countryside. Urban sprawl occurs in three ways. (See figure 12.4.) One type of growth involves the development of exclusive, wealthy suburbs adjacent to the city. These homes are usually on large individual lots in the more pleasing geographic areas of the city. Often they are located along water, on elevated sites, or in wooded settings. A second development pattern is tract development. **Tract development** is the construction of similar residential units over large areas. Initially, these tracts are often separated from each other by farmland. New roads are constructed to link new housing to the central city and other suburbs which stimulated the development of a third form of urban sprawl along transportation routes. This is referred to as **ribbon sprawl** and usually consists of commercial and industrial buildings. Ribbon sprawl results in high costs for the extension of utilities and other public services. It also makes the extent of urbanization seem much larger than it actually is, since the driver cannot see the undeveloped land hidden by the storefronts that face the highway.

As suburbs continued to grow, cities (once separated by farmland) began to merge, and it became difficult to tell where one city ended and another began. This

FIGURE 12.3 **Transportation and Suburbia** Traffic congestion is a common experience for people who work in cities but live in the suburbs. The popularity of automobiles and the desire to live in the suburbs are closely tied.

a.

b.

c.

FIGURE 12.4 **Urban Sprawl** Note the three different types of growth depicted in these photos. (*a*) The wealthy suburbs with large lots are adjacent to the city. (*b*) Ribbon sprawl develops as a commercial strip along highways. (*c*) Tract development results in neighborhoods consisting of large numbers of similar houses on small lots.

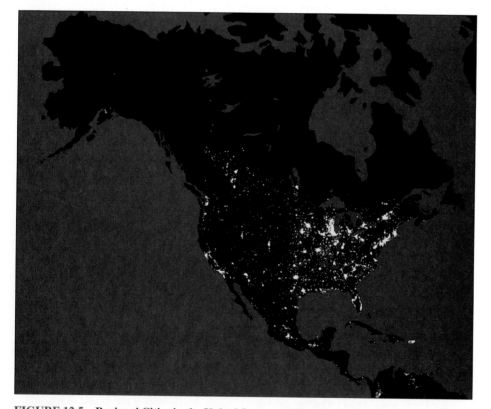

FIGURE 12.5 **Regional Cities in the United States and Canada** The lights from this satellite image show population concentrations. More than thirty major regional cities have developed with each having more than one million people. Many of these cities merge with their neighbors to form huge regional cities. Major urban regions are the Northeast coast of the United States (Boston to Washington, D.C.), the region south of the Great Lakes (Chicago to Pittsburgh), south Flordia (Jacksonville to Miami) the Toronto and Montreal regions of Canada, and the west coast of California (San Francisco to San Diego).

type of growth led to the development of regional cities. Although these cities maintain their individual names, they are really just part of one large urban area called a **megalopolis.** (See figure 12.5.) The eastern seaboard of the United States, from Boston, Massachusetts, to Washington, D.C., is an example of a continuous city. Other examples are London to Dover in England, the Toronto-Mississauga region of Canada, and the southern Florida coast from Miami northward.

In some areas throughout North America, the growth of suburbs has been slowed due to the increased cost of housing and transportation. People have migrated back to some cities on a limited scale because of the lower cost of urban houses and the fact that public transportation is generally more efficient in the city than in the suburbs, thus freeing urban residents from the cost of daily commuting. This reverse migration, however, is still greatly offset by the continual growth of the suburban communities.

SOME PROBLEMS ASSOCIATED WITH UNPLANNED URBAN GROWTH

Transportation Problems

Most cities experience continual problems with transportation. This is primarily because as cities grew there was little thought given to how people were going to move around and through the city. Furthermore, as housing patterns and commercial sectors change, transportation mechanisms must be changed to meet the shifting needs of the public. This often involves the abandoning of old transportation corridors and the establishment of new ones. Paradoxically the establishment of new transportation corridors stimulates increased growth in the areas served and the transportation corridors soon become inadequate. The reliance on the automobile as the primary method of transportation has required the constant building of new highways.

Loss of Open Space

One of the important features of a pleasing urban landscape is the presence of open space. Open fields, parks, boulevards, and similar land use allows people to visually escape from the congestion of the city. Unplanned urban growth does not take this important factor into account. Consequently, the quality of the urban environment is diminished or parks and other

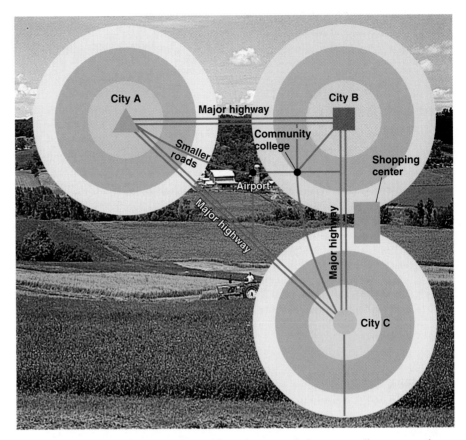

FIGURE 12.6 Loss of Farmland As cities grow outward, they eventually grow together to form a regional city. The land between them, once used for farming, becomes developed for residential and commercial purposes. Improved transportation routes and joint facilities (such as airports, shopping centers, and community colleges) hasten this loss of farmland.

green space must be retrofitted into the urban landscape at great expense.

Loss of Farmland

Most of the land that has recently been urbanized was previously used for high-value crops. Land that is flat, well drained, accessible to transportation, and close to cities is ideal farmland. However, it is also prime development land. Areas that once supported crops now support housing developments, shopping centers, and parking lots. (See figure 12.6.)

Urban development of farmland is proceeding at a rapid pace. One reason for this conversion is the way the land is taxed. Property is often taxed on what can be done with it, not necessarily on what is being done with it. For example, if land can be used for both farming and residential development, it is taxed as if it were residential. If a farmer sells a portion of the farm to a developer who builds five houses, local taxing authorities would consider these houses to be the "highest and best use" of the land. All of the farmer's land would then probably be reassessed and the taxes increased substantially. Farmers faced with this situation are often forced to sell their land because they are taxed on its commercial value rather than on its value as farmland. This policy encourages development and forces people out of farming. New policies that assess taxes based on current use of the land and not on the land's highest potential use must be explored.

Floodplain Problems

Because most cities were established along water, many cities are located in areas called **floodplains**. Floodplains are the low areas near rivers that are subject to flooding. Some may flood annually, while

ENVIRONMENTAL CLOSE-UP

Wetlands Loss in Louisiana

The state of Louisiana has extensive coastal wetlands along the Gulf of Mexico. Approximately 10,000 hectares (100 square kilometers) are lost per year for a variety of reasons, most of them the result of human activity.

Much of the coastline consists of poorly compacted muds that are easily eroded if exposed to wind and wave action. Dams and reservoirs upstream on the Mississippi River have reduced the amount of sediment delivered to the region, making replacement of lost sediments impossible. Furthermore, as the muds settle and compact, the land subsides and erosion is more likely.

Another major contributing factor to the loss of these coastal wetlands is oil and gas exploration and the development of other resources from beneath the Gulf of Mexico. Typically, production companies cut canals through the marshes to give barges and other equipment easy access to the drilling platforms and production facilities. When oil or gas is found, it must be transported to land through pipelines. Laying pipelines often requires cutting additional canals through the wetlands. These canals break the wall of vegetation that protects the soft muds, which accelerates the rate of erosion.

The nutria, a rodent introduced from South America, is also contributing to the problem. These animals reproduce rapidly and form such large populations that they eat all the vegetation in a local area. This exposes the muds to erosion and contributes to the loss of wetlands. Trapping these animals for their fur was at one time profitable for the local people and helped to control the animals. However, a shift in public opinion has reduced the acceptability of fur for clothing and depressed the fur market, making the trapping of nutria uneconomical.

---

other flood less regularly. They are generally flat, and so are inviting areas for residential development. A better use of these areas is for open space or for recreation, yet developers continue to build houses and light industry there.

When a floodplain is developed for residential or commercial use, a retaining wall usually is built to prevent the periodic flooding natural to the area. This increases the cost of the development, increases the cost of insurance protection, and creates high-water problems downstream. Frequently, tax monies are used to repair the damage that results from the unwise use of floodplains. Floodplain development is one of the insidious problems associated with a rapidly expanding population. As long as the population continues to increase, these less desirable areas are likely to be used for housing, whether they are subject to annual flooding or less frequent damage.

Floods are natural phenomena. Contrary to popular impressions, no evidence supports the premise that floods are worse today than they were one hundred or two hundred years ago, except perhaps on small, isolated watersheds. What has increased is the economic loss from the flooding. This loss reflects the fact that more cities, industries, railroads, and highways have been constructed on floodplains. Sometimes there are no alternatives to floodplain development, but too often, risks are simply ignored. Because flooding causes loss of life and property, floodplains should no longer be developed for uses other than agriculture and recreation.

Many communities have enacted **floodplain zoning ordinances** to restrict future building in floodplains. Although such ordinances may prevent further economic losses, what happens to individuals who already live in floodplains? Floodplain building ordinances usually allow current residents to remain. However, these residents often find it extremely difficult to obtain property insurance. Relocation, usually at a financial loss, is the only alternative. Such situations are unfortunate; perhaps proper planning in the future will prevent these problems.

Wetlands Misuse

Since access to water was and continues to be important to industrial development, many cities are located in areas with extensive wetlands. **Wetlands** include swamps, tidal marshes, coastal areas, and estuaries. Because wetlands breed mosquitoes and are sometimes barriers to the free movement of people, they have often been considered useless or harmful. Most of them have been drained, filled, or used as dumps. Many modern cities have completely covered over extensive wetland areas and may even have small streams running under streets, completely enclosed in concrete.

Each kind of wetland has unique qualities and serves as a home to many kinds of plants and animals. Wetlands are frequently associated with the reproductive portion of the life cycle of the organisms that inhabit them. They provide nesting sites and spawning sites, and commercial and sport fisheries depend upon these habitats to produce and protect the young of the species they harvest. Because most wetlands receive constant inputs of nutrients from the water that drains from the surrounding land, they are highly productive and excellent places for aquatic species to grow rapidly. Human impact on wetlands has severely degraded or eliminated these spawning and nursery habitats.

Besides providing a necessary habitat for fish and other organisms, wetlands provide natural filters for sediments and runoff. This filtration process allows time for water to be biologically cleaned before it enters larger bodies of water such as lakes and oceans and reduces the sediment load carried by runoff. Wetlands also protect shorelines from erosion. When destroyed, the natural erosion protection provided by wetlands must be replaced by costly artificial measures, such as breakwalls.

Other Land-Use Considerations

The geologic status of an area must also be considered in land-use decisions. Building cities on the sides of volcanos or on major earthquake-prone faults has led to much loss of life and property. Siting homes and villages on unstable hillsides or in areas subject to periodic fires is also unwise.

Another problem in some locations is lack of water. Southern California and metropolitan areas in Arizona must import water to sustain their communities. Wise planning would limit growth to whatever could be sustained by available resources. The risk of serious water shortages in Beijing, China, has forced the government to consider a massive 1200-kilometer diversion of water from the Yangtze River to the city. As these areas in California, Arizona, and China continue to grow, the strain on regional water resources will increase.

LAND-USE PLANNING: PRINCIPLES AND CHALLENGES

As we look back at the location and development of cities and metropolitan areas, we can see the mistakes that have been made. If we were in a position to start over, we might locate our population centers differently: we might regulate building and development for the least damaging but highest use of the land. But how does one go about determining the best use and least damaging use of an area? What goes into the process of land-use planning?

Land-use planning is the construction of an orderly list of priorities for the use of available land. Developing a plan involves gathering data on current use and geological, biological, and sociological information. From these data, projections are made about what human needs will be. All of the data collected are integrated with the projections, and each parcel of land is evaluated and assigned a best use under the circumstances. Finally, mechanisms for implementing the plan are divided into two categories: purchase of the land or regulation of its use.

Probably the simplest way to protect desirable lands is to purchase them. This method has often been used by government or by conservation-minded organizations or individuals. The major problem is the cost. Many communities are not in a financial position to purchase lands; therefore, they attempt to regulate land use by zoning laws.

Zoning is a common type of land-use regulation. When land is zoned, it is designated for specific potential uses. Common designations are agricultural, commercial, residential, recreational, and industrial. (See figure 12.7.)

Local governments often lack funds to hire professional planners. As a result, zoning regulations are frequently made by people who see only the short-term gain and not the possible long-term loss. The land is simply zoned for its current use. Even when well-designed land-use plans exist, they are usually modified to encourage local short-term growth rather than to provide for the long-range needs of the community. The community needs to be alert to variances from established land-use plans, because once the plan is compromised, it becomes easier to accept future deviations that may not be in the best interests of the community. Many times, individuals who make zoning decisions are real estate agents, developers, or local business people. These individuals wield significant local political power and are not always unbiased in their decisions. Concerned citizens must try to combat special interests by attending zoning commission meetings and by participating in the planning process.

Urban Transportation Planning

A growing concern of city governments is to develop comprehensive urban transportation plans. While the specifics of such plans might vary from region to region, urban transportation planning usually involves four major goals:

1. Conserve energy and land resources.
2. Provide efficient and inexpensive transportation within the city for the elderly, young, poor, and handicapped.

FIGURE 12.7 Zoning Most communities have a zoning authority that designates areas for particular use. This sign indicates that decisions have been made about the "best" use for the land.

3. Provide suburban people opportunities to commute efficiently.
4. Reduce urban pollution.

Any successful urban transportation plan should integrate all of these goals, but funding and intergovernmental cooperation are needed to achieve this. The problems associated with current urban transportation will certainly not disappear overnight, but comprehensive planning is the first step to solving them.

A primary method of transportation in most urban centers is the automobile. The automobile's advantages include convenience and freedom of movement. However, many cities are beginning to recognize that the automobile's disadvantages may outweigh its advantages. The disadvantages include increased air pollution, substantial land use for highways and parking areas, and greater urban sprawl. Recognizing these problems, some cities, such as Toronto, London, San Francisco, and New York, have attempted to dissuade automobile use by developing mass transit systems and by allowing automobile parking costs to increase substantially.

The major urban mass transit systems are railroads, subways, trolleys, and buses. Such systems, however, are not without their problems. Mass transit is:

1. Economically feasible only along heavily populated routes
2. Less convenient than the automobile
3. Extremely expensive to build and operate
4. Often crowded and uncomfortable

Although mass transit meets a substantial portion of the urban transportation needs in some parts of the world, such as in Europe and Russia, its use in North America has declined since the 1940s. (See figure 12.8.) A variety of forces has caused this decline. As people became more affluent, they could afford to own automobiles, which are a convenient, individualized method of transportation. Governments in North America encourage automobile use by financing highways and expressways, by maintaining a cheap energy policy, and by withdrawing support for most forms of mass transportation. Thus they encourage automobile transportation with hidden subsidies (highway construction and cheap gasoline) but maintain that rail and bus transportation should not be subsidized. North Americans will only seek alternatives to private automobile use when the cost of fuel, the cost of parking, or the inconvenience of driving becomes too high.

Urban Recreation Planning

Nearly three-fourths of the population of North America live in urban areas. These urban dwellers value open space because it breaks up the sights and sounds of the city and provides a place for recreation. Inadequate land-use planning in the past has rapidly converted urban open spaces to other uses. Until recently, creating a new park within a city was considered an uneconomical use of the land, but people are now beginning to realize the need for parks and open spaces.

Some cities recognized the need for open space a long time ago and allocated land for parks. London, Toronto, and Perth, Australia, have centrally located and well-used parks. New York City set aside approximately 200 hectares for Central Park in the late 1800s. (See figure 12.9.) Boston has developed a park system that provides a variety of urban open spaces. Other cities have not dealt with this need for open space because they have lacked either the foresight or the funding. Recreation is a basic human need. The most primitive tribes and cultures all engaged in games or recreational activities. New forms of recreation are continually being developed. In the congested urban center, cities often must

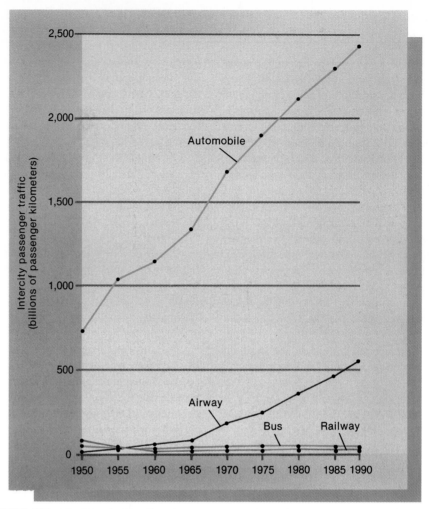

FIGURE 12.8 Decline of Mass Transportation Automobile use has increased consistently since 1950 while rail and bus transport has remained low.

construct special areas where recreation can take place.

A major problem with urban recreation is locating recreational facilities near residential areas. Facilities that are not conveniently located may be infrequently used. For example, the hundreds of thousands of square kilometers of national parks in Alaska and the Yukon will be visited every year by a tiny proportion of the population of North America. Large urban centers are discovering that they must provide adequate, low-cost recreational opportunities within their jurisdiction. Some of these opportunities take the form of commercial establishments, such as bowling centers, amusement parks, and theaters. Others must be subsidized by the community. (See figure 12.10.) Playgrounds, organized recreational activities, and open space have usually been combined into an arm of the municipal

government known as the parks and recreation department. Cities spend millions of dollars to develop and maintain recreation programs. Often, there is conflict over the allocation of financial and land resources. These are closely tied because open land is scarce in urban areas, and it is expensive. Riverfront property is ideal for park and recreational use, but it is also prime land for industry, commerce, or high-rise residential buildings. Although conflict is inevitable, many metropolitan areas are beginning to see that recreational resources may be as important as economic growth for maintaining a healthy community.

An outgrowth of the trend toward urbanization is the development of **nature centers.** In many urban areas, there is so little natural area left that the people who live there need to be given opportunities to learn about nature. Nature centers are basically teaching institutions that provide a

New York Central Park

Hyde Park, London

FIGURE 12.9 Urban Open Space These photos show open space in two urban areas, New York's Central Park and Hyde Park in London. If the land had not been set aside, it would have been developed.

variety of methods for people to learn about and appreciate the natural world. Zoos, botanical gardens, and some urban parks, combined with interpretative centers, also provide recreational experiences. Nature centers are usually located near urban centers, in places where some appreciation of the natural processes and phenomena can be developed. They may be operated by municipal governments or by school systems or other nonprofit organizations.

NATIONAL AND REGIONAL PLANNING

National and regional planning is often more effective than local land-use planning because political boundaries seldom reflect the geological and biological data base used in planning. Larger units contain more diverse collections of landscape resources and can afford to hire professional planners. A regional approach is also likely to prevent duplication of facilities and lead to greater efficiency. For example, airport locations should be based on a regional plan that incorporates all local jurisdictions. Three cities only

30 kilometers apart should not build three separate airports when one regional airport could serve their needs better and at a lower cost to the taxpayers.

Although regional planning is increasing in North America, the majority of regional governmental bodies are presently voluntary and lack any power to implement programs. Their only role is to advise the member governments. Unfortunately, members of local governments still seem unwilling to give up power. They may view policy from a narrow perspective and put their own interests above the goals of the region. An elected, multipurpose, regional government, on the other hand, would be ideal for implementing land-use policy. Such governments exist in only a handful of places and show few signs of spreading.

One way to encourage regional planning is to develop policies at the state, provincial, or national level. The first state to develop a comprehensive statewide land-use program was Hawaii. During the early 1960s, much of Hawaii's natural beauty was being destroyed to build houses and apartments for the increasing population. The same land that attracted tourists was being destroyed to provide hotels and supermarkets for them. Local

FIGURE 12.10 Urban Recreation In urban areas, recreation often takes the form of sports programs, playgrounds, and walking. Most cities recognize the need for such activities and develop extensive recreation programs for their citizens.

governments had failed to establish and enforce land-use controls. Consequently, the Hawaii State Land-Use Commission was founded in 1961. This commission designated all land as urban, agricultural,

FIGURE 12.11 Hawaii Land-Use Plan Hawaii was the first state to develop a comprehensive land-use plan. This development in an agricultural area was stopped when the plan was implemented in the 1960s.

TABLE 12.1 NUMBER OF PEOPLE WHO PARTICIPATED IN SELECTED OUTDOOR RECREATIONAL ACTIVITIES IN 1995

Activity	Percent of population over 7 years of age participating
Exercise walking	30
Swimming	27
Bicycling	24
Fishing	18
Camping	18
Golf	12
Hiking	10
Running	9
Hunting	7
Backpacking	4
Skiing	4

Source: Data from *Statistical Abstract of the United States, 1997.*

or conservational. Each parcel of land could be used only for its designated purpose. Other uses were allowed only by special permit. To date, the record for Hawaii's action shows that it has been successful in controlling urban growth and preserving the islands' natural beauty, even though the population continues to grow. (See figure 12.11.)

Several states and provinces are attempting to follow Hawaii's lead in state land-use regulation. Some have passed legislation dealing with special types of land use. Examples include wetland preservation, floodplain protection, and scenic and historic site preservation. Although direct state involvement in land-use regulation is relatively new, it is expected to grow. Only large, well-financed levels of government can afford

to pay for the growing cost of adequate land-use planning. State, provincial, and regional governments are also more likely to have the power to counter the political and economic influences of land developers, lobbyists, and other special-interest groups when conflicts over specific land-use policies arise.

National governments also have a role to play. Since national governments own and administer the use of large amounts of land, national policy will determine how those lands are used. The designation of lands as wilderness, forests, rangelands, or parks at the federal level often involves a balancing of national priorities with local desires. As with local land-use issues, the conflicts over the use of federal lands are often economic and involve compromise.

Multiple Land Use

Once a land-use decision is made, it may be irreversible. For example, when a highway is constructed, it is virtually impossible to use that land for any other purpose. If land is designated as a wilderness area, no human development is allowed. Other single exclusionary uses of land are agriculture, waste storage, housing, and industrial development.

Sometimes several different uses of the land can take place at the same time. They do not have to be exclusionary. Two or more uses of land occurring at the same time is called **multiple land use.** Perhaps the best example of multiple land use occurs in U.S. national forests. In 1960, the Multiple Use Sustained Yield Act divided use of national forests into four categories: wildlife habitat preservation, recreation, lumbering, and watershed protection. This act encouraged both economic and recreational use of the forests.

Another example of multiple land use is the development of parks on floodplains. These parks provide open space, recreational land, and storage reservoirs for storm sewer runoff. Storage reservoirs can even be blended into the park landscape as ponds or lakes.

The Use of Public Land for Outdoor Recreation

Not all people enjoy the same recreation. Some people enjoy reading or watching television. Others prefer commercial activities such as golf, tennis, bowling, amusement areas, racetracks, and skiing. Another major area of recreation, usually classified as **outdoor recreation,** involves using the natural out-of-doors for hiking, camping, canoeing, and so forth.

Many people want to use the natural world for recreational purposes because nature can provide challenges that may be lacking in their day-to-day lives. Whether the challenge is hiking in the wilderness, underwater exploration, climbing mountains, or driving a vehicle through an area that has no roads, these activities offer a sense of adventure. Look at table 12.1. All of these activities use the out-of-doors, but not in the same way. Conflicts develop because some of these activities cannot occur in the same place at the same time. For ex-

FIGURE 12.12 Conflict over Recreational Use of Land Land may be for both motorized and nonmotorized use. The people who participate in these two kinds of recreation are often antagonists over the allocation of land for recreational use.

ample, wilderness camping and backpacking often conflict with off-road vehicles.

There is a basic conflict between those who prefer to use motorized vehicles and those who prefer to use muscle power in their recreational pursuits. (See figure 12.12.) This conflict is particularly strong because both groups would like to use the same publicly owned land. Both have paid taxes, both "own" the land, and both feel that it should be available for them to use as they wish. When everyone "owns" something, there is often little desire to regulate activities.

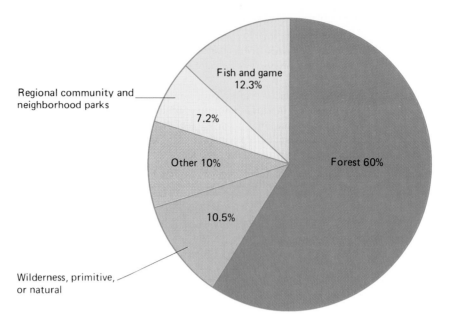

FIGURE 12.13 U.S. Federal Recreational Lands in 1995 Of the approximately 108 million hectares of federal recreational lands in the United States, approximately 10 percent is designated as wilderness, primitive, or natural.

Source: Data from the *Statistical Abstract of the United States, 1995.*

For example, much of the rangeland in the western United States is publicly owned. Based on "Animal Use Months" established by the Bureau of Land Management or the Forest Service, ranchers are allowed to graze cattle, but only on certain public lands. Technically, failure to comply can mean a loss of grazing rights. However, since the regulations are politically motivated and the regulatory agency is understaffed, there is no incentive for ranchers to limit the number of cattle or sheep on the land unless all ranchers agree to do so. The usual result is that a few individuals exploit the land in hopes of a short-term profit. This exploitation could cause permanent damage to the rangeland. Many people want to use the publicly owned rangelands for outdoor recreation and resent the control exercised by grazing interests. On the other hand, ranchers resent the intrusion of hikers and campers on land they have traditionally controlled.

An obvious solution to this problem is to allocate land to specific uses and to regulate the use once allocations have been made. Several U.S. governmental agencies, such as the National Park Service, the Bureau of Land Management, the U.S. Forest Service, and the U.S. Fish and Wildlife Service, allocate and regulate the lands they control. However, these agencies have con-

flicting roles. The U.S. Forest Service has a mandate to manage forested public lands for timber production. This mandate often comes in conflict with recreational uses. Similarly, the Bureau of Land Management has huge tracts of land that can be used for recreation, but it traditionally has been aligned with grazing interests.

A particularly sensitive issue is the designation of certain lands as wilderness areas. Obviously, if an area is to be wilderness, human activity must be severely restricted. This means that the vast majority of Americans will never see or make use of it. Many people argue that this is unfair because they are paying taxes to provide recreation for a select few. Others argue that if everyone were to use these areas, their charm and unique character would be destroyed and that, therefore, the cost of preserving wilderness is justifiable.

Areas designated as wilderness make up a very small proportion of the total public land available for recreation. (See figure 12.13.) The fact that there are relatively few wilderness areas has resulted in a further problem: The areas are being loved to death. People pressure on this resource has become so great that, in some cases, the wilderness quality is being tarnished. The designation of additional wilderness would relieve some of this pressure.

The following situation has happened thousands of times during the past 20 years:

A developer has just announced plans to build a large shopping mall on the outskirts of your city in what is now prime farmland. Many jobs will be created by the construction and operation of the proposed mall, which will include stores and three new theaters. Presently, your city has some unemployment, only one theater in the downtown area, and little variety in its retail businesses. On the surface, the proposed mall seems to be only good news. Is this the case? Before answering, look at the entire situation.

- What will happen to the downtown area when the mall is opened?

- If you owned a downtown business, would you favor building the mall?

- What will happen to the taxes on the farms near the new mall?

- If you were a farmer, would you favor the project?

- What effect will paving prime farmland have?

- How will storm water runoff be affected?

- How will future housing development be influenced?

- What other problems can be associated with building the complex?

- If you were the mayor or city manager, would you favor construction of the mall? Why?

- Is the proposed project all goods news after all?

Summary

Historically, waterways served as transportation corridors that allowed for the exploration of new land and for the transport of goods. Therefore, most large urban centers began as small towns located near water. Water served the needs of the towns in many ways, especially as transportation. Several factors resulted in the shift of the population from rural to urban. These included the Industrial Revolution which provided jobs in cities and the addition of foreign immigrants to the cities. As towns became larger, the farmland surrounding them became suburbs surrounding industrial centers. Unregulated industrial development in cities resulted in the degradation of the waterfront and stimulated the development of suburbs around the city as people sought better places to live and had the money to purchase new homes. The rise in automobile ownership further stimulated the movement of people from the cities to the suburbs.

Many problems have resulted from unplanned growth. Current taxation policies encourage residential development of farmland, which results in a loss of valuable agricultural land. Floodplains and wetlands are often mismanaged. Loss of property and life results when people build on floodplains. Wetlands protect our shorelines and provide a natural habitat for fish and wildlife. Transportation problems and lack of open space are also typical in many large metropolitan areas.

Land-use planning involves gathering data, projecting needs, and developing mechanisms for implementing the plan. Purchasing and zoning land are two ways to enforce land-use planning. The scale of local planning is often not large enough to be effective because problems may not be confined to political boundaries. Regional planning units can afford professional planners and are better able to withstand political and economic pressures. A growing concern of urban governments is to develop comprehensive urban transportation plans that seek to conserve energy and land resources, provide efficient and inexpensive transportation and commuting, and help to reduce urban pollution. Urban areas must also provide recreational opportunities for their residents.

Federal governments own and manage large amounts of land, therefore national policy must be developed. This usually involves designating land for particular purposes, such as forest, grazing, parks, or wilderness. The recreational use of public land often requires the establishment of rules that prevent conflict between potential users who have different ideas about what appropriate uses should be. Often federal policy is a compromise between competing uses and land is managed for multiple uses.

Review Questions

1. Why did urban centers develop near waterways? Are they still located near water?
2. Describe the typical changes that have occurred in cities from the time they were first founded until now.
3. Why do people move to the suburbs?
4. Why do some farmers near urban areas sell their land for residential or commercial development? If you were in this position, would you sell?
5. What is a megalopolis?
6. What land uses are suitable on floodplains?
7. What is multiple land use? Can land be used for multiple purposes?
8. Why is it important to provide recreational space in urban planning?
9. How can recreational activities damage the environment? Do you engage in any of those activities?
10. What is the monetary impact of recreational activities?
11. What are some strictly urban-related recreational activities?
12. List some conflicts that arise when an area is designated strictly as recreational.
13. Describe the steps necessary to develop a land-use plan.
14. What are the advantages of regional or state planning?
15. List three benefits of land-use planning.

Critical Thinking Questions

1. Choose a city where you live. Interview local residents and look at old city maps. What did the city look like 75 years ago? What were the city's boundaries? Where did people do their shopping? How did they get around? How does this compare with the current situation in the city?
2. What historical factors brought members of your family to the city? How does this compare to the factors that are currently contributing to the growth of cities in the developing world?
3. Consider the outer rim of the city closest to you. Which, if any, of the problems associated with unplanned growth are associated with your city? What factors make them a problem? What do you think can be done about them?
4. There has been tremendous development in the arid West of the United States over the past few decades, creating demands for water. How should these demands be met? Should there be limits to this type of development? What kinds of limits, if any?
5. Imagine you are a National Forest Supervisor who is creating a 10-year plan that is in the public comment stage. What interests would be contacting you? What power would each interest have?

How would you manage the competing interests of timber, mining, grazing, and recreation or between motorized and non-motorized recreation? What values, beliefs, and perspectives helped you form your recommendations?
6. Imagine that you lived in an area of the country that has the potential to be named a wilderness area. What conflicts do you think would arise from such a declaration? Who might be some of the antagonists? Which perspective do you think is most persuasive? How would you answer the objections of the other perspective?
7. Look at the Issues and Analysis section of this chapter and answer the questions. What values, beliefs, and perspectives lead you, as the imaginary city manager, to reach your decision about the shopping mall?
8. After reading the Environmental Close-up in this chapter concerning wetland loss in Louisiana, what kinds of recommendations would you make to help preserve wetlands? What do you suppose might happen if nothing is done? What resistance might wetland preservation generate?

Key Terms

floodplain 221	outdoor recreation 226
floodplain zoning ordinances 222	ribbon sprawl 220
land-use planning 223	tract development 219
megalopolis 220	urban sprawl 219
multiple land use 226	wetlands 222
nature centers 224	zoning 223

Related Internet Sites

Visit our website at http://www.mhhe.com/environmental science/ for specific information on each of the topics below.

Land management, fire, ecology, and the great basin

Bureau of Land Management

Outdoor recreation planning and management

Transportation, community, and the environment

Protection of the Wolf River and its floodplains

Worldwide threats to wildlife, wetlands, clean water, and more

U.S. Fish and Wildlife national wetlands inventory

Cold-climate-constructed wetlands

Efficient and sustainable management of land and water resources

Land use studies, habitat restoration, and geographic information systems

Habitat restoration projects in California

Soil and Its Uses

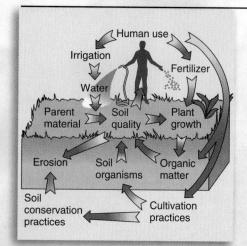

OBJECTIVES

After reading this chapter you should be able to:

- Describe the geologic processes that build and degrade the Earth's surface.
- List the physical, chemical, and biological factors responsible for soil formation.
- Explain the importance of humus to soil fertility.
- Differentiate between soil texture and soil structure.
- Explain how texture and structure influence soil atmosphere and soil water.
- Explain the role of living organisms in soil formation and fertility.
- Describe the various layers in a soil profile.
- Describe the processes of soil erosion by water and wind.
- Explain how contour farming, strip farming, terracing, waterways, windbreaks, and conservation tillage reduce soil erosion.
- Understand that the misuse of soil reduces soil fertility, pollutes streams, and requires expensive remedial measures.
- Explain how land not suited for cultivation may still be productively used for other purposes.

GEOLOGIC PROCESSES

We tend to think of the Earth as being stable and unchanging until we recognize such events as earthquakes, volcanic eruptions, floods and windstorms that change the surface of the places we live. There are forces that build new land and opposing forces that tear it down. Much of the building process involves shifting of large portions of the Earth's surface as plates. **Plate tectonics** is the concept that the outer surface of the Earth, known as the **crust,** consists of large plates that are slowly moving over the surface of the mantle. The **mantle** is a layer that makes up the majority of the Earth and surrounds a small core made up primarily of iron. The mantle consists of an inner solid portion and an outer portion that is capable of flow. (See figure 13.1.) The heat from the Earth causes slow movements of the outer layer of the mantle similar to what happens when you heat a liquid on the stove, only much slower.

The movements of the plates on this plastic outer layer of the mantle are independent of each other. Therefore, some of the plates are pulling apart from one another while others are colliding.

Where the plates are pulling apart from one another, the liquid mantle moves upward to fill the gap and solidifies. Thus new crust is formed from the liquid mantle. Approximately half of the surface of the Earth has been formed in this way in the past 200 million years. The bottom of the Atlantic and Pacific Oceans and the Rift Valley and Red Sea area of Africa are areas where this is occurring.

If plates are pulling apart on one portion of the Earth they must be colliding elsewhere. Where plates collide several things can happen. (See figure 13.2.) Often one of the plates slides under the other and is melted. When this occurs often some of the liquid mantle makes its way to the surface and volcanoes are formed which results in the formation of mountains. The west coasts of North and South America have many volcanoes and mountain ranges where the two plates are colliding. The volcanic activity adds new material to the crust. When a collision occurs between two plates under the ocean, the volcanoes may eventually reach the surface and form a chain of volcanic islands such as can be seen in the Aleutian Islands and many of the Caribbean Islands. When two continental plates collide neither plate slides under the other and the crust buckles to form mountains. The Himalayas, Alps, and Appalachian Mountain ranges are thought to have formed from the collision of two continental plates. All of these movements of the Earth's surface are associated with earthquakes. The movements of the plates are not slow and steady sliding movements but tend to occur in small jumps. However, what is a small movement between two plates on the Earth is a huge movement for the relatively small structures and buildings produced by humans, and they can cause tremendous amounts of damage.

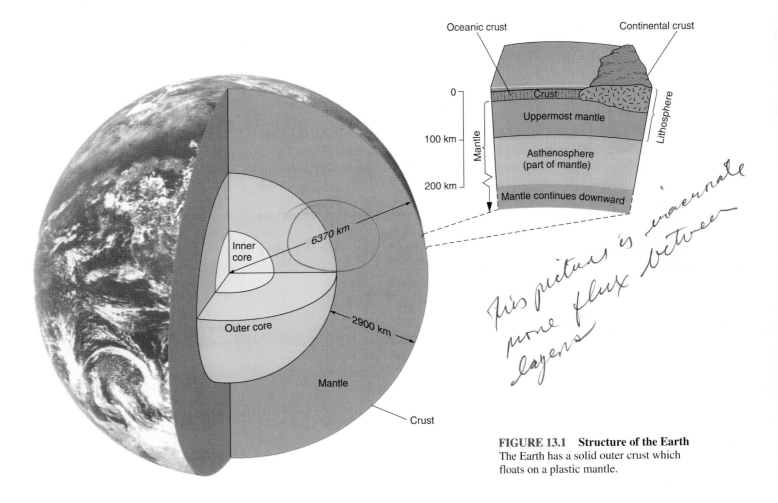

FIGURE 13.1   Structure of the Earth
The Earth has a solid outer crust which floats on a plastic mantle.

△ Direction of plate motion	Subduction zone (sawteeth point toward overriding plate)
	Spreading ridge
	Transform fault

FIGURE 13.2 Tectonic Plates The plates which make up the outer crust move with respect to one another. They pull apart from one another in some parts of the world and collide with one another in other parts of the world.

These building processes are counteracted by processes that tend to make the elevated surfaces lower. Gravity provides a force that tends to wear down the high places. Moving water and ice (glaciers) and wind assist in the process; however their effectiveness is related to the size of the particles. Several kinds of **weathering** processes are important in reducing the size of particles which can then be dislodged by moving water and air. **Mechanical weathering** results from physical forces that reduce the size of rock particles without changing the chemical nature of the rock. Common causes of mechanical weathering are: the freezing of water which expands and tends to split larger pieces of rock into smaller ones; changes in temperature that tend to result in fractures in rock; and the actions of plants and animals. **Chemical weathering** involves the chemical alteration of the rock in such a manner that it is more likely to fragment or to be dissolved. Because of gravity the prevailing movement of particles is from high elevations to lower ones. This process of loosening and

redistributing particles is known as **erosion.** Wind can move sand and dust and can also cause the wearing away of rocky surfaces by sandblasting their surfaces. Glaciers can move large rocks and also cause their surfaces to be rounded by being rubbed against each other and the surface of the Earth. Moving water transports much material in streams and rivers. In addition, wave action along the shores of lakes and the coasts of oceans constantly wears away and transports particles.

SOIL AND LAND

The geologic processes just discussed are involved in the development of both soil and land; however, soil and land are not the same. **Land** is the part of the world not covered by the oceans. **Soil** is a thin covering over the land consisting of a mixture of minerals, organic material, living organisms, air, and water that together support the growth of plant life. The proportions of the soil components vary with different

types of soils but a typical, "good" soil is about 45 percent mineral, 25 percent air, 25 percent water, and 5 percent organic matter (see figure 13.3). Farmers are particularly concerned with soil because the nature of the soil determines the kinds of crops that can be grown and which farming methods must be employed. Urban dwellers should also be concerned about soil because its health determines the quality and quantity of food they will eat. If the soil is so abused that it can no longer grow crops, or if it is allowed to erode, degrading air and water quality, both urban and rural residents suffer. To understand how soil can be protected, we must first understand its properties and how it is formed.

SOIL FORMATION

A combination of physical and biological events acting over time is responsible for the formation of soil. Soil building begins with the physical fragmentation of the **parent material,** which consists of ancient

1, Pedologist - Soil as a natural body

2. Edaphology = plant production

layers of rock or more recent geologic deposits from lava flows or glacial activity. The kind and amount of soil developed depends on the kind of parent material present, the plants and animals present, the climate, time, and the slope of the land. Factors that can bring about fragmentation or chemical change of the parent material are known as weathering. As was discussed earlier, the physical breakdown of parent material is known as mechanical weathering. Temperature changes and abrasion are two primary agents of mechanical weathering.

Because rock does not expand evenly, heating a large rock can cause it to fracture, causing pieces of the rock to flake off. These pieces can be further reduced in size by other processes, such as the repeated freezing and thawing of water. Water that has seeped into rock cracks and crevices expands as it freezes, causing the cracks to widen. Subsequent thawing allows more water to fill the widened cracks, which are enlarged further by another period of freezing. Alternating freezing and thawing fragments large rock pieces into smaller ones (see figure 13.4). The roots of plants growing in cracks can also exert enough force to break rock.

The physical breakdown of rock is also caused by forces that move and rub rock particles against each other. For example, a glacier causes rock particles to grind against one another, resulting in smaller fragments and smoother surfaces. These particles are deposited by the glacier when the ice melts. In many parts of the world, the parent material from which soil is formed consists of glacial deposits. Wind and moving water also cause small particles to collide, resulting in further weathering. The smoothness of rocks and pebbles in a stream or on the shore is evidence that moving water has caused them to rub together, removing their sharp edges. Similarly, particles carried by wind collide with objects, fragmenting both the objects and the wind-driven particles.

Wind and moving water also remove small particles and deposit them at new locations, exposing new surfaces to the weathering process. For example, the landscape of the Painted Desert in the southwest United States was created by a combination of wind and moving water

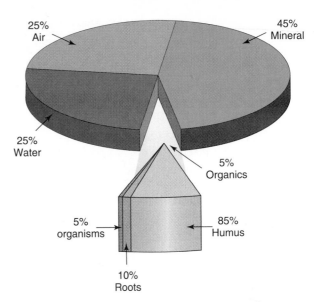

FIGURE 13.3 **The Components of Soil** Although soils vary considerably in composition, they all contain the same basic components: mineral material, air, water, and organic material. The organic material can be further subdivided into humus, roots, and other living organisms. The percentages shown are those that would be present in a good soil.

that removed easily transported particles, while rocks more resistant to weathering remained. (See figure 13.5.)

In addition to wind, moving water, glaciers, and changing temperature, certain chemical weathering activities also alter the size and composition of parent material and participate in the soil-building process. Some small rock fragments exposed to the atmosphere may be oxidized; that is, they combine with oxygen from the air and chemically change to different compounds. Other kinds of rock may combine with water molecules in a process known as hydrolysis. Often, the oxidized or hydrolyzed molecules are more readily soluble in water and, therefore, may be removed by rain or moving water. Since rain is normally slightly acid, the acid content helps dissolve rocks.

The role of organisms in the development of soil is very important. The first organisms to gain a foothold in this modified parent material also contribute to soil formation. Lichens often form a pioneer community that traps small particles and chemically alters the underlying rock, causing further fragmentation. The release of chemicals from the roots of plants causes further chemical breakdown of rock particles. As other kinds of organisms, like plants and small animals, become established, they contribute increasing amounts

FIGURE 13.4 **Physical Fragmentation by Freezing and Thawing** The crack in the rock fills with water. As the water freezes and becomes ice, the pressure of the ice enlarges the crack. The ice melts, and water again fills the crack. The water freezes again and widens the crack. Alternate freezing and thawing splits the rock into smaller fragments.

of organic matter, which are incorporated with the small rock fragments to form an important component of the soil known as humus.

Humus, which is the organic material resulting from the decay of organic matter, becomes mixed with the top layers of rock particles and constitutes a very important ingredient of soil. Humus supplies some of the nutrients needed by plants, increases the water-holding capacity of the soil, and also increases the acidity of the soil so that inorganic nutrients, which are more soluble under acidic conditions, become available to plants. For example, wheat and corn grow best in soils with a pH between 5.5 and 7.0. A soil with a pH above 7.0 would be more productive if humus were added to increase the acidity. One problem with an acidic soil is that because nutrients are more soluble they can be carried away if there is heavy rainfall or if there are few plants to capture them. Humus also helps to create a loose, crumbly soil texture that allows water to soak in and permits air to be incorporated into the soil. Compact soils have poor aeration, and the water runs off.

Burrowing animals, soil bacteria, fungi, and the roots of plants are also part of the biological process of soil formation. One of the most important burrowing animals is the earthworm. One hectare of soil may support a population of 500,000 earthworms that can process as much as 9 metric tons of soil a year. These animals literally eat their way through the soil, resulting in further mixing of organic and inorganic material, which increases the amount of nutrients available for plant use. They often bring nutrients from the deeper layers of the soil up into the area where plant roots are more concentrated, thus improving the soil's fertility. Soil aeration and drainage are also improved by the burrowing of earthworms and other small soil animals, such as nematodes, mites, pill bugs, and tiny insects. They also help to incorporate organic matter into the soil by collecting dead organic material from the surface and transporting it into burrows and tunnels.

Fungi and bacteria are decomposers and serve as important links in many mineral cycles. (See chapter 4.) They, along with animals, reduce organic material to smaller particles and, therefore, improve the quality of the soil.

Climate and time are also important in the development of soils. In general, extremely dry or cold climates develop soils very slowly, while humid and warm climates develop them more rapidly. Under ideal climatic conditions, soft parent material may develop into a centimeter of soil within 15 years. Under poor climatic

FIGURE 13.5   The Painted Desert This landscape was created by the action of wind and moving water. The particles removed by these forces were deposited elsewhere and may have become part of the soil in that new location.

conditions, a hard parent material may require hundreds of years to develop into that much soil. In any case, soil formation is a slow process.

The slope of the land also influences soil development. Soil formation on steep slopes is very slow because materials tend to be moved downslope with wind and water. Conversely, river valleys often have deep soils because they receive materials from elsewhere by these same erosive forces.

SOIL PROPERTIES

Soil properties include soil texture, structure, atmosphere, moisture, biotic content, and chemical composition. **Soil texture** is determined by the size of the rock particles within the soil. The largest soil particles are gravel, which consists of fragments larger than 2.0 millimeters in diameter. Particles between 0.05 and 2.0 millimeters are classified as sand. Silt particles range from 0.002 to 0.05 millimeters in diameter, and the smallest particles are clay particles, which are less than 0.002 millimeters in diameter.

Large particles, such as sand and gravel, have many tiny spaces between them, which allow both air and water to

flow through the soil. Water drains from this kind of soil very rapidly, often carrying valuable nutrients to lower soil layers, where they are beyond the reach of plant roots. Clay particles tend to be flat and are easily packed together to form waterproof layers. Soils with a lot of clay do not drain well and are poorly aerated. Because water does not flow through clay very well, clay soils tend to stay moist for longer periods of time and do not easily lose minerals to infiltrating water.

Rarely does a soil consist of a single size of particle. Various particles are mixed in many different combinations, resulting in many different soil classifications. (See figure 13.6.) An ideal soil for agricultural use is a **loam,** which combines the good aeration and drainage properties of large particles with the nutrient-retention ability of clay particles.

Soil structure is different from its texture. **Soil structure** refers to the way various soil particles clump together. The particles in sandy soils do not attach to one another and, therefore, sandy soils have a granular structure. The particles in clay soils tend to stick to one another to form large clods. Other soils that have a mixture of particle sizes tend to form smaller clods. A good soil forms small clods that crumble easily. This ability to

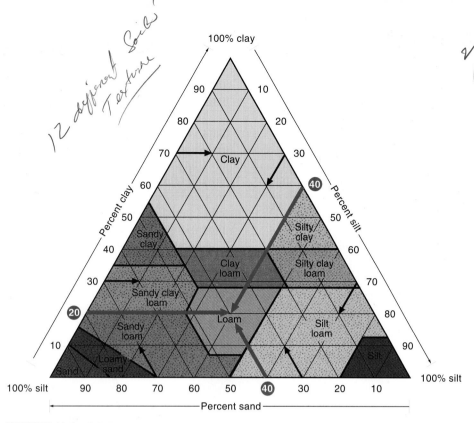

12 different soils'
Texture

FIGURE 13.6 **Soil Texture** Texture depends upon the percentage of clay, silt, and sand particles in the soil. Soils with the best texture for most crops are loams. As shown in the illustration, if a soil were 40 percent sand, 40 percent silt, and 20 percent clay, it would be a loam.

Source: Data from Soil Conservation Service

crumble is known as the soil's **friability.** The friability is determined by the soil structure and its moisture content. Sandy soils are very friable, while clay soils are not. If clay soil is worked when it is too wet, it can stick together in massive chunks that will not break up for years.

A good soil for agricultural use will crumble and has spaces for air and water. In fact, the air and water content depends upon the presence of these spaces. (See figure 13.7.) In good soil, about two-thirds of the spaces contain air after the excess water has drained. The air provides a source of oxygen for plant root cells and all the other soil organisms. The relationship between the amount of air and water is not fixed. After a heavy rain, most of the spaces may be filled with water and less oxygen is available to plant roots and other organisms. If some of the excess water does not drain from the soil, the plant roots may die from lack of oxygen. They are literally drowned. On the other hand, if there is not enough soil moisture, the plants wilt from lack of water. Soil moisture and air are also important in determining the numbers and kinds of soil organisms.

Protozoa, nematodes, earthworms, insects, algae, bacteria, and fungi are typical inhabitants of soil. (See figure 13.8.) The role of protozoa in the soil is not firmly established, but they seem to act as parasites and predators on other forms of soil organisms and, therefore, help to regulate the population size of those organisms. Nematodes, which are often called wireworms or roundworms, may aid in the breakdown of dead organic matter. Some nematodes are parasitic on the roots of plants. Insects and other soil arthropods contribute to the soil by forming burrows and consuming and fragmenting organic materials, but they are also major crop pests that feed on plant roots. Algae carry on photosynthesis and some are capable of fixing nitrogen from the air. They are also consumed by other soil organisms. Bacteria and fungi are particularly important in the decay and recycling of materials. Their chemical activities change complex organic materials into nutrients that can be used by plants. For example, some of these microorganisms can convert the nitrogen contained in the protein component of organic matter into ammonia or nitrate, which are nitrogen compounds that can be utilized by plants. The amount of nitrogen produced varies with the type of organic matter, type of microorganisms, drainage, and temperature. Finally it is important to recognize that the soil contains

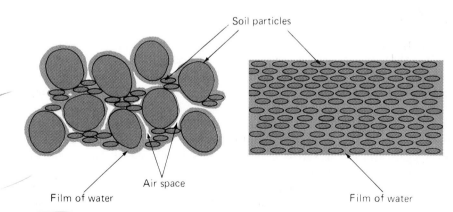

FIGURE 13.7 **Pore Spaces and Particle Size** The soil on the left, which is composed of particles of various sizes, has spaces for both water and air. The particles have water bound to their surfaces (represented by the colored halo around each particle), but some of the spaces are so large that an air space is present. The soil on the right, which is composed of uniformly small particles, has less space for air. Since roots require both air and water, the soil on the left would be better able to support crops than that on the right.

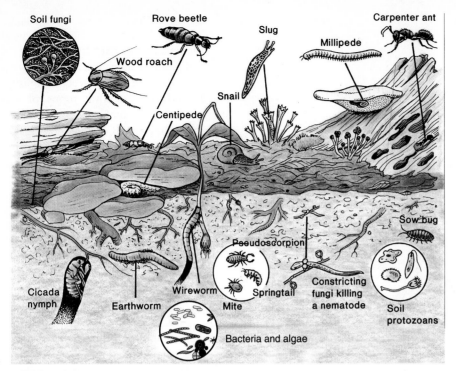

Soil fungi · Rove beetle · Slug · Carpenter ant · Millipede · Wood roach · Snail · Centipede · Pseudoscorpion · Sow bug · Cicada nymph · Earthworm · Wireworm · Mite · Springtail · Constricting fungi killing a nematode · Soil protozoans · Bacteria and algae

FIGURE 13.8 Soil Organisms All of these organisms occupy the soil and contribute to it by rearranging soil particles, participating in chemical transformation, and recycling dead organic matter.

From *Ecology and Field Biology*, 5th Edition, by Robert Leo Smith; copyright © 1996 by Harper Collins Publishers. Reprinted by permission.

a complicated food chain in which all organisms are subject to being consumed by others. All of these organisms are active within distinct layers of the soil, known as the soil profile.

SOIL PROFILE

The **soil profile** is a series of horizontal layers in the soil that differ in chemical composition, particle size, and amount of organic matter. Each recognizable layer is known as a **horizon.** (See figure 13.9.) The uppermost layer of the soil contains more nutrients and organic matter than do the deeper layers. This layer is known as the *A* horizon or topsoil. The thickness of the *A* horizon may vary from less than a centimeter on steep mountain slopes to over a meter in the rich grasslands of central North America. Most of the living organisms and nutrients are found near the top of the *A* horizon. The lower portions of the *A* horizon usually contain few nutrients because water flowing down through the

soil dissolves and transports nutrients to the *B* horizon. This process is known as **leaching.** The *B* horizon, often called the subsoil, contains less organic material and fewer organisms than the *A* horizon. It also contains accumulations of nutrients that were leached from higher levels. Because of this, the *B* horizon in many soils is a valuable source of nutrients for plants, and such subsoils support a well-developed root system. Because the amount of leaching depends on the available rainfall, areas of the world where rainfall is low may have a poorly developed *B* horizon.

The area below the subsoil is known as the *C* horizon, and it consists of weathered parent material. This parent material contains no organic materials, but it does contribute to some of the soil's properties. The chemical composition of the *C* horizon helps to determine the pH of the soil. The *C* horizon may also influence the soil's rate of water absorption and retention.

Soil profiles and the factors that contribute to soil development are extremely

varied. Over 15,000 separate soil types have been classified in North America. However, most of the cultivated land in the world can be classified as either grassland soil or forest soil. (See figure 13.10.)

The low amount of rainfall in grassland areas limits the amount of leaching from the topsoil. As a result, the *A* horizon is deep and supports most of the root growth. This lack of leaching also results in a thin layer of subsoil, which is low in mineral and organic content and supports little root growth.

Forest soils develop in areas of more abundant rainfall. The rainfall results in a layer of topsoil that is usually not as thick as that of grassland soil, but the material leached from the topsoil forms a subsoil that supports substantial root growth. One of the materials that accumulates in the *B* horizon is clay. In some soils, particularly forest soils, clay or other minerals may accumulate and form an impermeable "hardpan" layer that limits the growth of roots and may prevent water from reaching the soil's deeper layers.

Desert soils have very poorly developed horizons since there is little rainfall to leach materials from upper layers to lower layers. In addition, the low amount of plant growth results in little contribution of organic matter to the soil. In cold, wet climates, typical of the northern parts of Europe, Russia, and Canada, there may be considerable accumulations of organic matter since the rate of decomposition is reduced. The extreme acidity of these soils also reduces the rate of decomposition. Hot, humid climates also tend to have poorly developed soil horizons since the organic matter decays very rapidly, and soluble materials are carried away by the abundant rainfall.

In addition to the differences caused by the kind of vegetation and rainfall, topography influences the soil profile. (See figure 13.11.) On a relatively flat area, the topsoil formed by soil-building processes will collect in place and gradually increase in depth. The topsoil formed on rolling hills or steep slopes is often transported down the slope as fast as it is produced. On such slopes, the accumulation of topsoil may not be sufficient to support a cultivated crop. The topsoil

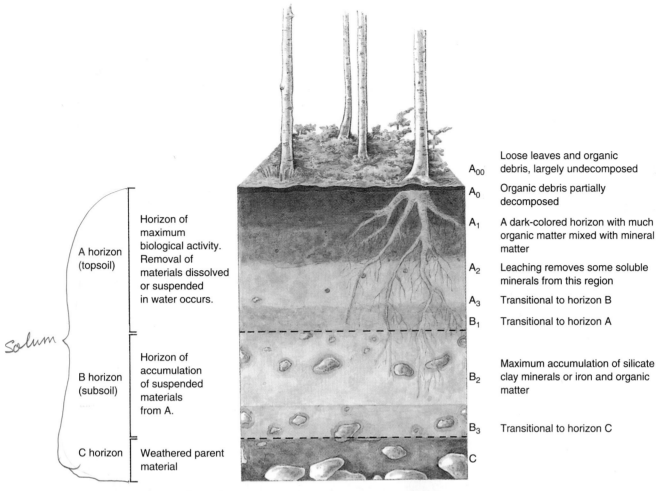

A horizon (topsoil) — Horizon of maximum biological activity. Removal of materials dissolved or suspended in water occurs.

B horizon (subsoil) — Horizon of accumulation of suspended materials from A.

C horizon — Weathered parent material

Solum

A_{00} — Loose leaves and organic debris, largely undecomposed

A_0 — Organic debris partially decomposed

A_1 — A dark-colored horizon with much organic matter mixed with mineral matter

A_2 — Leaching removes some soluble minerals from this region

A_3 — Transitional to horizon B

B_1 — Transitional to horizon A

B_2 — Maximum accumulation of silicate clay minerals or iron and organic matter

B_3 — Transitional to horizon C

C

FIGURE 13.9 **Soil Profile** A soil has layers that differ physically, chemically, and biologically. The top layer is known as the *A* horizon and contains most of the organic matter. The *B* horizon accumulates minerals and particles as water carries dissolved minerals downward from the *A* to the *B* horizon. The *B* horizon is often called the subsoil. Below the *B* horizon is a *C* horizon of weathered parent material.

removed from these slopes is eventually deposited in the flat floodplains. These regions serve as collection points for topsoil that was produced over extensive areas. As a result, these river-bottom and delta regions have a very deep topsoil layer and are highly productive agricultural land.

SOIL EROSION

Erosion is the wearing away and transportation of soil by water or wind. The Grand Canyon of the Colorado River, the floodplains of the Nile in Egypt, the little gullies on hillsides, and the deltas that develop at the mouths of rivers all attest to the ability of water to move soil. Anyone who has seen muddy water after a rainstorm has observed soil being moved by water. (See figure 13.12.) The force of moving water allows it to carry large amounts of soil. While erosion is a natural process it is greatly accelerated by agricultural practices that leave the soil exposed. Each year, the Mississippi River transports over 325 million metric tons of soil from the central regions of North America to the Gulf of Mexico. This is equal to the removal of a layer of topsoil approximately 1 millimeter thick from the entire region. This movement of soil by water occurs in every stream and river in the world. Dry Creek, a small stream in California, has only 500 kilometers of mainstream and tributaries; however, each year, it removes 180,000 metric tons of soil from a 340-square-kilometer area. (See chapter 15.)

Soil erosion takes place everywhere in the world, but some areas are more exposed than others. Erosion occurs wherever grass, bushes, and trees are disappearing. Every year erosion carries away far more topsoil than is created primarily because of agricultural practices that leave the soil exposed. (See figure 13.13.)

Worldwide, erosion removes about 25.4 billion tons of soil each year. Deforestation and desertification both leave land open to erosion. In deforested areas, water washes down steep, exposed slopes, taking the soil with it. In desertified

Litter

Topsoil

Subsoil

Decomposed
parent
material

A Horizon

B Horizon

C Horizon

Grassland soil
(mollisol)

A Horizon

B Horizon

C Horizon

Forest soil
(podzol)

A Horizon

B Horizon

C Horizon

Desert soil
(aridozol)

FIGURE 13.10 **Major Soil Types** There are thousands of different soil types, but many of them can be classified into three broad categories. Soils formed in the grasslands are known as mollisol soils and have a deep *A* horizon. The shallow *B* horizon does not have sufficient nutrients to support root growth. In forest soils, known as podzol soils, the *A* horizon is thinner, and leaching transfers many nutrients to the *B* horizon. Thus roots are found in both the *A* and *B* horizons. Desert soils known as aridozols have very thin *A* horizons.

FIGURE 13.11 **The Effect of Slope on a Soil Profile** The topsoil formed on a large area of the hillside is continuously transported down the slope by the flow of water. It accumulates at the bottom of the slope and results in a thicker *A* horizon. The resulting "bottomland" is highly productive because it has a deep, fertile layer of topsoil, while the soil on the slope is less productive.

FIGURE 13.12 **Water Erosion** The force of moving water is able to pick up soil particles and remove them. In cases of prolonged erosion, gullies (such as the one shown here) are likely to form.

regions, exposed soils, cleared for farming, building, or mining, or overgrazed by livestock, simply blow away. Wind erosion is most extensive in Africa and Asia. Blowing soil not only leaves a degraded area behind but can bury and kill vegetation where it settles. It will also fill drainage and irrigation ditches. When high-tech farm practices are applied to poor lands, soil is washed away and chemical pesticides and fertilizers pollute the runoff.

In Africa, soil erosion has reached critical levels, with farmers pushing farther onto deforested hillsides. In Ethiopia, for example, soil loss occurs at a rate of between 1.5 billion and 2 billion cubic meters a year, with some 4 million hectares of highlands considered irreversibly degraded. In Asia, in the eastern hills of Nepal, 38 percent of the land area is fields that have been abandoned because the topsoil has washed away. In the Western Hemisphere, Ecuador is losing soil at 20 times the acceptable rate.

According to the International Fund for Agricultural Development (IFAD), traditional labor-intensive, small-scale soil conservation efforts that combine maintenance of shrubs and trees with crop growing and cattle grazing work best at controlling erosion. In parts of Pakistan, a program begun by IFAD in 1980 to control rainfall runoff, erosion, and damage to rivers from siltation has increased crop yields and livestock productivity by 20 to 30 percent.

Badly eroded soil has lost all of the topsoil and some of the subsoil and is no longer productive farmland. Most current agricultural practices lose soil faster than it is replaced. Farming practices that reduce erosion, such as contour farming and terracing, are discussed later in the chapter.

Wind is also an important mover for soil. Under certain conditions, it can move large amounts. (See figure 13.14.) Wind erosion may not be as evident as water erosion, since it does not leave gullies. Nevertheless, it can be a serious problem. Wind erosion is most common in dry, treeless areas where the soil is exposed. In the Sahel region of Africa, much of the land has been denuded of vegetation because of drought, overgrazing, and improper farming practices. This has resulted in extensive wind erosion of the soil. (See figure 13.15.) In the Great Plains region of North America, there have been four serious periods of wind erosion since European settlement in the 1800s. If this area receives less than 30 centimeters of rain per year, there is not enough moisture to support crops. When this occurs for several years in a row, it is called a drought. Farmers plant crops, hoping for rain. When the rain does not come, they plow their fields again to prepare it for another crop. Thus, the loose, dry soil is left exposed, and wind erosion results. Because of the large amounts of dust in the air during those times, the region is known as the Dust Bowl. During the 1930s, wind destroyed 3.5 million hectares of farmland and seriously damaged an additional 30 million hectares in the Dust Bowl. Unfortunately, many agricultural practices adopted to combat erosion following the 1930s are being discontinued today for economic reasons, thus leaving the land vulnerable to erosion again.

SOIL CONSERVATION PRACTICES

Whenever soil is lost by water or wind erosion, the topsoil, the most productive layer, is the first to be removed. When the topsoil is lost, the soil's fertility decreases, and expensive fertilizers must be used to restore the fertility that was lost. This raises the cost of the food we buy. In addition, the movement of excessive amounts of soil from farmland into streams has several undesirable effects. (See chapter 14.) First, a dirty stream is less aesthetically pleasing than a clear stream. Second, a stream laden with sediment affects the fish population. Fishing may be poor because of unwise farming practices hundreds of kilometers upstream. Third, the soil carried by a river is eventually deposited somewhere. In many cases, this soil must be removed by dredging to clear shipping channels. We

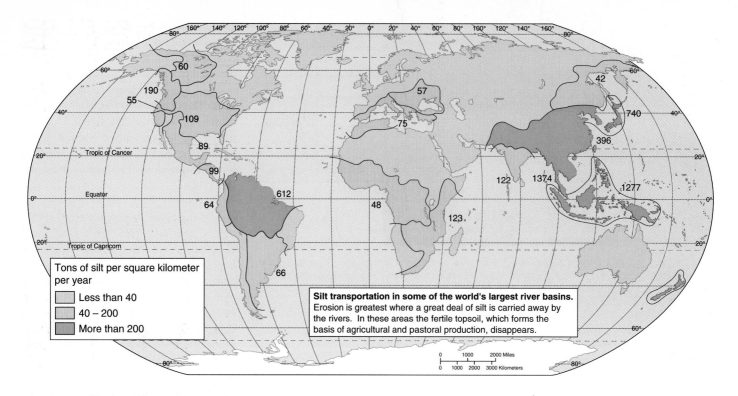

Tons of silt per square kilometer per year

- [ ] Less than 40
- [ ] 40 – 200
- [ ] More than 200

Silt transportation in some of the world's largest river basins. Erosion is greatest where a great deal of silt is carried away by the rivers. In these areas the fertile topsoil, which forms the basis of agricultural and pastoral production, disappears.

Silt-laden river

FIGURE 13.13 Soil Erosion Soil erosion is widespread throughout the world. Silty rivers are evidence of poor soil conservation practices upstream.

pay for dredging with our tax money, and it is a very expensive operation.

For all of these reasons, proper soil conservation measures should be employed to minimize the loss of topsoil. Figure 13.16 contrasts poor soil conservation practices with proper soil protection. When soil is not protected from the effects of running water, the topsoil is removed and gullies result. This can be prevented by slowing the flow of water over sloping land.

The kinds of agricultural activity that land can be used for are determined by soil structure, texture, friability, drainage, fertility, rockiness, slope of the land, amount and nature of rainfall, and other climatic conditions. A relatively large proportion—over 50 percent—of U.S. land is suitable for crop cultivation. (See figure 13.17.) However, only 2 percent of that land does not require some form of soil conservation practice. This means that nearly all of the soil in the United States must be managed in some way to reduce the effects of soil erosion by wind or water.

Not all parts of the world are as well supplied as the United States with land that has agricultural potential. (See table 13.1.) For example, worldwide, approximately 11 percent of the land surface is suitable for crops, and an additional 24 percent is in permanent pasture. In the United States, about 21 percent is cropland, and 26 percent is in permanent pasture. Contrast this with the continent of Africa, in which only 6 percent is suitable for crops, and 26 percent can be used for pasture. Canada has only 5 percent suitable for crops and under 3 percent for pasture. Europe has the highest percentage of cropland with 30 percent, but it has only 18 percent in permanent pasture.

Since there is very little land left that can be converted to agriculture we must use what we have wisely. There are many techniques that protect soil from erosion while allowing agriculture. Some of the more common methods are discussed here.

Contour Farming

Contour farming, which is tilling at right angles to the slope of the land, is one of the simplest methods to prevent soil erosion. This practice is useful on gentle slopes and produces a series of small ridges at right angles to the slope. (See figure 13.18.) Each ridge acts as a dam to hold water from running down the incline. This allows more of the water to soak into the soil. Contour farming reduces soil erosion by as much as 50 percent and, in drier

FIGURE 13.14 **Wind Erosion** The dry, unprotected topsoil from this field is being blown away. The force of the wind is capable of removing all the topsoil and transporting it several thousand kilometers.

a.

b.

FIGURE 13.16 **Poor and Proper Soil Conservation Practices** (*a*) This land is no longer productive farmland since erosion has removed the topsoil. (*b*) This rolling farmland shows strip contour farming to minimize soil erosion by running water. It should continue to be productive farmland indefinitely.

FIGURE 13.15 **Wind Erosion in the Sahel** The semiarid region just south of the Sahara Desert is in an especially vulnerable position. The rainfall is unpredictable, which often leads to crop failure. In addition, population pressure forces people in this region to try to raise crops in marginal areas. This often results in increased wind erosion.

regions, increases crop yields by conserving water.

Strip Farming

When a slope is too steep or too long, contour farming alone may not prevent soil erosion. However, a combination of contour and strip farming may work. **Strip farming** is alternating strips of closely sown crops like hay, wheat, or other small grains with strips of row crops like corn, soybeans, cotton, or sugar beets. (See figure 13.19.) The closely sown crops retard the flow of water, which reduces soil erosion and allows more water to be absorbed into the ground. The type of soil, steepness, and length of slope dictate the width of the strips and determine whether strip or contour farming is practical.

Terracing

On very steep land, the only practical method of preventing soil erosion is to construct terraces. **Terraces** are level areas constructed at right angles to the slope to retain water and greatly reduce the amount of erosion. (See figure 13.20.) Terracing has been used for centuries in nations with a shortage of level farmland. The type of terracing seen in figure 13.20*a* requires the use of small machines and considerable hand labor and is not suitable for the mechanized farming typical in much of the world. Terracing is an expensive method of controlling erosion since it

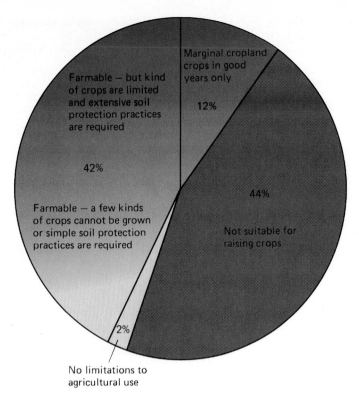

FIGURE 13.17 U.S. Land Suitable for Raising Crops Only 2 percent of the land in the United States can be cultivated without some soil conservation practices. This 2 percent is primarily flatland, which is not subject to wind erosion. On 42 percent of the remaining land, some special considerations for protecting the soil are required, or the kinds of crops are limited. Twelve percent of the land is marginal cropland that can only provide crops in years when rainfall and other conditions are ideal. Forty-four percent is not suitable for cultivating crops but may be used for other purposes, such as grazing cattle or as forests.

TABLE 13.1 PERCENTAGE OF LAND SUITABLE FOR AGRICULTURE

Country	Percent cropland	Percent pasture
World	11.0	26.0
Africa	6.3	28.8
Egypt	2.8	5.0
Ethiopia	12.7	40.7
Kenya	7.9	37.4
South Africa	10.8	66.6
North America	13.0	16.8
Canada	4.9	3.0
United States	19.6	25.0
South America	6.0	28.3
Argentina	9.9	51.9
Venezuela	4.4	20.2
Asia	15.2	25.9
China	10.3	30.6
Japan	12.0	1.7
Europe	29.9	17.1

Source: Data from *World Resources 1996.*

requires the moving of soil to construct the level areas, protecting the steep areas between terraces, and constant repair and maintenance. Many factors, such as length and steepness of slope, type of soil, and amount of precipitation, determine whether terracing is feasible.

Waterways

Even with such soil conservation practices as contour farming, strip farming, and terracing, farmers must often provide protected channels for the movement of water. **Waterways** are depressions on sloping land where water collects and flows off the land. When not properly maintained, these areas are highly susceptible to erosion. (See figure 13.21.) If a waterway is maintained with a permanent sod covering, the speed of the water is reduced, and soil erosion is decreased.

Windbreaks

Contour farming, strip farming, terracing, and maintaining waterways are all important ways of reducing water erosion, but wind is also a problem with certain soils, particularly in dry areas of the world. Wind erosion can be reduced if the soil is protected. (See chapter 17.) The best protection is a layer of vegetation over the surface. However, the process of preparing soil for planting and the method of planting often leave the soil exposed to wind. **Windbreaks** are plantings of trees or other plants that protect bare soil from the full force of the wind. Windbreaks reduce the velocity of the wind, thereby decreasing the amount of soil that it can carry away. (See figure 13.22.) In some cases, rows of trees are planted at right angles to the prevailing winds to reduce their force, while in other cases, a kind of strip farming is practiced in which strips of hay or grains are alternated with row crops that leave large amounts of the soil exposed. In some areas of the world, the only way to protect the soil is not to cultivate it at all, but to leave it in a permanent cover of grasses.

Worldwide Soil Degradation

As humans have used the land and its soil for agriculture, grazing, fuelwood production and forestry, we have had profound effects on the rates at which soils are formed and lost. Removal of the vegetation has accelerated soil erosion so that most of the world is losing soil faster than it is being formed. The degradation is most severe where the need for food is greatest, since starving people are forced to overexploit their already overused soils. The map and graphs illustrate the magnitude of the soil degradation problem and the relative importance of the causes.

Percent of soil degraded by human activity

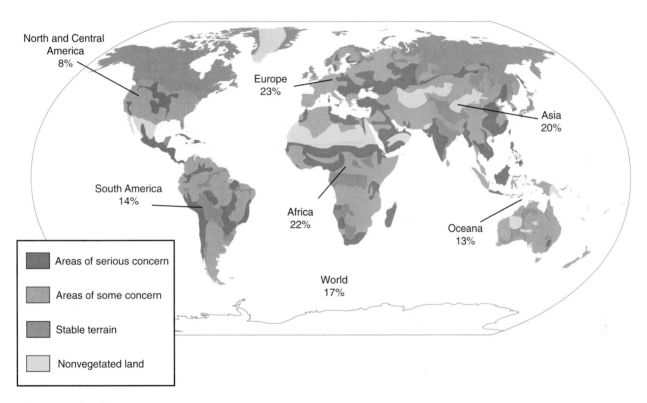

Causes of soil degradation

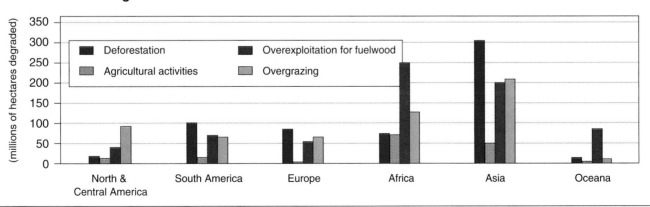

Source: World Resources Institute in collaboration with the United Nations Environment Programme and the United Nations Development Programme, *World Resources 1992–93* (Oxford University Press, New York, 1992), Table 19.4, p. 290.

FIGURE 13.18 Contour Farming Tilling at right angles to the slope creates a series of ridges that slows the flow of the water and prevents soil erosion. This soil conservation practice is useful on gentle slopes.

FIGURE 13.19 Strip Farming On rolling land, a combination of contour and strip farming prevents excessive soil erosion. The strips are planted at right angles to the slope, with bands of closely sown crops, such as wheat or hay, alternating with bands of row crops, such as corn or soybeans.

Conservation Tillage

Conventional tillage methods in much of the world require extensive use of farm machinery to prepare the soil for planting and to control weeds. Typically, a field is plowed and then disked or harrowed one to three times before the crop is planted.

The plowing, which turns the soil over, has several desirable effects: any weeds or weed seeds are buried, thus reducing the weed problem in the field. Crop residue from previous crops is incorporated into the soil where it will decay faster and

contribute to soil structure. Nutrients that had been leached to deeper layers of the soil are brought near the surface. And the dark soil is exposed to the sun so that it warms up faster. This last effect is most critical in areas with short growing seasons. In many areas, fields are plowed in the fall, after the crop has been harvested, and the soil is left exposed all winter.

After plowing, the soil is worked by disks or harrows to break up any clods of earth, kill remaining weeds, and prepare the soil to receive the seeds. After the seeds are planted, there may still be weed problems. Farmers often must cultivate row crops to kill the weeds that begin to grow between the rows. Each trip over the field costs the farmer money, while at the same time increasing the amount of time the soil is exposed to wind or water erosion.

In recent years, several innovations in chemical herbicides and farm equipment have led to the development of tilling practices that protect the soil by reducing the amount of time it is exposed to erosion forces. Reduced tillage uses less cultivation to control weeds and to prepare the soil to receive seeds. Selective herbicides are used to kill unwanted vegetation prior to planting the new crop and to control weeds afterward. Several variations of conservation tillage are used:

1. Plowing followed by reduced secondary tillage. This usually involves plowing followed immediately by planting.

2. Strip tillage, a method that involves tilling only in the narrow strip that is to receive the seeds. The rest of the soil and the crop residue from the previous crop is left undisturbed.

3. No-till farming involves special planters that place the seeds in slits cut in the soil that still has on its surface the crop residue from the previous crop.

All these types of reduced tillage use less fuel and less time but may require more herbicides. (See table 13.2.) For many kinds of crops, yields are comparable to that produced by conventional tillage methods.

a.

b.

FIGURE 13.20 **Terraces** Since the construction of terraces requires the movement of soil and the protection of the steep slope between levels, terraces are expensive to build. (*a*) The terraces seen here are extremely important for people who live in countries that have little flatland available. They require much energy and hand labor to maintain but make effective agricultural use of the land without serious erosion. (*b*) This modification of the terracing concept allows the use of the large farm machines typical of farming practices in Canada, Europe, and the United States.

a.

b.

FIGURE 13.21 **Protection of Waterways Prevents Erosion** (*a*) An unprotected waterway has been converted into a gully. (*b*) A well-maintained waterway is not cultivated; a strip of grass retards the flow of water and protects the underlying soil from erosion.

Other positive effects of reduced tillage, in addition to reducing erosion, are:

1. The amount of winter food and cover available for wildlife increases, which can lead to increased wildlife populations.

2. Since there is less runoff, siltation in streams and rivers is reduced. This results in clearer water for recreation and less dredging to keep waterways open for shipping.

3. Row crops can be planted on hilly land that cannot be converted to such crops under conventional tilling methods. This allows a farmer to convert low-value pasture land into cropland that gives a greater economic yield.

4. Since fewer trips are made over the field, petroleum is saved, even if the petrochemical feedstocks necessary to produce the herbicides are taken into account.

5. Two crops may be grown on a field in areas that had been restricted to growing one crop per field per year. In some areas, immediately after harvesting wheat, farmers have used reduced tillage methods to plant soybeans in the wheat stubble.

6. Because reduced tillage reduces the number of trips made over the field by farm machinery, the soil does not become compacted as quickly.

a.

b.

FIGURE 13.22 Windbreaks (*a*) In sections of the Great Plains, trees provide protection from wind erosion. The trees along the road are planted in a north-south direction to protect the land from the prevailing westerly winds. (*b*) In this field, temporary strips of vegetation serve as windbreaks.

TABLE 13.2 COMPARISON OF VARIOUS TILLING METHODS

Tillage method	Hours required per 100 hectares	Liters of fuel per 100 hectares	Cost of herbicide per 100 hectares (1993)
Conventional	200	2915	$3,717
Reduced secondary tillage	125	1390	$3,717
Strip tillage	95	1020	$3,717
No tillage	62	375	$5,000

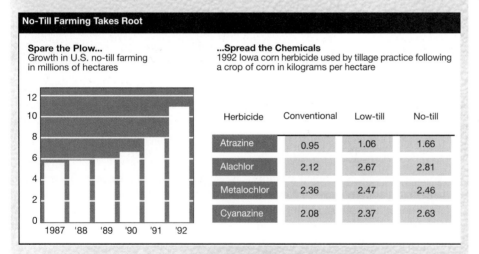

No-Till Farming Takes Root

Spare the Plow...
Growth in U.S. no-till farming in millions of hectares

...Spread the Chemicals
1992 Iowa corn herbicide used by tillage practice following a crop of corn in kilograms per hectare

Herbicide	Conventional	Low-till	No-till
Atrazine	0.95	1.06	1.66
Alachlor	2.12	2.67	2.81
Metalochlor	2.36	2.47	2.46
Cyanazine	2.08	2.37	2.63

Source: (top) Data from J. E. Beuerlein and S. W. Bone, "Selecting a Tillage System" Extension Publication, Ohio State University; and D. H. Doster, "Economics of Alternative Tillage Systems" in *Bulletin of the Entomological Society of America* 22 (1976): 297. (bottom left) Data from Conservation Technology Information Center, West Lafayette, IN; (bottom right) Data from U.S. Department of Agriculture.

However, there are also some drawbacks to reduced-tillage methods:

1. The residue from previous vegetation may delay the warming of the soil, which may, in turn, delay planting some crops for several days.

2. The crop residue reduces evaporation from the soil and the upward movement of water and soil nutrients from deeper layers of the soil, which may retard the growth of plants.

3. The accumulation of plant residue can harbor plant pests and diseases that will require more insecticides and fungicides.

This is particularly true if the same crop is planted repeatedly in the same location. Reduced tillage is not the complete answer to soil erosion problems but may be useful in reducing soil erosion on well-drained soils. It also requires that farmers pay close attention to the condition of the soil and the pests to be dealt with.

In 1972, 12 million hectares in the United States were under some form of conservation tillage. About 1.3 million hectares were being farmed using no-till methods. By 1992, this had risen to 60 million hectares of conservation-tilled land, of which 6.2 million hectares were being farmed using no-till methods. It is estimated that, by the year 2010, 95 percent of U.S. cropland will be under some form of reduced-tillage practice.

PROTECTING SOIL ON NONFARM LAND

Not all land is suitable for crops or continuous cultivation. Some must never be plowed for use as cropland, but people can still use it to provide for some of their needs. By using appropriate soil conservation practices, much of the land not usable for crops can be used for grazing, wood production, wildlife production, or scenic and recreational purposes. Figure 13.23 shows land that is not suitable for cultivation but that is capable of producing grass for cattle or sheep. Such land, if properly used, will produce food for many generations.

The land shown in figure 13.24 is not suitable for either crops or grazing. However, it is still a valuable and productive piece of land since it can be used to furnish lumber, wildlife habitats, and recreational opportunities.

Not all land is equal. Each section has its own soil characteristics, climate, and degree of slope. When all these are taken into consideration, a proper use can be determined for each portion of the planet. Wise planning and careful husbandry of the soil is necessary if the land and its soil are to provide food and other necessities of life.

FIGURE 13.23 **Noncrop Use of Land to Raise Food** As long as it is properly protected and managed, this land can produce food through grazing, but it should never be plowed to plant crops because the topsoil is too shallow and the rainfall is too low.

FIGURE 13.24 **Forest and Recreational Use** Although this land is not capable of producing crops or supporting cattle, it furnishes lumber, a habitat for wildlife, and recreational opportunities.

ENVIRONMENTAL CLOSE-UP

Land Capability Classes

Not all land is suitable for raising crops, or urban building. Such factors as the degree of slope, soil characteristics, rockiness, erodibility, and other characteristics determine the best use for a parcel of land. In an attempt to encourage people to use land wisely, the U.S. Soil Conservation Service has established a system to classify land-use possibilities. The table shows the eight classes of land and lists the characteristics and capabilities of each.

Unfortunately, many of our homes and industries are located on type I and II land, which has the least restrictions on agricultural use.

This does not make the best use of the land. Zoning laws and land-use management plans should consider the land-use capabilities and institute measures to assure that land will be used to its best potential.

- Can you provide examples from your community where you feel there was not proper land use?
- In your opinion, should there be more or less land-use planning?
- What are some of the consequences of building or farming on land that is not suitable?

	Land class	Characteristics	Capability	Special conservation measures
Land suitable for cultivation	I	Excellent, flat, well-drained land	Cropland	Normal good practices adequate
	II	Good land; has minor limitations, such as slope, sandy soil, or poor drainage	Cropland Pasture	Strip cropping Contour farming
	III	Moderately good land with important limitations of soil, slope, or drainage	Cropland Pasture Watershed	Contour farming Strip cropping Terraces Waterways
	IV	Fair land with severe limitations of soil, slope, or drainage	Pasture Orchards Urban Industry Limited cropland	Crops on a limited basis Contour farming Strip cropping Terraces Waterways
Land not suitable for cultivation	V	Use for grazing and forestry; slightly limited by rockiness, shallow soil, or wetness	Grazing Forestry Watershed Urban Industry	No special precautions if properly grazed or logged; must not be plowed
	VI	Moderate limitations for grazing and forestry because of moderately steep slopes	Grazing Forestry Watershed Urban Industry	Grazing or logging may be limited at times
	VII	Severe limitations for grazing and forestry because of very steep slopes vulnerable to erosion	Grazing Forestry Watershed Recreation Wildlife Urban Industry	Careful management is required when used for grazing or logging
	VIII	Unsuitable for grazing and forestry because of steep slope, shallow soil, lack of water, or too much water	Watershed Recreation Wildlife Urban Industry	Not to be used for grazing or logging; steep slope and lack of soil present problems

In a court case in Virginia, a city attorney filed suit for damages against a land developer. The city contended that the developer was responsible for property damages as a result of erosion from his construction site. During the trial, it was shown that the developer had constructed his subdivision without considering the possible effects of soil erosion. As a result, during a heavy rain, mud from his project had covered the yards, sidewalks, and drives of homes located nearby. As much as 10 meters of soil had eroded from the exposed land of his development and been carried downhill, covering portions of the adjacent community.

Speaking to the developer, the city attorney stated: "After gambling and principally winning over a period of fifteen months, it is apparent that you are attempting to go with your winning streak; and flood, storms, and weather hold no terror for you." The city claimed that the developer had acted in a negligent manner and that he was,

therefore, responsible for the damages caused by the deposited mud and should be ordered to pay for the cleanup. The developer claimed no responsibility. His position was that grading large tracts of land was a common construction procedure and that the weather was "an act of God" for which he could not be held accountable.

- If you were a member of the jury hearing the case, would you think the developer acted in a negligent manner by leaving the soil exposed for a long period of time?

- Do you believe that property owners can hold a person responsible for actions taken in an area some distance from their property?

- Do you believe that the erosion and subsequent deposit of mud was "an act of God"?

- Would you find the developer guilty or not guilty? Why?

Summary

The surface of the Earth is in constant flux. The movement of tectonic plates results in the formation of new land as old land is worn down by erosive activity. Soil is an organized mixture of minerals, organic material, living organisms, air, and water. Soil formation begins with the breakdown of the parent material by such physical processes as changes in temperature, freezing and thawing, and movement of particles by glaciers, flowing water, or wind. Oxidation and hydrolysis can chemically alter the parent material. Organisms also affect soil building by burrowing into and mixing the soil, by releasing nutrients, and by their decomposition.

Topsoil contains a mixture of humus and inorganic material, both of which supply soil nutrients. Soil fertility is determined by the inorganic matter, organic matter, water, and air spaces in the soil. The mineral portion of the soil consists of various mixtures of sand, silt, and clay particles.

A soil profile consists of the A horizon, which is rich in organic matter; the B horizon, which accumulates materials leached from the A horizon; and the C horizon, which consists of slightly altered parent material. Forest soils typically have a shallow A horizon and a deep, nutrient-rich B horizon with much root development. Grassland soils usually have a thick A horizon and very few nutrients in the thin B horizon. Therefore, most of the roots of the grasses are in the A horizon.

Soil erosion is the removal and transportation of soil by water or wind. Proper use of such conservation practices as contour farming, strip farming, terracing, waterways, windbreaks, and conservation tillage can reduce soil erosion. Misuse reduces the soil's fertility and causes air- and water-quality problems. Land unsuitable for crops may be used for grazing, lumber, wildlife habitats, or recreation.

Review Questions

1. How are soil and land different?
2. Name the five major components of soil.
3. Describe the process of soil formation.
4. Name five physical and chemical processes that break parent material into smaller pieces.
5. In addition to fertility, what other characteristics determine the usefulness of soil?
6. How does soil particle size affect texture and drainage?
7. Describe a soil profile.
8. Define erosion.
9. Describe three soil conservation practices that help to reduce soil erosion.
10. Besides cropland, what are other possible uses of soil?

Critical Thinking Questions

1. Minimum tillage soil conservation often uses greater amounts of herbicides to control weeds. What do you think about this practice? Why?
2. As populations grow should we try to bring more land into food production, or should we use technology to aid in producing more food on the land we already have in production of food? What are the tradeoffs?
3. Given what you know about soil formation, how might you explain the presence of mollisol soil in the North American Midwest?
4. Why should non-farmers be interested in soil conservation?
5. Imagine that you are a scientist hired to consult on a project to evaluate land use practices at the edge of a small city. The area in question has deep ravines and hills. What kinds of agricultural, commercial, and logging practices would you recommend in this area to help preserve the environment?
6. Look at your own community. Can you see examples of improper land use (urban or rural)? What are the consequences of these land use practices? What recommendations would you make to improve land use?

Key Terms

chemical weathering *232*

contour farming *240*

crust *231*

erosion *232*

friability *235*

horizon *236*

humus *234*

land *232*

leaching *236*

loam *234*

mantle *231*

mechanical weathering *232*

parent material *232*

plate tectonics *231*

soil *232*

soil profile *236*

soil structure *234*

soil texture *234*

strip farming *241*

terrace *241*

waterways *242*

weathering *232*

windbreak *242*

Related Internet Sites

Visit our website at http://www.mhhe.com/environmentalscience/ for specific information on each of the topics below.

Preservation and control of erosion

Hydrodynamic and ocean engineering

On soil and its conservation

Working to protect soil, water, air, and animals

Index for environmental science research materials

Soil characteristics laboratory

National Society of Consulting Soil Scientists

On-line journal for soil science

Soil fertility, erosion, and the impact of agriculture on land, water, and air

Soil science resources

Soils and crop production, environmental quality, and ecosystem sustainability

14

Agricultural Methods and Pest Management

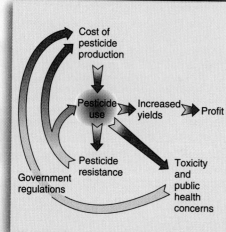

DIFFERING AGRICULTURAL METHODS

Around the world, people who grow food use a variety of agricultural methods. In some parts of the world with poor soil and low populations, a form of agriculture known as "slash and burn" is practiced. The forest in a small area is felled and the trees and other vegetation on the site are burned. (See figure 14.1.) This releases nutrients that are tied up in the biomass and allows a crop or two to be raised before the soil is exhausted. Once the soil is no longer suitable for raising crops, the site is abandoned. The surrounding forest quickly recolonizes the area, which returns over time to the original forest through the process of succession. This method is particularly useful on thin tropical soils and on steep slopes. The small size of the opening in the forest and its temporary existence prevent widespread damage to the soil, and erosion is minimized. The gardens that are planted are typically a mixture of plants, rather than rows of single species.

The traditional practices of the people who engage in slash-and-burn agriculture have been developed over hundreds of years and often are more effective than other methods of gardening for their local conditions. For example, mixing plants may have a beneficial effect, since shade-requiring species may be helped by taller plants, or nitrogen-fixing legumes may provide needed nutrients for species that have a nitrogen requirement. In addition, mixing species may reduce insect pest problems because some plants produce molecules that are natural insect repellents. The small, isolated, temporary nature of the gardens also reduces the likelihood of insect plagues. This form of agriculture is limited to a very small number of people in specialized habitats and is not the most efficient way to farm all kinds of habitats. Slash-and-burn farming also requires a long recovery time before the forest can be cleared for another cycle of agriculture.

In other parts of the world, extensive areas are farmed with a great deal of manual labor. Two situations favor this kind of farming: (1) when the growing site does not allow for mechanization, and (2) when the kind of crop does not allow it. Soils or terrain that require that fields be small discourage mechanization, since large tractors and other machines cannot be used efficiently on small, oddly shaped fields. Many mountainous areas of the world fit into this category. In addition, some crops require such careful handling in planting, weeding, or harvesting that large amounts of hand labor are required. The planting of paddy rice and the harvesting of many fruits and vegetables are examples.

However, the primary reason for labor-intensive farming is economic. Many densely populated countries have numerous small farms that can be effectively managed with human labor, supplemented by that of draft animals and a few small gasoline-powered engines. (See figure 14.2.) In addition, in the less-developed regions of the world, the cost of

FIGURE 14.1   Slash-and-Burn Agriculture In many areas of the world where the soils are poor and human populations are low, crops can be raised by disturbing small parts of the ecosystem followed by several years of recovery. The burning of vegetation releases nutrients that can be used by crops for one or two years before the soil is exhausted. The return of the natural vegetation prevents erosion and repairs the damage done by temporary agricultural use.

FIGURE 14.2   Labor-Intensive Agriculture In many of the less-developed countries of the world the extensive use of hand labor allows for impressive rates of production with a minimal input of fossil fuels and fertilizers. This kind of agriculture is also necessary in areas that have only small patches of land suitable for farming.

labor is low, which encourages the use of hand labor rather than mechanization to do planting, weeding, and other activities. Merchanization requires large tracts of land that could only be accumulated by the expenditure of large amounts of money or the development of larger cooperative farms from many small units. Even if social and political obstacles to such large landholdings could be overcome, there is still the problem of obtaining the necessary capital to purchase the machines. Large parts of the developing world fit into this category, including much of Africa, many areas in Central and South America, and many areas in Asia.

Mechanized agriculture is typical of North America, much of Europe, the republics of the former USSR, South America, and other parts of the world where money and land are available to support this form of agriculture. In large measure, machines and fossil-fuel energy replace the energy formerly supplied by human and animal muscles. Mechanization requires large expanses of fairly level land for the machines to operate efficiently. In addition, large tracts of land must be planted in the same crop for efficient planting, cultivating, and harvesting. Planting large fields in the same crop is known as **monoculture.** (See figure 14.3.) Small sections of land with many kinds of crops require many changes of farm machinery, which takes time. Also, many crops interspersed with one another reduces the efficiency of farming operations because farmers must skip parts of the field, which increases travel time and uses expensive fuel.

Even though monoculture is an efficient method of producing food, it is not without serious drawbacks. When large farm machines are used to plow, plant, and harvest vast expanses of land at the same time, large tracts of land are left uncovered by vegetation, and soil erosion increases. (See chapter 13.) Mechanized farming also removes much of the organic matter each year. When the crop is harvested, a major portion of the plant typically is removed for food or some other purpose. This means that little organic matter is returned to the soil unless the farmer specifically plants a crop that is later plowed under to increase the soil's organic content.

To ensure that a crop can be planted, tended, and harvested efficiently, farmers rely on hybrid seed that provides uniform plants with characteristics suitable for mechanized farming. These hybrid plants have little genetic variety. When all the farmers in an area plant the same hybrids, pest control becomes a serious problem. If diseases or pests begin to spread, the magnitude of the problem becomes devastating because all the plants have the same characteristics and, thus, are susceptible to the same diseases. If genetically diverse crops are planted, or crops are rotated from year to year, this problem is not as great.

Because farm equipment is expensive, farmers tend to specialize in a few crops. This means that the same crop may be planted in the same field several years in a row. This lack of crop rotation may deplete certain essential soil nutrients, thereby requiring special attention to soil chemistry. In addition, planting the same crop repeatedly encourages the growth of insect and fungus pest populations because they have a huge food supply at their disposal. This requires the frequent use of insecticides and fungicides or some other methods of pest control.

Even though there are problems associated with mechanized, monoculture agriculture, it has greatly increased the amount of food available to the world over the past one hundred years. Yields per hectare of land being farmed have increased over much of the world, particularly in the developed world, which includes the United States. (See figure 14.4.) This increase has come about because of improved varieties of crops, irrigation, better farming methods, the use of agricultural chemicals, more efficient machines, and the use of energy-intensive as opposed to labor-intensive technology.

Throughout the 1950s, 1960s, and 1970s, the introduction of new plant varieties and farming methods resulted in increased agricultural production worldwide. This has been called the Green Revolution. Both the developed world, which uses highly mechanized farming methods, and the developing world, where labor-intensive farming is typical, have benefited from these advances and food production increased significantly. However, the population of the world continues to increase and more food is needed. More recently agricultural scientists have been working in the area of genetic engineering and other forms of biotechnology. Over the next 20 to 30 years, scientists hope to use biotechnology to produce high-yield plant strains that are more resistant to insects and disease, thrive on less fertilizer, make their own nitrogen fertilizer, do well in slightly salty soils, withstand drought, and use solar energy more efficiently during photosynthesis. Generally biotechnology involves genetic engineering procedures that insert genes into plants that should allow them to function more efficiently. To date,

FIGURE 14.3 Monoculture This wheat field is an example of monoculture, a kind of agriculture that is highly mechanized and requires large fields for the efficient use of machinery. In this kind of agriculture, machines and fossil-fuel energy have replaced the energy of humans and draft animals.

FIGURE 14.4 **Increased Yields Resulting from Modern Technology** The increased yields in the United States and many other parts of the world are the result of a combination of factors, including better genetic qualities in the seed, improved agricultural methods, the application of fertilizers and pesticides, and more efficient machinery.

Source: Data from Department of Agriculture, *Agriculture Statistics*, 1998 and previous years.

however, several factors have limited the success of these undertakings and most genetically engineered plant varieties produce yields no higher and often lower than those from traditional strains. Second, the cost of genetically engineered crop strains is too high for most of the world's subsistence farmers. There are also ethical issues regarding genetic engineering that farmers, consumers, and scientists are only beginning to debate. Will genetically engineered crops have hidden effects such as that experienced by the use of pesticides? Will the special genes inserted into plants to improve them, negatively affect other important plant properties? Will the genes inserted into crop plants "escape" to other plants and alter weed species so that they become more difficult to control?

ENERGY VERSUS LABOR

Mechanized agriculture has substituted the energy stored in petroleum products for the labor of humans. For example, in the United States in 1913, it required 135 hours of labor to produce 2500 kilograms of corn. In 1980, it required about 15 hours of labor to produce 2500 kilograms of corn. The energy supplied by petroleum products replaced the equivalent of 120 hours of labor. Energy is needed for tilling, planting, harvesting, and pumping irrigation water. Much of the energy is in the form of manu-factured fertilizers, herbicides, fungicides, and insecticides that encourage crop growth while inhibiting weeds and reducing the impact of pests that feed on the crop or cause disease. For example, about 5 metric tons of fossil fuel are required to produce about 1 metric ton of fertilizer. Since the developed world is dependent on oil to provide energy to manufacture pesticides to support its agriculture, any change in the availability or cost of oil will have a major impact on the world's ability to feed itself.

THE IMPACT OF FERTILIZER

Table 14.1 shows that, in some respects, fertilizer can be used to replace human labor. The United States, with only 3 percent of its workforce in agriculture, uses nearly 30 percent more fertilizer than Brazil and over twice as much as India and has a cereal grain yield more than twice that of Brazil or India. The yields are similar for India and Brazil; however, Brazil uses less labor to accomplish the same yield. In this case, fertilizer appears to directly substitute for labor.

Approximately 25 percent of the world's crop yield is estimated to be directly attributed to the use of chemical fertilizers. The use of fertilizer has increased significantly over the last few decades and is projected to increase even more (see figure 14.5). However, since fertilizer pro-duction relies on energy from fossil fuels, the price and availability of chemical fertilizers is strongly influenced by world energy prices. If the price of oil increases, the price of fertilizer goes up, as does the cost of food. This is felt most acutely in parts of the world where money is in short supply, since the farmers are unable to buy fertilizer, and crop yields fall accordingly.

Fertilizers are valuable because they replace the soil nutrients removed by plants. Some plant nutrients, such as carbon, hydrogen, and oxygen, are easily replaced by carbon dioxide from the air and water from the soil, but others are less easily replaced. The three primary soil nutrients often in short supply are potassium, phosphorus, and nitrogen compounds. They are often referred to as **macronutrients** and are the common ingredient of chemical fertilizers. Their replacement is important because, when the crop is harvested, the chemical elements that are a part of the crop are removed from the field. Since many of those elements originated from the soil, they need to be replaced if another crop is to be grown. Certain other elements are necessary in extremely small amounts and are known as **micronutrients.** Examples are boron, zinc, and manganese. As an example of the difference between macronutrients and micronutrients, harvesting a metric ton of potatoes removes 10 kilograms of nitrogen (a macronutrient) but only 13 grams of boron (a micronutrient). When the same crop is grown repeatedly in the same field, certain micronutrients may be depleted, resulting in reduced yields. These necessary elements can be returned to the soil in sufficient amounts by incorporating them into the fertilizer the farmer applies to the field.

Although chemical fertilizers replace inorganic nutrients, they do not replace soil organic matter. Organic material is important because its decay produces humus. Humus modifies the structure of the soil, preventing compaction and maintaining pore space, which allows water and air to move to the roots. Organic matter also maintains proper soil chemistry because it tends to lower the pH of the soil, which helps to release certain soil nutrients for use by plants. Soil bacteria and other organisms use organic matter as a source of energy. Since these organisms serve as important links in the carbon and nitrogen cycles, the presence of organic matter is important to their function. Thus, total dependency upon chemical fertilizers usually reduces the amount of organic matter and can change the physical, chemical, and biotic properties of the soil.

As water moves through the soil, it dissolves soil nutrients and carries them into streams and lakes, where they may encourage the growth of unwanted plants and algae. This is particularly true when fertilizers are applied at the wrong time of the year, just before a heavy rain, or in such large amounts that the plants cannot efficiently remove them from the soil before they are lost. These ideas are covered in greater detail in chapter 15 on water pollution.

PESTICIDES

In addition to chemical fertilizers, mechanized monoculture requires large amounts of other agricultural chemicals, such as pesticides, growth regulators, and preservatives. These chemicals have specific scientific names but are usually categorized into broad groups based on their effects. A **pesticide** is any chemical used to kill or control populations of unwanted fungi, animals, or plants, often called **pests.** The term pest is not scientific but refers to any organism that is unwanted. Insects that

TABLE 14.1	FERTILIZER AND FOOD PRODUCTION		
Country	Percentage of the workforce in agriculture (1991)	Fertilizer used (kilograms/hectare) (1993)	Yield in kilograms/hectare (cereal grains) (1994)
United States	3	108	5092
Brazil	25*	85	2256
India	70*	46	2062

*Estimates based on various data.
Source: Data from *World Resources Institute, 1994–95.*

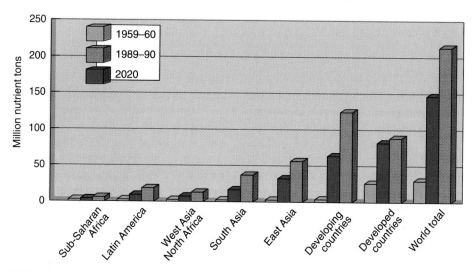

Estimated growth in worldwide fertilizer use, 1960–2020

(Legend: 1959–60, 1989–90, 2020)

FIGURE 14.5 **Increasing Fertilizer Use** The use of fertilizer is increasing rapidly. Growth in use will be most rapid in the developing countries over the next 20 years.

Source: Balu Bumb and Carlos Baanaunte, *World Trends in Fertilizer Use and Projections to 2020,* 2020 Brief No. 38 (International Food Policy Research Institute, Washington, D.C., 1996), Table 1.

feed on crops are pests, while others, like bees, are beneficial for pollinating plants. Unwanted plants are generally referred to as **weeds.**

Pesticides can be subdivided into several categories based on the kinds of organisms they are used to control. **Insecticides** are used to control insect populations by killing them. Unwanted fungal pests that can weaken plants or destroy fruits are controlled by **fungicides.** Mice and rats are killed by **rodenticides,** and plant pests are controlled by **herbicides.** A perfect pesticide is one that kills or inhibits the growth of only the specific pest organism causing a problem. The pest is often referred to as the **target organism.** However, most pesticides are not very specific and kill many **nontarget**

organisms as well. For example, most insecticides kill both beneficial and pest species, rodenticides kill other animals as well as rodents, and most herbicides kill a variety of plants, both pests and nonpests. Pesticides do not just kill pests but can kill a large variety of living things, including humans. Because of this, these chemicals might be more appropriately called **biocides,** since they kill many kinds of living things.

Insecticides

If insects are not controlled they consume a large proportion of the crops produced by farmers. In small garden plots, insects can be controlled by manually removing them and killing them. However, in large fields, this is not practical, so people have sought other ways to control pest insects. In addition, many insects harm humans because they spread diseases, such as sleeping sickness, bubonic plague, and malaria. Mosquitoes are known to carry over 30 diseases harmful to humans. In 1965, mosquitoes were estimated to be responsible for spreading diseases like malaria and sleeping sickness to 100 million people. The discovery of chemicals that could kill insects was celebrated as a major advance in the control of disease and the protection of crops.

Nearly 3000 years ago, the Greek poet Homer mentioned the use of sulfur to control insects. For centuries, it was known that natural plant products could repel or kill insect pests. Plants with insect-repelling abilities were interplanted with crops to help control the pests. Nicotine from tobacco, rotenone from tropical legumes, and pyrethrum from chrysanthemums were extracted and used to control insects. In fact, these compounds are still used today. However, because plant products are difficult to apply and have short-lived effects, other compounds were sought. In 1867, the first synthetic inorganic insecticide, Paris green, was formulated. It was a mixture of acetate and arsenide of copper and was used to control Colorado potato beetles.

The first synthetic organic insecticide to be used was DDT. It was originally thought to be the perfect insecticide. It was long-lasting, relatively harmless to humans, and very deadly to insects. During the first 10 years of its use (1942–1952), DDT is estimated to have saved 5 million lives, primarily because of its use to control disease-carrying mosquitoes. However, after a period of use, many mosquitoes and other insects became tolerant of DDT. The early success of DDT and its loss of effectiveness promoted a search for new synthetic organic compounds that could replace it. Since then, over 60,000 different compounds have been synthesized that have potential as insecticides. However, most of these have never been put into production because of cost, human health effects, or other drawbacks that make them unusable. Several categories of these compounds

have been developed. Three that are currently used are chlorinated hydrocarbons, organophosphates, and carbamates.

Chlorinated Hydrocarbons

Chlorinated hydrocarbons are a group of pesticides of complex, stable structure that contain carbon, hydrogen, and chlorine. DDT (dichlorodiphenyltrichloroethane) was the first such pesticide manufactured, but several others have been developed. The chemical structure of DDT is shown in figure 14.6. Other chlorinated hydrocarbons are chlordane, aldrin, heptachlor, dieldrin, and endrin. It is not fully understood how these compounds work, but they are believed to affect the nervous systems of insects, resulting in their death.

One of the major characteristics of these pesticides is that they are very stable chemical compounds. This is both an advantage and a disadvantage. Since they are long-lasting, they are called **persistent pesticides.** They can be applied once and be effective for a long time. However, since they do not break down easily, they tend to accumulate in the soil and in the bodies of animals in the food chain. Thus, they affect many nontarget organisms, not just the original target insects. In temperate regions of the world, DDT has a half-life (the amount of time required for half of the chemical to decompose) of 10 to 15 years. This means that if 1 metric ton of DDT were sprayed over an area, 500 kilograms would still be present in the area 10 to 15 years later; 30 years from the date of application, 250 kilograms would still be present. The half-life of DDT varies depending on soil type, temperature, the kinds of soil organisms present, and other factors. In tropical parts of the world, the half-life may be as short as six months. An additional complication is that persistent pesticides may break down into products that are still harmful.

Because of their negative effects, most of the chlorinated hydrocarbons are no longer used in most of the world. DDT, aldrin, dieldrin, toxaphene, chlordane, and heptachlor have been banned in the United States and some other countries. However, many developing countries still use chlori-

FIGURE 14.6 **The Chemical Structure of DDT** This diagram shows the arrangement of the atoms in a molecule of DDT. The two chlorophenyl portions of the molecule are shown in blue. The other chlorines are shown in red, and the remainder of the ethane molecule appears black.

nated hydrocarbons for insect control to protect crops and public health.

Organophosphates and Carbamates

Because of the problems associated with persistent pesticides, a new generation of **nonpersistent pesticides** that decompose to harmless products in a few hours or days was developed. However, like other insecticides, these are not species specific; they kill beneficial insects as well as harmful ones. Although the short half-life prevents the accumulation of toxic material in the environment, it is a disadvantage for farmers, since more frequent applications are required to control pests. This requires more labor and fuel and, therefore, is more expensive.

Both **organophosphates** and **carbamates** work by interfering with the ability of the nervous system to conduct impulses normally. Under normal conditions, a nerve impulse is conducted from one nerve cell to another by means of a chemical known as acetylcholine. When this chemical is produced at the end of one nerve cell, it causes an impulse to be passed to the next cell, thereby transferring the nerve message. As soon as this transfer is completed, an enzyme known as cholinesterase destroys acetylcholine, so the second nerve cell in the chain is stimulated for only a short time. Organophosphates and carbamates attach to the enzyme cholinesterase, preventing it from

ENVIRONMENTAL
CLOSE-UP

Regulation of Pesticides

The Environmental Protection Agency (EPA) is charged by Congress to protect the nation's land, air, and water systems. Under a mandate of national environmental laws focused on air and water quality, solid waste management, and the control of toxic substances, pesticides, noise, and radiation, the agency strives to formulate and implement actions that lead to a compatible balance between human activities and the ability of natural systems to support and nurture life.

To fulfill this mandate, the EPA is charged with the enforcement of various federal laws and acts. One such act governs the registration of pesticides. Section 3 (c) (1) of the Federal Insecticide, Fungicide, and Rodenticide Act (FIFRA) states:

Procedure for Regulation

1. Statement required. Each applicant for registration of a pesticide shall file with the Administrator a statement which includes:

   A. the name and address of the applicant and of any other person whose name will appear on the labeling;

   B. the name of the pesticide;

   C. a complete copy of the labeling of the pesticide, a statement of all claims to be made for it, and any directions for its use;

   D. except as otherwise provided in subsection (c) (2) (D) of this section, if requested by the Administrator, a full description of the tests made and the results thereof upon which the claims are based, or alternately, a citation to data that appears in the public literature or that previously had been submitted to the Administrator and that the Administrator may consider in accordance with the following provisions;

   E. the complete formula of the pesticide; and

   F. a request that the pesticide be classified for general use, or restricted use, or for both.

If a corporation wants to manufacture and market a pesticide within the United States, the pesticide must not adversely affect the environment. There is no specific format to test a product's effects on the environment, but some of the tests conducted include the following:

1. **Degradation**—A determination of the physical and chemical methods of decay and an identification of the decay products.

2. **Metabolism**—A determination of the organisms that act on the pesticide and the metabolic products released.

3. **Mobility**—A determination of where the molecules are likely to go and what route they will follow in the ecosystem.

4. **Accumulation**—Do the pesticide molecules accumulate in living tissue? If so, in what form and in what quantities?

5. **Hazard**—Evaluation of the possible hazards to plants, microorganisms, wildlife, and humans, whether target or nontarget organisms.

After receiving the information submitted in the procedure for registration, the Administrator of EPA may register the pesticide for use under the provisions of Section 3 (c) (5) of FIFRA:

Approval of Registration
The Administrator registers a pesticide if he or she determines that, when considered with any restrictions imposed under subsection (d),

1. its composition is such as to warrant the proposed claims for it;

2. its labeling and other material required to be submitted comply with the requirements of this Act;

3. it will perform its intended function without unreasonable adverse effects on the environment; and

4. when used in accordance with widespread and commonly recognized practice, it will not generally cause unreasonable adverse effects on the environment.

The Administrator also publishes all applications and supporting data so that various governmental agencies and the public have an opportunity to comment.

The FIFRA is designed to protect the environment from being damaged by the use of pesticides. The law places the burden upon the manufacturer to prove that a pesticide is safe. One source estimates that eight to twelve years and $50 million are needed to bring a major new pesticide from discovery to first registration. These figures do not include the capital cost of constructing a facility to manufacture the pesticide.

---

destroying acetylcholine. This results in nerve cells being continuously stimulated, causing uncontrolled spasms of nervous activity and uncoordination that result in death.

Although these pesticides are less persistent in the environment than are chlorinated hydrocarbons, they are generally much more toxic to humans and other vertebrates. Persons who use such pesticides must use special equipment and should receive special training because improper use can result in death. Since

organophosphates bind more strongly to cholinesterase than do carbamates, they are considered more dangerous and for many applications have been replaced by carbamates.

Common organophosphates are malathion, parathion, and diazinon. Malathion is widely used for such projects as mosquito control, but parathion is a restricted organophosphate because of its high toxicity to humans. Diazinon is widely used in gardens. Sevin, aldicarb, and propoxur are examples of carbamates.

Herbicides

Herbicides are another major class of chemical control agents. In fact, about 60 percent of the pesticides used in U.S. agriculture are herbicides. They are widely used to control unwanted vegetation along power-line rights-of-way, railroad rights-of-way, and highways, as well as on lawns and cropland, where they are commonly referred to as weed killers.

Weeds are plants growing in the wrong place. Weed control is extremely important

A New Generation of Insecticides

The perfect insecticide would harm only target insect species, not be toxic to humans, not be persistent, and would break down into harmless materials. Most currently used insecticides were developed by screening a wide variety of compounds and modifying compounds that showed insect-killing properties. Most are highly toxic to many kinds of organisms and must be used with great care. However, a new generation of insecticides is being developed that starts with knowledge of insect biology and develops a compound that interferes with some essential aspect of the life cycle. Some mimic normal hormones and prevent insects from maturing into adults. Therefore, the insects do not reproduce and the populations are greatly reduced.

Other insecticides attack specific chemical processes in cells and cause death. Nicotine is a normal product of some plants like tobacco,

which is toxic to insects. It binds to specific receptors on nerve cells, causing them to fire uncontrollably. However, it is not stable in sunlight. By tinkering with the chemical structure of nicotine, a modified molecule, called imidacloprid, was developed. It is more stable, not toxic to mammals, because they do not have as many of the receptors as insects, and very effective against sucking insects.

Another new insecticide works by turning off the energy-producing processes of mitochondria in cells. All cells of higher organisms have mitochondria and would be injured by this insecticide, but a nontoxic form of the molecule was produced that is converted to a toxic form in the bodies of insects but is not converted in mammals. It is highly toxic to birds and some aquatic organisms, however, which would require strict control of its use.

---

for agriculture since weeds take nutrients and water from the soil, making them unavailable to the crop species. In addition, weeds may shade the crop species and prevent it from getting the sunlight it needs for rapid growth. At harvest time, weeds reduce the efficiency of harvesting machines. Also, weeds generally must be sorted from the crop before it can be sold, which adds to the time and expense of harvesting.

Traditionally, farmers have expended much energy trying to control weeds. Initially, weeds were eliminated with manual labor and the hoe. Tilling the soil also helps to control weeds. Once the crop is planted, row crops like corn or soybeans may be cultivated to remove weeds from between the rows. All of these activities are expensive in time and fuel. Selective use of herbicides can have a tremendous impact on a farmer's profits.

Many of the recently developed herbicides can be very selective if used appropriately. Some are used to kill weed seeds in the soil before the crop is planted, while others are used after the weeds and the crop begin to grow. In some cases, a mixture of herbicides can be used to control several weed species simultaneously. Figure 14.7 shows the effects of using a herbicide that kills grasses but not other kinds of plants.

Several major types of herbicides are in current use. One type is synthetic plant-growth regulators that mimic natural-

growth regulators known as **auxins.** Two of the earliest herbicides were of this type: 2, 4-dichlorophenoxyacetic acid (2, 4-D) and 2, 4, 5-trichlorophenoxyacetic acid (2, 4, 5-T). When applied to broadleaf plants, these chemicals caused the metabolism of the plants to exceed their food-producing potential. The plants grew so rapidly that they ran out of stored food and died. These herbicides have been replaced

by newer kinds that use a variety of mechanisms to control or kill weeds. Some disrupt photosynthetic activity of plants, causing their death. Others inhibit enzymes, precipitate proteins, stop cell division, or destroy cells directly. Depending on the concentration of the herbicide used, some are toxic to all plants, while others are very selective as to which species of plants they affect. One such

FIGURE 14.7　The Effect of Herbicides The grasses in this photograph have been treated with herbicides. The soybeans are unaffected and grow better without competition from grasses.

herbicide is diuron. In proper concentrations and when applied at the appropriate time, it can be used to control annual grasses and broadleaf weeds in over 20 different crops. However, at higher concentrations, it kills all vegetation in an area. Fenuron is a herbicide that kills woody plants. In low concentration, it is used to control woody weed plants in cropland. In high concentrations, it is used on noncroplands, such as power-line rights-of-way. (See figure 14.8.)

Other Agricultural Chemicals

In addition to herbicides, other agrochemicals are used for special applications. For example, a synthetic auxin sprayed on cotton plants prior to harvest causes the leaves to drop off, which facilitates the harvesting process by reducing clogging of the mechanical cotton picker.

NAA (naphthaleneacetic acid) is used by fruit growers to prevent apples from dropping from the trees and being damaged. This chemical can keep the apples on the trees for up to 10 extra days, which allows for a longer harvest period and fewer lost apples.

Under other conditions, it may be valuable to get fruit to fall more easily. Cherry growers use ethephon to promote loosening of the fruit so that the cherries will fall more easily from the tree when shaken by the mechanical harvester. This method lowers the cost of harvesting the fruit. (See figure 14.9.)

Fungicides and Rodenticides

Fungus pests can be divided into two categories. Some are natural decomposers of organic material, but when the organic material being destroyed happens to be a crop or other product useful to humans, the fungus is considered a pest. Other fungi are parasites on crop plants and weaken or kill the plants, thereby reducing the yield. Fungicides are used as fumigants (gases) to protect agricultural products from spoilage, as sprays and dusts to prevent the spread of diseases among plants, and as seed treatments to protect seeds from rotting in the soil before they have a chance to germinate. Methylmercury is often used on seeds to protect them

from spoilage prior to germination. However, since methylmercury is extremely toxic to humans, these seeds should never be used for food. To reduce the chance of a mix-up, treated seeds are usually dyed a bright color.

Like fungi, rodents are harmful because they destroy food supplies. In addition, they can carry disease and damage crops in the field. In many parts of the world, such as India, the government pays a bounty to people who kill rats because this is an inexpensive way to protect the food supply. Several kinds of rodenticides have been developed to control rodents. One of the most widely used is warfarin, a chemical that causes internal bleeding in animals that consume it. It is usually in-

FIGURE 14.8 Herbicide Use to Maintain Rights-of-Way Power-line rights-of-way are commonly maintained by herbicides that kill the woody vegetation, which might grow so tall that it interferes with the power lines.

FIGURE 14.9 Chemical Loosening of Cherries to Allow Mechanical Harvesting By using chemicals and machinery, this farmer can rapidly harvest the cherry crop. This practice reduces the amount of labor required to pick the cherries but requires the application of chemicals to loosen the fruit.

Politics and the Control of Ethylene Dibromide (EDB)

Since the 1940s, ethylene dibromide (EDB) has been used as a fumigant to protect stored grain from insect pests. More recently, it has been used to fumigate milling machinery and kill nematodes in the soil. It has also been used as a fumigant in fruit shipments to prevent the transportation of insect pests from one part of the world to another.

It was originally thought that EDB dissipated and was not a health hazard. However, residues were found in grain and fruits, and were also reported in drinking water from wells in areas where EDB was used as a soil fumigant. In the 1970s, it was shown that EDB causes cancer in rats and mice at levels of twenty parts per million.

In 1977, the U.S. Environmental Protection Agency first sought to control use of EDB, but manufacturers and users effectively fought its

prohibition until September 1984, when the EPA succeeded in banning the use of EDB as a soil fumigant.

When the EPA banned the use of EDB, it was required by law to purchase the remaining stocks of the material from the manufacturers and to dispose of them at its own expense. (This feature of the law was lobbied for very strongly by pesticide manufacturers.) In March 1988, the EPA incinerated the remaining approximately 1.3 million liters of the EDB it had acquired when it banned the substance. The EPA had already paid $2.5 million for these stocks and disposing of them properly cost several million more. The requirement that the EPA acquire and dispose of banned pesticides has had a significant impact on the agency's ability to function.

corporated into a food substance so that rodents eat warfarin along with the bait. Because it is effective in all mammals, including humans, it must be used with care to prevent nontarget animals from having access to the chemical. As with many kinds of pesticides, some populations of rodents have become tolerant of warfarin, while others avoid baited areas. In many cases, rodent problems can be minimized by building storage buildings that are rodent proof, rather than relying on rodenticides to control them.

PROBLEMS WITH PESTICIDE USE

A perfect pesticide would have the following characteristics:

1. It would be inexpensive.
2. It would affect only the target organism.
3. It would have a short half-life.
4. It would break down into harmless materials.

However, the perfect pesticide has not been invented. Many of the more recently developed pesticides have fewer drawbacks than the early pesticides, but none are without problems.

Persistence

Although there has been a trend away from using persistent pesticides in the United States, some are still allowed for special purposes, and they are still in common use in other parts of the world. Because of their stability, these chemicals have become a long-term problem. Persistent insecticides become attached to small soil particles, which are easily moved by wind or water to any part of the world. Persistent pesticides and other pollutants have been discovered in the ice of the poles and are present in detectable amounts in the body tissues of animals, including humans, throughout the world. Thus, chemicals originally sprayed on a cornfield in Iowa or a sugarcane field in Brazil may become a problem in Arctic or Antarctic ecosystems.

Bioaccumulation and Bioamplification

Another problem associated with persistence is that pesticides may accumulate in the bodies of animals. If an animal receives small quantities of persistent pesticides or other persistent pollutants in its food and is unable to eliminate them, the concentration within the animal increases. This process of accumulating higher and higher amounts of material within the

body of an animal is called **bioaccumulation.** Many of the persistent pesticides and their breakdown products are fat soluble and build up in the fat of animals. When an affected animal is eaten by a carnivore, these toxins are further concentrated in the body of the carnivore, causing disease or death, even though lower-trophic-level organisms are not injured. This phenomenon of acquiring increasing levels of a substance in the bodies of higher-trophic-level organisms is known as **biological amplification.**

The well-documented case of DDT is an example of how biological amplification occurs. DDT is not very soluble in water, but dissolves in oil or fatty compounds. When DDT falls on an insect or is consumed by the insect the DDT is accumulated in the insect's fatty tissue. Large doses kill insects but small doses do not and their bodies may contain as much as one part per billion of DDT. This is not very much, but it can have a tremendous effect on the animals that feed on the insects.

If an aquatic habitat is sprayed with a small concentration of DDT, or received DDT from runoff, small aquatic organisms may accumulate a concentration that is up to 250 times greater than the concentration of DDT in the surrounding water. These organisms are eaten by shrimp, clams, and small fish, which are, in turn, eaten by larger fish. DDT concentrations of large fish can be as much as 2000 times the original concentration sprayed on the area. What was a very small initial concentration has now become so high that it could be fatal to animals at higher trophic levels. This has been of particular concern for birds, since DDT interferes with the production of eggshells, making them much more fragile. This problem is more common in carnivorous birds be-cause they are at the top of the food chain. Although all birds of prey have probably been affected to some degree, those that rely on fish for food seem to have been affected most severely. Eagles, osprey, cormorants, and pelicans are particularly susceptible species. (See figure 14.10.)

DDT is a well-known case of bio-logical amplification in ecosystems. Other persistent molecules are known to be-have in similar fashion. Mercury, aldrin, chlordane, and other chlorinated hydro-carbons, such as polychlorinated biphenyls (PCBs) used as insulators in electric transformers, are all known to ac-cumulate in ecosystems.

Because of their persistence, their effects on organisms at higher trophic levels, and concerns about long-term human health problems, most chlorinated hydrocarbon pesticides have been banned from use in the United States and some

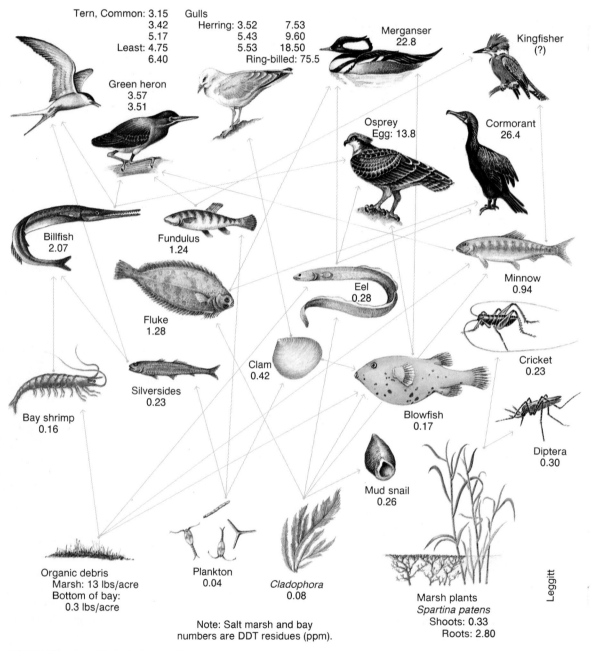

Tern, Common: 3.15
3.42
5.17
Least: 4.75
6.40

Gulls
Herring: 3.52   7.53
5.43   9.60
5.53   18.50
Ring-billed: 75.5

Merganser
22.8

Kingfisher
(?)

Green heron
3.57
3.51

Osprey
Egg: 13.8

Cormorant
26.4

Billfish
2.07

Fundulus
1.24

Minnow
0.94

Eel
0.28

Fluke
1.28

Clam
0.42

Cricket
0.23

Silversides
0.23

Blowfish
0.17

Diptera
0.30

Bay shrimp
0.16

Mud snail
0.26

Organic debris
Marsh: 13 lbs/acre
Bottom of bay:
0.3 lbs/acre

Plankton
0.04

Cladophora
0.08

Marsh plants
Spartina patens
Shoots: 0.33
Roots: 2.80

Leggitt

Note: Salt marsh and bay numbers are DDT residues (ppm).

FIGURE 14.10 Biological Amplification Note how the concentration of DDT increases as it passes through the food chain. At each successive trophic level, the amount of DDT increases because the animals are accumulating the DDT from the bodies of animals they are eating as food.

other countries. The use of DDT was prohibited in the United States in the early 1970s. Aldrin, dieldrin, heptachlor and chlordane have also been prohibited from use on crops, although heptachlor and chlordane were still used for termite control until recently. In 1987, Velsicol Chemical Corporation agreed to stop selling chlordane in the United States.

Pesticide Resistance

Another problem associated with insecticides is the ability of insect populations to become resistant to them. Not all organisms within a given species are identical. Each individual has a slightly different genetic composition and slightly different characteristics. When an insecticide is used for the first time on an insect, it kills all the individuals that are susceptible. Individuals with characteristics that allow them to tolerate the insecticides may live to reproduce.

If, in a population of insects, only 5 percent of the individuals possess genes that make them resistant to an insecticide, the first application of the insecticide will kill 95 percent of the population and so will be of great benefit in controlling the size of the insect population. However, the surviving individuals that are tolerant of the insecticide will constitute the majority of the breeding populations. Since these individuals possess genetic characteristics for tolerating the insecticide, so will most of their offspring. Therefore, in the next generation, the number of individuals able to tolerate the insecticide will increase, and the second use of the insecticide will not be as effective as the first. Since some species of insect pests can produce a new generation each month, this process of selecting individuals capable of tolerating the insecticide can result in resistant populations in which 99 percent of the individuals are able to tolerate the insecticide within five years. As a result, that particular insecticide is no longer as effective in controlling insect pests, and increased dosages or more frequent spraying may be necessary. Figure 14.11 indicates that over 500 species of insects have populations resistant to insecticides.

Cotton growers have become extremely dependent on the use of insecticides. About 40 percent of the insecticides used in the United States are for controlling pests in cotton. Because of this intensive use, many populations of pest insects have become resistant to the commonly used insecticides. Table 14.2 shows the effect of continual use of insecticides on two pests, the bollworm and the tobacco budworm. The size of the dose necessary to kill the pest increased greatly in a five-year period. In some cases, the dose increased tenfold and in other a hundredfold or more. By 1965, these insecticides were no longer able to control bollworm or tobacco budworm.

Effects on Nontarget Organisms

Most pesticides kill beneficial species as well as pest species. In fact, they may kill predator and parasitic insects that normally control the pest insects. This allows pest species to increase rapidly following the use of a pesticide because there are no natural checks to their population growth. Additional applications of insecticides are necessary to prevent the pest population from rebounding to levels even higher than the initial one. Once the decision is made to use pesticides, it often becomes

Pest species resistant to insecticides

FIGURE 14.11 Resistance to Insecticides The continued use of insecticides has constantly selected for genes that give resistance to a particular insecticide. As a result, many species of insects and other arthropods are now resistant to many kinds of insecticides, and the number continues to increase.

Source: Data from George P. Georghiou, University of California at Riverside.

TABLE 14.2 AVERAGE DOSE NECESSARY TO KILL TWO COTTON PESTS

Compound	Average dose necessary to kill (milligrams per gram of larva)			
	Bollworm		Tobacco budworm	
	1960	1965	1961	1965
DDT	0.03	1.000+	0.13	16.51
Endrin	0.01	0.13	0.06	12.94
Carbaryl	0.12	0.54	0.30	54.57
Strobane and DDT	0.05	1.04	0.73	11.12
Toxaphene and DDT	0.04	0.46	0.47	3.52

Reprinted with permission from P.L. Adkisson, "Controlling Cotton's Insect Pests: A New System" in *Science*, 216:19–22, April 1982. Copyright © 1982 American Association for the Advancement of Science.

an irreversible tactic, because stopping their use would result in rapid increases in pest populations and extensive crop damage.

A related concern is that the use of insecticides may cause an insect that is not a problem to become one. For example, when synthetic organic insecticides came into common use with cotton in the 1940s, the insect parasite and predators were eliminated, and the bollworm and tobacco budworm became major pests. The use of insecticides caused a different pest problem to develop.

Human Health Concerns

Short-term and long-term health effects to the persons using the pesticide and the public that consumes the food grown by using pesticides are also concerns. If properly applied, most pesticides can be applied with little danger to the applicator. However, in many cases, people applying pesticides are unaware of how they work and what precautions should be used in their application. In many parts of the world, farmers may not be able to read the caution labels on the packages or do not have access to the protective gear specified for use with the pesticide. Therefore, many incidences of acute poisoning occur each year. In most cases the symptoms disappear after a period free from exposure. Estimates of the number of poisonings are very difficult to obtain since many go unreported, but in the United States pesticide poisonings requiring medical treatment are in the thousands per year. In the developing world the number of poisonings is in the tens of millions per year. Consequently, some deaths due to misuse of pesticides happen each year.

For most people, however, the most critical health problem related to pesticide use is inadvertent exposure to very small quantities of pesticides. Many pesticides have been proven to cause mutations, produce cancers, or cause abnormal births in experimental animals. There are questions about the effects of chronic, minute exposures to pesticide residues in food or through contamination of the environment.

Although the risks of endangering health by consuming tiny amounts of pesticides is very small compared to other risk factors such as automobile accidents, smoking, or poor eating habits, many people find pesticides unacceptable and seek to prohibit their sale and use. The U.S. Environmental Protection Agency requires careful studies of the effectiveness and possible side effects of each pesticide licensed for use. It has banned several pesticides from further use because new information suggests that they are not as safe as originally thought. For example, it banned the use of dinoseb, a herbicide, because tests by the German chemical company Hoechst AG indicated that dinoseb causes birth defects in rabbits. Other studies indicate that dinoseb causes sterility in rats. Similarly, in 1987, Velsicol Chemical Corporation signed an agreement with the U.S. EPA to stop producing and distributing chlordane in the United States. Chlordane had been banned previously for all applications except for termite control. After it was shown that harmful levels of chlordane could exist in treated homes, its use was finally curtailed. Also in 1987, the Dow Chemical Company announced that it would stop producing the controversial herbicide 2, 4, 5-T. In both of these cases, new pesticides were developed to replace the older types, so that discontinuing them did not result in an economic hardship for farmers or other consumers.

WHY ARE PESTICIDES SO WIDELY USED?

Figure 14.12 shows changes in the use of pesticides throughout the world. Use is projected to continue to increase signifi-

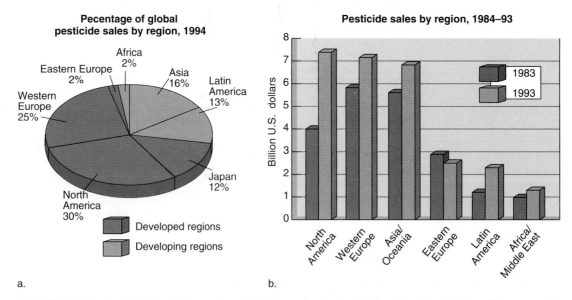

Percentage of global pesticide sales by region, 1994

Africa 2%
Eastern Europe 2%
Western Europe 25%
North America 30%
Asia 16%
Latin America 13%
Japan 12%

Developed regions
Developing regions

a.

Pesticide sales by region, 1984–93

Billion U.S. dollars

1983
1993

North America
Western Europe
Asia/Oceania
Eastern Europe
Latin America
Africa/Middle East

b.

FIGURE 14.12 Pesticide Use The use of pesticides is greatest in the developed world with about 67 percent of the total. However, all parts of the world except for Eastern Europe are experiencing increased use of pesticides.

Source: (a) "Upturn in World Agrochemical Sales in 1994," *AGRO: World Crop Protection News,* No. 238 (August 1995), p. 20. (b) The Fredonia Group, World Pesticides Report No. 636 (The Fredonia Group, Cleveland, Ohio, 1994) as cited in *AGRO,* No. 225 (1995), p. 16.

China's Ravenous Appetite

History has shown that when people rise above subsistence level, one of the first things they want is to eat a little better. They tend to leave the basic grains to the animals and eat the animals themselves or animal products in the form of meat, fish, eggs, and butter. They eat more vegetables and use more edible oils. In short, they move a bit higher up the food chain.

A sizable percentage of China's more than 1.2 billion citizens are moving up the food chain. In the early 1980s, the typical urban Chinese diet consisted of rice, porridge, and cabbage. By the middle 1990s the diet had dramatically changed to include meat, eggs, or fish at least once a day.

Diet is not the only thing changing in China. Even with population control, China's population grows by 14 million people a year, and young people are leaving the countryside and moving to the cities. China's urban population, currently over 300 million, should double by 2010. China has 21 percent of the world's population but only 7 percent of the arable land; in other words, it has less arable land than the United States but five times the population. China's farm sector is simply unable to keep up with the surging demand for what people regard as better food.

Total meat consumption in China is growing 10 percent a year; feed consumption is growing 15 percent. Demand for poultry, which requires 2 to 3 pounds of feed per pound of bird, has doubled in five years. China's new diet will make it more dependent on the United States, Canada, and Australia for feed grains. By 1997 China had gone from being a net exporter of grain to importing 16 million tons. The switch in corn is even more dramatic. As recently as 1995, China was the second-largest corn exporter in the world. But with the chickens and pigs eating so much corn, in one

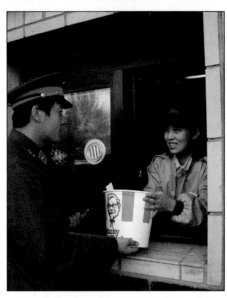

KFC and McDonald's in Beijing. Demand for poultry and beef is soaring.

year China moved from exporting 12 million tons of corn to importing 4 million tons.

China has also become the world's largest importer of fertilizer. In addition, while the government is reclaiming some wasteland in the north for agriculture, this is more than offset by the loss of fertile, multiple-cropped farmland in the southern coastal provinces. Overall, China's farmland is shrinking by at least 0.5 percent per year.

China also faces a severe water shortage in the arid north. In the Yellow River Valley's large grain belt, irrigation projects tripled crop yields during the 1950s and 1960s, resulting in severe overpumping of groundwater. Today, the water table is falling, aquifers are vanishing, and farmers now have to compete with industry and households for water. The era of the flush toilet has arrived.

China is only a part—but certainly the biggest part—of a larger story. What's true for China is also happening throughout much of Asia. Large, populous countries such as Indonesia, Thailand, and the Philippines are rapidly urbanizing. They are gaining purchasing power and losing farmland, and the people are adding more animal proteins and processed foods to their diets. Given the magnitude of the changes, the world's ability to feed itself in the future is difficult to predict.

- What changes do you anticipate in the food consumption pattern in China in the next decade?
- Is China capable of maintaining a diet similar to that of North Americans? Is such a diet desirable?

cantly. If pesticides have so many drawbacks, why are they used so extensively? There are three primary reasons. First, the use of pesticides has increased, at least in the short term, the amount of food that can be grown in many parts of the world. In the United States, pests are estimated to consume 33 percent of the crops grown. On a worldwide basis, pests consume approximately 35 percent of crops. This represents an annual loss of $18.2 billion in the United States alone. Farmers, grain-storage operators, and the food industry continually seek to reduce this loss. A retreat from dependence on pesticides would certainly reduce the amount of food produced. Agricultural planners in most countries are not likely to suggest changes in pesticide use that would result in malnutrition and starvation for many of their inhabitants.

GLOBAL PERSPECTIVE

Contaminated Soils in the Former Soviet Union

The newly independent republics of the former Soviet Union have inherited many kinds of environmental problems generated by the failed policies of the Soviet government. One of these policies was a push for increased agricultural production. New land was brought under cultivation, irrigation projects were developed, and extensive use of fertilizer and pesticides stimulated production.

This single-minded approach has resulted in serious environmental damage. In many areas, the soil has been contaminated by excessive fertilizer and pesticide use, poisoning groundwater and tainting food products. It is estimated that 30 percent of the food produced throughout the former Soviet Union is contaminated and should not be eaten. Forty-two percent of all baby food is contaminated by nitrates and pesticides. Latvian ecologists estimate that excessive pesticide use currently results in 14,000 deaths per year and that 700,000 people become ill from pesticides throughout the former Soviet Union annually.

One problem pesticide is DDT. It was officially banned in the Soviet Union in the 1970s, but the drive to improve agricultural production allowed its continued use. Waivers to the ban were allowed, or DDT was simply renamed by government bureaucrats and used under its new label. It is estimated that 10 million hectares of agricultural land in the former Soviet Union are contaminated with DDT.

The widespread use of agricultural chemicals has also affected the groundwater. Persistent pesticides, nitrates, and heavy metals have entered the groundwater, which many rely on for drinking water. Some areas have cadmium levels three to ten times higher than the amount permitted.

It appears that the fall of Communism in the former Soviet Union did not solve all the problems of the people. These environmental problems are surfacing after years of secrecy and denial under the short-sighted, production-oriented agricultural policy of the failed political system.

A second reason pesticides are used so extensively is economic. The cost of pesticides is more than offset by increased yields and profits for the farmer. In addition, the production and distribution of pesticides is big business. Companies that have spent millions of dollars developing a pesticide are going to argue very strongly for its continued use. Since farmers and agrochemical interests have a powerful voice in government, they have successfully lobbied for continued use of pesticides.

A third reason for extensive pesticide use is that many health problems are currently impossible to control without insecticides. This is particularly true in areas of the world where insect-borne diseases would cause widespread public health consequences if insecticides were not used.

ALTERNATIVES TO CONVENTIONAL AGRICULTURE

Before the invention of synthetic fertilizers, herbicides, fungicides, and other agrochemicals, animal manure and crop rotation provided soil nutrients; a mixture of crops prevented regular pest problems; and manual labor killed insects and weeds. With the development of mechanization, larger areas could be farmed, draft animals were no longer needed, and many farmers changed from mixed agriculture, in which animals were an important ingredient, to monoculture. Chemical fertilizers replaced manure as a source of soil nutrients and crop rotation was no longer important since hay and grain were no longer grown for draft animals and cattle. The larger fields of crops like corn, wheat, and cotton presented opportunities for pest problems to develop and chemical pesticides were used to "solve" this problem.

Today many people feel current agricultural practices are not sustainable and that the degradation of the soil and excessive use of fertilizers and pesticides are not in the best interest of humankind and that we should look for alternative methods for assuring good yields and controlling the pests that compete with us for the food that we raise. Several different approaches have developed that have somewhat different but overlapping goals. Some of the terms used to describe these approaches are: *alternative agriculture, sustainable agriculture,* and *organic agriculture.* Alternative agriculture is the broadest term which includes all nontraditional agricultural methods and encompasses sustainable agriculture, organic agriculture, alternative uses of traditional crops, alternative methods for raising crops, and producing crops for industrial use. The sustainable agriculture movement maintains that current practices are degrading natural resources and seeks methods to produce adequate, safe food in an economically viable manner while enhancing the health of agricultural land and related ecosystems. Organic agriculture advocates avoiding the use of chemical fertilizers and pesticides in the production of food, thus preventing damage to related ecosystems and the food-consuming public. A portion of the proposed definition of organic agriculture by the National Organic Standards Board established by the 1990 Organic Foods Production Act points out a central theme of these three movements.

Organic agriculture is an ecological production management system that promotes and enhances biodiversity, biological cycles, and soil biological activity. It is based on minimal use of off-farm inputs and management practices that restore, maintain, and enhance ecological harmony.

ENVIRONMENTAL CLOSE-UP

Food Additives

Food additives are chemicals added to food before its sale. They have several purposes:

1. To prolong the storage life of the food.
2. To make the food more attractive by adding color or flavor.
3. To modify nutritive value.

Many kinds of molecules are added to foods to prolong shelf life. Calcium propionate is added to baked goods because they can become contaminated with airborne spores of molds and bacteria, which grow on the food and spoil it. BHT, TBHQ, and other commonly seen alphabetic mixtures have a similar function.

Sometimes additives are used just to make the food appear more attractive. For years, Red Dye II was used to color a variety of foods. Its use was discontinued when it was found to be carcinogenic. Other food colorings are still widely used to increase appeal to the consumer. Many kinds of artificial flavors are added to products as well. Commonly used flavor enhancers are monosodium glutamate, table salt, and citric acid.

Other food additives are used to modify the nutritive value of the food product. Iodine in table salt is a good example. Iodine is a trace element required for proper thyroid functioning. Individuals suffering from a lack of iodine often develop an enlargement of the thyroid gland known as goiter. The addition of iodine to table salt has eliminated goiter in the United States. Most cereals and baked goods and many other products have various vitamins and minerals added to improve their nutritive value. Some additives, like Nutrasweet™, reduce the calories while giving the consumer the sensation of tasting something sweet.

Some additives are an unavoidable residue of some step of the food production industry and could more properly be called contaminants. Pesticide residues are an example. Pesticides are used to grow foods, but they are also used to eliminate pests in the storing, processing, and transportation steps of the food industry.

Diethylstilbestrol (DES) was at one time used in the poultry industry to produce fatter birds. Because there were indications that DES is carcinogenic, the U.S. Food and Drug Administration banned the use of DES in chickens and declared that it was potentially hazardous to humans. Further studies were conducted to determine if DES was safe to use in raising beef. It was used with cattle because an animal gains weight more rapidly with DES in its feed. Because DES was eventually linked to breast cancer in women, it has been banned from all animal feed use in the United States and Europe.

In order to reduce negative impacts of conventional agriculture while making a profit, farmers must deal with the twin problems of protecting the quality and fertility of the soil while controlling pests of their crops. A wide variety of techniques can be used to reduce the negative impact of conventional agricultural activity while maintaining economic viability for the farmer.

Techniques for Protecting Soil and Water Resources

Conventional farming practices have several negative effects on soil and water. Soil erosion is a problem throughout the world. (See chapter 13.) However, two other problems are also important—compaction of the soil and reduction in soil organic matter. Several changes in agricultural production methods can help to reduce these problems. Reducing the number of times farm equipment travels over the soil will reduce the degree of compaction. Leaving crop residue on the soil and incorporating it into the soil reduces erosion and increases soil organic matter. In addition, the introduction of organic matter into the soil makes compaction less likely.

Conventional agriculture has several negative impacts on watersheds: fertilizer runoff stimulates aquatic growth and degrades water resources, pesticides can accumulate in food chains, and groundwater resources can be contaminated by fertilizer, pesticides, or animal waste. Reducing or eliminating these sources of contamination would enhance ecological harmony and reduce a threat to human health. Fertilizer runoff can be lessened by reducing the amount of fertilizer applied and the conditions under which it is applied. Applying fertilizer as plants need it will assure that more of it is taken up by plants and less runs off. Increased organic matter in the soil also tends to reduce runoff. More careful selection, timing, and use of pesticides would decrease the extent to which these materials become environmental contaminants.

True organic agriculture which uses neither chemical fertilizers nor pesticides is the most effective in protecting soil and water resources from these forms of pollution but require several adjustments in the way in which farming is done. Crop rotation is an effective way to enhance soil fertility, reduce erosion, and control pests. The use of nitrogen-fixing legumes, such as clover, alfalfa, beans, or soybeans, in crop rotation increases soil nitrogen, but places other demands on the farmer. For example, it typically requires that cattle be a part of the farmer's operation. Crop rotation also requires a greater investment in farm machinery, since certain crops require specialized equipment. Also, the raising of cattle requires additional expenditures for feed supplements and veterinary care. Critics say that organic farming cannot produce the amount of food required for today's population and that it can be economically successful only in specific cases. Proponents disagree and stress that when the hidden costs of soil erosion and pollution are included, organic agriculture or some modification of it is a viable alternative approach to conventional means of food production. Furthermore, organic farmers are willing to accept lower yields because they do not have to pay for expensive chemical fertilizers and pesticides. In addition, organic farmers often receive premium prices for products that are organically grown. Thus, even with lower yields they can still make a profit.

A study by Lockeretz, Shearer, and Kohl lists some comparisons between conventional and organic farming. (See table 14.3.) Conventional farms had a higher yield than organic farms and also netted $384 per hectare, as compared to $333 per hectare for organic farms. Organic farms used only 40 percent of the energy used on conventional farms, thus reducing their costs and raising their profit margins. However, by 1994 only 0.1 percent of agricultural land was being used for "certified" organic farming and most of these were small farms involved in the production of vegetables.

Integrated Pest Management

Another way to reduce the impact of conventional agriculture is to use integrated pest management which uses a variety of methods to control pests, rather than relying on pesticides alone. **Integrated pest management** is a technique that depends on a complete understanding of all ecological aspects of the crop and the particular pests to which it is susceptible to establish pest control strategies. It requires information about the metabolism of the crop plant, the biological interactions between pests and their predators or parasites, the climatic conditions that favor certain pests, and techniques for encouraging beneficial insects. It may involve the selective use of pesticides. Much of the information necessary to make integrated pest management work goes beyond the knowledge of the typical farmer. The metabolic and ecological studies necessary to pinpoint weak points in the life cycles of pests can usually be carried out only at universities or government research institutions. These studies are expensive and must be completed for each kind of pest, since each pest has a unique biology. Once a viable technique has been developed, an educational program is necessary to interest farmers in using integrated pest management rather than

the "spray and save" techniques that they understand and that pesticide salespersons continually encourage.

Several methods are employed in integrated pest management. These include sex attractants, male sterilization, the release of natural predators or parasites, the development of resistant crops, the use of natural pesticides, the modification of farming practices, and the selective use of pesticides. In some species of insects, a chemical called a **pheromone** is released by females to attract males. Males of some species of moths can detect the presence of a female from a distance of up to 3 kilometers. Since many moths are pests, synthetic odors can be used to control them. Spraying an area with the pheromone confuses the males and prevents them from finding females which results in a reduced moth population the following year. In a similar way, a synthetic sex attractant molecule known as Gyplure is used to lure gypsy moths into traps, where they become stuck. Since the females cannot fly and the males are trapped, the reproductive rate drops, and the insect population may be controlled.

Another technique that reduces reproduction is male sterilization. In the southern United States and Central America, the screwworm fly weakens or kills large grazing animals, such as cattle, goats, and deer. The female screwworm fly lays eggs in open wounds on these animals, where the larvae feed. However, it was discovered that the female mates only once in her lifetime. Therefore, the fly population can be controlled by raising and releasing large numbers of sterilized male screwworm flies. Any female that mates with a sterile male fails to produce fertilized eggs

and cannot reproduce. In Curaçao, an island 65 kilometers north of Venezuela, a program of introducing sterile male screwworm flies eliminated this disease from the 25,000 goats on the island. In parts of the southwestern United States, the sterile male technique has also been very effective. The screwworm fly has been eliminated from the United States and northern Mexico, and much of Central America may become free of them as well. In 1990, sterile males were released in Libya to begin eliminating screwworm flies that had been introduced with a South American cattle shipment.

During an epidemic of Mediterranean fruitflies in southern California and northern Mexico, in the early 1980s, a similar technique was employed. Unfortunately the X-ray technique used to sterilize the males was ineffective, and most of the flies released were not sterile, which made the problem worse rather than better. Pesticides were eventually used to control the fruitflies. Recent concern about the Mediterranean fruitfly in California has resulted in the controversial aerial spraying of malathion.

The manipulation of predator-prey relationships can also be used to control pest populations. For instance, the ladybird beetle, commonly called a ladybug, is a natural predator of aphids and scale insects. (See figure 14.13.) Artificially increasing the population of ladybird beetles reduces aphid and scale populations. In California during the late 1800s, scale insects on orange trees damaged the trees and reduced crop yields. The introduction of a species of ladybird beetle from Australia quickly brought the pests under control. Years later, when chemical pesticides were first used in

TABLE 14.3 YIELD DIFFERENCE BETWEEN ORGANIC AND CONVENTIONAL FARMING

Crop	Metric tons per hectare	
	Organic farming	**Conventional farming**
Corn	6.45	7.00
Soybeans	2.44	2.57
Wheat	1.88	3.28

Source: Data from William Lockeretz, et al., "Organic Farming in the Corn Belt," in *Science*, 211:540–47, February 6, 1981.

the area, so many ladybird beetles were accidentally killed that scale insects once again became a serious problem. When pesticide use was discontinued, ladybird beetle populations rebounded, and the scale insects were once again brought under control. (See figure 14.14.)

Herbivorous insects can also be used to control weeds. Purple loosestrife (*Lythrum salicaria*) is a wetlands plant accidentally introduced from Europe in the mid-1800s. It takes over sunny wetlands and eliminates native vegetation such as cattails and, therefore, eliminates many species that rely on cattails as food, nesting places, or hiding places (see figure 14.15). The plant is known to exist in most states and provinces in the United States and Canada. In Europe the plant is not a pest because there are several insect species that attack it in various stages of the plant's life cycle. Since purple loose- strife is a European plant a search was made to identify candidate species of insects that would control it. The criteria are that the insects must live only on purple loosestrife and not infest other plants and must have the capacity to do major damage to purple loosestrife. Five species of beetles have been identified that will attack the plant in various ways. After extensive studies to determine if introductions of these insects from Europe would be likely to cause other problems, several were selected as candidates to help control purple loosestrife in the United States and Canada. Some combination of these beetles have been released in a large proportion of the states and provinces infested with purple loosestrife. Two species (*Galerucella calmariensis, Galerucella pusilla*) feed on the leaves, shoots, and flowers of the newly growing purple loosestrife; one (*Hylobius transversovittatus*) has larvae that feed on the roots. These species have been introduced and appear to be forming colonies. Two other species (*Nanophyes brevis, Nanophyes marmoratus*) feed on the flowers. Their release was scheduled for the summer of 1998. This multi-pronged attack by several species of beetles is projected to reduce the number of purple loosestrife plants by 90 percent.

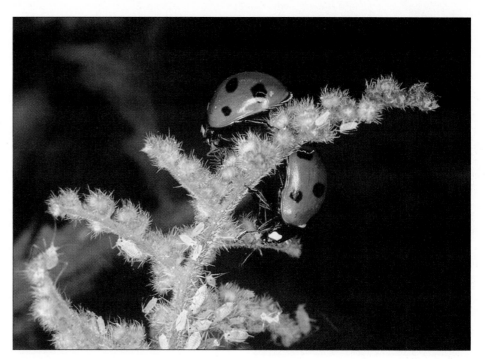

FIGURE 14.13 **Beneficial Insect** The ladybird beetle is a predator of many kinds of pest insects, including aphids.

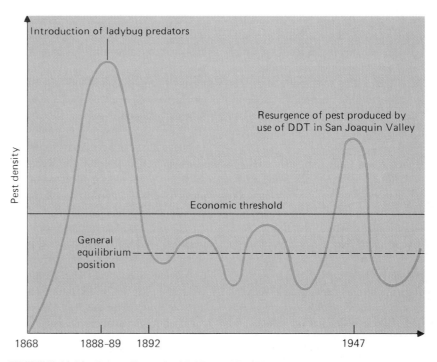

FIGURE 14.14 **Insect Control with Natural Predators** In 1889, the introduction of ladybird beetles (ladybugs) brought the cottony cushion scale under control in the orange groves of the San Joaquin Valley. In the 1940s, DDT reduced the ladybird beetle population and the cottony cushion scale population increased. Stopping the use of DDT allowed the ladybird population to increase, reducing the pest population and allowing the orange growers to make a profit.

From *Man and the Environment*, 2/e by Arthur S. Boughey, © 1975. Reprinted by permission of Prentice-Hall, Inc., Upper Saddle River, N.J.

CHAPTER 14   Agricultural Methods and Pest Management   269

a. Purple loosestrife

b. *Galerucella calmariensis*

c. *Galerucella pusilla*

d. *Hylobius transversovittatus*

FIGURE 14.15   Biological Control of Purple Loosestrife Purple loosetrife is a European plant that invades wetlands and prevents the growth and reproduction of native species of plants. Several kinds of European beetles that feed on purple loosestrife have been released as biological control agents. Two species (*b. Galerucella calmariensis, c. Galerucella pusilla*) feed on the leaves, shoots, and flowers of the newly growing purple loosestrife, one (*d. Hylobius transversovittatus*) has larvae that feed on the roots. It appears that they are actually slowing the spread of purple loosestrife.

The use of specific strains of the bacterium *Bacillus thuringiensis* to control mosquitoes and moths is another example of the use of one organism to control another. A crystalline material produced by the bacterium destroys the lining of the gut of the feeding insect, resulting in its death. One strain of *B. thuringiensis* is used to control mosquitoes while another is primarily effective against the caterpillars of leaf-eating moths, including the gypsy moth.

A method receiving increasing attention is that of producing resistant crop species through selective breeding or genetic engineering. These techniques seek to develop characteristics in major crop plants that allow them to resist fungi, insects, and other pests. For a long time, wheat rust, a fungus parasite that weakens wheat and reduces yields, was a major problem. Modern strains of wheat, and other crops, have now been developed that are able to resist such common plant pests. Another development is strains of crop plants with shorter growing seasons that complete their entire life cycle before pest populations become large enough to cause problems. Another recent development could allow herbicide-resistant

genes to be introduced into crop plants by genetic engineering techniques. This would allow herbicides to be used on the crop without harming it, while the competing weeds would be destroyed. Field trials of crop plants with herbicide-resistant genes have not been as successful as originally hoped.

Naturally occurring pesticides found in plants also can be used to control pests. For example, marigolds are planted to reduce the number of soil nematodes, and garlic plants are used to check the spread of Japanese beetles.

Often, modification of farming practices can reduce the impact of pests. In some cases, all crop residues are destroyed to prevent insect pests from finding overwintering sites. For example, shredding and plowing under the stalks of cotton in the fall reduces overwintering sites for boll weevils and reduces their numbers significantly, thereby reducing the need for expensive insecticide applications. Many farmers are also returning to crop rotation, which tends to prevent the buildup of specific pests that typically occurs when the same crop is raised in a field year after year.

Pesticides can also play a part in integrated pest management. Identifying the precise time when the pesticide will have the greatest effect at the lowest possible dose has several advantages: it reduces the amount of pesticide used, and may still allow the parasites and predators of pests to survive. Such precise applications often require the assistance of a trained professional who can correctly identify the pests, measure the size of the population, and time pesticide applications for maximum effect. In several instances, pheromone-baited traps capture insect pests from fields, and an assessment of the number of insects caught can be a guide to when insecticides should be applied.

Integrated pest management will become increasingly popular as the cost of pesticides increases and knowledge about the biology of specific pests becomes available. However, as long as humans raise crops, there will be pests that will outwit the defenses we develop. Integrated pest management is just another approach to a problem that began with the dawn of agriculture.

Herring gulls nest on islands and other protected sites throughout the Great Lakes region. Since they feed primarily on fish, they are near the top of aquatic food chains and tend to accumulate toxic materials from the food they eat. Eggs taken from nests can be analyzed for a variety of contaminants.

Since the early 1970s, the Canadian Wildlife Service has operated a monitoring program to assess trends in the levels of contaminants in the eggs of herring gulls. In general, the contaminant levels have declined as both the Canadian and U.S. governments have taken action to stop new contaminants from entering the Great Lakes. The figure shows the trends for PCBs and DDE. PCBs are a group of organic compounds that were used as fire retardants, lubricants, insulation fluids in electrical transformers, and in some printing inks. Some forms of PCB are much more toxic than others. Both Canada and the United States have eliminated most uses of PCBs, reducing the levels found in herring gull eggs. DDE is a breakdown product of DDT. DDT use was discontinued in much of the world in the early 1970s. The illustration shows a general decline in the amount of DDE present following the ban on DDT.

- Should the health of a bird species be used to develop policy? Why or why not?
- If PCBs and DDT are no longer being used, why are herring gulls still being contaminated?

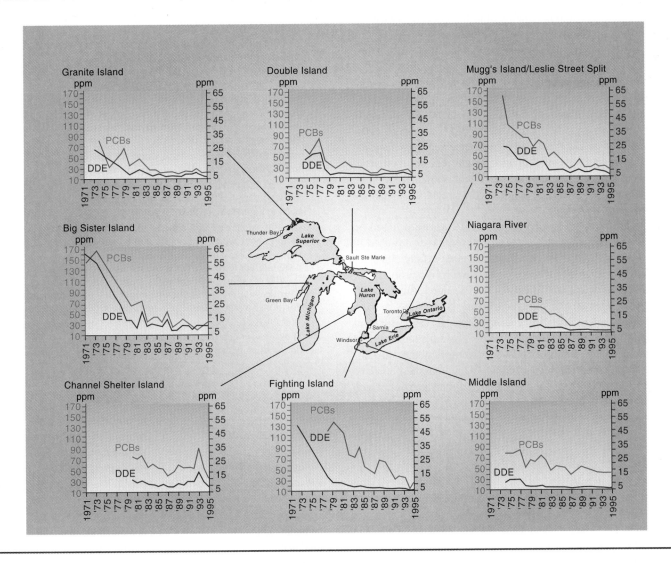

Summary

Although small slash-and-burn garden plots are common in some parts of the world, most of the food in the world is raised on more permanent farms. In countries where population is high and money is in short supply, much of the farming is labor intensive, making use of human labor for many of the operations necessary to raise crops. However, much of the world's food is grown on large, mechanized farms that use energy rather than human muscle for tilling, planting, and harvesting crops, and for the production and application of fertilizers and pesticides.

Monoculture involves planting large areas of the same crop year after year. This causes problems with plant diseases, pests, and soil depletion. Although chemical fertilizers can replace soil nutrients that are removed when the crop is harvested, they do not replace the organic matter necessary to maintain soil texture, pH, and biotic richness.

Mechanized monoculture is heavily dependent on the control of pests by chemical means. Persistent pesticides are stable and persist in the environment, where they may be amplified in ecosystems. Consequently, many of the older persistent pesticides have been quickly replaced by nonpersistent pesticides that decompose much more quickly and present less of an environmental hazard. However, most nonpersistent pesticides are more toxic to humans and must be handled with greater care than the older persistent pesticides.

Pesticides can be divided into several categories based on the organism they are used to control. Insecticides are used to control insects, herbicides are used for plants, fungicides for fungi, and rodenticides for rodents. Because of the problems of persistence, biological amplification, resistance of pests to pesticides, and human health concerns, many people are seeking pesticide-free alternatives to raising food. Several different philosophies that seek the same ends—less use of chemicals and better stewardship of soil—are alternative agriculture, sustainable agriculture, and organic agriculture. One ingredient in all of these approaches is the use of integrated pest management, which uses a complete understanding of an organism's ecology to develop pest-control strategies that use no or few pesticides.

Review Questions

1. What is monoculture?
2. List three reasons why fossil fuels are essential for mechanized agriculture.
3. Describe why pesticides are commonly used in mechanized agriculture.
4. Why are fertilizers used? What problems are caused by fertilizer use?
5. How do persistent and nonpersistent pesticides differ?
6. What is biological amplification? What problems does it cause?
7. How do organic farms differ from conventional farms?
8. Name three nonchemical methods of controlling pest populations.
9. What are the advantages and disadvantages of integrated pest management?
10. List three uses of food additives.
11. List three actions farmers could use to reduce the effect of pesticides on the environment.

Critical Thinking Questions

1. If you were a public health official in a developing country, would you authorize the spraying of DDT to control mosquitoes that spread malaria? What would be your reasons?
2. Look at Table 14.2. What reasons can you advance for the evidence that is presented here? What might you recommend as an alternative to this kind of pest control if you were an agricultural extension agent?
3. Imagine that you are a scientist examining fish in Lake Superior and you find toxaphene, in the fish you are studying. Toxaphene was used primarily in cotton farming and has been banned since 1982. How can you explain this?
4. Are the risks of pesticide use worth the benefits? What values, beliefs, and perspectives lead you to this conclusion?
5. Do you think that current agricultural practices are sustainable? Why or why not? What changes in agriculture do you think will need to happen in the next 50 years?
6. Imagine you are an EPA official who is going to make a recommendation about whether an agricultural pesticide can remain on the market or should be banned. What are some of the facts you would need in order to make your recommendation?

   Who are some of the interest groups interested in the outcome of your decision? What arguments might they present regarding their positions? What political pressures might they be able to bring to bear on you?
7. Why are few consumers demanding alternative methods of crop production and why are farmers not using those methods?

Key Terms

auxin *259*	monoculture *254*
bioaccumulation *261*	nonpersistent pesticide *257*
biocide *256*	nontarget organism *256*
biological amplification *261*	organophosphate *257*
carbamate *257*	persistent pesticide *257*
chlorinated hydrocarbon *257*	pest *256*
fungicide *256*	pesticide *256*
herbicide *256*	pheromone *269*
insecticide *256*	rodenticide *256*
integrated pest management *268*	target organism *256*
macronutrient *255*	weed *256*
micronutrient *255*	

Related Internet Sites

Visit our website at http://www.mhhe.com/environmentalscience/
for specific information on each of the topics below.

Organic foods and farming for human and environmental health

Vegetable pest fact sheet and pest management data

Pesticide contamination threats to humans and the environment

Sustainable agriculture and ecological alternatives to pesticides

Rural development agencies

Information and links related to agriculture

Information, research, and links on agriculture

Biological controls, sustainable systems, and integrated pest
management

Pesticide regulations, toxicology, usage, and other pesticide issues

Evaporation

Navigation/
hydroelectric
use

Rain water

Agricultural
use

Salinization

Groundwater

Industrial and
domestic use

Pollution

OBJECTIVES

After reading this chapter, you should be able to:

- Explain how water is cycled through the hydrologic cycle.
- Explain the significance of groundwater, aquifers, and runoff.
- Explain how land use affects infiltration and surface runoff.
- List the various kinds of water use and the problems associated with each.
- List the problems associated with water impoundment.
- List the major sources of water pollution.
- Define biochemical oxygen demand (BOD).
- Explain how nutrients cause water pollution.
- Differentiate between point and nonpoint sources of pollution.
- Explain how heat can be a form of pollution.
- Differentiate between primary, secondary, and tertiary sewage treatments.
- Describe some of the problems associated with storm-water runoff.
- List sources of groundwater pollution.
- Explain how various federal laws control water use and prevent misuse.
- List the problems associated with water-use planning.
- Explain the rationale behind the federal laws that attempt to preserve certain water areas and habitats.
- List the problems associated with groundwater mining.
- Explain the problem of salinization associated with large-scale irrigation in arid areas.
- List the water-related services provided by local governments.

THE WATER ISSUE

Water is the material that makes life possible on Earth. All living organisms are composed of cells that contain at least 60 percent water and their metabolic activities take place in a water solution. Organisms can exist only where there is access to adequate supplies of water. Water is also unique because it has remarkable physical properties. Water molecules are polar, that is, they have positive and negative ends. Because of this the water molecules tend to stick together and they also have a great ability to separate other molecules from each other. Water's ability to act as a solvent and its capacity to store heat are a direct consequence of its polar nature and these abilities make water extremely valuable for human societal and industrial activities. Water dissolves and carries substances ranging from nutrients to industrial and domestic wastes. A glance at any urban sewer will quickly point out the importance of water in dissolving and transporting wastes. Because water heats and cools more slowly than most other substances, it is used in large quantities for cooling in electric power generation plants and in other industrial processes. Water's ability to retain heat also modifies local climatic conditions in areas near large bodies of water. These areas do not have the wide temperature changes characteristic of other areas.

For most human uses, as well as some commercial ones, the *quality* of the water is as important as its *quantity.* Water must be substantially free of salinity, plant and animal waste, and bacterial contamination to be suitable for human consumption. The oceans contain most of the water but its saltiness makes it unusable for most purposes. Freshwater lacks the salts found in ocean waters. Unpolluted freshwater that is suitable for drinking is known as **potable water.** Early human migration routes and settlement sites were influenced by the availability of drinking water. Today, despite advances in drilling, irrigation, and purification, the location, quality, quantity, ownership, and control of potable waters remains an important human concern.

Over 70 percent of the earth's surface is covered by water. The vast majority of

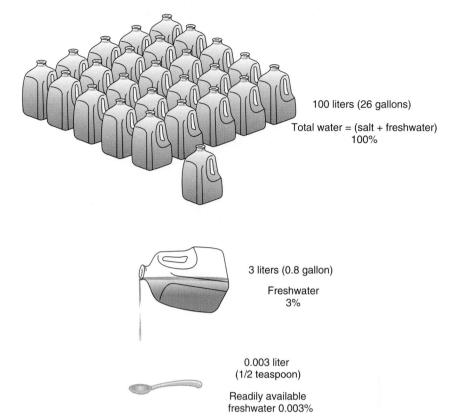

100 liters (26 gallons)

Total water = (salt + freshwater)
100%

3 liters (0.8 gallon)

Freshwater
3%

0.003 liter
(1/2 teaspoon)

Readily available
freshwater 0.003%

FIGURE 15.1   Freshwater Resources Although water covers 75 percent of the Earth's surface, 97 percent is salt water. Of the 3 percent that is freshwater only a tiny fraction is available for human use.

it, however, is ocean saltwater, which has limited use. Only about 3 percent is freshwater and of that only a small fraction is available for use. (See figure 15.1.) Saltwater cannot be consumed by humans or used for many industrial processes. Clean freshwater supplies were once considered inexhaustible. Only recently have we begun to understand that we will probably exhaust our usable water supplies in some areas of the world because of increases in human populations and limitations on the supply available. Some areas of the world have abundant freshwater resources while others have few. In addition, there is an increased demand for freshwater for industrial, agricultural, and personal needs.

Shortages of potable freshwater throughout the world can also be directly attributed to human abuse in the form of pollution. Water pollution has negatively affected water supplies in almost all of the world's densely populated industrialized nations, including Japan, Europe, the former Soviet Union, Canada, and the United States.

Unfortunately, the outlook for the world's freshwater supply is not very promising. According to studies by the United Nations and the International Joint Commission, many sections of the world are currently experiencing a shortage of freshwater and the problem will only intensify. Although the world's supply of water is continually being replenished by rainfall, this rainfall varies significantly in different parts of the world. (See figure 15.2.) Furthermore, changes in the amount of rain from year to year results in periodic droughts for some areas and devastating floods for others. However, rainfall is needed to regenerate freshwater and, therefore, is an important link in the cycling of water.

THE HYDROLOGIC CYCLE

All water is locked into a constant recycling process called the **hydrologic cycle.** (See figure 15.3.) Two important processes

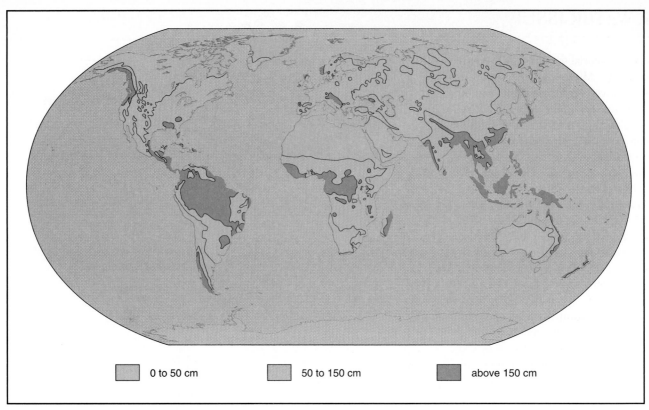

| 0 to 50 cm | 50 to 150 cm | above 150 cm |

FIGURE 15.2   Global Average Annual Precipitation   The world's precipitation is not uniformly distributed. Although some areas of the world, such as Southeast Asia, Central Africa, and Central and South America, have extremely high levels of precipitation, other areas, such as northern Africa, the Middle East, central Asia, and western North America have very little annual precipitation.

involved in the cycle are the evaporation and condensation of water. Evaporation involves adding energy to molecules of a liquid so that they become farther apart to form a gas. Condensation is the reverse process in which molecules of a gas give up energy, become closer together, and become a liquid. Solar energy provides the energy that evaporates water from the ocean surface. Water also evaporates from the soil, bodies of freshwater, and from the surfaces of plants. The water loss from plants is called **transpiration** and involves plants taking up water from the soil and releasing it from the leaves. The air containing water vapor moves across the surface of the earth. As this warm, moist air cools, water droplets form and fall to the land as precipitation. Although some precipitation may simply stay on the surface until it evaporates, most will either sink into the soil or flow downhill until it eventually returns to the ocean. The water that infiltrates the soil may be stored for long periods in underground reservoirs. This

water is called **groundwater,** as opposed to the surface water that enters a river system as **runoff.**

Water that enters the soil and is not picked up by plant roots moves slowly downward through the spaces in the soil and subsurface material until it reaches an impervious layer of rock. The porous layer which becomes saturated with water is called an **aquifer.** There are two basic kinds of aquifers: **unconfined** and **confined.** An unconfined aquifer usually occurs near the land's surface and may be called a water-table aquifer because the upper surface of its water is the **water table.** The lower boundary is an impermeable layer of clay or rock. Unconfined aquifers are replenished (recharged) primarily by rain that falls on the ground directly above the aquifer and percolates down to the water table. The water in such aquifers is at atmospheric pressure and flows in the direction of the water table's slope. Above the water table and below the land surface is a layer that is not saturated

with water known as the **vadose zone** (also known as the unsaturated zone or zone of aeration). (See figure 15.4.)

A confined aquifer is bounded on both the top and bottom by confining layers and is saturated with water under greater-than-atmospheric pressure. A confined aquifer is primarily replenished by rain and surface water from a recharge zone that may be many kilometers from where the aquifer is tapped for use. If water cannot pass through the confining layer the layer is called an **aquiclude.** If water can pass in and out of the confining layer, the layer is called an **aquitard.** If the recharge area is at a higher elevation than the place where an aquifer is tapped, water will flow up the pipe until it reaches the same elevation as the recharge area. Such well are called **artesian wells.** The nature of the substrate in the aquifer influences the amount of water the aquifer can hold and the rate at which water moves through it. **Porosity** is a measure of the size and number of the spaces in the substrate. The

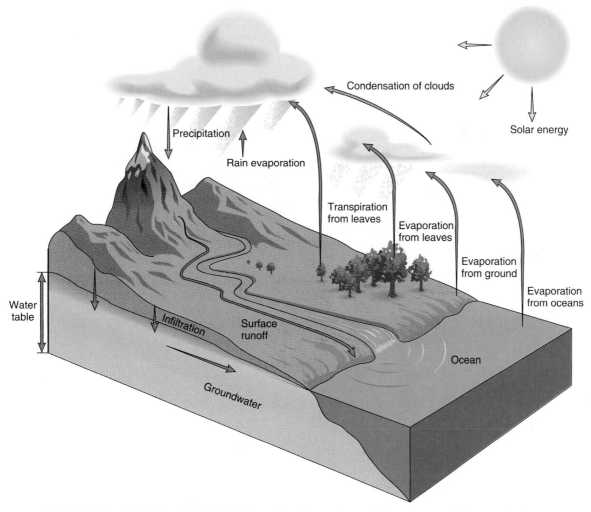

FIGURE 15.3 **The Hydrologic Cycle** The cycling of water through the environment follows a simple pattern. Moisture in the atmosphere condenses into droplets that fall to the earth as rain or snow, supplying all living things with its life-sustaining properties. Water, flowing over the earth as surface water or through the soil as groundwater, returns to the oceans, where it evaporates back into the atmosphere to begin the cycle again.

greater the porosity the more water it can contain. **Permeability** is a measure of how rapidly water can flow through the layer and determines how much water can be pumped from a well per minute.

HUMAN INFLUENCES ON THE HYDROLOGIC CYCLE

The way in which land is used has significant impact on evaporation, runoff, and infiltration. When water is used for cooling in power plants or to irrigate crops, the rate of evaporation is increased. Water impounded in reservoirs also evaporates

rapidly. This rapid evaporation can affect local atmospheric conditions. Water infiltration also is greatly influenced by human activity. Urban complexes with paved surfaces and storm sewers increase runoff and reduce infiltration. Demand for underground water in urban areas is usually high, but urban developments actually increase the gap between supply and demand.

Aquifers are extremely important in supplying water for industrial, agricultural, and municipal use. Most of the larger urban areas in the western part of the United States depend upon underground water for their water supply. This groundwater supply exists as long as it is not used faster than it can be replaced. Determining how

much water can be used and what the uses should be is often a problem.

Water uses can be measured by either the amount withdrawn or the amount consumed. Water withdrawn for use is diverted from its natural course. It may be withdrawn and then later returned to its source so it can be used again in the future. For example, when a factory removes water from a river for cooling purposes, it returns the water to the river and it can be used later. Water that is incorporated into a product or lost to the atmosphere through evaporation and transpiration cannot be reused in the same geographic area and is said to be consumed. Much of the water used for irrigation evaporates, is transpired by plants, or is removed with the crop when it is

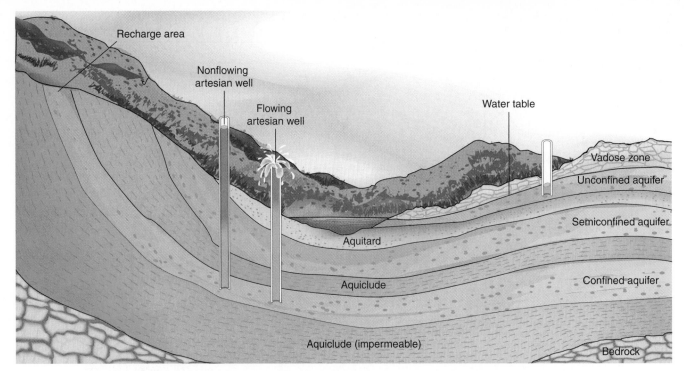

FIGURE 15.4 Aquifers and Groundwater Groundwater is found in the pores in layers of sediment or rock. The various layers of sediment and rock determine the nature of the aquifer and how it can be used.

harvested. Therefore, much of the water withdrawn for irrigation is consumed.

KINDS OF WATER USE

Water use varies considerably around the world depending on availability and degree of industrialization. However, use can be classified into four broad categories: (1) domestic use, (2) agricultural use, (3) industrial use, and (4) in-stream use. It is also important to make a distinction between consumptive and nonconsumptive uses. Consumptive uses do not return the water after use while nonconsumptive uses do.

Domestic Use of Water

Over 90 percent of the water used for domestic purposes in North America is supplied by municipal water systems which typically include complex, costly water purification facilities. However, many rural residents can obtain safe water from untreated private wells. Wells are also important sources of water for municipal systems. Nearly 40 percent of municipal water supplies come from wells.

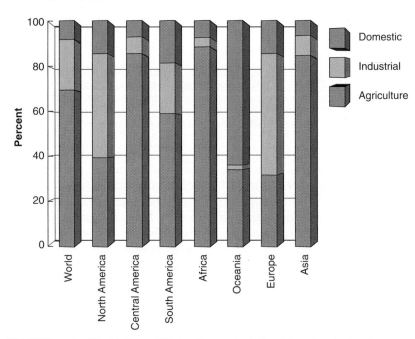

How Water Is Used

FIGURE 15.5 World Uses of Water Domestic, industrial, and agricultural uses dominate the allocation of water resources. However there is considerable variety in how these resources are used in different parts of the world.

Source: *World Resources 1998–99* Data Table 12.1.

Urban growth has created problems in the development, transportation, and maintenance of quality water supplies. Domestic use of water is a relatively small component of the total water-use picture (see figure 15.5). In regions experiencing rapid population growth, such as Asia, domestic use is expected to increase sharply.

Domestic activities in highly developed nations require a great deal of water. This domestic use includes drinking, air conditioning, bathing, washing clothes, washing dishes, flushing toilets, and watering lawns and gardens. On average, each person in a North American home uses 300 to 400 liters of water each day. Most of this **domestic water** (about 70 percent) is used as a solvent to carry away wastes (nonconsumptive use), with only a small amount used for drinking and about 30 percent for lawns and gardens (consumptive use). (See figure 15.6.) Natural processes cannot cope with the highly concentrated wastes typical of a large urban area. The unsightly and smelly waste presents a potential health problem which cities and towns must treat before releasing it to a local water source. Yet all water that enters the house has been purified and treated to make it safe for consumption. Until recently, the cost of water in almost every community has been so low that there was very little incentive to conserve, but shortages of water and increasing purification costs have raised the price of domestic water in many parts of the world and it is becoming evident that increased costs do tend to reduce use. (See figure 15.7.)

The major problem associated with domestic use of water is maintaining an adequate, suitable supply for growing metropolitan areas. (See figure 15.8.) Demand for water in urban areas sometimes exceeds the immediate supply, particularly when the supply is local surface water. This is especially true during the summer, when water demand is high and precipitation is often low. Many communities have begun public education campaigns designed to help reduce water usage. (See figure 15.9.)

In addition to encouraging the public to conserve water, municipalities need to pay attention to losses that occur within the distribution system. Leaking water pipes and mains account for significant losses of water. Even in the developed world losses may be as high as 20 percent. Poorer countries typically exceed this and some may lose over 50 percent of the water to leaks. Another major cause of

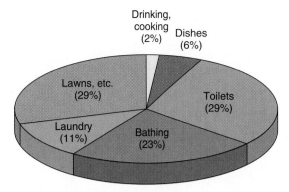

Water Use by a Typical North American Family of Four

FIGURE 15.6 Urban Domestic Water Uses Over 150 billion liters of water are used each day for urban domestic purposes in North America.

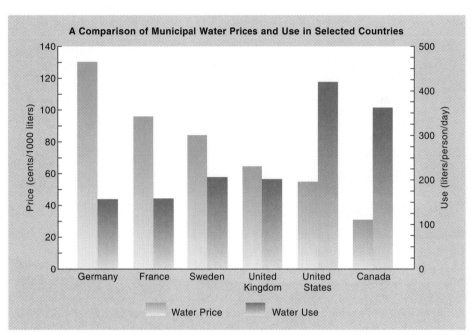

FIGURE 15.7 Water Use Decreases as Water Price Increases There is a direct correlation between the amount of water that is used and the increase in price.

water loss has been public attitudes. As long as water is considered a limitless, inexpensive resource, little effort will be made to conserve it. As the cost of water rises and attitudes toward water change, so will usage and efforts to conserve.

Agricultural Use of Water

In North America groundwater accounts for about 40 percent of the water used in agriculture and surface water about 60 percent. **Irrigation** is the major consumptive use of water in most parts of the world and accounts for about 80 percent of all the water consumed in North America. The amount of water used for irrigation and livestock continues to increase throughout the world. Future agricultural demand for water will depend on the cost of water for irrigation; the demand for agricultural products, food, and fiber; governmental policies; and the development of new technology.

Since irrigation is typical in arid and semiarid areas, local water supplies are often lacking and it is often necessary to transport water great distances to water crops. This is particularly true in the western United States, where about 14

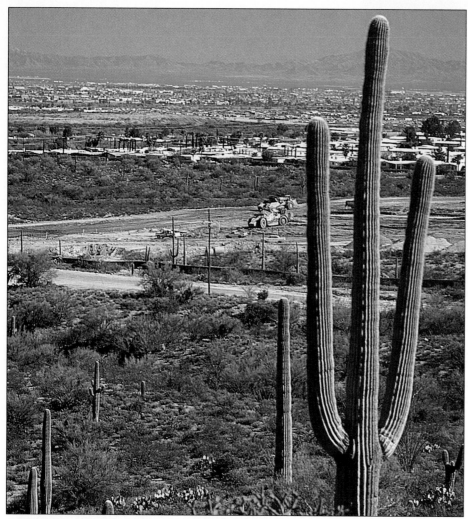

FIGURE 15.8 Urban Expansion Maintaining a suitable supply of water for growing metropolitan areas can pose major problems, especially in arid areas.

Many forms of irrigation require a great deal of energy. This is particularly true when pumps are used to deliver the water to the crop. It is estimated that 40 percent of the energy devoted to agriculture in Nebraska is used for irrigation. Increasing energy costs may force some farmers to reduce or discontinue irrigation. In addition, much of western Nebraska relies on groundwater for irrigation, and the water table is dropping rapidly. If a water shortage develops, land values will decline. Land use and water use are interrelated and cannot be viewed independently.

Industrial Use of Water

Industrial use of water accounts for nearly half of total water withdrawals in the United States, about 70 percent in Canada, and about 23 percent worldwide. Ninety percent of the water used by industry is for cooling and is returned to the source, so only a small amount is actually consumed. Most industrial processes involve heat exchanges. Water is a very effective liquid for carrying heat away from these processes. For example, electric power generating plants use water to cool steam so that it changes back into water. If the water heated in an industrial process is dumped directly into a watercourse, it significantly changes the water temperature. This affects the aquatic ecosystem by increasing the metabolism of the organisms and reducing the water's ability to hold dissolved oxygen.

Industry also uses water to dissipate and transport waste materials. In fact, many streams are now overused for this purpose, especially in urban centers. The use of watercourses for waste dispersal degrades the quality of the water and may reduce its usefulness for other purposes. This is especially true if the industrial wastes are toxic.

Many nations in the industrialized world have passed laws that severely restrict industrial discharges of wastes into watercourses. In the United States, the federal role in maintaining water quality began in 1948 with the passage of the Federal Water Pollution Control Act. This act provided federal funds and technical assistance to strengthen local, state, and interstate water-quality programs. Through

million hectares of land are irrigated. The construction and maintenance of dams, canals, pipes, and pumps is expensive. Therefore, it is becoming increasingly important to modify irrigation practices to use less water. (See figure 15.10.) Costs for irrigation water have traditionally been low since many of the dams and canals were constructed with federal assistance and farmers have often used water wastefully. Increasing the cost of water will stimulate farmers to conserve, just as it does homeowners. Another way to reduce the demand for irrigation water is to reduce the quantity of water-demanding crops grown in dry areas, or change from high water-demanding to lower water-demanding crops. For example, wheat or soybeans require less water than do potatoes or sugar beets.

There are four commonly used methods for irrigation. Surface or flood irrigation involves supplying the water to crops by having the water flow over the field or in furrows. This requires extensive canals and is not suitable for all kinds of crops. Spray irrigation involves the use of pumps to spray water on the crop. Trickle irrigation uses a series of pipes with strategically placed openings so that water is delivered directly to the roots of the plants. Subirrigation involves supplying water to plants through underground pipes. Often this method is used where soils require draining at certain times of the year. The underground pipes can be used to drain excess water at one time of the year and supply water at others. Each of these methods has its drawbacks and advantages as well as conditions under which it works well.

Water savings guide

Conservative use will save water		Normal use will waste water
Wet down, soap-up, rinse off 15 liters (4 gal)	**Shower**	Regular shower 95 liters (25 gal)
May we suggest a shower?	**Tub bath**	Full tub 135 liters (36 gal)
Minimize flushing Each use consumes 20–25 liters (5–7 gal) New toilets use 6 liters (1.6 gal)	**Toilet**	Frequent flushing is very wasteful
Fill basin 4 liters (1 gal)	**Washing hands**	Tap running 8 liters (2 gal)
Fill basin 4 liters (1 gal)	**Shaving**	Tap running 75 liters (20 gal)
Wet brush, rinse briefly 2 liters (1/2 gal)	**Brushing teeth**	Tap running 38 liters (10 gal)
Take only as much as you require	**Ice**	Unused ice goes down drain
Please report immediately	**Leaks**	A small drip wastes 95 liters (25 gal)
Turn off light, TV, heaters, and air conditioning when not in room	**Energy**	Wasting energy also wastes water

Thank you for using this column...and not this one

FIGURE 15.9 Ways to Conserve Water Minor changes in the way people use water could significantly reduce domestic water use. Note: 1 gallon equals approximately 3.785 liters.

amendments to the act in 1956, 1965, 1972, and 1987, the federal role in water-pollution control was increased to include establishing area-wide waste-treatment management plans, and establishing the framework for a national program of water-quality regulation. However, many countries in the developing world have done little to control industrial pollution and water quality is significantly reduced by careless use. See Global Perspective: Comparing Water Use and Pollution in Industrialized and Developing Countries.

In-Stream Use of Water

In-stream uses of water do not remove water but make use of it in its channels and basins. Therefore all in-stream uses are nonconsumptive. Major **in-stream uses** of water are for hydroelectric power, recreation, and navigation. Although in-stream uses do not remove water, they may require modification of the direction, time, or volume of flow and can negatively affect the watercourse.

Electricity from hydroelectric power plants is an important energy resource. Presently, hydroelectric power plants produce about 13 percent of the total electricity generated in the United States. (See figure 15.11.) They do not consume water and do not add waste products to it. However, the dams needed for the plants have definite disadvantages, including the high cost of construction and the resulting destruction of the natural habitat in streams and surrounding lands. The sudden discharge of impounded water from a dam can seriously alter the downstream environment. If the discharge is from the top of the reservoir, the stream temperature rapidly increases. Discharging the colder water at the bottom of the reservoir causes a sudden decrease in the stream's water temperature. Either of these changes is harmful to aquatic life. The impoundment of water also reduces the natural scouring action of a flowing stream. If water is allowed to flow freely, the silt accumulated in the river is carried downstream during times of high water. This maintains the river channel and carries nutrient materials to the river's mouth. But if a dam is constructed, the silt deposits behind the dam, eventually filling the reservoir with silt. Many other dams were constructed to control floodwaters. While dams reduce flooding, they do not eliminate it. In fact, the building of a dam often encourages people to develop the floodplain. As a result, when flooding occurs, the loss of property and lives may be greater.

Because dams create lakes that have a large surface area, evaporation is increased. In areas where water is scarce, the amount of water lost can be serious. This is particularly evident in hot climates. Furthermore, flow is often intermittent below the dam, which alters the water's oxygen content and interrupts fish migration. The populations of algae and other small organisms are also altered. Because of all these impacts, dam construction requires careful planning.

Water tends to be a focal point for recreational activities. Dam construction creates new recreational opportunities because reservoirs provide sites for boating, camping, and related recreation. (See figure 15.12.) However, these opportunities come at the expense of a previously free-flowing river. Some recreational pursuits, such as river fishing, are lost. Sailing, waterskiing, swimming, fishing, and camping all require water of reasonably good quality. Water is used for recreation in its natural setting and often is not physically affected. Even so, it is necessary to plan for recreational use, because overuse or inconsiderate use can degrade water quality. For example, waves generated by powerboats can accelerate shoreline erosion and cause siltation.

a.

b.

c.

FIGURE 15.10 **Types of Irrigation.** Many arid areas require irrigation to be farmed economically. *(a)* The use of irrigation canals and ditches is a common method of irrigation. The land is graded so that water flows from the source into the fields. Water is siphoned from a canal into ditches between rows of crops. *(b)* Central pivot irrigation uses a pump to spray water into the air above the plants as a long pipe on wheels slowly rotates about a central point. *(c)* Trickle irrigation conserves water by delivering water directly to the roots of the plants but requires an extensive network of pipes.

Most major rivers and large lakes are used for navigation. North America currently has more than 40,000 kilometers of commercially navigable waterways. These waterways must have sufficient water to ensure passage of transport vessels. Canals, locks, and dams guarantee this volume. Often, dredging is necessary to maintain the proper channel depth. Dredging can resuspend in the water contaminated sediments that had been covered over. In addition, the flow within the hydrologic system is changed, which, in turn, affects the water's value for other uses.

Most large urban areas rely on water to transport resources. During recent years, the inland waterway system has carried about 10 percent of goods such as grain, coal, ore, and oil. In North America, expenditures for the improvement of the inland waterway system have totaled billions of dollars.

In the past, almost any navigation project was quickly approved and funded, regardless of the impact on other uses. Today, however, such decisions are not made until the impacts on other uses are carefully analyzed.

KINDS AND SOURCES OF WATER POLLUTION

Water pollution occurs when the use of water by one segment of society interferes with the health and well-being of other members. In an industrialized society, maintaining completely unpolluted water in all drains, streams, rivers, and lakes is probably impossible. (See table 15.1.) But we can evaluate a body of water and its water quality can be matched to its intended use. Adding material to water may make it unfit for some uses but not others. If silt is added to a lake, the water may still be drinkable, but the lake may no longer be an acceptable place to swim. If salts are added to a lake, the water may be less acceptable for drinking but not for recreation. It may not be necessary to maintain absolutely pure water.

There are also economic considerations. The cost of removing the last few percentages of some materials from the

Is It Safe to Drink the Water?

Roughly 1000 contaminants have been detected in the public water supply in the United States, and virtually every major water source is vulnerable to pollution. About 60 percent of the U.S. population relies on surface water from rivers, lakes, and reservoirs that may contain industrial and agricultural wastes and pesticides washed off fields by rain. The other 40 percent uses groundwater that may be tainted by chemicals slowly seeping in from toxic-waste dumps, agricultural activities, and leaking sewage and septic systems. In some areas where groundwater supplies are being gradually depleted, the chemical pollutants are becoming more concentrated.

Most pollutants are probably not concentrated enough to pose significant health hazards; however, there are exceptions. The most widespread danger in water is lead, which can cause high blood pressure and an array of other health problems. Lead is especially hazardous to children, since it impairs the development of brain cells. The U.S.

Environmental Protection Agency estimates that at least 42 million Americans are exposed to unacceptably high levels of lead, and the U.S. Public Health Service estimates that perhaps 9 million children are at least slightly affected by it.

The contamination comes from old lead pipes and solder that have been used in plumbing for years. These materials are gradually being replaced in homes and water systems. Individuals may want to have their water tested for lead by an official lab. If the level is too high, they can investigate ways to deal with the problem or switch to bottled water for drinking and cooking. Even then, caution is called for—some bottled waters contain many of the same contaminants that tap water does.

The other four types of contamination in the U.S. water supply, along with their source and risk, are shown in the table. Regardless of the problems, however, the water supply in the United States is among the cleanest in the world.

COMMON CONTAMINANTS OF DRINKING WATER

Substance	Source	Health effects
Persistent chlorinated organic compounds	Used as solvents in industry; past use as pesticides	Various; including reproductive problems and cancer
Trihalomethanes	Produced by chemical reactions when water is disinfected by chlorination	Liver and kidney damage and possible cancer
Nitrates	Primarily from fertilizer and effluent from concentrated livestock raising	Can react to reduce oxygen uptake by blood; particularly a problem with children
Lead	Old piping and solder in public water distribution systems, homes, and other buildings	Nerve damage, learning difficulties in children, birth defects, possible cancer
Pathogenic bacteria, protozoa, and viruses	Leaking septic tanks and sewers, contamination of water supply from birds and mammals, and inadequate disinfection	Acute gastrointestinal illness and other serious health problems

Source: Data, in part, from the U.S. Environmental Protection Agency.

water may not be justified. This is certainly true of organic matter, which is biodegradable. However, radioactive wastes and toxins that may accumulate in living tissue are a different matter. Vigorous attempts to remove these materials are often justified because of their potential harm to humans and other living organisms.

Attempts to control water pollution have been focused on point sources of pollution. When a source of pollution can be readily identified because it has a definite source and place where it enters the water, it is said to come from a **point source.** Municipal and industrial discharge pipes are good examples of point sources. Economic pressure and adverse publicity can affect

companies that continue to pollute from point sources. Diffuse pollutants, such as agricultural runoff, road salt, and acid rain, are said to come from **nonpoint sources** and are much more difficult to identify and control. In addition, pollution legislation relating to them is very difficult to enforce.

Municipal Water Pollution

Municipalities are faced with the double-edged problem of providing suitable drinking water for the population and disposing of wastes. These wastes consist of storm-water runoff, wastes from industry, and wastes from homes and commercial establishments. Wastes from homes consist

primarily of organic matter from garbage, food preparation, cleaning of clothes and dishes, and human wastes. Human wastes are mostly undigested food material and a concentrated population of bacteria, such as *Escherichia coli* and *Streptococcus faecalis*, collectively called coliform bacteria. These particular bacteria normally grow in the large intestine of humans, where they are responsible for some food digestion and for the production of vitamin K. Low numbers of these bacteria in water are not harmful to healthy people. Because they can be easily identified, their presence in the water is used to indicate the amount of pollution from human waste. The numbers of these types of bacteria present in water

FIGURE 15.11   Dams Interrupt the Flow of Water   The flow of water in most large rivers is controlled by dams. Most of these dams provide electricity. In addition, they prevent flooding and provide recreational areas. However, dams destroy the natural river system.

FIGURE 15.12   Recreational Use of Water   Marinas provide recreation; however, wetlands destruction and large dredging operations may be necessary to build them.

are directly related to the amount of human waste entering the water. When human wastes are disposed of in water systems,

other potentially harmful bacteria from humans may be present in amounts too small to detect by sampling. Even in small

numbers these harmful bacteria may cause disease epidemics. The greater the amount of wastes deposited in the water, the more likely it is that there will be populations of disease-causing bacteria. Therefore, coliform bacteria are used as indicator that other more harmful organisms may be present.

The nonliving organic matter in sewage presents a different kind of pollution problem because it decays in the water. As microorganisms metabolize the organic matter, they use up available oxygen. The amount of oxygen required to decay a certain amount of organic matter is called the **biochemical oxygen demand (BOD).** (See figure 15.13.) Measuring the BOD of a body of water is one way to determine how polluted it is. If too much organic matter is added to the water, all of the available oxygen will be used up. Then anaerobic (not requiring oxygen) bacteria begin to break down wastes. Anaerobic respiration produces chemicals that have a foul odor and an unpleasant taste and that generally interfere with the well-being of humans.

Although the wastewater from cleaning dishes and clothing may contain some organic material, the more important group of contaminants found in this water are soaps and detergents. Soaps and detergents are useful because one end of the molecule dissolves in dirt or grease and the other end dissolves in water. When the soap or detergent molecules are rinsed away by the water, the dirt or grease goes with them.

Until recently, many detergents contained phosphates as a part of their chemical structure. Plants and algae need phosphate as a nutrient. Phosphates are generally present in very limited amounts in freshwater and, therefore, are a limiting factor on their growth. (A **limiting factor** is a necessary material that is in short supply, and because of the lack of it, an organism cannot reach its full potential growth. See chapter 4.) Thus, when phosphates from detergents are added to the surface water, they act as a fertilizer and promote the growth of undesirable algae populations. The excessive growth of algae and aquatic plants due to added nutrients is called **eutrophication.** Algae and larger aquatic plants may interfere with the use of the water by fouling boat

TABLE 15.1 SOURCES AND IMPACTS OF SELECTED POLLUTANTS

Pollutant	Source	Effects on humans	Effects on aquatic ecosystem
Acids	Atmospheric deposition; mine drainage; decomposing organic matter	Reduced availability of fish and shellfish. Increase heavy metals in fish.	Death of sensitive aquatic organisms; increased release of trace metals from soils, rock, and metal surfaces such as water pipes
Chlorides	Runoff from roads treated for removal of ice or snow; irrigation runoff; brine produced in oil extraction; mining	Reduced availability of drinking water supplies; reduced availability of shellfish	At high levels, toxic to freshwater organisms
Disease-causing organisms	Dumping of raw and partially treated sewage; runoff of animal wastes from feed lots	Increased costs of water treatment; death and disease; reduced availability and contamination of fish, shellfish, and associated species	Reduced survival and reproduction of aquatic organisms due to disease
Elevated temperatures	Heat trapped by cities that is transferred to water; unshaded streams; solar heating of reservoirs; warm-water discharges from power plants and industrial facilities	Reduced availability of fish	Elimination of cold-water species of fish and shellfish; less oxygen; heat-stressed animals susceptible to disease. Inappropriate spawning behavior.
Heavy metals	Atmospheric deposition; road runoff; discharges from sewage treatment plants and industrial sources; creation of reservoirs; acidic mine effluents	Increased costs of water treatment; disease and death; reduced availability and healthfulness of fish and shellfish. Bioamplification.	Lower fish population due to failed reproduction; death of invertebrates leading to reduced prey for fish. Bioamplification.
Nutrient enrichment	Runoff from agricultural fields, pastures, and livestock feedlots; landscaped urban areas; dumping of raw and treated sewage and industrial discharges; phosphate detergents	Increased water treatment costs; reduced availability of fish, shellfish, and associated species; color and odor associated with algal growth; impairment of recreational uses	Algal blooms occur. Death of algae results in low oxygen levels and reduced diversity and growth of large plants; reduced diversity of animals; fish kills.
Organic molecules	Runoff from agricultural fields and pastures; landscaped urban areas; logged areas; discharges from chemical manufacturing and other industrial processes; combined sewers	Increased costs of water treatment; reduced availability of fish, shellfish, and associated species; odors	Reduced oxygen; fish kills; reduced numbers and diversity of aquatic life
Sediment	Runoff from agricultural pastures and livestock feed lots; logged hillsides; degraded stream banks; road construction; and other improper land use	Increased water treatment costs; reduced availability of fish, shellfish, and associated species; filling in of lakes, streams, and artificial reservoirs and harbors requiring dredging.	Covering of spawning sites for fish; reduced numbers of insect species; reduced plant growth and diversity; reduced prey for predators; clogging of gills and filters
Toxic chemicals	Urban and agricultural runoff; municipal and industrial discharges; leachate from landfills and mines; atmospheric deposits	Increased costs of water treatment; increased risk of certain cancers; reduced availability and healthfulness of fish and shellfish	Reduced growth and survivability of fish eggs and young; fish diseases; death of carnivores due to bioamplification in the food chain

Source: Data, in part, from World Resources 1994–95.

propellers, clogging water intake pipes, changing the taste and odor of water, and causing the buildup of organic matter on the bottom. As this organic matter decays, oxygen levels decrease, and fish and other aquatic species die. Because these problems are all associated with the addition of plant nutrients such as phosphates to the water, many states have banned the sale of detergents with high phosphate content. Detergent manufacturers point out that nutrients from **agricultural runoff** are probably more significant than detergents in adding phosphate to lakes and streams. While such runoff is important in adding to the total phosphate load of surface water, this defense does not alter the fact that

detergents also add to this surface water load. Most domestic wastewater goes through a sewage treatment plant, so it is easier to measure and control the phosphate content. In areas where agriculture is uncommon, detergents can be the major source of phosphate pollution.

Agricultural Water Pollution

Agricultural activities are the primary cause of water pollution problems. Excessive use of fertilizer results in eutrophication in many aquatic habitats, because precipitation carries dissolved materials into streams and lakes. In addition, groundwater may become contaminated with

fertilizer and pesticides. The exposure of land to erosion results in increased amount of sediment being added to water courses. Runoff from animal feedlots carries nutrients, organic matter, and bacteria. Irrigation water that leaves fields and enters a body of water typically carries a heavy load of salt that degrades the water body. And the use of agricultural chemicals results in contamination of sediments and aquatic organisms. One of the largest water pollution problems is agricultural runoff from large expanses of open fields.

Farmers can reduce runoff in several ways. One is to leave a zone of undisturbed land near drains or stream banks. This retards surface runoff because soil

ENVIRONMENTAL CLOSE-UP

New Optimism for the New River

The New River that flows from Mexico into the United States is pea-green in color and texture, laden with fecal matter and carcinogens, topped with detergent foam, and carrying the virus that causes polio and the bacteria that cause typhoid and cholera. For half a century, the river has been a binational disaster, carrying the human and industrial wastes of Mexicali, Mexico, into the Imperial Valley of California and earning an ignominious distinction as the dirtiest river in the United States.

In Mexico, the river runs for 35 kilometers past schools and neighborhoods before reaching the U.S. border at Calexico. Then it winds an additional 10 kilometers north through some of the richest farmland in the world before emptying its load—now including agricultural pesticides—into the Salton Sea in California. Five U.S. presidential administrations and their counterparts in Mexico City have promised to do something about the New River, but until recently little has been done. (See insert.)

Although it is too early to say that the river is getting cleaned up, there are encouraging signs that help is finally on the way, for in this era of the North American Free Trade Agreement, border environmental problems are being taken more seriously in both the United States and Mexico. The U.S. Congress has appropriated funding for repair jobs on the Mexicali sewer system aimed at significantly reducing the flow of raw sewage into the river. In addition, the International Boundary and Water Commission has made the New River a priority in its anti-pollution efforts along the border.

Part of the credit for the burst of official interest in the river lies with NAFTA and the binational agencies created by it. But some of the credit also belongs to the Imperial County Board of Supervisors and its strategy to get the attention of the Environmental Protection Agency (EPA) in Washington, D.C.

At first, the supervisors thought of suing the federal government for letting the New River fester for so long. Instead, they hired a Los Angeles attorney who had once been the general counsel to the EPA. The attorney recommended a bureaucratic procedure called a "petition for rule-making" in which Imperial County laid out its grievances against the polluted river and asked that something be done.

A Tide of Broken Promises

Here are some highlights of the half-century history of accords that have done little to clean up the New River:

1944: Washington and Mexico City agree to work cooperatively to clean up the river.

1950: Congress authorizes the secretary of state to work with the Mexican government on cleanup.

1972: After meeting with President Richard Nixon on the matter, Mexico's President Luis Echeverria Alvarez orders an end to the dumping of raw sewage in the New River, promising that most of it will be curbed within the year.

1979: U.S. President Jimmy Carter and Mexican President Jose Lopez Portillo, meeting in Mexico City, discuss the continuing problems of the river. The same year, a Mexican federal official is quoted in the Los Angeles Times: "We have the solution already planned. We just need the time to finish it."

1986: U.S. President Ronald Reagan and Mexican President Roberto de la Madrid, meeting in Mexicali, discuss river problems.

1987: A Mexican federal official quoted in the Los Angeles Times says "give us one year" to make major improvements.

1992: The International Boundary and Water Commission adopts a plan to clean up the river. A U.S. commissioner promises to swim in it in 1995.

1994: The Imperial County Board of Supervisors sends a plea for help to the Environmental Protection Agency: "The New River has been an environmental disgrace for over 50 years."

1995: Imperial County Supervisor Wayne Van De Graaff says, "When all is said and done about the New River, there's been a lot more said than done."

From *Los Angeles Times,* November 9, 1995. Copyright, 1995, *Los Angeles Times.*

The EPA responded by promising to deal more aggressively with the New River. Officials and citizens in Imperial County are optimistic that this time Washington and Mexico City really mean what they say about cleaning up the river.

Copyright © 1995, Los Angeles Times.

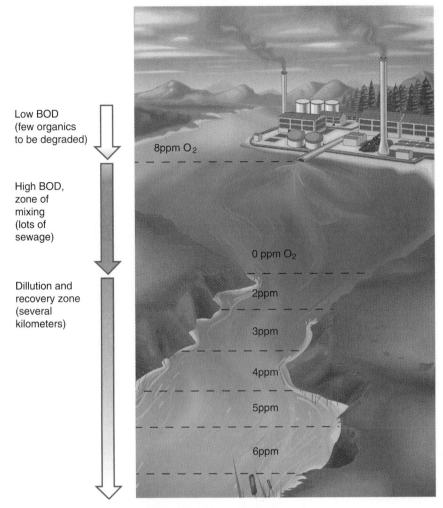

Low BOD
(few organics
to be degraded)

8ppm O₂

High BOD,
zone of
mixing
(lots of
sewage)

0 ppm O₂

2ppm

Dillution and
recovery zone
(several
kilometers)

3ppm

4ppm

5ppm

6ppm

FIGURE 15.13   Effect of Organic Wastes on Dissolved Oxygen Sewage contains a high concentration of organic materials. When these are degraded by organisms, oxygen is removed from the water. This is called the biochemical oxygen demand (BOD). Therefore, there is an inverse relationship between the amount of organic matter and oxygen in the water. The greater the BOD, the less desirable the water is for human use. The more the pollution, the greater the BOD.

covered with vegetation tends to slow the movement of water and allows the silt to be deposited on the surface of the land rather than in the streams. This can be costly because farmers may need to remove valuable cropland from cultivation. To retard leaching, farmers can keep the soil covered with a crop as long as possible and carefully control the amount and the timing of fertilizer application. This makes good economic sense because any fertilizer that runs off or leaches out of the soil has to be replaced.

Industrial Water Pollution

Factories and industrial complexes frequently dispose of some or all of their wastes into municipal sewage systems. Depending on the type of industry involved, these wastes may be a combination of organic materials, petroleum products, metals, acids, and so forth. The organics and oil add to the BOD of the water. The metals, acids, and other ions need special treatment, depending on their nature and concentration. As a result, municipal sewage treatment plants must be designed with their industrial customers in mind. In most cases, cities prefer that industries take care of their own wastes. This allows industries to segregate and control toxic wastes and design wastewater facilities that meet their specific needs.

Since industries are point sources of pollution they have been relatively easy to identify as pollution sources, have been vigorously regulated, and have responded to mandates that they clean up their effluent. Most companies, when they remodel their facilities, include wastewater treatment as a necessary part of an industrial complex. However, some older facilities continue to pollute. These companies discharge acids, particulates, heated water, and noxious gases into the water. While industrial water pollution in the industrialized world has been significantly regulated, in much of the developing world this is not the case and many lakes, streams, and harbors are severely polluted with heavy metals and other toxic materials, organic matter, and human and animal waste.

A special source of industrial water pollution is mining. By its very nature mining disturbs the surface of the earth and increases the chances that sediment and other materials will pollute surface waters. Hydraulic mining is practiced in some countries and involves spraying hillsides with high pressure water jets to dislodge valuable ores. Often chemicals are used to separate the valuable metals from the ores and the waste from these processes are released into streams as well. When coal is mined, water that drains from the mines is often very acidic. In addition, fine coal dust particles are suspended in the water, which makes the water chemically and physically less valuable as a habitat. Dissolved ions of iron, sulfur, zinc, and copper also are present in mine drainage. Control involves containing mine drainage so that it does not mix with surface water.

Thermal Pollution

Amendments to the U.S. Federal Water Pollution Control Act of 1972 (PL 92-500) have mandated changes in how industry treats water. Industries are no longer allowed to use water and return it to its source in poor condition. One of the standards regulates the temperature of the water that is returned to its source. Because many industries use water for cooling, thermal pollution can be a problem. **Thermal pollution** occurs when an industry removes water from a source, uses the water for cooling purposes, and then returns the heated water to its source.

The Cleanup of the Holy Ganges

Every day, thousands of Hindus flock to the banks of the Holy Ganges River in India. There they drink and bathe in what they believe to be holy water, as partially cremated corpses float past them and nearby drains emit millions of liters of raw sewage. Clean water is one of India's most scarce resources, but as in many developing nations, the money and technology needed to properly treat sewage are not available in most cities and villages. With the country's 900 million population expected to double in 34 years, officials are concerned that Indians will have no choice but to continue dumping raw waste into local waterways, contributing to epidemics of diarrhea and other diseases that kill thousands of people annually.

Keeping the Ganges clean is made especially difficult because faith in the river's incorruptible purity has generated complacency and ambivalence about its pollution among many of the 300 million people who live in the Ganges Basin. More than 1600 million liters of untreated municipal sewage, industrial waste, agricultural runoff, and other pollutants are discharged into the river every day. At the same time, officials estimate that more than 1 million people a day bathe or take a "holy dip" in the Ganges, and thousands drink straight from its banks.

The Hindu belief in cremation has led to several environmental problems. A wood cremation takes more than 50 pounds of wood, costing two week's wages for the typical Indian. There are complaints that wood cremations are helping to devour India's forests. Given the high cost of wood, many bodies are not completely cremated, and the partially burned corpses are disposed of in the Ganges.

One solution to this problem has been the introduction of 25,000 specially raised snapping turtles that are attracted to the rotten smell of corpses. While this solution may seem extreme, there is another that is more acceptable in the long run. In 1992, as part of the Ganges Action Plan, an electric crematorium was built at the city of Varanasi that charges less than $2 per body and does a thorough job of turning the bodies to ash. There is some concern however, that many Hindus will not want to abandon the traditional ritual of a wood cremation for the more efficient and less costly electric cremation.

Although this entire scenario may seem somewhat unusual to North Americans, it is important to keep in mind how culture and religion affect our environment. To many Indians, the cultural and religious practices of North Americans are equally puzzling.

Power plants heat water to convert it into steam, which drives the turbines that generate electricity. For steam turbines to function efficiently, the steam must be condensed into water after it leaves the turbine. This condensation is usually accomplished by taking water from a lake or stream to absorb the heat. This heated water is then discharged. The least expensive and easiest method of discharging heated water is to return the water to the aquatic environment, but this can create problems for the inhabitants of the area. Although an increase in temperature of only a few degrees may not seem significant, some aquatic ecosystems are very sensitive to minor temperature changes. Many fish are triggered to spawn by increases in temperature while others may be inhibited from spawning if the temperature rises. For example, lake trout will not spawn in water above 10°C. If a lake has a temperature of 8°C, the lake trout will reproduce, but an increase of 3°C would prevent spawning and result in this species' eventual elimination from that lake.

Ocean estuaries are very fragile. The discharge of heated water into an estuary may alter the type of plant food available. As a result, animals with specific food habits may be eliminated because the warm water supports different food organisms. The entire food web in the estuary may be altered by only slight temperature increases.

Comparing Water Use and Pollution in Industrialized and Developing Countries

Water Use		
Characteristic	**Industrialized countries**	**Developing countries**
Domestic water use per capita	1. Heavy per capita use 2. Highest usage in Australia, New Zealand, U.S., & Canada 3. Usage is stabilizing.	1. Small per capita use 2. Water usage increases as living standards go up.
Where water is used	1. Irrigation and industry total about 85 percent. 2. Domestic about 15 percent.	1. Irrigation is over 80 percent, particularly high in Asia and Africa. 2. Industry and domestic uses each less than 10 percent.
Access to safe drinking water and wastewater treatment	1. Safe drinking water generally available 2. Wastewater treatment generally available 3. Only small population increases expected	1. Large numbers of people lack safe drinking water. 2. Effective wastewater treatment generally *not* available 3. Rapidly growing urban population will create greater need for safe drinking water and wastewater treatment.
Pollution Control		
Domestic wastewater treatment	1. Most countries treat domestic waste. 2. Central and Eastern European countries lack effective sewage treatment.	1. Almost all sewage is discharged without effective treatment.
Industrial wastes	1. Industrial discharges strictly regulated in most countries 2. Some accidental discharges 3. Discharges from disused industrial sites are a problem. 4. Industrial air pollution has caused acidification of lakes in parts of North America and Europe. 5. Eastern and Central Europe have serious industrial waste problems from historically unregulated industry. Black Sea and Baltic Sea heavily polluted.	1. Largely untreated 2. Little attention to regulating industrial waste 3. Acidification of lakes becoming important in China and tropical Africa
Land use runoff	1. Fertilizers and pesticides are a continuing problem. 2. Soil conservation practices used but agricultural runoff is still a significant problem. 3. Runoff from urban areas causes some particle, oil, and other chemicals to enter water.	1. Heavy fertilizer and pesticide use causes serious water-quality problems. 2. Deforestation and poor farming practices causes soil erosion and degrades water quality. 3. Runoff from urban areas causes serious pollution problems; trash, human waste, animal waste, chemical waste all present.

Cooling water used by industry does not have to be released into aquatic ecosystems. Today in the industrialized world most cooling water is not released in such a way that aquatic ecosystems are endangered. There are three other methods of discharging the heat. One method is to construct a large, shallow pond. Hot water is pumped into one end of the pond, and cooler water is removed from the other end. The heat is dissipated from the pond into the atmosphere and substrate.

A second method is to use a cooling tower. In a cooling tower, the heated water is sprayed into the air and cooled by evaporation. The disadvantage of cooling towers and shallow ponds is that large amounts of water are lost by evaporation. The release of this water into the air can also produce localized fogs.

The third method of cooling, the dry tower, does not release water into the atmosphere. In this method, the heated water is pumped through tubes, and the heat is released into the air. This is the same principle used in an automobile radiator. The dry tower is the most expensive to construct and operate.

Marine Oil Pollution

Marine oil pollution has many sources. One source is accidents, such as oil-drilling blowouts or oil-tanker accidents. The *Exxon Valdez,* which ran aground in Prince William Sound, Alaska, in 1989, released over 42 million liters of oil and affected nearly 1500 kilometers of Alaskan coastline. However, by 1994, there was little visible sign of the oil that had covered the coastline in 1989. A U.S. National Oceanic and Atmospheric Administration (NOAA) study estimates that 50 percent of the oil biodegraded on beaches or in the water; 20 percent evaporated; 14 percent was recovered; 12 percent is at the bottom of the sea, mostly in the Gulf of Alaska; 3 percent lies on shorelines; and less than 1 percent still drifts in the water column. Although accidents such as the *Exxon Valdez* are

spectacular events, much more oil is released as a result of small, regular releases from other less visible sources. Nearly two-thirds of all human-caused marine oil pollution comes from three sources: (1) runoff from streets, (2) improper disposal of lubricating oil from machines or automobile crankcases, and (3) intentional oil discharges that occur during the loading and unloading of tankers. With regard to the latter, pollution occurs when the tanks are cleaned or oil-contaminated ballast water is released. Oil tankers use seawater as ballast to stabilize the craft after they have discharged their oil. This oil-contaminated water is then discharged back into the ocean when the tanker is refilled.

As the number of offshore oil wells and the number and size of oil tankers have grown, the potential for increased oil pollution has also grown. Many methods for controlling marine oil pollution have been tried. Some of the more promising methods are recycling and reprocessing used oil and grease from automobile service stations and industries, and enforcing stricter

regulations on the offshore drilling, refining, and shipping of oil. As a result of oil spills from shipping tankers, an international agreement was reached in 1992 which required that all future oil tankers be constructed with two hulls—one inside the other. Such double-hulled vessels would be much less likely to rupture and spill their contents. Today approximately 15 percent of oil tankers are double-hulled.

Groundwater Pollution

A wide variety of activities, some once thought harmless, have been identified as potential sources of groundwater contamination. In fact, possible sources of human-induced groundwater contamination span every facet of social, agricultural, and industrial activities. (See figure 15.14.) Once groundwater pollution has occurred it is extremely difficult to remedy. Pumping groundwater and treating it is very slow and costly. A much better way to deal with the issue of groundwater pollution is to work very hard to prevent the pollution from occurring in the first place.

FIGURE 15.14 Sources of Groundwater Contamination A wide variety of activities have been identified as sources of groundwater contamination.

Major sources of groundwater contamination include:

1. **Agricultural products**

   Pesticides contribute to unsafe levels of organic contaminants in groundwater. Seventy-three different pesticides have been detected in the groundwater in Canada and the United States. Accidental spills or leaks of pesticides pollute groundwater sources with 10 to 20 additional pesticides. Other agricultural practices contributing to groundwater pollution include animal feeding operations, fertilizer applications, and irrigation practices.

2. **Underground storage tanks**

   Up to 350,000 of the 1.4 million underground storage tanks containing gasoline and other hazardous substances in North America may already be leaking. Four liters of gasoline can contaminate the water supply of a community of 50,000 people.

3. **Landfills**

   Even though recently constructed landfills have special liners and water collection systems, approximately 90 percent of the landfills in North America have no liners to stop leaks to underlying groundwater, and 96 percent have no system to collect the leachate that seeps from the landfill. Sixty percent of landfills place no restrictions on the waste accepted, and many landfills are not inspected even once a year.

4. **Septic tanks**

   Poorly designed and inadequately maintained septic systems have contaminated groundwater with nitrates, bacteria, and toxic cleaning agents. Over 20 million septic tanks are in use, and up to a third have been found to be operating improperly.

5. **Surface impoundments**

   Over 225,000 pits, ponds, and lagoons are used in North America to store or treat wastes. Seventy-one percent are unlined, and only 1 percent use a plastic or other synthetic, nonsoil liner. Ninety-nine percent of these impoundments have no leak-detection systems. Seventy-three percent have no restriction on the waste placed in the impoundment. Sixty percent are not even inspected annually. Many of these ponds are located near groundwater supplies.

Other sources of groundwater contamination include mining wastes, salting for snow control, land application of treated wastewater, open dumps, cemeteries, radioactive disposal sites, urban runoff, construction excavation, fallout from the atmosphere, and animal feedlots.

WATER-USE PLANNING ISSUES

In the past, wastes were discharged into waterways with little regard to the costs imposed on other users by the resulting decrease in water quality. Furthermore, as population has grown and the need for irrigation and domestic water has intensified, in many parts of the world there has not been enough water to satisfy everyone's needs. With today's increasing demands for high-quality water, unrestrained waste disposal and unlimited withdrawal of water could lead to serious conflicts about water uses causing social, economic, and environmental losses at both local and international levels. (See table 15.2.)

Metropolitan areas must deal with a variety of issues and maintain an extensive infrastructure in order to provide three basic water services:

1. Water supply for human and industrial needs

2. Wastewater collection and treatment

3. Storm-water collection and management

Water sources must be identified and preserved for use. Some cities obtain all their municipal water from groundwater and must have a thorough understanding of the size and characteristics of the aquifer they use. Some cities such as New York City obtain potable water by preserving a watershed that supplies the water needed by the population. Other cities have abundant water in the large rivers that flow by them but must deal with pollution problems caused by upstream users. Water for human and industrial use must be properly treated and purified. It is then pumped through a series of pipes to consumers. After the water is used, it flows

TABLE 15.2 INTERNATIONAL WATER DISPUTES

River	Countries in dispute	Issues
Nile	Egypt, Ethiopia, Sudan	Siltation, flooding, water flow/diversion
Euphrates & Tigris	Iraq, Syria, Turkey	Reduced water flow, salinization
Jordan, Yarmuk Litani & West Bank aquifer	Israel, Jordan, Syria, Lebanon	Water flow/diversion
Indus & Sutlei	India, Pakistan	Irrigation
Ganges-Brahmaputra	Bangladesh, India	Siltation, flooding, water flow
Salween	Myanmar, China	Siltation, flooding
Mekong	Cambodia, Laos, Thailand, Vietnam	Water flow, flooding
Paraná	Argentina, Brazil	Dam, land inundation
Lauca	Bolivia, Chile	Dam, salinization
Rio Grande & Colorado	Mexico, United States	Salinization, water flow, agrochemical pollution
Rhine	France, Netherlands, Switzerland, Germany	Industrial pollution
Maas & Schelde	Belgium, Netherlands	Salinization, industrial pollution
Elbe	Czech Republic, Germany	Industrial pollution
Szamos	Hungary, Romania	Industrial pollution

From Michael Renner, "National Security: The Economic and Environmental Dimensions" in *Worldwatch Paper 89*, 1989. Reprinted by permission of Worldwatch Institute.

through a network of sewers to a waste-water treatment plant before it is released. Maintaining the infrastructure of pipes, pumps, and treatment plants is expensive.

Metropolitan areas must also deal with great volumes of excess water during storms. Because urban areas are paved and little rainwater can be absorbed into the ground, managing storm water is a significant problem. Cities often have severe local flooding because the water is channeling along streets to storm sewers. If these sewers are overloaded or blocked with debris, the water cannot escape and flooding occurs.

Many cities have a single system to handle both sewage and **storm-water runoff.** During heavy precipitation or spring thaws the runoff from streets may be so large that the wastewater treatment plant cannot handle the volume. The wastewater is then diverted directly into the receiving body of water without being treated. Some cities have areas in which to store this excess water until it can be treated. This is expensive and, therefore, is done only if federal or state funding is available.

Many cities along the Mississippi River have recently completed sewer separation projects. Rainwater runoff now empties directly into the river without passing through treatment plants. This system is easier on the treatment facilities, but city residents must be more careful about their use of lawn chemicals and disposal of wastes to avoid polluting the river.

Water services are expensive. They must be provided with an understanding that water supplies are limited. Also, water's ability to dilute and degrade pollutants is limited. Proper land-use planning is essential if metropolitan areas are to provide services and limit pollution.

In pursuing these objectives, city planners encounter many obstacles. Large metropolitan areas often have hundreds of local jurisdictions (governmental and bureaucratic areas) that divide responsibility for management of basic water services. The Chicago metropolitan area is a good example. This area is composed of six counties and approximately two thousand local units of government. It has 349 separate water-supply systems and 135 separate wastewater disposal systems. Efforts to implement a water-management plan when so many layers of government are involved are complicated and frustrating.

To meet future needs urban, agricultural, and national interests will need to deal with a number of issues, such as the following:

- Increased demand for water will encourage continued efforts to divert water to highly populated areas or areas capable of irrigated agriculture.

- Increased demand for water will force increased treatment of wastewater and reuse of existing water supplies.

- In many areas where water is used for irrigation, both the water and the soil become salty because of evaporation. When this water returns to a stream, the quality of the water is lowered.

- In some areas, wells provide water for all categories of use. If the groundwater is pumped out faster than it is replaced, the water table is lowered.

- In coastal areas, seawater may intrude into the aquifers and ruin the water supply.

- The demand for water-based recreation is increasing dramatically and requires high-quality water, especially for water recreation involving total body contact, such as swimming.

Water Diversion

Water diversion is the physical process of transferring water from one area to another. An early example of water diversion is the aqueducts of ancient Rome. Thousands of diversion projects have been constructed since then. New York City, for example, diverts water from Pennsylvania (250 kilometers away), and Los Angeles obtains part of its water supply from the Colorado River. (See Issues and Analysis: The California Water Plan.)

While diversion is necessary in many parts of the world, it can be misused or detrimentally affect an area. An example of this is the proposed Garrison Diversion Unit, which is designed to divert water from the Missouri River system for irrigation. (See figure 15.15.) This action would have enormous environmental consequences. Wildlife refuges, native grasslands, forests, and waterfowl breeding marshes would be damaged or destroyed. In addition, hundreds of kilometers of streams could be seriously damaged.

There is considerable opposition to this plan, both by private groups and governmental agencies. In fact, the Canadian government has fought against the project. In contrast, the people who need the water for irrigation see the Garrison Diversion Unit plans as beneficial. If this plan was implemented, they could raise corn (a more profitable crop), rather than the dryland crop of wheat. The corn could then be used for cattle feed.

The initial plan to irrigate this portion of the Great Plains was thought of during the Dust Bowl era of the 1930s. The Federal Flood Control Act of 1944 authorized the construction of the Garrison Diversion Unit, but controversy concerning it continues over 50 years after its proposal.

Wastewater Treatment

Because water must be cleaned before it is released, most companies and municipalities maintain wastewater treatment facilities. The percentage of sewage that is treated, however, varies greatly throughout the world. (See table 15.3.) Treatment of sewage is usually classified as primary, secondary, or tertiary. **Primary sewage treatment** is primarily a physical process that removes larger particles by filtering water through large screens and then allowing smaller particles to settle in ponds or lagoons. Water is removed from the top of the settling stage and either released to the environment or to a subsequent stage of treatment. If the water is released to the environment it does not have any sand or grit, but it still carries a heavy load of organic matter, dissolved salts, bacteria, and other microorganisms. The organisms use the organic material for food, and as long as there is sufficient oxygen, they will continue to grow and reproduce. If the receiving body of water is large enough and the organisms have enough time, the organic matter will be degraded. In crowded areas, where several municipalities take water and return it to a lake or stream, within a few kilometers

ENVIRONMENTAL
CLOSE-UP

Is It Too Late for the Everglades?

Close to a million people visit the Everglades Park each year to view a landscape that is part African veldt and part tropical swamp. While the Everglades may look healthy, parts of its finely tuned ecosystem are breaking down. According to some scientists, the park is dying.

Years of mismanagement of the water supply have desiccated much of the park's wetlands. Agriculture and urban development have increasingly diverted the flow that sustains the park. Water policies favoring farmers and urban dwellers have forced wading birds to relocate their colonies and cut their population in the southern Everglades from 300,000 to 15,000. Alligators, whose population in the drying park is also drastically shrinking, are cannibalizing their young. Exotic plants, thriving on nutrients from agricultural runoff, are forcing out natural vegetation. Bass and catfish in the park should not be eaten because they are laced with natural mercury leached into ponds from soil dried to dust. Panthers, snail kites, and wood storks, three of the park's 13 endangered wildlife species, are on the verge of extinction.

After Everglades Park was established in 1947, about 800,000 hectares of wetlands to the north were drained for farms and urban development. South Florida boomed. Some 4.5 million people now live in the horseshoe crescent around the Everglades region, and 600 new residents arrive each day. The once-clear water of Lake Okeechobee, which is north of the park, are now being polluted with dairy-farm runoff and choked with algae. Cattails that proliferate in the phosphorus-laden runoff from the sugar plantations are choking the waterways. Some 175,000 hectares of drained Everglades are in sugar production. Since the water is free, South Florida's farmers take as much as they need and have little incentive to conserve.

There are plans to help the Everglades. The Kissimmee River, which flows into Lake Okeechobee, was channelized into an arrow-straight river in the late 1960s. This project destroyed the marshes and allowed nutrients to pollute the lake and thus the park. Beginning in 1990, the Kissimmee was being returned to its natural state, with twisting oxbow curves and extensive wetlands. In 1989, the U.S. Congress approved the purchase of 43,000 hectares to be added to the east section of the park. The state of Florida has obtained an additional 60,000 hectares as a buffer zone for the park. In addition, canals, spillways, and pumping stations are being modified to assure a steady supply of cleaner water for the park. These ambitious attempts to put a little wild back in the wilderness of the Everglades project have a good chance to succeed.

In 1993, the U.S. Department of the Interior proposed an agreement with the state of Florida and the state's vegetable farmers and sugar industry to restore the Everglades. The agreement would begin an ambitious plan to reduce phosphorus pollution and return natural water flow. The cost of the restoration program would be paid for out of a $465 million settlement with the sugar industry. A scaled down plan was approved in 1996 over the objections of many in the agricultural community. Plans for the expansion of airport facilities in the area have been the latest threat to surface. The growing population of the Miami area places continual pressure on the park and requires constant activity to preserve what remains of the Everglades.

Everglades wading birds once numbered more than a million, but drought and pollution have decimated their number.

Construction threatens to destroy the fragile ecosystem of the Everglades.

of each other, primary water treatment is not adequate.

Secondary sewage treatment is a biological process that usually follows primary treatment which involves holding the wastewater until the organic material has been degraded by the bacteria and other microorganisms. Most secondary treatment facilities try to promote the growth of microorganisms. To encourage this action, the wastewater is mixed with large quantities of highly oxygenated water, or the water is aerated directly, as in a trickling filter system or an activated sludge system. In a **trickling filter system,** the wastewater is sprayed over the surface of rock or other substrate to increase the amount of dissolved oxygen. The rock also provides a place for the bacteria and other microbes to attach so they are exposed simultaneously to the organic material in the water and to oxygen. These microorganisms feed on the dissolved organic matter and small suspended particles, which then become incorporated into their bodies as part of their cell structure. The bodies of the microorganisms are larger than the dissolved and suspended organic matter, so this process concentrates the organic wastes into particles that are large enough to settle out. This mixture of organisms and other particular matter is called **sewage sludge.** The sludge that settles consists of living and dead microorganisms and their waste products.

In **activated sludge sewage treatment** plants, the wastewater is held in tanks and has air continuously bubbled through it. The sludge eventually is moved to settling tanks where the water and sludge can be separated. Some of the sludge is returned to aeration tanks, where it is mixed with incoming wastewater. This kind of process uses less land than a trickling filter. (See figure 15.16.) Both processes produce a sludge that settles out of the water.

The sludge that remains is concentrated and often dried before disposal. Sludge disposal is a major problem in large population centers. In the San Francisco Bay area, 2500 metric tons of sludge are produced each day. Most of this is carried to landfills and lagoons, and some is composted and returned to the land as fertilizer.

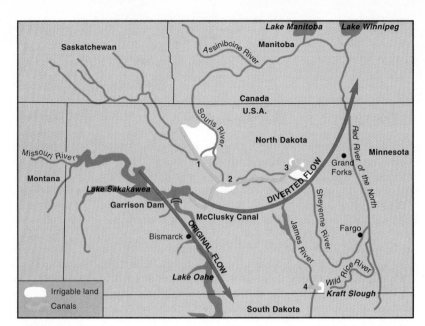

FIGURE 15.15 **The Garrison Diversion Unit.** This unit would divert water that flows into the Missouri River to the McClusky Canal and the Sheyenne River and eventually into Hudson Bay.

TABLE 15.3	PERCENT OF SEWAGE TREATED IN SELECTED AREAS
Area	**Percent**
Europe	72
Mediterranean Sea	30
Caribbean Basin	Less than 10
Southeast Pacific	Almost zero
South Asia	Almost zero
South Pacific	Almost zero
West and Central Africa	Almost zero

Source: Data from *World Resources Institute 1994–95.*

In other areas the sewage sludge is applied directly to agricultural land as a fertilizer and soil conditioner.

In North America and much of the developed world wastewater receives both primary and secondary sewage treatment. The water discharged from these sewage treatment plants must be disinfected. The least costly method of disinfection is chlorination of the wastewater after it has been filtered and the organic materials have been allowed to settle. Using ultrasonic energy to mechanically break down waste may be less harmful and more effective, but it is also more expensive.

A growing number of larger sewage treatment plants use additional processes called tertiary sewage treatment. **Tertiary sewage treatment** involves a variety of techniques to remove dissolved pollutants left after primary and secondary treatments. (See table 15.4.) The tertiary treatment of municipal wastewater is often used to remove phosphorus and nitrogen that could increase aquatic plant growth. Some municipalities are using natural or constructed wetlands to serve as tertiary sewage treatment systems. Tertiary treatment of industrial and other specialized wastewater streams is very costly because it requires specific chemical treatment of the water to eliminate specific problem materials. Certain industries are beginning to maintain their own secondary or tertiary

a.

c.

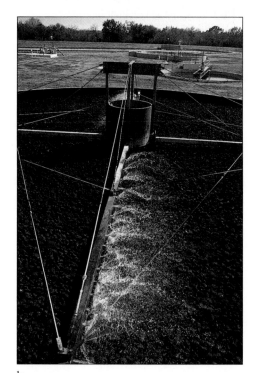

b.

FIGURE 15.16 Primary and Secondary Wastewater Treatment Primary treatment is physical; it includes filtrating and settling of wastes. Photograph (*a*) is a settling tank in which particles settle to the bottom. Secondary treatment is mostly biological and includes the concentration of dissolved organics by microorganisms. Two major types of secondary treatment are trickling filter and activated sludge methods. Photograph (*b*) shows a trickling filter system and photograph (*c*) shows an activated sludge system.

wastewater facilities because of the specific nature of their waste products.

As water has become scarce in many parts of the world people have looked at wastewater as a source of water for other purposes. Water from wastewater plants can be used for irrigation water, industrial purposes, cooling, and many other activities. Ultimately it is possible to have a closed loop system for domestic water in which the output of the wastewater plant becomes the input for the drinking water supply.

Salinization

Another water-use problem results from **salinization,** an increase in salinity caused by increasing salt concentrations in soil. This is primarily a problem in areas where irrigation is practiced. When plants extract the water they need, the salts present in all natural waters become concentrated. Irrigation of arid farmland makes this problem more acute because so much water is lost due to evaporation. Irrigation is most common in hot, dry areas, which normally have high rates of evaporation. This results in a concentration of salts in the soil and in the water that runs off the land. (See figure 15.17.) Every river increases in salinity as it flows to the ocean. The salinity of the Colorado River water increases 20 times as it passes through irrigated cropland between Grand Lake in north central Colorado and the Imperial Dam in southwest Arizona. The problem of salinity will continue to increase as irrigation increases.

Groundwater Mining

Groundwater mining means that water is removed from an aquifer faster than it is replaced. When this practice continues for a long time, the water table eventually declines. Groundwater mining is common in areas of the western United States due to growing cities and increasing irrigation. In aquifers with little or no recharge, virtually any withdrawal constitutes mining, and sustained withdrawals will eventually exhaust the supply. This problem is particularly serious in communities that depend heavily upon groundwater for their domestic needs.

Groundwater mining can also lead to problems of settling or subsidence

GLOBAL PERSPECTIVE

Death of a Sea

The Aral Sea lies on the border between Uzbekistan and Kazakhstan in the former Soviet Union. It was once larger than any of the Great Lakes except Superior, and now it is disappearing. Since the 1920s, Soviet agricultural planners have used up the Aral Sea, diverting its waters for irrigation. The two rivers feeding the Aral were drawn off to irrigate millions of hectares of cotton. The irrigation canal, the world's longest, stretches over 1300 kilometers, paralleling the boundaries of Afghanistan and Iran. The cotton production plan worked, and by 1937, the Soviet Union was a net exporter of cotton. The success of the cotton program, however, spelled the end for the Aral Sea.

For a long time, the ecological impact on the sea and surrounding area was largely hidden from public view. Since the 1960s, however, the Aral has lost about 40 percent of its surface area, or almost 20,000 square kilometers of what are now largely dry, salt-encrusted wastelands. The once-thriving fishing industry that depended on the water is all but gone. Some 20 of the 24 fish species there have disappeared. The fish catch, which totaled 44,000 tons a year in the 1950s and supported some 60,000 jobs, has dropped to zero. Abandoned fishing villages dot the sea's former coastline.

Another apparent consequence of the dried-up sea is a host of human illnesses. A high rate of throat cancer is attributed to dust from the drying sea. Each year, winds pick up 40 million to 150 million tons of a toxic dust–salt mixture from the dry sea bed and deposit them on the surrounding farmland, harming or killing crops. The low river flows have concentrated salts and toxic chemicals, making water supplies hazardous to drink and contributing to disease. In the northwest part of the Republic of Uzbekistan, the infant mortality rate is the highest in the former Soviet Union.

The former fishing center of the sea was a town named Muynak. The town is now landlocked more than 30 kilometers from the water. Less than 25 years ago, Muynak was a seaport. The population of Muynak is down from 40,000 citizens in 1970 to 12,000 today. In 1990, the mayor and last harbormaster of Muynak commented:

> The water continued to go away while the salinity increased. The weather changed for the worse, with the summers getting hotter and the winters colder. The people feel salt on their lips and in their eyes all the time. It's getting harder to open your eyes here.

What has happened in the Aral Sea basin shows vividly how damage to the economy, community, and human health can follow close on the heels of ecological destruction.

Death of a sea

TABLE 15.4 TERTIARY TREATMENT METHODS

Kind of tertiary treatment	Problem chemicals	Methods
Biological	Phosphorus and nitrogen compounds	1. Large ponds are used to allow aquatic plants to assimilate the nitrogen and phosphorus compounds from the water before the water is released. 2. Columns containing denitrifying bacteria are used to convert nitrogen compounds into atmospheric nitrogen.
Chemical	Phosphates and industrial pollutants	1. Water can be filtered through calcium carbonate. The phosphate substitutes for the carbonate ion and the calcium phosphate can be removed. 2. Specific industrial pollutants, which are nonbiodegradable, may be removed by a variety of specific chemical processes.
Physical	Primarily industrial pollutants	1. Distillation 2. Water can be passed between electrically charged plates to remove ions. 3. High-pressure filtration through small-pored filters. 4. Ion-exchange columns

FIGURE 15.17 Salinization As water evaporates from the surface of the soil, the salts it was carrying are left behind. In some areas of the world, this has permanently damaged cropland. Other areas flush the soil to rid it of salt and maintain its fertility.

of the ground surface. Removal of the water allows the ground to compact, and large depressions may result. For example, in the San Joaquin Valley of California, groundwater has been withdrawn for irrigation and cultivation since the 1850s and groundwater levels have fallen over 100 meters. More than 1000 hectares of ground have subsided, some as much as 6 meters. Currently, the ground surface in that area is sinking 30 centimeters per year. In 1981, in Winter Park, Florida, a

large area of subsidence—a "sinkhole"— occurred because of excessive water withdrawal. (See figure 15.18.) London, Mexico City, Venice, Houston, and Las Vegas are some other cities that have experienced subsidence as a result of groundwater withdrawal. (See table 15.5.)

Groundwater mining poses a special problem in coastal areas. As the fresh groundwater is pumped from wells along the coast, the saline groundwater moves inland, replacing fresh groundwater with

unusable saltwater. (See figure 15.19.) Saltwater intrusion is a serious problem in heavily populated coastal areas throughout the world.

Preserving Scenic Water Areas and Wildlife Habitats

As mentioned earlier, the use of land influences water quality. This is particularly true along shorelines and riverbanks where the land and water meet. Some bodies of water have unique scenic value. To protect these resources, the way in which the land adjacent to the water is used must be consistent with preserving these scenic areas.

The U.S. Federal Wild and Scenic Rivers Act of 1968 established a system to protect wild and scenic rivers from development. All federal agencies must consider the wild, scenic, or recreational values of certain rivers in planning for the use and development of the rivers and adjacent land. The process of designating a river or part of a river as wild or scenic is complicated. It often encounters local opposition from businesses dependent on growth. Following reviews by state and federal agencies, rivers may be designated as wild and scenic by action of either Congress or the Secretary of the Interior. Sections of over 65 streams in the United States have been designated as wild or scenic.

Many unique and scenic shorelands have also been protected from future development. Until recently, estuaries and shorelands have been subjected to significant physical modifications, such as dredging and filling, which may improve conditions for navigation and construction but destroy fish and wildlife habitats. Recent actions throughout North America have attempted to restrict the development of shorelands. Development has been restricted in some particularly scenic areas, such as Cape Cod National Seashore in Massachusetts and the Bay of Fundy in the Atlantic provinces of Canada.

Historically, poorly drained areas were considered worthless. Subsequently, many of these wetlands were filled or drained and used for development. The natural and economic importance of wetlands has been recognized only recently.

TABLE 15.5 GROUNDWATER DEPLETION IN MAJOR REGIONS OF THE WORLD, CIRCA 1995

Region/aquifer	Estimates of depletion
California	Groundwater overdraft exceeds 1.7 billion cubic meters per year. The majority of the depletion occurs in the Central Valley, which is referred to as the vegetable basket of the United States.
Southwestern United States	In parts of Arizona, water tables have dropped more than 120 meters. Projections for parts of New Mexico indicate that water tables will drop an additional 22 meters by 2020.
High Plains aquifer system, United States	This aquifer underlies nearly 20 percent of all the irrigated land in the United States. To date the net depletion of the aquifer is in excess of 350 billion cubic meters, or roughly 15 times the average annual flow of the Colorado River. Most of the depletion has been in the Texas high plains, which has witnessed a 26 percent decline in irrigated land from 1979 to 1989. Current depletion is estimated to be in excess of 13 billion cubic meters a year.
Mexico City and Valley of Mexico	Use exceeds natural recharge by 60 to 85 percent, causing land subsidence and falling water tables.
African Sahara	North Africa has vast non-recharging aquifers where current depletion exceeds 12 billion cubic meters a year.
India	Water tables are declining throughout much of the most productive agricultural land in India. In parts of the country, groundwater levels have declined 90 percent during the past two decades.
North China	The water table underneath portions of Beijing has dropped 40 meters during the past 40 years. A large portion of northern China has significant groundwater overdraft.
Arabian peninsula	Groundwater use is nearly three times greater than recharge. At the depletion rates projected into the beginning of the next century, exploitable groundwater reserves could be exhausted within the next 50 years. Saudi Arabia depends on nonrenewable groundwater for roughly 75 percent of its waters. This includes irrigation of 2 million to 4 million tons of wheat per year.

FIGURE 15.18   Development of a "Sinkhole" If the water table is lowered as a result of groundwater mining or a severe drought, the land surface can subside. The space formerly occupied by water can collapse and create sinkholes. This photo shows only one of many sinkholes that occurred in Florida in 1981 as a result of the combined effects of excessive groundwater use and a long drought.

In addition to providing spawning and breeding habitats for many species of wildlife, wetlands act as natural filtration systems by trapping nutrients and pollutants and preventing them from entering adjoining lakes or streams. Wetlands also slow down floodwaters and permit nutrient-rich particles to settle out. In addition, wetlands can act as reservoirs and release water slowly into lakes, streams, or aquifers, thereby preventing floods. (See figure 15.20.) Coastal estuarine zones and adjoining sand dunes also provide significant natural flood control. Sand dunes act as barriers and absorb damaging waves caused by severe storms.

FIGURE 15.19 Saltwater Intrusion When saltwater intrudes on fresh groundwater, the groundwater becomes unusable for human consumption and for many industrial processes.

The value of wetlands

- •Filter toxic wastes, excess nutrients, sediments and other pollutants
- •Help prevent erosion
- •Reduce flooding by storing storm water

- •Reduce storm damage by absorbing waves
- •Are feeding, resting spot for migratory waterfowl

- •Provide food, habitat for other aquatic species

- •Provide nesting grounds for a number of species, including oysters, clams, crabs, and shrimp

FIGURE 15.20 The Value of Wetlands Wetlands are areas covered by water enough of the year that they support aquatic plant and animal life. Wetlands can be either fresh or saltwater and may be isolated potholes or extensive areas along rivers, lakes, and oceans. We once thought of wetlands as only a breeding site for mosquitoes. Today we are beginning to appreciate their true value.

ECOPARQUE

The city of Tijuana, Mexico, is home to approximately 1 million people. Rapid population growth has strained local water resources in this arid region. Both the city and the adjacent U.S.–Mexico border region also have severe water-pollution problems. One source of pollution is the direct discharge of residential wastewater into the Tijuana River and the nearby coastal waters. Because of the water scarcity and the pollution, scientists in Tijuana began thinking of alternative technologies for treating wastewater stating that "water is too valuable to be called waste." One answer is an alternative treatment facility called ECOPARQUE.

ECOPARQUE began in 1986 as a study of decentralized wastewater treatment by El Colegio de la Frontera Norte (COLEF), a college in Tijuana. By 1996, ECOPARQUE had developed into a successful model of sustainable and integrated urban wastewater treatment. The central feature of ECOPARQUE is the basic treatment unit for recapturing wastewater. The unit includes a stainless steel fine screen, a plastic biofilter, and a passive sediment deposit clarifier. These three elements operate without electrical energy or any mechanical parts. For this reason, the process falls under the category of appropriate technology for construction and use in developing countries.

ECOPARQUE and its treated wastewater are being used for a variety of sustainable purposes, including:

- Urban reforestation on hillsides and canyons in Tijuana.

- A nursery that produces a variety of plants, including vegetables and fruits.

- Food production in an urban center that includes an expanding fruit harvest, soon to be joined by beekeeping and an agricultural production unit. There are also plans for raising some species of fish as a food source.

- A meteorological station that will provide local authorities with useful urban planning information.

- An experimental solar field for use at ECOPARQUE and the surrounding neighborhoods.

- A site for environmental and technological education for schoolchildren and scientists.

Given the range of activities at ECOPARQUE, it is easy to see why the scientists who planned the facility believed that water is too valuable to be called waste. While it may not be feasible to build facilities similar to ECOPARQUE in every city in the developing world, the design plans for ECOPARQUE are being made available to other countries. There is hope that many other cities will turn what is a water pollution and health problem into a multifaceted and sustainable environmental and community success story.

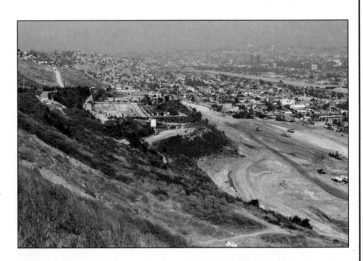

Summary

Water is a renewable resource that circulates continually between the atmosphere and the earth's surface. The energy for the hydrologic cycle is provided by the sun. Water loss from plants is called transpiration. Water that infiltrates the soil and is stored in underground reservoirs is called groundwater, as opposed to surface water that enters a river system as runoff. The two basic kinds of aquifers are unconfined and confined. The way in which land is used has a significant impact on rates of evaporation, runoff, and infiltration.

The four human uses of water are domestic, agricultural, instream, and industrial. Water use is measured by either the amount withdrawn or the amount consumed. Domestic water is in short supply in many metropolitan areas. Most domestic water is used for waste disposal and washing, with only a small amount used for drinking. The largest consumptive use of water is for agricultural irrigation. Major in-stream uses of water are for hydroelectric power, recreation, and navigation. Most industrial uses of water are for cooling and for dissipating and transporting waste materials.

Water can be heavily used in many ways without permanent damage. However, adding material to water may make it unfit for many uses though not for all.

Major sources of water pollution are municipal sewage, industrial wastes, and agricultural runoff. Organic matter in water requires oxygen for its decomposition and therefore has a large biochemical oxygen demand (BOD). Oxygen depletion can result in fish death and changes in the normal algae community, which leads to visual and odor problems. Nutrients, such as nitrates and phosphates from detergents and agricultural runoff, enrich water and stimulate algae and aquatic plant growth.

Point sources of pollution are easy to identify and resolve. Nonpoint sources of pollution, such as agricultural runoff and mine drainage, are more difficult to detect and control than those from municipalities or industries.

Thermal pollution occurs when an industry returns heated water to its source. Temperature changes in water can alter the kinds and numbers of plants and animals that live in it. The methods of controlling thermal pollution include cooling ponds, cooling towers, and dry cooling towers.

Wastewater treatment consists of primary treatment, a physical settling process; secondary treatment, biological degradation of the wastes; and tertiary treatment, chemical treatment to remove specific components. Two major types of secondary wastewater treatments are the trickling filter and the activated sludge sewage methods.

Groundwater pollution comes from a variety of sources, including agriculture, landfills, and septic tanks. Marine oil pollution results from oil drilling and oil-tanker accidents, runoff from streets, improper disposal of lubricating oil from machines and car crankcases, and intentional discharges from oil tankers during loading or unloading.

Reduced water quality can seriously threaten land use and in-place water use. In the United States and other nations, legislation helps to preserve certain scenic water areas and wildlife habitats. Shorelands and wetlands provide valuable services as buffers, filters, reservoirs, and wildlife areas. Water management concerns of growing importance are groundwater mining, increasing salinity, water diversion, and managing urban water use. Urban areas face several problems, such as providing water suitable for human use, collecting and treating wastewater, and handling storm-water runoff in an environmentally sound manner. Water planning involves many governmental layers, which makes effective planning difficult.

Review Questions

1. Describe the hydrologic cycle.
2. List several uses of water that are nonconsuming and nonwithdrawing.
3. What is the major industrial use of water?
4. What are the similarities between domestic and industrial water use? How are they different from in-stream use?
5. How is land use related to water quality and quantity? Can you provide local examples?

6. Is the definition of water pollution related to the water's intended use? Give some examples to support your answer.
7. What is biochemical oxygen demand? How is it related to water quality?
8. How is water pollution related to agricultural activities? Why is this a growing concern?
9. Differentiate between point and nonpoint sources of water pollution.
10. How are most industrial wastes disposed of? How has this changed over the past 25 years?
11. What is thermal pollution? How can it be controlled?
12. Describe primary, secondary, and tertiary sewage treatment.
13. What are the types of wastes associated with agriculture?
14. Why is storm-water management more of a problem in an urban area than in a rural area?
15. What is the Federal Wild and Scenic Rivers Act? Why is it important?
16. Define groundwater mining.
17. How does irrigation increase salinity?
18. What are the three major water services provided by metropolitan areas?

Critical Thinking Questions

1. Leakage from freshwater distribution systems accounts for significant losses. Is water so valuable that governments should require systems that minimize leakage in order to preserve the resource? Under what conditions would you change your evaluation?
2. Do non-farmers have an interest in how water is used for irrigation? Under what conditions should the larger public be involved in making these decisions along with the farmers who are directly involved?
3. Should the United States allow Mexico to have water from the Rio Grande and the Colorado Rivers, both of which originate in the United States and flow to Mexico?
4. Do you believe that large scale hydroelectric power should be promoted as a renewable alternative to power plants that burn fossil fuels? What criteria do you use for this decision?
5. What are the costs and what are the benefits of the proposed Garrison Diversion Unit? What do you think should happen with this project?
6. How might you be able to help save freshwater in your daily life? Would the savings be worth the costs?
7. Look at the hydrologic cycle in Figure 15.3. If global warming increases the world-wide temperature, what do you think the effects of global warming on the hydrologic cycle would be?
8. Why might it be difficult to clean up a river like the New River that flows from Mexico into the Imperial Valley of California? (See the Environment Close-up in this chapter.)
9. After reading the Issues and Analysis concerning the California Water Plan, do you believe water should be diverted from Northern California to Southern California?
10. What do you think about increased development in arid regions? If development continues and water resources decline, what do you think the political effects will be?

The management of freshwater is often a controversial subject, involving social, ecological, and economic aspects. A good example is the California Water Plan.

In the early 1900s, it became clear that the growth of Los Angeles, which was then a small coastal town, would be encouraged by irrigating the surrounding land. Los Angeles looked to the Owens Valley, 40 kilometers north, for a source of water. The Los Angeles Aqueduct connecting these two areas was completed in 1913.

After the Owens Valley project, California developed a statewide water program known as the California Water Plan. This plan is necessary because most of the state's population and irrigated land are found in the central and southern regions, but most of the water is in the north. The plan details the construction of aqueducts, canals, dams, reservoirs, and power stations to transport water from the north to the south. In addition to supplying water for southern regions, the aqueducts provide irrigation for the San Joaquin Valley. Eventually, the new land made available for agriculture will amount to about 400,000 hectares.

The California Water Plan has been one of the most controversial programs ever undertaken in California. Its adoption instigated a sectional feud between the moist "north" and the dry "south." Southern California was accused of trying to steal northern water, but because the large population in the southern part of the state carried the vote, the plan was adopted. Environmentalists still claim that the project has irreparably scarred the countryside and upset natural balances of streams, estuaries, vegetation, and wildlife. They argue that providing water to southern California promotes population growth, which leads to further urbanization and land development.

Many questions and controversies center on whether the water is really needed. Ninety percent of the water used in southern California is for irrigation; there is evidently abundant water for domestic and industrial use. A few of the crops raised in the San Joaquin Valley account for a large percentage of the irrigation water. Most notable is rice. Rice is not a native crop and demands intensive irrigation. The cost is borne by all the rate payers, most of whom are urban. California is now one of the nation's most productive agricultural areas.

Allocation of water resources is a matter of economics as well as technology. The California water project has been criticized for using public funds to increase the value of privately held farmland. Furthermore, technological advances in desalination plants may give a new dimension (unforeseen when the water plan was devised) to the problem of water resources. Although more aqueducts, canals, and pumping plants are planned, whether they will be completed is uncertain.

Southern California has been in a drought emergency for several years. Water rates have increased, certain municipal uses have been curtailed, and irrigation has been reduced. Increasing population creates a demand for water that is now causing people to recognize that they will need to choose between using water for municipal purposes or irrigation. There is not enough water to increase both.

- What are the major advantages of having a water plan?
- What problems develop when water is transported to arid regions?
- Should water from northern California be sent to southern California?
- Should crops like rice be raised in California?
- Aside from lack of rainfall, why do you think southern California is in a water crisis?

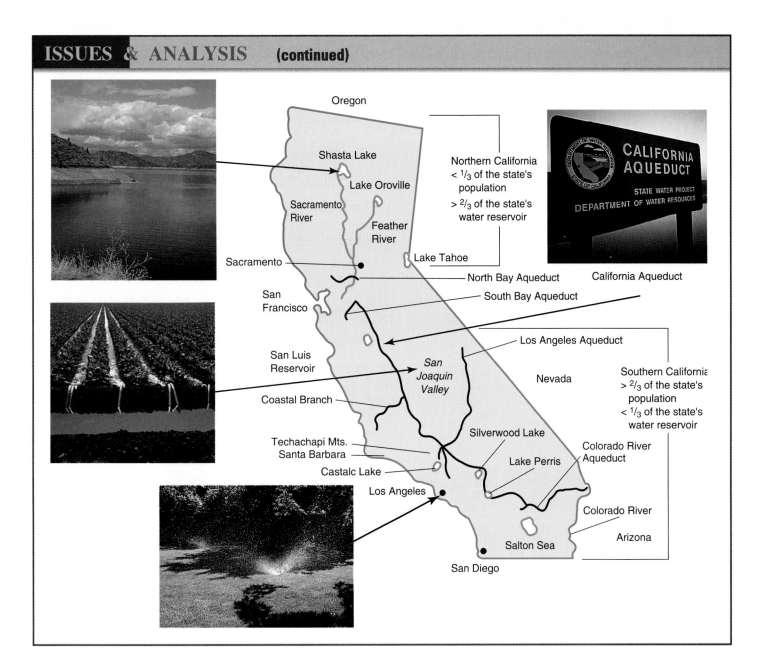

Oregon

Shasta Lake

Lake Oroville

Sacramento River

Feather River

Sacramento

Lake Tahoe

Northern California
< 1/3 of the state's population
> 2/3 of the state's water reservoir

CALIFORNIA AQUEDUCT
STATE WATER PROJECT
DEPARTMENT OF WATER RESOURCES

California Aqueduct

North Bay Aqueduct

San Francisco

South Bay Aqueduct

Los Angeles Aqueduct

San Luis Reservoir

San Joaquin Valley

Nevada

Southern California
> 2/3 of the state's population
< 1/3 of the state's water reservoir

Coastal Branch

Silverwood Lake

Colorado River Aqueduct

Techachapi Mts.
Santa Barbara

Lake Perris

Castalc Lake

Los Angeles

Colorado River

Salton Sea

Arizona

San Diego

Key Terms

activated sludge sewage treatment *294*
agricultural products *291*
agricultural runoff *285*
aquiclude *276*
aquifer *276*
aquitard *276*
artesian well *276*
biochemical oxygen demand (BOD) *284*
confined aquifer *276*
domestic water *279*
eutrophication *284*
groundwater *276*
groundwater mining *295*
hydrologic cycle *275*
industrial uses *280*
in-stream uses *281*
irrigation *279*
landfills *291*
limiting factor *284*
nonpoint source *283*
permeability *277*
point source *283*
porosity *276*
potable waters *275*
primary sewage treatment *292*
runoff *276*
salinization *295*
secondary sewage treatment *294*
septic tanks *291*
sewage sludge *294*
storm-water runoff *292*
surface impoundments *291*
tertiary sewage treatment *294*
thermal pollution *287*
transpiration *276*
trickling filter system *294*
unconfined aquifer *276*
underground storage tanks *291*
vadose zone *276*
water diversion *292*
water table *276*

Related Internet Sites

Visit our website at http//www.mhhe.com/environmentalscience/ for specific information on each of the topics below.

Water treatment and drinking water

Water treatment

Humanity's attempt to adapt to desert environments

Environmental geochemistry and mineralogy

Pacific Islanders' struggle against phosphate mining

Protecting and conserving Idaho's forestry, grazing, mining, water, and endangered species

Worldwide river management

Watershed sustainability

Water conservation, irrigation and landscaping, and conservation sites and programs

Water data, publications, and technical resources U.S.

Plans to protect water supplies and information on water quality, biosolids and American Heritage Rivers

Facts on drinking water contaminants, water treatment, and home water treatment and testing

Electronic access to information regarding water and agriculture

PART FIVE

Pollution and Policy

Most people have strong feelings about the quality of the environment. However, these feelings vary greatly because people do not agree on what is good or bad for environmental quality. Economics, perception of risk, and politics have become interwoven into the environmental quality equation. There are great differences of opinion about what environmental quality should be or what the current quality of the environment is. The amount of pollution can often be quantified, and pollution may be more or less serious depending on local conditions. Although pollution, by definition, is harmful, some amounts of pollution can be tolerated, depending on the situation.

Chapter 16 provides background information on risk assessment and economic principles useful in evaluating the costs of pollution and the value gained from eliminating pollution.

Chapters 17, 18, and 19 discuss air pollution, solid waste disposal, and hazardous and toxic wastes, respectively. These chapters also describe approaches used to deal with these problems. Throughout the selection, economic and political realities are discussed as a part of the pollution equation.

Chapter 20 discusses how environmental decisions are made. It is a complex process, integrating public input, economic demands, political posturing, and legal requirements. As complex as the process is, the decisions we have made have significantly improved the quality of our environment over the past several decades.

chapter 16

Risk and Cost: Elements of Decision Making

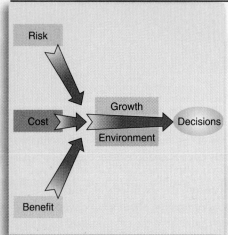

OBJECTIVES

After reading this chapter, you should be able to:

- Describe why the analysis of risk has become an important tool in environmental decision making.
- Understand the difference between risk assessment and risk management.
- Describe the issues involved in risk management.
- Understand the difference between true and perceived risks.
- Define what an economic good or service is.
- Understand the relationship between the available supply of a commodity or service and its price.
- Understand how and why cost-benefit analysis is used.
- Understand the concept of sustainable development.
- Understand environmental external costs and the economics of pollution prevention.
- Understand the market approach to curbing pollution.

CHAPTER OUTLINE

MEASURING RISK

Two factors are primary in many decisions in life: risk and cost. We commonly ask such questions as "How likely is it that someone will be hurt?" and "What is the cost of this course of action?" Environmental decision making is no different. If a new air-pollution regulation is contemplated, industry will be sure to point out that it will cost a considerable amount of money to put these controls in place and will reduce profitability. Citizens will point out that their tax money will have to support another governmental bureaucracy. On the other side, advocates will point out the reduced risk of illnesses and the reduced cost of health care for people who live in areas of heavy air pollution.

Risk analysis has become an important decision-making tool at all levels of society. In the area of environmental concerns, assessing and managing risks help us determine what environmental policies are appropriate. The analysis of risk generally involves a probability statement. **Probability** is a mathematical statement about how likely it is that something will happen. Probability is often stated in terms like "The probability of developing a particular illness is 1 in 10,000," or "The likelihood of winning the lottery is 1 in 5,000,000." It is important to make a distinction between *probability* and *possibility*. When we say something is *possible*, we are just saying that it could occur. It is a very inexact term. *Probability* defines how likely *possible* events are.

Another important consideration is the consequences of an event. If a disease is likely to make 50 percent of the population ill (the probability of becoming ill is 50 percent) but no one dies, that is very different from analyzing the safety of a dam, which if it failed would cause the deaths of thousands of people downstream. We would certainly not accept a 50 percent probability that the dam would fail. Even a 1 percent probability in that case would be unacceptable. The assessment and management of risk involve an understanding of probability and the consequences of decisions. (See figure 16.1.)

Risk assessment involves analyzing a risk to determine the probability of an

FIGURE 16.1 Decision-Making Process The assessment, cost, and consequences of risks are all important to the decision-making process.

adverse effect. Risk management is a much broader task that includes assessment and the consequences of risks in the decision-making process. We will look at both these tasks in the next two sections.

Risk Assessment

Environmental **risk assessment** is the use of facts and assumptions to estimate the probability of harm to human health or the environment that may result from exposures to pollutants, toxic agents, or management decisions. What risk assessment provides for environmental decision makers is an orderly, clearly stated, and consistent way to deal with scientific issues when evaluating whether a hazard exists and what the magnitude of the hazard may be.

Calculating the hazardous risk to humans of a particular activity, chemical, or technology is difficult. If a technology is well known, scientists use probabilities based on past experience to estimate risks. For example, the risk of developing black lung disease from coal dust in mines is well established. To predict the risks associated with new technology, much less accurate statistical probabilities, based on models rather than real-life experiences, must be used.

Risks associated with new chemicals are difficult to quantify. While animal tests are widely accepted in predicting whether or not a chemical will cause cancer in humans, their use in predicting how many cancers will be caused in a group of exposed people is still very controversial. Most risk assessments are *estimates* of the probability that a person will develop cancer or other negative effects. (See figure 16.2.)

Such estimates typically are based on broad assumptions to ensure that a lack of complete knowledge does not result in an underestimation of the risk. For example, people may be more or less sensitive to the effects of certain chemicals than the laboratory animals studied. Also, people vary in their sensitivity to cancer-causing compounds. Thus, what may present no risk to one person may be a high risk to others. Persons with breathing difficulties are more likely to be adversely affected by high levels of air pollutants than are healthy individuals. In addition, the estimate of human risk is based on extrapolation from animal tests in which high, chronic doses are used. Human exposure is likely to be lower or infrequent. Because of all these uncertainties, government regulators have decided to err on the side of safety to protect the public health.

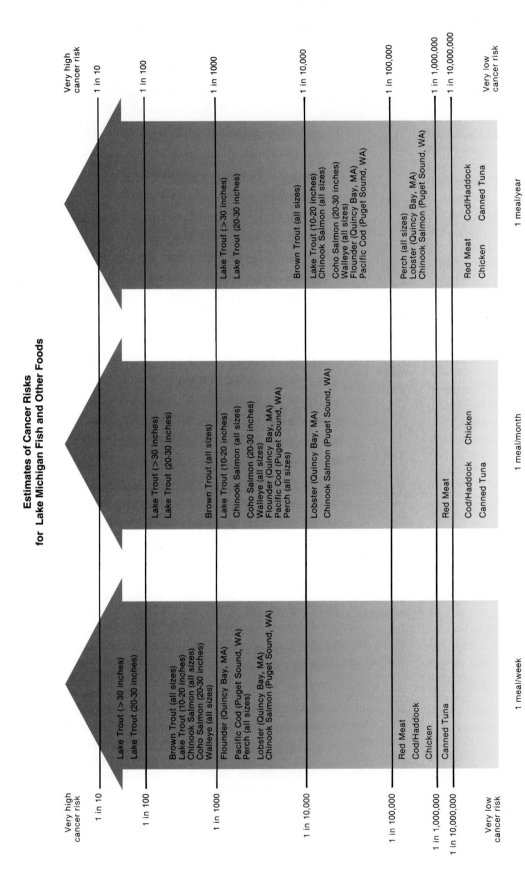

FIGURE 16.2 **Cancer Risk and Fish Consumption** A study by the National Wildlife Federation provides estimates for cancer risks associated with the consumption of sport fish.

What's In a Number?

Risk values are often stated, shorthand-fashion, as a number. When the risk concern is cancer, the risk number represents a probability of occurrence of additional cancer cases. For example, such an estimate for Pollutant X might be expressed as 1×10^{-6}, or simply 10^{-6}. This number can also be written as 0.000001, or one in a million—meaning one additional case of cancer projected in a population of one million people exposed to a certain level of Pollutant X over their lifetimes. Similarly, 5×10^{-7}, or 0.0000005, or five in *100 million,* indicates a potential risk of five additional cancer cases in a population of 100 million people exposed to a certain level of the pollutant. These numbers signify incremental cases above the background cancer incidence in the general population. American Cancer Society statistics indicate that the background cancer incidence in the general population is one in three over a lifetime.

If the effect associated with Pollutant X is not cancer but another health effect, perhaps neurotoxicity (nerve damage) or birth defects, then numbers are not typically given as probability of occurrence, but rather as levels of exposure estimated to be without harm. This often takes the form of a reference dose (RfD). An RfD is typically expressed in terms of milligrams (of pollutant) per kilogram of body weight per day, e.g., 0.004 mg/kg/day. Simply described, an RfD is a rough estimate of daily exposure to the human population (including sensitive subgroups) that is likely to be without appreciable risk of deleterious noncancerous effects during a lifetime. The uncertainty in an RfD may be one or several orders of magnitude (i.e., multiples of 10).

What's in a number? The important point to remember is that the numbers by themselves don't tell the whole story. For instance, even though the numbers are identical, a cancer risk value of 10^{-6} for the "average exposed person" (perhaps someone exposed through the food supply) is not the same thing as a cancer risk of 10^{-6} for a "most exposed individual" (perhaps someone exposed from living or working in a highly contaminated area). It's important to know the difference. Omitting the qualifier "average" or "most exposed" incompletely describes the risk and would mean a failure in risk communication.

A numerical estimate is only as good as the data it is based on. Just as important as the *quantitative* aspect of risk characterization (the risk numbers), then, are the *qualitative* aspects. How extensive is the data base supporting the risk assessment? Does it include human epidemiological data as well as experimental data? Does the laboratory data base include less data on more than one species? If multiple species were tested, did they all respond similarly to the test substance? What are the "data gaps," the missing pieces of the puzzle? What are the scientific uncertainties? What science policy decisions were made to address these uncertainties? What working assumptions underlie the risk assessment? What is the overall confidence level in the risk assessment? All of these qualitative considerations are essential to deciding what reliance to place on a number and to characterizing a potential risk.

Source: Data from EPA Journal.

That approach has been criticized by those who say it carries protection to the extreme, usually at the expense of industry.

Over the past decade, risk assessment has had its largest impact in regulatory practices involving cancer-causing chemicals called **carcinogens.** In the United States, for example, the decisions to continue registration of pesticides, to list substances as hazardous air pollutants under the Clean Air Act, and to regulate water contaminants under the Safe Drinking Water Act depend to a large degree on the risk assessments for the substances in question.

Risk assessment analysis is also being used to help set regulatory priorities and support regulatory action. Those chemicals or technologies that have the highest potential to cause damage to health or the environment receive attention first, while those perceived as having minor impacts receive less immediate attention. Medical waste is perceived as having high risk, and laws have been enacted to minimize the risk, while the risk associated with the use of fertilizer on lawns is considered minimal and is not regulated.

The science supporting environmental regulatory decisions is complex and rapidly evolving. Many of the most important threats to human health and the environment are highly uncertain. Risk assessment quantifies risk and states the uncertainty that surrounds many environmental issues. This can help institutions research and plan in a way that is consistent with scientific and public concern for environmental protection.

Risk Management

Risk management is a decision-making process that involves risk assessment, technological feasibility, economic impacts, public concerns, and legal requirements. Risk management includes:

1. Deciding which risks should be given the highest priority

2. Deciding how much money will be needed to reduce each risk to an acceptable level

3. Deciding where the greatest benefit would be realized by spending limited funds

4. Deciding how much risk is acceptable

5. Deciding how the plan will be enforced and monitored

Risk management raises several issues. With environmental concerns such as acid rain, ozone depletion, and hazardous waste, the scientific basis for regulatory decisions is often controversial. As was previously mentioned, hazardous substances can be tested, but only on animals. Are animal tests appropriate for determining impacts on humans? There is not an easy answer to this question. Dealing with global warming, ozone depletion, and acid rain require projecting into the future and estimating the magnitude of future effects. Will sea level rise? How many lakes will become acidified? How many additional skin cancers will be caused by depletion of the ozone layer? Estimates from equally reputable sources vary widely. Which ones do we believe?

The politics of risk management frequently focus on the adequacy of the scientific evidence. The scientific basis can be thought of as a kind of problem definition. Science determines that some threat or hazard exists, but because scientific facts are open to interpretation, there is controversy. For example, it is a fact that dioxin is a highly toxic material known to cause cancer in laboratory animals. It is also very difficult to prove that human exposure to dioxin has led to the development of cancer, although high exposures have resulted in acne in exposed workers. Acne is a common result of exposure to molecules like dioxin.

This is why problem definition is so important. Defining the problem helps to determine the rest of the policy process (making rules, passing laws, or issuing statements). If a substance poses little or no risk, then policy action is unnecessary. For example, some observers believe chemicals pose many threats that need to be addressed. Others believe chemicals pose little threat; instead, they see scare tactics and government regulations as unnecessary attacks on businesses. Logging forests poses risks of soil erosion and the loss of some animal species. The timber industry sees these risks as minimal and as a threat to its economic well-being, while many environmentalists consider the risks unacceptable. These and similar disagreements are often serious public-relations problems for both government and business because most of the public has a poor understanding of the risks they accept daily.

True and Perceived Risks

People often overestimate the frequency and seriousness of dramatic, sensational, well-publicized causes of death and underestimate the risks from more familiar causes that claim lives one by one. Risk estimates by "experts" and by the "public" on many environmental problems differ significantly. This discrepancy and the reasons for it are extremely important because the public generally does not trust

TABLE 16.1 RISKS OF DEATH

Deaths per million hours of exposure

Mountain climbing	40,000
Canoeing	10,000
Cigarette smoking	3000
Swimming	2560
Automobile travel	1200
Hunting	1000
Air travel	500
Being struck by lightning	100
Vaccination	1.5
Living beside a nuclear power plant	0.5

experts to make important risk decisions alone.

While public health and environmental risks can be minimized, eliminating all risks is impossible. Almost every daily activity—driving, walking, working—involves some element of risk. (See table 16.1.)

From a risk management standpoint, whether one is dealing with a site-specific situation or a national standard, the deciding question ultimately is: What degree of risk is acceptable? In general, we are not talking about a "zero risk" standard, but rather a concept of **negligible risk:** At what point is there really no significant health or environmental risk? At what point is there an adequate safety margin to protect public health and the environment?

Risk management involves comparing the estimated true risk of harm from a particular technology or product with the risk of harm perceived by the general public. The public generally perceives involuntary risks, such as nuclear power plants or nuclear weapons, as greater than voluntary risks, such as drinking alcohol or smoking. In addition, the public perceives newer technologies, such as genetic engineering or toxic-waste incinerators, as greater risks than more familiar technologies, such as automobiles and dams. Many people are afraid of flying for fear of crashing; however, automobile accidents account for a far greater number of deaths—about 45,000 in the United States each year, compared to less than a thousand annually from plane crashes. (See figure 16.3.)

A fundamental problem facing governments today is how to satisfy people who are concerned about a problem that

experts state presents less hazard than another less visible problem, especially when economic resources to deal with the problems are limited. Debate continues over the health concerns raised by asbestos, dioxin, radon, and Alar.

Some researchers argue that the public is frequently misled by the politics of public health and environmental safety. This is understandable since many prominent people become involved in such issues and use their public image to encourage people to look at issues from a particular point of view.

Whatever the issue, it is hard to ignore the will of the people, particularly when sentiments are firmly held and not easily changed. A fundamental issue surfaces concerning the proper role of a democratic government and other organizations in a democracy when it comes to matters of risk. Should the government focus available resources and technology where they can have the greatest tangible impact on human and ecological well-being, or should it focus them on problems about which the public is most upset? What is the proper balance? For example, adequate prenatal health care for all pregnant women would have a greater effect on the health of children than would removing asbestos from all school buildings.

Obviously, there are no clear answers to these questions. However, experts and the public are both beginning to realize that they each have something to offer concerning how we view risk. Many risk experts who have been accustomed to looking at numbers and probabilities are now conceding that there is rationale for looking at risk in broader terms. At the same time, the public is being supplied

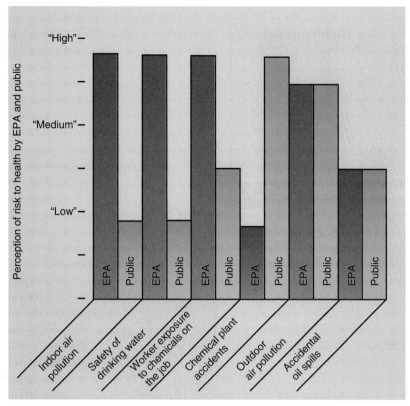

FIGURE 16.3 **Perception of Risk** Professional regulators and the public do not always agree on what risks are.

with more data to enable them to make more informed judgments.

Throughout this discussion of risk assessment and management we have made numerous references to costs and economics. It is not economically possible to eliminate all risk. As risk is eliminated, the cost of the product or service increases. Many environmental issues are difficult to evaluate from a purely economic point of view, but economics is one of the tools useful to analyzing any environmental problems.

ECONOMICS AND THE ENVIRONMENT

Environmental problems are primarily economic problems. While this may be an overstatement, it often is difficult to separate economics from environmental issues or concerns. Basically, economics deals with resource allocation. It is a description of how we value goods and services. We are willing to pay for things or services we value highly and are unwilling to pay for things we think there is plenty of. For example, we will readily pay for a warm, safe place to live but would be offended if someone suggested that we pay for the air we breathe.

Our goal as a society is to seek long-term economic growth that creates jobs while improving and sustaining the environment. Achieving this goal requires an environmental strategy that repairs past environmental damage; helps us shift from waste management to pollution prevention; and uses valuable resources more efficiently.

In the three decades since the first Earth Day, many nations have made considerable progress in responding to threats to public health and the environment. Yet major challenges remain. We can put in place a set of policies and programs that will establish a new course for the development and use of environmental technologies into the next century. We must, however, broaden our environmental tool kit, replacing those instruments that are no

longer effective with a new set of tools designed to meet today's challenges and tomorrow's needs.

Economic Concepts

An economic good or service can be defined as anything that is scarce. Scarcity exists whenever the demand for anything exceeds its supply. We live in a world of general scarcity. **Resources** are anything that contributes to making desired goods and services available for consumption. Resources are limited, relative to the desires of humans to consume. The **supply** is the amount of a good or service available to be purchased. **Demand** is the amount of a product that consumers are willing and able to buy at various prices. In economic terms, supply depends on:

1. The raw materials available to produce a good or service using present technology

2. The amounts of those materials available

3. The costs of extracting, shipping, and processing the raw materials

4. The degree of competition for those materials among users

5. The feasibility and cost of recycling already used material

6. The social and institutional arrangements that might have an impact. (See figure 16.4.)

The relationship between available supply of a commodity or service and its price is known as a **supply/demand curve.** (See figure 16.5.) The price of a product or service reflects the strength of the demand for and the availability of the commodity. When demand exceeds supply, the price rises. Cost increases cause people to seek alternatives or to decide not to use a product or service, which results in lower demand.

For example, food production depends heavily on petroleum for the energy to plant, harvest, and transport food crops. In addition, petrochemicals are used to make fertilizer and chemical pest-control agents. As petroleum prices rise, farmers reduce their petroleum use. Perhaps they farm less land or use less fertilizer or pesticide. Regardless, the price of food must rise as

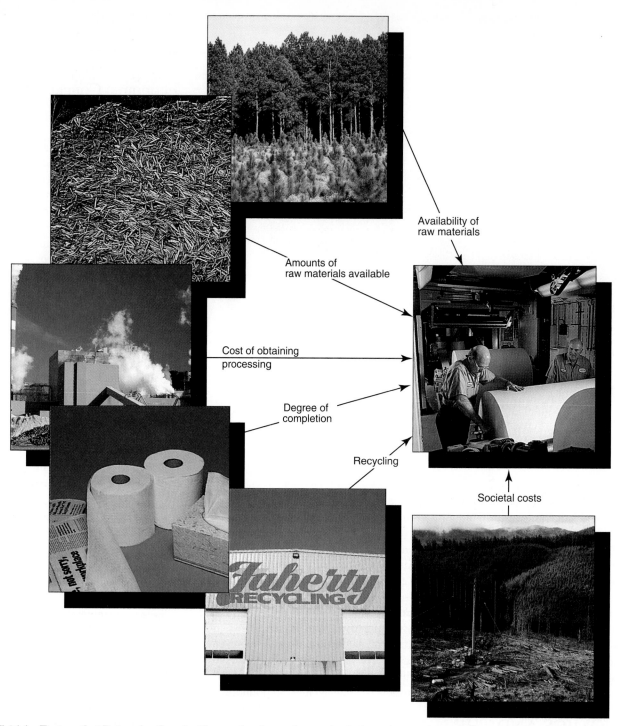

FIGURE 16.4 **Factors that Determine Supply** The supply of a good or service is dependent on the several factors shown here.

the price of petroleum rises. As the prices of certain foods rise, consumers seek less costly forms of food.

When the supply of a commodity exceeds the demand, producers must lower their prices to get rid of the product, and eventually, some of the producers go out of business. Ironically, this happens to farmers when they have a series of good

years. Production is high, prices fall, and some farmers go out of business.

Market-Based Instruments

With the growing interest in environmental protection during the past decade, policy makers are examining new methods to reduce harm to the environment. One area

of growing interest is market-based instruments (MBIs). MBIs provide an alternative to the common command-and-control legislation because they use economic forces and the ingenuity of entrepreneurs to achieve a high degree of environmental protection at a low cost. Instead of dictating how industry should conduct its activities, MBIs provide incentives by imposing costs

a.

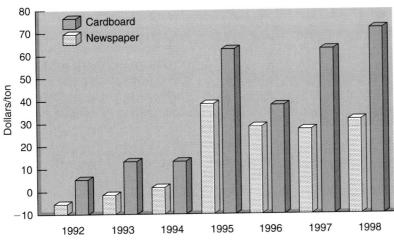

b.

FIGURE 16.5 **Supply and Demand** (*a*) As a result of increased interest in recycling, the supply of recycled newspapers has increased substantially between 1970 and 1993. The demand changed very little; therefore, the price in 1993 was extremely low. (*b*) Beginning in 1994, the demand for recycled newspaper and cardboard increased, thus the price rose significantly. Prices were more stable through 1998.

Source: (*b*) Data from *Recycling Times*.

on pollution-causing activities. This approach allows companies to decide for themselves how best to achieve the required level of environmental protection.

MBIs can be grouped into five basic categories:

1. **Information programs**

   These programs rely on informed consumers' market choices to reduce environmental problems. Information about the environmental or risk consequences of choices make clear to consumers that it is in their personal interest to change their decisions or behavior. Examples include radon or lead testing or labeling pesticide products. Another type of program, such as the Toxic Release Inventory in the United States, discloses information on environmental releases by polluters. This provides corporations with incentives to improve their environmental performance to enhance their public image.

2. **Tradable emissions permits**

   These permits give companies the right to emit specified quantities of pollutants. Companies that emit less than the specified amounts can sell their permits to other firms or "bank" them for future use. Businesses responsible for pollution have an incentive to internalize the external cost they were previously imposing on society: If they clean up their pollution sources, they can realize a profit by selling their permit to pollute. Once a business recognizes the possibility of selling its permit, it sees that pollution is not costless. The creation of new markets (the permits) to reduce pollution reduces the external aspects to waste disposal and makes them internal costs, just like costs of labor and capital.

3. **Emission fees, taxes, and charges**

   These fees provide incentives for environmental improvement by making environmentally damaging activity or products more expensive. Businesses and individuals control the level of pollution to where it is cheaper to abate the pollution than to pay the charges. Emissions fees can be useful

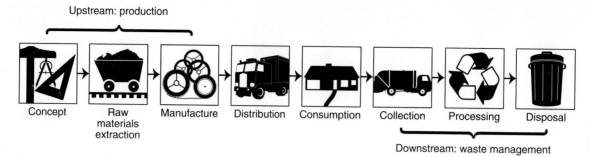

FIGURE 16.6   The Life Cycle of a Typical Product

Source: *Environment*, Vol. 39, No. 7, September 1997.

when pollution is coming from many small sources, such as vehicular emissions or agricultural runoff, where direct regulation or trading schemes are impractical. Taxes contribute to government revenue; charges are used to fund environmental cleanup programs.

4. Deposit-refund programs

These programs place a surcharge on the price of a product, which is refunded when the used product is returned for reuse or recycling.

5. Subsidies

Subsidies may include consumer rebates for purchases of environmentally friendly goods, soft loans for businesses planning to implement environmental products, and other monetary incentives designed to reduce the costs of improving environmental performance.

A **subsidy** is a gift from government to a private enterprise that is considered important to the public interest. This is particularly true when the private enterprise is having economic difficulty. Agriculture, transportation, space technology, and communication are frequently subsidized by governments. These gifts, whether loans, favorable tax situations, or direct grants, are all paid for by taxes on the public.

Subsidies are costly in two ways: First, the bureaucracy necessary to administer a subsidy costs money, and the subsidy is an indirect way of keeping the market price of a product low. The actual cost is higher because these subsidy costs must be added to the market price to arrive at the product's true cost. Second, in many cases, subsidies encourage activities that

in the long term may be detrimental to the environment. For example, the transportation subsidies for highway construction encourage use of inefficient individual automobiles. Higher taxes on automobile use to cover the cost of building and repairing highways would encourage the use of more energy-efficient public transport.

While research is still needed to determine how MBIs should be structured and used, there has been sufficient experience to justify their further deployment as a means of attaining environmental objectives at least possible cost. Successful programs so far include a sulfur dioxide trading market, lead phase-down banking and trading, and hundreds of pay-by-the-bag trash collection programs in many countries.

When used correctly, market-based approaches will allow us to reach a level of environmental protection at lower total cost than would be possible with the traditional means of command and control. Market forces tend to drive decisions toward least-cost solutions. Offer the right incentives, and business will develop and adopt better pollution-control technology, rather than stagnating at "commanded" technology.

Market-oriented policies, however, will not always work. For instance, utilities and other producers in southern California have complained that the area's tradable permit system has been so poorly planned that it will cost them more than traditional approaches. It could also be stated that no single one of these economic incentive approaches will be a panacea for all problems. It could be a mistake to start with a policy instrument, then go in search of applications. But this kind of flexible approach, using several

solutions, is gaining favor among the regulated and the regulators.

Although the various economic instruments discussed have their own niches, they can be used effectively in combination. For example, trading or pricing approaches work better if supported by information programs: communities that adopted pay-by-the-bag systems of trash disposal had fewer problems if households were given adequate information well in advance. Environmental tax systems can incorporate trading features (e.g., taxes can be levied on net emissions after trades). It will be increasingly important to use the various MBI methods together. The challenge is to design the most appropriate instruments to deal with environmental problems, bearing in mind the relevant policy objectives: steady progress in reducing risks, cost-effectiveness, encouragement of technological innovation, fairness, and administrative simplicity.

Extended Product Responsibility

Extended product responsibility (EPR) is an emerging principle for a new generation of pollution prevention policies that focus on product systems instead of production facilities. It relies for its implementation on life cycle analysis to identify opportunities to prevent pollution and reduce resource and energy use in each stage of the product chain through changes in product design and process technology. All factors along the product chain share responsibility for the life cycle environmental impacts of products, from the upstream impacts inherent in selection of materials and impacts from

Georgia-Pacific Corporation: Recycled Urban Wood—A Case Study in Extended Product Responsibility

Georgia-Pacific manufactures particleboard from multi-species wood recovered from commercial disposal or general urban solid waste. The company has agreements with five recycling and processing companies that accept or collect wood at various sites. The wood is cleaned of contaminants and sent to a Georgia-Pacific particleboard manufacturing plant in Martell, California, or to other end users.

The project involves five stakeholder groups: (1) wood waste producers (e.g., operations involved with construction and demolition debris, cut-to-size lumber, commercial wood waste from furniture), (2) collection agents, (3) processors of wood waste, which make the waste into a product that can be reused, (4) transportation contractors, shippers, and haulers, and (5) end users (e.g., Georgia-Pacific's Martell plant). The project has a variety of goals, including increasing the availability of the wood supply for particleboard production, contributing to Georgia-Pacific's goals of product stewardship, and contributing to California's mandated reduction in solid waste (e.g., 50 percent reduction by 2000).

Business factors driving the project include the shortage of fiber for the particleboard plant, rising costs of landfilling, and mandated solid waste reductions. Benefits include an expanded fiber supply in the Northwest United States. Contamination is one of the most significant barriers to the wood recovery program. Often the collected wood is mixed in with metal, plastic, and paper and must be cleared of these contaminants to be usable. The captured paper, plastic, and nonferrous metals are sent to a landfill. Wood by-products that cannot be used in particleboard processing are sold for use as animal bedding, playground cover, soil additive, and lawn or garden mulch. Currently virgin fiber, a by-product from sawmills is often less expensive than recovered fiber for use in particleboard. As wood becomes more scarce, however, the economics will reverse.

---

the manufacturing process itself to downstream impacts from the use and disposal of the products. (See figure 16.6.)

EPR had its origins in Western Europe. The term *extended product responsibility* is derived from the term *extended producer responsibility*, which is often applied to the German packaging ordinance and similar policies in other Western European countries. The second term is something of a misnomer, however, because most extended responsibility schemes allocate the burden of environmental protection all along the product chain rather than placing it entirely on producers. Under the German packaging ordinance, for instance, consumers, retailers, and packaging manufacturers all share this responsibility, with the financial burden of waste management falling on the last two. In the United States, extended product responsibility has gained greater currency because it highlights systems of shared responsibility.

Specific benefits of EPR include:

Cost savings, particularly through the process of producers taking back used products, allow manufacturers to recover valuable materials, reuse them, and save money.

Companies are looking at designing for recycling and disassembly— innovations necessitated by end-of-life management of products, which in many cases helps companies realize how to assemble products more efficiently.

There are more efficient environmental protections, since product-based environmental strategies often are a more cost-effective method of complying with environmental regulations and avoiding environmental liabilities than existing facility-based programs.

Despite these benefits, however, obstacles to EPR exist. There include:

Cost of EPR

Lack of information and tools to access overall product system impacts

Difficulty in building relationships among actors in different life cycle stages

Hazardous waste regulations that require hazardous waste permits for collection and take-back of certain products

Antitrust laws that make it difficult for companies to cooperate

These are the kinds of questions that further explorations of EPR and, most important, that real-world experience with EPR programs can help to understand and address. Ultimately, EPR is an opportunity to explore new models of environmental policy that are less costly and more flexible.

Cost-Benefit Analysis

People use **cost-benefit analysis** to determine whether a policy generates more social costs than social benefits, and if benefits outweigh costs, how much spending would obtain optimal results. Steps in cost-benefit analysis include:

1. Identification of the project to be evaluated

2. Determination of all impacts, favorable and unfavorable, present and future, on all of society

3. Determination of the value of those impacts, either directly through market values or indirectly through price estimates

4. Calculation of the net benefit, which is the total value of positive impacts less the total value of negative impacts

For example, the cost of reducing the amount of lead in drinking water in the United States to acceptable limits is estimated to be about $125 million a year. The benefits to the nation's health from such a program are estimated at nearly $1 billion

TABLE 16.2 COSTS AND BENEFITS OF IMPROVING AIR QUALITY

Costs	Benefits
Installation and maintenance of new technology	Reduced deaths and disease
1. Scrubbers on smokestacks	Fewer respiratory problems
2. Automobile emissions control	
Redesign of industries and machines	Reduced plant and animal damage
Additional energy costs to industry and public	Lower cleaning costs for industry and public
Unemployment as some industries go out of business	More clear, sunny days; better visibility
Retraining of employees to use new technology	Less eye irritation
Costs associated with monitoring and enforcement	Fewer odor problems

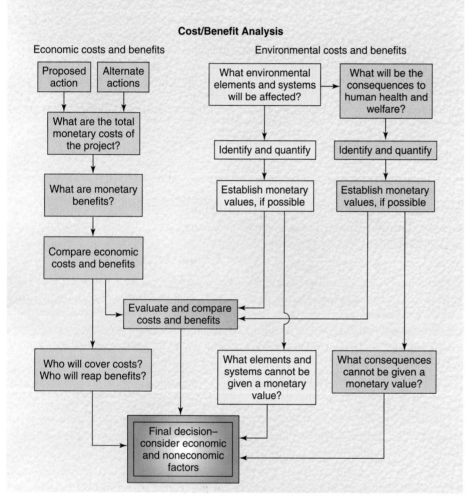

per year. Thus under a cost-benefit analysis, the program is economically sound. Recycling of solid waste, which was once cost-prohibitive, is now cost-effective in many communities because the costs of landfills have risen dramatically. Table 16.2 gives examples of the kinds of costs and benefits involved in improving air quality. Although not a complete list, the table indicates the kinds of considerations that go into a cost-benefit analysis. Some of these are easy to measure in monetary terms; others are not.

Concerns about the Use of Cost-Benefit Analysis

Does everything have an economic value? Critics of cost-benefit analysis raise this point, among others. Some people argue that, if economic thinking pervades society, many simple noneconomic values like beauty or cleanliness can survive only if they prove to be "economic." (See figure 16.7.)

It has long been the case in many developed countries that major projects, especially those undertaken by the government, require some form of cost-benefit analysis with respect to environmental impacts and regulations. In the United States, for example, such requirements were established by the National Environmental Policy Act of 1969, which requires environmental impact statements for major government-supported projects. Increasingly, similar analyses are required for projects supported by national and international lending institutions such as the World Bank.

There are clearly benefits to requiring such analysis. Although environmental issues must be considered at some point during project evaluation, efforts to do so are hampered by the difficulty of assigning specific value to environmental resources. In cases of Third World development projects, these already difficult environmental issues are made more difficult by cultural and socioeconomic differences. A less-developed country, for example, may be less inclined to insist on or be able to afford expensive emissions-treatment technology on a project that will provide jobs and economic development.

One particularly compelling critique of cost-benefit analysis is that for analysis to be applied to a specific policy, the analyst must decide which preferences count—that is, which preferences have "standing" in cost-benefit analysis. In theory, cost-benefit analysis should count all benefits and costs associated with the policy under review, regardless of who benefits or bears the costs. In practice, however, this is not always done. For example, if a cost is spread thinly over a great many people, it may not be recognized as a cost at all. The cost of air pollution in many parts of the world could fall into such a category. Debates over how to count benefits and costs for future generations, inanimate objects such as rivers, and nonhumans, such as endangered species, are also common.

A noted critic of cost-benefit analysis is E. F. Schumacher, who states in his book *Small Is Beautiful* (Harper and Row, 1973):

FIGURE 16.7 **Does Everything Have an Economic Value?** The use of water and land is often based on the economic benefits obtained. Are there other benefits that cannot be measured economically?

Cost-benefit analysis is a procedure by which the higher is reduced to the level of the lower and the priceless is given a price. All it can do is lead to self-deception or the deception of others; for to undertake to measure the immeasurable is absurd and constitutes but an elaborate method of moving from preconceived notions to foregone conclusions; all one has to do to obtain the desired results is to impute suitable values to the immeasurable costs and benefits. The logical absurdity, however, is not the greatest fault of the undertaking; what is worse, and destructive of civilization, is the pretense that everything has a price or, in other words, that money is the highest of all values.

While Schumacher's objectives to cost-benefit analysis may be somewhat idealistic, he does raise some issues that should be discussed. Is it possible to separate economic issues from environmental issues? (See table 16.3.)

Economics and Sustainable Development

The past president of the Japan Economic Research Center, Saburo Okita, once stated

that a slowdown of economic growth is needed to prevent further deterioration of the environment. Whether or not a slowdown is necessary provokes sharp differences of opinion.

One school of thought argues that economic growth is essential to finance the investments necessary to prevent pollution and to improve the environment by a better allocation of resources. A good school of thought, which is also progrowth, stresses the great potential of science and technology to solve problems and advocates relying on technological advances to solve environmental problems. Neither of these schools of thought sees any need for fundamental changes in the nature and foundation of economic policy. Environmental issues are viewed mainly as a matter of setting priorities in the allocation of resources.

A newer school of economic thought believes that economic and environmental well-being are mutually reinforcing goals that must be pursued simultaneously if either one is to be reached. Economic growth will create its own ruin if it continues to undermine the healthy functioning of Earth's natural systems or to exhaust natural resources. It is also true that healthy economies are most likely to provide the necessary financial investments

in environmental protection. For this reason, one of the principal objectives of environmental policy must be to ensure a decent standard of living for all. The solution, at least in the broad scope, would be for a society to manage its economic growth in such a way as to do no irreparable damage to its environment. This concept is referred to as **sustainable development.** The term has been criticized as ambiguous and open to a wide range of interpretations, many of which are contradictory. The confusion arises because "sustainable growth" and "sustainable use" have been used interchangeably, as if their meanings were the same. They are not. "Sustainable growth" is a contradiction in terms: Nothing physical can grow indefinitely. "Sustainable use" is applicable only to renewable resources: it means using them at rates within their capacity for renewal.

By balancing economic requirements with ecological concerns, the needs of the people are satisfied without jeopardizing the prospects of future generations. While this concept may seem to be common sense, the history of the world shows that it has not been a common practice. A major obstacle to sustainable development in many countries is a social structure that gives most of the nation's wealth to a tiny

Sustainable Development

The most commonly used definition of the term "sustainable development" is one that originated with the 1987 report, *Our Common Future,* by the World Commission on Environment and Development (known as the Bruntland Commission). By that formulation, sustainable development is "development that meets the needs of the present without compromising the ability of future generations to meet their own needs."

Since the release of the Bruntland Commission report, the phrase has been broadened and modified. The term "sustainable" has gained usage because of increasing concern over exploitation of natural resources and economic development at the expense of environmental quality. Although disagreement exists as to the precise meaning of the term beyond respect for the quality of life of future generations, most definitions refer to the viability of natural resources and ecosystems over time, and to maintenance of human living standards and economic growth. The popularity of the term stems from the melding of the dual objectives of environmental protection and economic growth. A sustainable agricultural system, for example, can be defined as one that can indefinitely meet the demands for food and fiber at socially acceptable economic costs and environmental impacts.

Finally, as pointed out by Tan Sri Razali, former chairman of the United Nation's Commission on Sustainable Development, the transfer of modern, environmentally sound technology to developing nations is the "key global action to sustainable development."

Source: From Science and Technology Council, 1995, Office of the President, Washington, D.C.

TABLE 16.3 THE U.S. ENVIRONMENTAL INDUSTRY 1980–1998

Environmental Industry Segment	1988	80–88 Growth	1990	88–90 Growth	1992	90–92 Growth	1994	92–94 Growth	1996	94–96 Growth	1998	96–98 Growth
Services												
Analytical services	1.2	225%	1.5	28%	1.4	−7.8%	1.3	−6.4%	1.2	−10.6%	1.3	−7.0%
Wastewater treatment works	18.2	67%	20.4	12%	21.5	5.4%	22.7	5.6%	24.0	5.8%	23.0	6.0%
Solid waste management	21.4	153%	26.1	22%	28.2	8.0%	31.0	9.9%	33.9	9.4%	32	8.2%
Hazardous waste management	4.7	662%	6.3	34%	6.6	5.2%	6.4	−3.5%	6.0	−6.6%	6.0	−6.6%
Remediation/industrial services	6.7	1431%	8.0	20%	7.8	−2.6%	8.4	8.7%	8.6	2.1%	9.0	2.2%
Consulting and engineering	8.4	479%	12.5	49%	14.3	14.4%	15.3	7.0%	15.2	−0.6%	15	−0.7%
Equipment												
Water equipment and chemicals	12.0	73%	13.5	13%	14.7	8.4%	15.6	6.6%	17.5	12.1%	18.0	12.1%
Instruments and information systems	1.3	500%	2.0	53%	2.6	29.1%	2.9	10.9%	3.1	10.1%	3.1	10%
Air pollution control equipment	3.7	13%	13.1	254%	13.8	5.0%	14.5	5.4%	15.7	8.0%	16.0	8.1%
Waste management equipment	8.8	120%	10.4	18%	11.1	6.7%	11.2	0.9%	12.0	6.9%	12.3	7.0%
Process and prevention technology	0.2	178%	0.4	86%	0.6	46.3%	0.8	26.7%	0.8	10.5%	0.9	10.4%
Resources												
Water utilities	17.7	49%	19.8	12%	21.9	10.6%	24.2	10.5%	26.4	8.9%	27	9.0%
Resource recovery	11.5	161%	13.1	14%	12.2	−6.9%	15.4	26.1%	14.3	−6.7%	15.1	−5.9%
Environmental energy sources	1.4	−8%	1.8	25%	2.0	12.6%	2.2	11.7%	2.4	8.2%	2.5	8.3%
Totals:	**117.1**	**115%**	**148.9**	**27%**	**158.5**	**6.5%**	**171.9**	**8.4%**	**181.1**	**5.4%**	**181.2**	**5.4%**

Source: From *Environmental Business Journal,* Environmental Business International, Environmental Business Publishing, San Diego, CA. Reprinted by permission.

majority of its people. It has been said that a person who is worrying about his next meal is not going to listen to lectures on protecting the environment. What to residents in the Northern Hemisphere seem like some of the worst environmental outrages—cutting rain forests to make charcoal for sale as cooking fuel, for example—are often committed by people who have no other form of income.

The disparities that mark individual countries are also reflected in the planet as a whole. Most of the wealth is concentrated in the Northern Hemisphere. From the Southern Hemisphere's point of view, it is the rich world's growing consumption patterns—big cars, refrigerators, and climate-controlled shopping malls—that are the problem. The problem for the long term is that people in developing countries now want those consumer items that make life

ENVIRONMENTAL CLOSE-UP

"Green" Advertising Claims—Points to Consider

Like many consumers, you may be interested in buying products that are less harmful to the environment. You have probably seen products with such "green" claims as "environmentally safe," "recyclable," "degradable," or "ozone friendly." But what do these claims really mean? How can you tell which products really are less harmful to the environment? Here are some pointers to help you decide.

1. Look for environmental claims that are specific. Read product labels to determine whether they have specific information about the product or its packaging. For example, if the label says "recycled," check how much of the product or packaging is recycled. Labels with "recyclable" claims mean that these products can be collected and made into useful products. This is relevant to you, however, only if this material is collected for recycling in your community.

2. Be wary of overly broad or vague environmental claims. These claims provide little information to help you make purchasing decisions. Labels with unqualified claims that a product is "environmentally friendly," "eco-safe," or "environmentally safe" have little meaning, for two reasons. First, all products have some environmental impact, though some may have less impact than others. Second, these phrases alone do not provide the specific information needed to compare products and packaging on their environmental merits.

3. Some products claim to be "degradable." Degradable materials will not help save landfill space. Biodegradable materials, like food and leaves, break down and decompose into elements found in nature when exposed to air, moisture, and bacteria or other organisms. Photodegradable materials, usually plastics, disintegrate into smaller pieces when exposed to enough sunlight. Either way, however, degradation of any material occurs very slowly in landfills, where most solid waste is sent. That is because modern landfills are designed to minimize the entry of sunlight, air, and moisture. Even organic materials like paper and food may take decades to decompose in a landfill.

GENUINE RECYCLED PAPER

30

30% POST CONSUMER WASTE

Three types of recycling symbols are commonly used in the United States. *(a)* This symbol simply means that the object is potentially recyclable, not that it has been or will be recycled. *(b)* This symbol indicates that a product contains recycled material, but it does not indicate how much recycled material is in the product (it could be only a very small amount). *(c)* This symbol states explicitly the percentage of recycled content found in the product.

---

in the industrial world so comfortable as well as environmentally costly. If the standard of living in China and India were to rise to that of Germany or the United States, the environmental impact on the planet would be significant.

If sustainable development is to become feasible, it will be necessary to transform our approach to economic policy. Steps in that direction would include changing the definition of gross national product (GNP) to include environmental improvement or decline. The concept of sustainable development may seem simple, but implementing it will be a very complex process.

Historically, rapid exploitation of resources has provided only short-term economic growth, and the environmental consequences in some cases have been incurable. For example, 40 years ago, forests covered 30 percent of Ethiopia.

Today, forest covers only 1 percent, and deserts are expanding. One-half of India once was covered by trees; today, only 14 percent of the land is in forests. As the Indian trees and topsoil disappear, the citizens of Bangladesh drown in India's runoff. (See figure 16.8.)

Sustainable development requires choices based on values. Both depend upon information and education, especially regarding the economics of decisions that affect the environment. A. W. Clausen, in his final address as president of the World Bank, noted the

> increasing awareness that environmental precautions are essential for continued economic development over the long run. Conservation, in its broadest sense, is not a luxury for people rich enough to vacation in scenic parks. It is not just a

motherhood issue. Rather, the goal of economic growth itself dictates a serious and abiding concern for resource management.

High-income developed nations, such as the United States, Japan, and much of Europe, are in a position to promote sustainable development. They have the resources to invest in research and the technologies to implement research findings. Some believe that the world should not impose environmental protection standards upon poorer nations without also helping them move into the economic mainstream.

Gaylord Nelson, the founder of the first Earth Day, lists five characteristics that define sustainability:

1. *Renewability:* A community must use renewable resources, like water, topsoil, and energy sources, no faster

Pollution Prevention Pays!

Extended product responsibility (EPR) is an emerging principle of resource conservation and pollution prevention. EPR advocates using a life cycle perspective to identify pollution prevention and resource conservation opportunities that maximize eco-efficiency. Under this principle, there is assumed responsibility for the environmental impacts of a product throughout its life cycle, including impacts on the selection of materials for the product, impacts from the manufacturer's production process, and downstream impacts from the use, recycling, or disposal of the product.

While this concept is relatively new, successful examples of it are in operation. For example, several years ago the Minnesota Mining and Manufacturing (3M) Company's European chemical plant in Belgium switched from a polluting solvent to a safer but more expensive water-based substance to make the adhesive for its Scotch™ Brand Magic™ Tape. The switch was not made to satisfy any environmental law in Belgium or the 12-nation European Community. 3M managers were complying with company policy to adopt the strictest pollution-control regulations that any of its subsidiaries is subject to—even in countries that have no pollution laws at all.

Part of the policy is founded on corporate public relations, a response to growing customer demand for "green" products and environmentally responsible companies. But as many North American multinationals with similar global environmental policies are discovering, cleaning up waste, whether voluntarily or as required by law, can cut costs dramatically.

Since 1975, 3M's "Pollution Prevention Pays" program—or 3P—has cut the company's air, water, and waste pollution around the world in half and, at the same time, has saved nearly $600 million in the last 18 years on changes in its manufacturing process, including $100 million overseas. Less waste has meant less spending to comply with pollution-control laws. But, in many cases, 3M actually has made money selling wastes it formerly hauled away. And, because of recycling prompted by the 3P program, it has saved money by not having to buy as many raw materials.

AT&T followed a similar path. In 1990, it set voluntary goals for the company's 40 manufacturing and 2500 nonmanufacturing sites worldwide. According to its latest estimates, AT&T has (1) reduced toxic air emissions, many caused by solvents used in the manufacture of computer circuit boards, by 73 percent; (2) reduced emissions of chlorofluorocarbons—gases blamed for destroying the ozone layer in the Earth's atmosphere—by 76 percent; and (3) reduced manufacturing waste 39 percent.

Xerox Corporation had focused on recycling materials in its global environment efforts. It provides buyers of its copiers with free United Parcel Service pickup of used copier cartridges, which contain metal-alloy parts that otherwise would wind up in landfills. The cartridges and other parts are now cleaned and used to make new ones.

CONTROL AND PREVENTION TECHNOLOGIES—SOME EXAMPLES

Control/Treatment/Disposal

- Sewage treatment
- Industrial wastewater treatment
- Refuse collection
- Incineration
- Off-site recovery and recycling of wastes
- Landfilling
- Catalytic conversion and oxidation
- Particulate controls
- Flue-gas desulfurization
- Nitrogen oxides control technology
- Volatile organic compound control and destruction
- Contaminated site remediation

Prevention

- Improved process control to use energy and materials more efficiently
- Improved catalysis or reactor design to reduce byproducts, increase yield, and save energy in chemical processes
- Alternative processes (e.g., low or no-chlorine pulping)
- In-process material recovery (e.g., vapor recovery, water reuse, and heavy metals recovery)
- Alternatives to chlorofluorocarbons and other organic solvents
- High-efficiency paint and coating application
- Substitutes for heavy metals and other toxic substances
- Cleaner or alternative fuels and renewable energy
- Energy-efficient motors, lighting, heat exchangers, etc.
- Water conservation
- Improved "housekeeping" and maintenance in industry

Source: Data from EPA Journal.

Other recycled Xerox copier parts include power supplies, motors, paper transport systems, printed wiring boards, and metal rollers. In all, 1 million parts per year are remanufactured. The initial design and equipment investment was $10 million. Annual savings total $200 million.

The philosophy of pollution prevention is that pollution should be prevented or reduced at the source whenever feasible. It is increasingly being shown that preventing pollution can cut business costs and thus increase profits. Pollution prevention, then, does make cents!

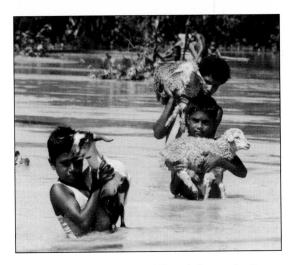

FIGURE 16.8 Indian Deforestation Causes Floods in Bangladesh Because the Ganges River drains much of India and the country of Bangladesh is at the mouth of the river, deforestation and poor land use in India can result in devastating floods in Bangladesh.

than they can replace themselves. The rate of consumption of renewable resources cannot exceed the rate of regeneration.

2. *Substitution:* Whenever possible, a community should use renewable resources instead of nonrenewable resources. This can be difficult, because there are barriers to substitution. To be sustainable, a community has to make the transition before the nonrenewable resources become prohibitively scarce.

3. *Interdependence:* A sustainable community recognizes that it is a part of a larger system and that it cannot be sustainable unless the larger system is also sustainable. A sustainable community does not import resources in a way that impoverishes other communities, nor does it export its wastes in a way that pollutes other communities.

4. *Adaptability:* A sustainable community can absorb shocks and can adapt to take advantage of new opportunities. This requires a diversified economy, educated citizens, and a spirit of solidarity. A sustainable community invests in, and uses, research and development.

5. *Institutional commitment:* A sustainable community adopts laws and political processes that mandate sustainability. Its economic system supports sustainable production and

consumption. Its educational systems teach people to value and practice sustainable behavior.

External Costs

Many of the important environmental problems facing the world today arise because modern production techniques and consumption patterns transfer waste disposal, pollution, and health costs to society. Such expenses, whether they are measured in monetary terms or in diminished environmental quality, are borne by someone other than the individuals who use a resource. They are referred to as **external costs.** For example, consider an individual who attempts to spend a day of leisure fishing in a lake. Transportation, food, fishing equipment, and bait for the day's activities can be purchased in readily accessible markets. But suppose that upon arrival at the lake, the individual finds a lifeless, polluted body of water. Let us further suppose that a chemical plant situated on the lakeshore is responsible for degrading the water quality on the lake to the point that all or most of the fish are destroyed. In this example, fishers collectively bear the external costs of chemical production in the form of lost recreational opportunities.

Pollution-control costs include pollution-prevention costs and pollution costs. **Pollution-prevention** costs are those incurred either in the private sector or by government to prevent, either

entirely or partially, the pollution that would otherwise result from some production or consumption activity. The cost incurred by local government to treat its sewage before dumping it into a river is a pollution-prevention cost; so is the cost incurred by a utility to prevent air pollution by installing new equipment.

Pollution-prevention costs can often be factored into a life cycle analysis. Life cycle analysis can help us understand the full cost, potential, and impact on new products and their associated technologies. As a systems approach, life cycle analysis examines the entire set of environmental consequences of a product, including those that result from its manufacture, use, and disposal. Because the relationships among industrial processes are complex, life cycle analysis requires understanding of material flows, resource reuse, and product substitution. Shifting to an approach that considers all resources, products, and waste as an interdependent system will take time, but governments can facilitate the shift by encouraging the transition to a systems approach.

Pollution costs can be broken down into two categories:

1. The private or public expenditures to avoid pollution damage once pollution has already occurred

2. The increased health costs and loss of the use of public resources because of pollution

TABLE 16.4 ECONOMIC SOLUTIONS TO POLLUTION AND RESOURCE WASTE

Solution	Internalizes external costs	Innovation	International competitiveness	Administrative costs	Increases government revenue
Regulation	Partially	Can encourage	Decreased*	High	No
Subsidies	No	Can encourage	Increased	Low	No
Withdrawing harmful subsidies	Yes	Can encourage	Decreased*	Low	Yes
Tradable rights	Yes	Encourages	Decreased*	Low	Yes
Green taxes	Yes	Encourages	Decreased*	Low	Yes
User fees	Yes	Can encourage	Decreased*	Low	Yes
Pollution-prevention bonds	Yes	Encourages	Decreased*	Low	No

Unless more cost-effective and productive technologies are developed.

The large cost of cleaning up spills, such as that from the *Exxon Valdez* and from the war in the Persian Gulf, is an example of a pollution cost, as is the increased health risk to humans from eating seafood contaminated from the oil. By the middle 1990s there were several new economic solutions being utilized to both internalize external costs and to prevent pollution. (See table 16.4.)

Common Property Resource Problems

Economists have stated that, when everybody shares ownership of a resource, there is a strong tendency to overexploit and misuse that resource. Thus, common public ownership could be better described as effectively having no owner.

For example, common ownership of the air makes it virtually costless for any industry or individual to dispose of wastes by burning them. The air-pollution cost is not reflected in the economics of the polluter but becomes an external cost to society. Common ownership of the ocean makes it inexpensive for cities to use the ocean as a dump for their wastes. (See figure 16.9.)

Similarly, nobody owns the right to harvest whales. If any one country delays in getting its share of the available supply of whales, other countries may beat that country to it. Thus, there is a strong incentive to overharvest the whale population and create a consequent threat to the survival of the species. What is true on an international scale for whales is true also for other species within countries. Note that

endangered species are wild and undomesticated; the survival of privately owned livestock is not a concern.

Finally, common ownership of land resources, such as parks and streets, is the source of other environmental problems. People who litter in public parks do not generally dump trash on their own property. The lack of enforceable property rights to commonly owned resources explains much of what economist John Kenneth Galbraith has termed "public squalor amid private affluence."

Economic Decision Making and the Biophysical World

For most natural scientists, current crises like biodiversity loss, climate change, and many other environmental problems are symptoms of an imbalance between the socioeconomic system and the natural world. While it is true that humans have always changed the natural world, it is also clear that this imprint is currently much greater than anything experienced in the past. One reason for the profound effect of human activity on the natural world is the fact that there are so many of us.

One of the most serious consequences of the growing human impact on the natural world is the loss of biodiversity. Biological diversity is thought to affect the stability of ecosystems and the ability to cope with crises. In the 570-million-year history of complex life on earth, there have been several major extinction events. The loss of biodiversity after major extinction episodes ranged from 20 percent to more than 90 percent. The loss in biodiversity

caused by human activity since the Industrial Revolution alone is somewhere between 10 and 20 percent. If current trends continue, losses are likely to reach 50 percent by the end of the next century. After each extinction event, it took between 20 million and 100 million years for biodiversity to recover to previous levels, a length of time between 100 and 500 times longer than the 200,000-year history of *Homo sapiens.*

The greatest single cause of the loss of biodiversity is habitat destruction, that is, the destruction of the web of organisms and functions that support individual species. The recognition that individual species are supported by others within the ecosystem is foreign to the way markets view the world.

The example of biodiversity illustrates the conflicting frameworks of economics and ecology. Market decisions fail to account for the context of a species or the interconnections between resource quality and ecosystem functions. For example, the value of land used for beef production is measured according to its contribution to output. Yet long before output and the use value of land decreases, the diversity of grass varieties, microorganisms in the soil, or groundwater quality may be affected by intensive beef production. As long as yields are maintained, these changes go unnoticed by markets and are unimportant to land-use decisions.

Another obvious difference between economics and ecology is the relevant time frame in markets and ecosystems. The biophysical world operates in tens of thousands and even millions of years. The

617 million

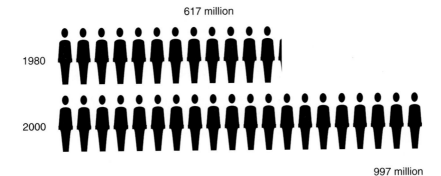

1980

2000

997 million

a.

b.

c.

FIGURE 16.9 **Abuse of the Ocean** Since the oceans of the world are a shared resource that nobody owns, there is a tendency to use the resource unwisely. (*a*) Growing population in coastal areas leads to more marine pollution and destruction of coastal habitats. (*b* and *c*) Many countries use the ocean as a dump for unwanted trash. This barge will dump its waste in the ocean, and much of this waste will ultimately end up on the shore as currents and waves move it from where it was dumped. Some 6.5 million tons of litter finds its way into the sea each year.

Source: (*a*) Data from UN Environment Programme; World Resources Institute.

time frame for market decisions is short. Particularly where economic policy is concerned, two- to four-year election cycles are the frame of reference; for investors and dividend earners, performance time frames of three months to one year are the rule.

Space or place is another issue. For ecosystems, place is critical. Take groundwater as an example. Soil quality, hydrogeological conditions, regional precipitation rates, plants that live in the region, and losses from evaporation, transpiration, and groundwater flow all contribute to the size and location of groundwater reservoirs. These capacities are not simply transferable from one location to another. For economic activities, place is increasingly irrelevant. Topography, location and function within a bioregion, or local ecological features do not enter into economic calculations except as simple functions of transportation costs or comparative advantage. Production is transferable, and the preferred location is anywhere production costs are the lowest.

Another difference between economics and ecology is that they are measured in different units. The unifying measure of market economics is money. Progress is measured in monetary units that everyone uses and understands to some degree. Biophysical systems are measured in physical units such as calories of energy, CO_2 absorption, centimeters of rainfall, or parts per million of nitrate contamination. Focusing only on the economic value of resources while ignoring biophysical health may mask serious changes in environmental quality or function.

Economics, Environment, and Developing Nations

As previously mentioned, the earth's "natural capital," on which humankind depends for food, security, medicines, and for many industrialized products, is its biological diversity. The majority of this diversity is in the developing world and much of it is threatened by exploitation and development. In order to pay for development projects, many economically poorer nations are forced to borrow money from banks in the developed world.

So great is the burden of external debt that many developing nations see little option but to overexploit their natural resource base. By 1998, the debt in the developing nations had risen to over US$1645 billion, a figure almost half their

Placing a Value on Ecosystem Services

There are many services provided by functioning ecosystems that are taken for granted. Protection of watersheds by forested land has long been known to be of great value. New York City found that it could provide water to its residents less expensively by protecting the watershed from which the water comes rather than by building expensive water purification plants to clean water from local rivers. Ducks Unlimited, an organization that supports waterfowl hunting, uses money provided by its members to protect nesting habitat for ducks and geese. Many countries have planted trees to help remove carbon dioxide from the air. All of these services can be converted into monetary terms, since it takes money to purify water, purchase land, and purchase and plant trees.

Since choices between competing uses for ecosystems often are determined by financial values assignable to ecosystems, it is important to have some kind of idea about the value of the "free" services provided by functioning ecosystems. Many environmental thinkers have begun to try to put a value on the many services provided by intact, functioning ecosystems. Obviously this is not an easy task and many will belittle these initial attempts to put monetary values on ecosystem services, but it is an important first step in forcing people to consider the importance of ecosystem services when making economic decisions about how ecosystems should be used. The following table represents approximate values for ecosystems services assigned by a panel of experts including ecologists, geographers, and economists. The total of $33 trillion per year is an estimate which many consider to be low. The current world GNP is about $18 trillion per year. Therefore the "free" services of ecosystems must not be overlooked when decisions are made about land use and how natural resources should be managed.

Categories of services	Examples of services	Estimated yearly value (trillion 1994 U.S. dollars)
Soil formation	Weathering, organic matter	17.1
Recreation	Outdoor recreation, eco-tourism, sport fishing	3.0
Nutrient cycling and waste treatment	Nitrogen, phosphorus, organic matter	2.4
Water services	Irrigation, industry, transportation, watersheds, reservoirs, aquifers	2.3
Climate regulation	Greenhouse gas regulation	1.8
Refuges	Nursery areas for animals, stopping places during migration, overwintering areas	1.4
Disturbance regulation and erosion control	Protection from storms and floods, drought recovery	1.2
Food and raw materials	Hunting and gathering, fishing, lumber, fuel wood, food for animals	0.8
Genetic resources	Medicines, reservoir of genes for domesticated plants and animals	0.8
Atmospheric gas balance	Carbon dioxide, oxygen, ozone, sulfur oxides	0.7
Pollination and pest control	Increased production of fruits	0.5
Total		33*

*Disparity is due to rounding.

Source: Data from Costanza, R. et al. "The Value of the World's Ecosystem Services and Natural Capital," in *Nature* Vol. 387 (1997).

collective gross national product. The debt burdens have led what investment there is in many developing countries to projects with safe, short-term returns and programs absolutely necessary for immediate survival. Environmental impacts are often neglected, the view being that severely indebted countries cannot afford to pay attention to environmental costs until other problems are resolved. This strategy suggests that environmental problems can be "corrected" once a country has reached a higher income level, but it ignores the growing realization that environmental impacts frequently cause international problems. Many countries under pressure from their debt crisis feel forced to over-exploit their natural resources, rather than manage them sustainably.

One new method of helping manage a nation's debt crises is referred to as debt-for-nature exchange. Debt-for-nature exchanges are an innovative mechanism for addressing the debt issue while encouraging investment in conservation and sustainable development. The exchanges, or swaps, allow debt to be bought at discount but redeemed at a premium, in local currency, for use in conservation and sustainable development projects. Debt-for-nature originated in 1987, when a nonprofit organization, Conservation International, bought $650,000 of Bolivia's foreign debt in exchange for Bolivia's promise to establish a national park. By 1998, at least 16 debtor countries—in the Caribbean, Africa, Eastern Europe, and Latin America—had made similar deals with official and nongovernmental organizations. By 1996, nearly US$120 million of debt around the world had been purchased at a cost of some US$21 million

The Tragedy of the Commons

The problems inherent in common ownership of resources were outlined by biologist Garrett Hardin in a now classic essay entitled "The Tragedy of the Commons" (1968). The original "commons" were areas of pastureland in England that were provided free by the king to anyone who wished to graze cattle.

There are no problems on the commons as long as the number of animals is small in relation to the size of the pasture. From the point of view of each herder, however, the optimal strategy is to enlarge his or her herd as much as possible: If my animals do not eat the grass, someone else's will. Thus, the size of each herd grows, and the density of stock increases until the commons becomes overgrazed. The result is that everyone eventually loses as the animals die of starvation. The tragedy is that, even though the eventual result should be perfectly clear, no one acts to avert disaster. In a democratic society, there are few remedies to keep the size of herds in line.

The ecosphere is one big commons stocked with air, water, and irreplaceable mineral resources—a "people's pasture," but a pasture with very real limits. Each nation attempts to extract as much from the commons as possible while enough remains to sustain the herd. Thus, the United States and other industrial nations consume far more than their share of the total world resource harvest each year, much of it imported from less-developed nations. The nations of the world compete frantically for all the fish that can be taken from the sea before the fisheries are destroyed. Each nation freely uses the commons to dispose of its wastes, ignoring the dangers inherent in overtaxing the waste-absorbing capacity of rivers, oceans, and the atmosphere.

The tragedy of the commons also operates on an individual level. Most people are aware of air pollution, but they continue to drive their automobiles. Many families claim to need a second or third car. It is not that these people are antisocial; most would be willing to drive smaller or fewer cars if everyone else did, and they could get along with only one small car if public transport were adequate. But people frequently get "locked into" harmful situations, waiting for others to take the first step, and many unwittingly contribute to tragedies of the commons. After all, what harm can be done by the birth of one more child, the careless disposal of one more beer can, or the installation of one more air conditioner?

---

but redeemed for the equivalent of US$64 million. This money was used to establish biosphere reserves and national parks, develop watershed protection programs, build inventories of endangered species, and develop environmental education.

In debt-for-nature exchanges, debtor countries benefit from the reduction of their foreign-currency debt obligations and add to their expenditure invested at home. The conservation investor receives a premium on the investment. This can be used for conservation and to establish sustainable development projects. Creditor banks gain by converting their nonpaying debts. Although they receive only part of their initial loan, some return is better than a total loss.

The primary goal of debt-for-nature exchanges has not been debt reduction but the funding of natural-resource-management investment. The contribution made by exchanges could increase, as in the case of the Dominican Republic, where 10 percent of the country's outstanding foreign commercial debt is to be redeemed by exchanges. Although eliminating the debt crisis alone is no guarantee of investment in environmentally sound projects, instruments like debt-for-nature exchanges can, on a small scale, reduce the mismanagement of natural resources and encourage sustainable development.

Attitudes of banks in the industrialized nations also seem to be changing. For example, the World Bank, which lends money for Third World development projects, has long been criticized by environmental groups for backing large, ecologically unsound programs, such as a cattle-raising project in Botswana that led to overgrazing. During the past few years, however, the World Bank has been factoring environmental concerns into its programs. One product of this new approach is an environmental action plan for Madagascar. The 20-year plan, which has been drawn up jointly with the World Wildlife Fund, is aimed at heightening public awareness of environmental issues, setting up and managing protected areas, and encouraging sustainable development.

Lightening the Load

Ship captains pay careful attention to a marking on their vessels called the Plimsoll line. If the water level rises above the Plimsoll line, the boat is too heavy and is in danger of sinking. When the line is submerged, rearranging items on the ship will not help much. The problem is the total weight, which has surpassed the carrying capacity of the ship.

This analogy points out that human activity can reach a scale that the earth's natural systems can no longer support. In 1992, more than 1600 scientists, including 102 Nobel laureates, underscored this point by collectively signing a "Warning to Humanity." Their warning stated in part that "a new ethic is required, a new attitude towards discharging our responsibility for caring for ourselves and for the earth. . . . This ethic must motivate a great movement, convincing reluctant leaders and reluctant governments and reluctant peoples themselves to effect the needed changes."

Such a new successful global effort to lighten humanity's load on the earth would need to directly address three major driving forces of environmental decline: the inequitable distribution of income, resource consumptive economic growth, and rapid population growth. It would redirect technology and trade to buy time for this great change to occur. Although there is much to say about each of these challenges, some key points bear noting.

Wealth inequality may be the most difficult problem, since it has existed for

The nets used for trawling for shrimp unfortunately do not only catch shrimp. One historic victim of such nets has been turtles, including the endangered Kemp's Ridley sea turtle as well as threatened loggerhead and endangered green sea turtles. Until the mid-1990s shrimp boats traveling the South Atlantic and Gulf of Mexico could accidentally catch an estimated 45,000 sea turtles in their nets, of which some 12,000 would drown.

The U.S. National Marine Fisheries Service (NMFS), looking for a technological innovation to stop the accidental netting and killing of the turtles, developed a device to keep turtles out of shrimp nets. The turtle excluder device, or TED, attaches to standard shrimp nets. A TED is a grid of bars with an opening either at the top or bottom. The grid is fitted into the neck of a shrimp trawl. Small animals like shrimp slip through the bars and are caught in the bag end of the trawl. Large animals such as turtles and sharks, when caught at the mouth of the trawl, strike the grid bars and are ejected through the opening. Data compiled by the NMFS show that TEDs can reduce turtle captures in shrimp nets by 97 percent with minimal loss of shrimp.

TEDs were first introduced to the shrimp industry on a voluntary basis in 1982. TEDs, however, were not widely accepted and many fishermen claimed that it was not fair for only United States fishermen to use them while their counterparts in Mexico did not. The U.S. fishermen argued that they were operating under an unfair economic policy. The shrimpers who opposed the TEDs also claimed that the devices were expensive and dangerous, and cut down on their catches.

TEDs were redesigned with input from the fishermen and there soon followed legislation replacing the voluntary program with a mandated one. The NMFS ensured that the TED requirements were phased in gradually to minimize the impact on the fishery. By 1998 inspectors from the NMFS uncovered only 98 TED violations in 2724 visits aboard shrimp boats.

In 1993 Mexico also mandated the use of TEDs. By the late 1990s the use of TEDs had spread to many parts of the world. The NMFS and the U.S. Department of State have worked closely with Mexico and the other shrimp-supplying nations in Latin America to help them develop comparable TED programs. TEDs have also been successfully implemented in Thailand, Malaysia, and the Philippines. After actual demonstration and dissemination of results of experiments with TEDs, shrimp fishermen did not resist their use and were understanding of the necessity for using them.

- Can you think of other examples where the development of a new technology has ensured economic growth without endangering one or more species?
- Is it possible to always have a "technological fix"?

Trawling for shrimp also kills several endangered and threatened species of turtles in the South Atlantic and Gulf of Mexico.

centuries. The difference today, however, is that the future of rich and poor alike depends on reducing poverty and thereby eliminating this driving force of global environmental decline. In this way, self-interest joins ethics as a motive for redistributing wealth, and raises the chances that it might be done.

Important actions to narrow the income gap must include reducing Third World debt. This was talked about a great deal in the 1980s, but little was accomplished. In addition, the developed nations must focus more foreign aid, trade, and international lending policies directly on improving the living standards of the world's poor.

A key description for reducing the kinds of economic growth that harm the environment is the same as that for making technology and trade more sustainable: internalizing environmental costs. If this is done through the adoption of environmental taxes, such as taxing based on pollution emitted, governments could avoid imposing heavier taxes overall by lowering income taxes accordingly. In addition, establishing better measures of economic accounting is critical. Since the calculations used to produce the gross national product do not account for the destruction or depletion of natural resources, this popular economic measure is extremely misleading. It tells us we are making progress even as our ecological foundations are being diminished. A better guide toward a sustainable path is essential. The United Nations and several governments have been working to develop better accounting methods, and while the progress has been slow, there is growing hope in the heightened awareness that a change is necessary.

As our discussion has shown, the economics of environmental problems is complex and difficult. The single most difficult problem to overcome is the assignment of an appropriate economic value to resources that have not previously been examined from an economic perspective. When air, water, scenery, and wildlife are assigned an economic value, they are looked at from an entirely different point of view.

Summary

Risk assessment is the use of facts and assumptions to estimate the probability of harm to human health or the environment that may result from exposures to pollutants, toxic agents, or management decisions. While it is difficult to calculate risks, risk assessment is used in risk management, which analyzes factors in decision making. The politics of risk management focus on the adequacy of scientific evidence, which is often open to divergent interpretations. In assessing risk, people frequently overestimate new and unfamiliar risks, while underestimating familiar ones.

To a large degree, environmental problems can be viewed as economic problems. Economic policies and concepts, such as supply/demand and subsidies, play important roles in environmental decision making. Another important economic tool is cost-benefit analysis. Cost-benefit analysis is concerned with whether a policy generates more social benefits than social costs. Criticism of cost-benefit analysis is based on the question of whether everything has an economic value. It has been argued that, if economic thinking dominates society, then even non-economic values, like beauty, can survive only if a monetary value is assigned to them.

A newer school of economic thought is referred to as sustainable development. Sustainable development has been defined as actions that address the needs of the present without compromising the ability of future generations to meet their own needs. Sustainable development requires choices based on values.

Pollution is extremely costly. When the costs are imposed on society, they are referred to as external costs. Costs of pollution control include pollution-prevention costs and pollution costs. Prevention costs are less costly, especially from a societal perspective.

Economists have stated that, when everyone shares ownership of a resource, there is a strong tendency to overexploit and misuse the resource. This concept was developed by Garrett Hardin in "The Tragedy of the Commons."

Recently, a market approach to curbing pollution has been proposed that would assign a value to not polluting, thereby introducing a profit motive to pollution reduction. Critics of this approach argue that pollution "permits" are nothing more than an acceptance of pollution of public resources.

Economic concepts are also being applied to the debt-laden developing countries. One such approach is the debt-for-nature swap. This program, which involves transferring loan payments for land that is later turned into parks and wildlife preserves, is gaining popularity.

Review Questions

1. How is risk assessment used in environmental decision making?
2. What is incorporated in a cost-benefit analysis? Develop a cost-benefit analysis for a local issue.
3. What are some of the concerns about the use of cost-benefit analysis in environmental decision making?
4. What concerns are associated with sustainable development?
5. What are some examples of environmental external costs?
6. Define what is meant by pollution-prevention costs.
7. Define the problem in common property resource ownership. Provide some examples.
8. Describe the concept of debt-for-nature.

Critical Thinking Questions

1. If you were a regulatory official, what kind of information would you require in order to make a decision about whether a certain chemical was "safe" or not? What level of risk would you deem acceptable for society? For yourself and your family?
2. Why do you suppose some carcinogenic agents, like those in cigarettes, are so difficult to regulate?
3. Imagine you were assessing the risk of a new chemical plant being built along the Mississippi River in Louisiana. Identify some of the risks that you would want to assess. What kinds of data would you need to assess whether the risk was acceptable, or not? Do you think that some risks are harder to quantify than others? Why?
4. Granting polluting industries or countries the right to buy and sell emissions permits is a controversial idea. Some argue that the market is the best way to limit pollution. Others argue that trade in permits allows polluting industries to continue to pollute and concentrates that pollution. What do you think?
5. Imagine you are an independent economist who is conducting a cost/benefit analysis of a hydroelectric project. What might be the costs of this project? The benefits? How would you quantify the costs of the project? The benefits? What kinds of costs and benefits might be hard to quantify, or might be too tangential to the project to figure into the official estimates?
6. Do you think environmentalists should stretch traditional cost-benefit analysis to include how development impacts the environment or shouldn't they? What are the benefits to this? The risks?
7. Looking at your own life, what kinds of risks do you take? What kinds would you be unwilling to take? What criteria do you use to make a decision about acceptable and unacceptable risk?
8. Is current worldwide growth and development sustainable? If there was less growth, what would be the effect on developing countries? How could we achieve a just distribution of resources and still limit growth?
9. Should our policies reflect an interest in preserving resources for future generations? If so, what level of resources should be preserved? What would you be willing to do without in order to save for the future?

Key Terms

carcinogens *311*

cost-benefit analysis *317*

demand *313*

external costs *323*

negligible risk *312*

pollution costs *323*

pollution prevention *323*

probability *309*

resources *313*

risk assessment *309*

risk management *311*

subsidy *316*

supply *313*

supply/demand curve *313*

sustainable development *319*

Related Internet Sites

Visit our website at http://www.mhhe.com/environmentalscience/ for specific information on each of the topics below.

Stratospheric ozone and human health

Domestic and international policy-making

International development and the environment

Television programs about science and the environment

Atlanta's experiment with alternative fuels

Information about the recycling industry

Opinion and policymakers and community and business leaders

In-depth information on wildlife and animal life

Animal rights news and information

Environmental science and public policy

Air Pollution

OBJECTIVES

After reading this chapter, you should be able to:

- Explain why air can accept and disperse significant amounts of pollutants.
- List the major sources and effects of the five primary pollutants.
- Describe how photochemical smog is formed and how it affects humans.
- Explain how PCV valves, APC valves, catalytic converters, scrubbers, precipitators, filters, and changes in fuel types reduce air pollution.
- Explain how acid rain is formed.
- Understand that humans can alter the atmosphere in such a way that the climate may change.
- Understand the link between chlorofluorocarbon use and ozone depletion.
- Recognize that enclosed areas can trap air pollutants that are normally diluted in the atmosphere.

CHAPTER OUTLINE

The Atmosphere

Primary Air Pollutants

Global Perspective: *Air Pollution in Mexico City*

    Carbon Monoxide (CO)
    Hydrocarbons (HC)
    Particulates
    Sulfur Dioxide (SO_2)
    Oxides of Nitrogen (NO and NO_2)

Photochemical Smog

Other Significant Air Pollutants

Control of Air Pollution

Acid Deposition

Environmental Close-Up: *The 1990 Clean Air Act*

Environmental Close-Up: *Secondhand Smoke*

Global Warming and Climate Change
    Worsening Health Effects
    Rising Sea Level
    Disruption of the Water Cycle
    Changing Forests and Natural Areas
    Challenges to Agriculture and the Food
      Supply

Environmental Close-Up: *Aesthetic Pollution*

Ozone Depletion

Indoor Air Pollution

Environmental Close-Up: *Radon*

Environmental Close-Up: *Noise Pollution*

Issues and Answers: *International Air Pollution*

THE ATMOSPHERE

The atmosphere, or air, is normally composed of 79 percent nitrogen, 20 percent oxygen, and a 1 percent mixture of carbon dioxide, water vapor, and small quantities of several other gases. Most of the atmosphere is held close to the earth by the pull of gravity, so it gets thinner with increasing distance from the earth. (See figure 17.1.)

Even though gravity keeps the air near the earth, the air is not static. As it absorbs heat from the earth, it expands and rises. When its heat content is radiated into space, the air cools, becomes more dense, and flows toward the earth. As the air circulates due to heating and cooling, it also moves horizontally over the surface of the earth because the earth rotates on its axis. The combination of all air movements creates the wind patterns characteristic of different regions of the world. (See figure 17.2.)

As we discussed in chapter 11, pollution is something produced by humans that interferes with our well-being. Because we cause pollution, we may be able to do something to prevent it. There are several natural sources that degrade the quality of the air, such as gases and particles from volcanoes, dust from natural sources, or odors from decomposition of dead plants and animals. However, since these activities are not controlled by humans, they do not fit our definition of pollution. Automobile emissions, chemical odors, and factory smoke are considered air pollution, however, and we will focus on these and similar examples.

The problem of air pollution is directly related to the number of people living in an area and the kinds of activities they engage in. When a population is small and its energy use is low, the impact of people is minimal. Their pollution is diluted and the overall negative effect is slight. However, our urbanized, industrialized civilization has a growing population and a history of increasing use of fossil fuels and technological aids. We release large quantities of polluting by-products into our environment as we manufacture products demanded by the population.

Gases or small particles released into the atmosphere are likely to be mixed,

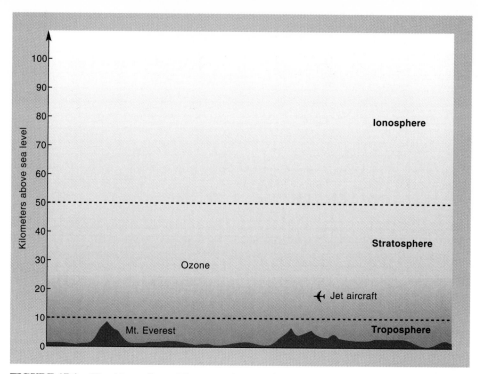

FIGURE 17.1　The Atmosphere The atmosphere is divided into the troposphere, the relatively dense layer of gases close to the surface of the earth; the stratosphere, more distant with similar gases but less dense; and the ionosphere, composed of ionized gases.

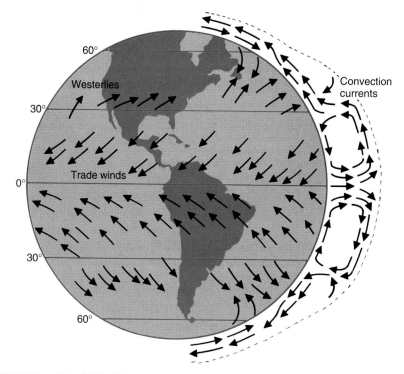

FIGURE 17.2　Global Wind Patterns Wind is the movement of air caused by temperature differences and the rotation of the earth. Both of these contribute to the patterns of world air movement. In North America, most of the winds are westerlies (from the west to the east).

diluted, and circulated, but they are likely to stay near the earth due to gravity. When we put a material into the air, we do not get rid of it; we just dilute it and move it out of the immediate area. When people lived in small groups, the smoke from their fires was diluted. It was produced in such low concentrations that it did not interfere with neighboring groups downwind. In industrialized urban areas, pollutants cannot always be diluted before the air reaches another city. The polluted air from Chicago is further polluted by Gary, Indiana, supplemented by the wastes of Detroit and Cleveland, and finally moves over southeastern Canada and New England to the ocean. (See figure 17.3.) While not every population center adds the same kind or amount of waste, each adds to the total load carried.

Air pollution is not just an aesthetic problem. It also causes health problems. Hundreds of deaths have been directly related to poor air quality in cities. A well-documented case of pollution that was harmful to human health occurred in Donora, Pennsylvania. (See figure 17.4.) The city of Donora is located in a valley. In October 1948, the pollutants from a zinc plant and steel mills became trapped in the valley, and a dense smog formed. Within five days, seventeen people died and 5910 persons became ill. The polluted atmosphere affected nearly 50 percent of the city's 12,300 inhabitants. The technology that provided jobs for the people was also killing them.

Many of the megacities of the developing world have extremely poor air quality. Beijing, Seoul, Mexico City, and Cairo exceed World Health Organization guidelines for air quality for at least two pollutants. The causes of this air pollution are open fires, large numbers of poorly maintained motor vehicles, and poorly regulated industrial plants. The World Health Organization estimates that particulates in Mexico City contribute to 6400 deaths each year. Not only does poor air quality in such cities increase the death rate, but the general health of the populace is lowered. Chronic coughing and susceptibility to infections are common in these cities. Deaths from air pollution occur primarily among the elderly, the infirm, and the very young. Bronchial inflammations, allergic

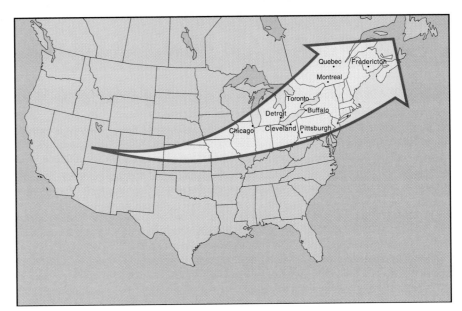

FIGURE 17.3   Accumulation of Pollutants As an air mass moves across the continent from west to east, each population center adds its pollutants to the total load in the atmosphere.

FIGURE 17.4   Air Pollution Donora, Pennsylvania was the scene of a serious air-pollution incident. The pollutants from industry were trapped by an inversion, and almost half of the population became ill.

reactions, and irritation of the mucous membranes of the eyes and nose all indicate that air pollution must be reduced.

PRIMARY AIR POLLUTANTS

Around the world, five major types of materials are released directly into the atmosphere in their unmodified forms and in sufficient quantities to pose a health risk. They are carbon monoxide, hydrocarbons, particulates, sulfur dioxide, and nitrogen compounds. This group of pollutants is known as **primary air pollutants.** (See table 17.1.) These materials may interact with one another in the presence of an energy source to form new **secondary air pollutants** such as ozone and other

Air Pollution in Mexico City

Mexico City has been labeled the city with the worst air pollution ever recorded. The air over Mexico City exceeded ozone limits set by the World Health Organization on more than 300 days in one year. With about 20 million people, Mexico City is one of the largest cities in the world and grows larger each day. Open sewers and garbage dumps contribute dust and bacteria to the atmosphere. The city has about 35,000 factories and 3.6 million vehicles. Most of these are older models that are poorly tuned and, therefore, pollute the air with a mixture of hydrocarbons, carbon monoxide, and nitrogen oxides. A large proportion of the air pollution is the result of automobiles. A major source of hydrocarbons is from leakage of LP gas. Fixing the leaks would significantly reduce the hydrocarbon level in the atmosphere. The high altitude (over 2000 meters) results in even greater air pollution from automobiles because automobile engines do not burn fuel efficiently at such high altitudes. Mexico City's location in a valley also allows for conditions suitable for thermal inversions during the winter.

Many foreign companies and governments give special "hazard pay" for working in Mexico City because of the air pollution. Pediatricians estimate that 85 percent of childhood illnesses are related to air pollution and say that the only way to improve the health of many of the children is to get them out of the city. The government has responded with a comprehensive program to clean up the air in the city. Twenty-five million trees have been planted to help clean the air, and 4400 hectares of land have been purchased to provide green space for the city.

Public information campaigns encourage people to keep their automobiles tuned to reduce air pollution, lead-free gasoline has been made available and is being used, and catalytic converters are now required on all automobiles manufactured after 1991. Taxis manufactured before 1985 have been banned, and by the year 2000, all pre-1991 taxis must be replaced.

Personal automobiles and taxis are prohibited from being driven on the streets one day a week. This encourages people to carpool or use public transportation. The government is planning to improve public transportation to make it more attractive for people to switch from private automobiles to public transport. Both actions should reduce the number of automobiles releasing pollutants into the air. A polluting, government-owned oil refinery was shut down, and power plants and many industries have switched from oil to natural gas, which pollutes less.

These actions have had significant effects on the quality of the air. Although ozone in the air continues to be a major problem, as it is in metropolitan areas around the world, lead levels, carbon monoxide, and sulfur dioxide have been brought under control. In order to further improve air quality, increasingly strong restrictions on polluting industries and further restrictions on the use of private automobiles will be necessary. The Mexican government is considering such actions.

---

very reactive materials. Secondary air pollutants also form from reactions with natural chemicals in the atmosphere.

Carbon Monoxide (CO)

Carbon monoxide (CO) is produced when organic materials, such as gasoline, coal, wood, and trash, are incompletely burned. The single largest source of carbon monoxide is the automobile. (See figure 17.5.) Although increased fuel efficiency and the use of catalytic converters have reduced carbon monoxide emissions per kilometer driven, carbon monoxide remains a problem because the number of automobiles and the number of kilometers driven have increased. In many parts of the world, automobiles are poorly maintained and may have inoperable pollution control equipment, resulting in even greater amounts of carbon monoxide.

The next largest source of carbon monoxide is smoking tobacco. Currently in the United States and some other countries, there is a great deal of pressure to restrict areas where smoking is permitted to

TABLE 17.1 SOURCES OF PRIMARY AIR POLLUTANTS	
Pollutant	**Sources**
Carbon monoxide	Incomplete burning of fossil fuels
	Tobacco smoke
Hydrocarbons	Incomplete burning of fossil fuels
	Tobacco burning
	Chemicals
Particulates	Burning fossil fuels
	Farming operations
	Construction operations
	Industrial wastes
	Building demolition
Sulfur dioxide	Burning fossil fuels
	Smelting ore
Nitrogen compounds	Burning fossil fuels

minimize exposure to secondhand cigarette smoke. Restaurants designate nonsmoking sections (some even advertise themselves as smoke-free), public buildings have designated nonsmoking areas, and corporations and colleges are designating their buildings as smoke-free. Smoking is decreasing in the industrialized world today, but in the developing nations, smoking retains its image of

glamour and sophistication as a result of extensive marketing campaigns by cigarette companies.

Several hours of exposure to air containing 0.001 percent of carbon monoxide can cause death. Because carbon monoxide remains attached to hemoglobin for a long time, even small amounts tend to accumulate and reduce the blood's oxygen-carrying capacity. The amount of carbon monoxide produced in heavy traffic can cause headaches, drowsiness, and blurred vision. A heavy smoker in congested traffic is doubly exposed and may experience severely impaired reaction time compared to nonsmoking drivers.

Fortunately, carbon monoxide is not a persistent pollutant. Natural processes convert carbon monoxide to other compounds that are not harmful. Therefore, the air can be cleared of its carbon monoxide if no new carbon monoxide is introduced into the atmosphere.

Hydrocarbons (HC)

In addition to carbon monoxide, automobiles emit a variety of **hydrocarbons (HC).** Hydrocarbons are a group of organic compounds consisting of carbon and hydrogen atoms. They are either evaporated from fuel supplies or are remnants of fuel that did not burn completely. The internal combustion engine is the major source of hydrocarbons, although refineries and other industries add hydrocarbons to the total atmospheric burden. Hydrocarbons in the atmosphere are no great problem. Most of them are washed out of the air when it rains and run off into surface water. They cause an oily film on surfaces, but hydrocarbons do not generally cause more than nuisance problems, except when they react to form secondary pollutants.

Many modifications to automobile engines have reduced the loss of hydrocarbons to the atmosphere. Recycling some gases through the engine, using higher oxygen concentrations in the fuel-air mixture, and using valves to prevent the escape of gases are three of these modifications. In addition, catalytic converters burn exhaust gases more completely so that fewer hydrocarbons leave the tail pipe.

Particulates

Particulates, small pieces of solid materials dispersed into the atmosphere, constitute the third largest category of air pollutants. Smoke particles from fires, bits of asbestos from brake linings and insulation, dust particles, and ash from industrial plants contribute to the particulate load. Particulates cause problems ranging from the annoyance of soot settling on a backyard picnic table to the **carcinogenic**

FIGURE 17.5 Carbon Monoxide The major source of carbon monoxide, hydrocarbons, and nitrogen oxides is the internal combustion engine, which is used to provide most of our transportation. The more concentrated the number of automobiles, the more concentrated the pollutants. Carbon monoxide concentrations of a hundred parts per million are not unusual in rush-hour traffic in large metropolitan areas. These concentrations are high enough to cause fatigue, dizziness, and headaches.

(cancer-causing) effects of asbestos. Particulates frequently get attention because they are so readily detected by the public. Heavy black smoke from a factory can be seen without expensive monitoring equipment and generally causes an outcry, whereas the production of colorless gases like carbon monoxide and sulfur dioxide goes unnoticed.

Particulates can accumulate in the lungs and interfere with the ability of the lungs to exchange gases. However, this lung damage usually happens to people who are repeatedly exposed to large amounts of particulate matter on the job. Miners and others who work in dusty conditions are most likely to be affected. For most of the population, particulates affect health by acting as centers for the deposition of moisture and gases from the atmosphere. As we breathe air containing particulates, we come in contact with concentrations of other potentially more harmful materials that have accumulated on the particulates. Sulfuric, nitric, and carbonic acids, which irritate the lining of our respiratory system, frequently form on particulates.

Sulfur Dioxide (SO$_2$)

Sulfur dioxide (SO$_2$) is a compound of sulfur and oxygen that is produced when sulfur-containing fossil fuels are burned. Coal and oil were produced from organisms that had sulfur in their living structure. When the coal or oil was formed, some of the sulfur was incorporated into the fossil fuel. The sulfur is released as sulfur dioxide when the fuel is burned. Sulfur dioxide has a sharp odor and irritates respiratory tissue. It also reacts with water, oxygen, and other materials in the air to form sulfur-containing acids. The acids can become attached to particles, which, when inhaled, are very corrosive to lung tissue. In 1306, Edward I of England banned the burning of "sea coles," coal found on the sea shore, in the city of London. This coal was high in sulfur content and was, in part, responsible for the city's noxious odors. Edward's ban might very well be the earliest environmental legislation concerning air quality.

London, England, was also the site of one of the earliest killer fogs. In 1952, the city was covered with a dense fog for several days. During this time, air over the city failed to mix with layers of air in the upper atmosphere due to temperature conditions. The factories continued to release smoke and dust into this stagnant layer of air, and it became so full of the fog, smoke, and dust that people got lost in familiar surroundings. This combination of smoke and fog has become known as smog. Many London residents developed respiratory discomfort, headaches, and nausea. Four thousand people died in a few weeks. Their deaths have been associated with the high levels of sulfur compounds in the smog. Thousands of others suffered from severe bronchial irritation, sore throats, and chest pains. The 1948 Donora, Pennsylvania, incident already mentioned also involved symptoms related to the particles and sulfur dioxide in the air.

Oxides of Nitrogen (NO and NO₂)

Oxides of nitrogen (NO and NO₂) are the fifth category of primary air pollutants. Several compounds contain nitrogen and oxygen in different combinations; nitrogen oxide (NO) and nitrogen dioxide (NO₂) are the most common. When combustion takes place in air, nitrogen and oxygen molecules from the air may react with each other, and oxides of nitrogen result.

$$N_2 + O_2 \rightarrow 2NO \text{ (nitrogen oxide)}$$
$$2NO + O_2 \rightarrow 2NO_2 \text{ (nitrogen dioxide)}$$

A mixture of nitrogen oxide and nitrogen dioxide is called NO_X. The nitrogen dioxide in the mixture reacts with other compounds to produce photochemical smog, discussed in the next section.

The primary source of nitrogen oxides is the automobile engine. Catalytic converters reduce the amount of nitrogen oxides released from the internal combustion engine, but increased automobile traffic has resulted in significant levels of NO_X in many metropolitan areas. Nitrogen oxides are noteworthy because they are involved in the production of secondary air pollutants.

PHOTOCHEMICAL SMOG

Secondary air pollutants are compounds that result from the interaction of various primary air pollutants. **Photochemical smog** is a mixture of pollutants resulting from the interaction of nitrogen oxide with ultraviolet light. (See figure 17.6.) The two most destructive components of

FIGURE 17.6 Photochemical Smog The interaction among hydrocarbons, oxides of nitrogen, and sunlight produces new compounds that are irritants to humans. The visual impact of smog is shown in these photographs taken in Los Angeles, CA.

photochemical smog are ozone (O₃) and peroxyacetylnitrates. Both of these materials are excellent oxidizing agents, which means that they react readily with many other compounds, including those found in living things, causing destructive changes. Ozone is particularly harmful because it destroys chlorophyll in plants and injures lung tissue in humans and other animals. Peroxyacetylnitrates, in addition to being oxidizing agents, are eye irritants. Both ozone and peroxyacetylnitrates are secondary air pollutants that result from chemical reactions assisted by light.

A typical photochemical smog incident involves an interesting series of events.

1. Morning rush-hour traffic produces large amounts of nitrogen oxide (NO):

$$N_2 + O_2 \rightarrow 2NO$$

2. The nitrogen oxide reacts with molecular oxygen (O₂) from the atmosphere to form nitrogen dioxide (NO₂). It is the nitrogen dioxide in the atmosphere that gives photochemical smog its reddish-brown haze.

$$2NO + O_2 \rightarrow 2NO_2$$

3. Later in the morning, nitrogen dioxide reacts with ultraviolet light to form atomic oxygen (O):

$$NO_2 \xrightarrow[\text{light}]{\text{ultraviolet}} NO + O$$

4. Abundant molecular oxygen in the atmosphere reacts with atomic oxygen to form ozone:

$$O_2 + O \rightarrow O_3$$

5. Hydrocarbons in the atmosphere react with ozone to produce peroxyacetylnitrates. Various pollution-control devices now reduce the amount of hydrocarbons escaping to the atmosphere, thereby decreasing the amount of peroxyacetylnitrates formed.

6. As the ozone and peroxyacetylnitrates react with living things, they cause damage and are converted to less reactive molecules, and the smog eventually clears.

Due to their climate and the geographic features of their locations, such large metropolitan areas as Los Angeles, Salt Lake City, Phoenix, and Denver have more trouble with photochemical smog than do East Coast metropolitan areas. Each of these cities is ringed by mountains. The prevailing winds are from the west. As cool air flows into these valleys, it pushes the warm air upward. This warm air becomes sandwiched between two layers of cold air and acts like a lid on the valley, a condition known as a **thermal inversion.** The air is trapped in the valley. (See figure 17.7.) The lid of warm air cannot rise further because it is covered by a layer of cooler air pushing down on it. It cannot move out of the area because of the ring of mountains. Without normal air circulation, smog accumulates. Harmful chemicals continue to increase in concentration until a major weather change causes the lid of warm air to move up and over the mountains. Then the underlying cool air can begin to circulate, and the polluted air is diluted.

Smog problems could be substantially decreased by reducing the use of internal combustion engines (perhaps eliminating them completely) or by moving population centers away from the valleys that produce thermal inversions. Both of these solutions would require expenditures of billions of dollars and major changes in lifestyle; therefore, people will probably continue to live with the problem.

Normal situation

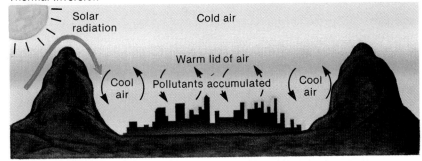

Thermal inversion

FIGURE 17.7   Thermal Inversion Under normal conditions, the air at the earth's surface is heated by the sun and rises to mix with the cooler air above it. When a thermal inversion occurs, a layer of warm air is formed above the cooler air at the surface. The cooler air is then unable to mix with the warm air above and cannot escape because of surrounding mountains. The cool air is trapped, sometimes for several days, and accumulates pollutants. If the thermal inversion continues, the levels of pollution can become dangerously high.

OTHER SIGNIFICANT AIR POLLUTANTS

In recent years, three other air pollutants have been recognized as significantly affecting the health and welfare of people: radon (see Environmental Close-Up, p. 352), lead, and toxic chemicals. The primary sources of lead are gasoline and paint. For many years lead was added to gasoline to help engines run more effectively. Recognition that lead emissions were hazardous resulted in the lead additives being removed from gasoline in North America and Europe. This has resulted in a decline in the amount of lead in the atmosphere. (See figure 17.8.) However, most other countries in the world still use leaded gasoline.

Another major source of lead is paints. Many older homes have paints that contain lead, since various lead compounds are colorful pigments. Dust from flaking paint or remodeling or demolition is released into the atmosphere. Although the amount of lead may be small, its presence in the home can result in significant exposure to inhabitants, particularly young children who chew on painted surfaces and often eat paint chips.

Air toxics are harmful chemicals that are released into the atmosphere on purpose or are released accidentally as a result of leaks or poorly designed manufacturing processes. Materials such as pesticides are purposely released to kill insects or other pests. However, the majority of air toxics are released as a result of manufacturing processes. Although air toxics are important to the entire public, they are most critical for people who are exposed on the job since they are likely to be exposed often and to higher concentrations. There are literally hundreds of different air toxics.

CONTROL OF AIR POLLUTION

All of the air pollutants we have examined thus far are produced by humans. That means their release into the atmosphere can be controlled. Methods of controlling air pollution depend upon the type of pollutant and the willingness or ability of industries, governments, and individuals to make changes. It is important to also note that positive change is possible. (See table 17.2.)

Eliminating photochemical smog completely would require large-scale changes in driving habits and other aspects of our lifestyle and culture. That is not likely to happen, but the problem can be lessened by reducing the particular primary air pollutants that contribute to photochemical smog: NO_X and hydrocarbons. United States government regulations have pressured the automobile industry to reduce emissions. The positive crankcase ventilation valve (PCV) and gas caps with air-pollution control valves (APC) reduce hydrocarbon loss. Catalytic converters reduce carbon monoxide, oxides of nitrogen, and hydrocarbons in emissions and necessitate the use of lead-free fuel. This lead-free fuel requirement, in turn, significantly reduces the amount of lead (and other metal additives) in the atmosphere.

Particulates are produced primarily in automobiles or by industries burning fuels. In industry, particulate release can be controlled with scrubbers, precipitators, and filters. (See box figure.) These devices are effective, but expensive. They can be retrofitted to the smokestack, or they can be designed into the combustion system. They filter the products of combustion and remove some, while allowing others to be released into the atmosphere. Electric charge differentials may also be used to remove the unwanted materials from the emission.

Another major source of particulates are dusts produced by a variety of industrial activities. Transfer of materials such as grain or coal from one container to another generates dust. Mining and other earth-moving activities, including farming operations, cause dust.

Industry and automobiles are not the only producers of particulate pollutants. Many people in the world use wood as their primary source of fuel for cooking and heat. In developed countries like the United States and Canada, some people use fireplaces and wood-burning stoves as a primary source of heat, but most use them for supplemental heat or for aesthetic

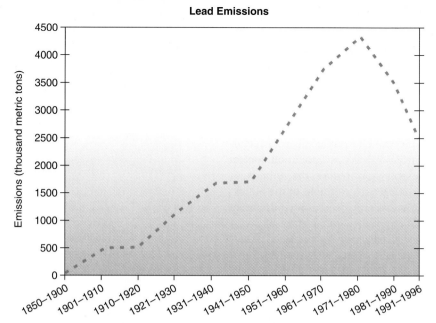

FIGURE 17.8   Lead Emissions The amount of lead released into the atmosphere has declined significantly since the early 1980s when the lead additive was removed from gasolines in North America and much of Europe.

Source: Data from *Science,* Vol. 272, April 12, 1996.

TABLE 17.2   TOP TEN CITIES FOR SMOG IMPROVEMENT IN UNITED STATES FROM 1987–1996

	Percent Reduction	Smog days 1987–1989	Smog days 1994–1996
1. Seattle	100.0	14.0	0.0
2. Buffalo	100.0	7.7	0.0
3. Tucson	100.0	4.0	0.0
4. Tacoma	95.5	7.3	0.3
5. Albuquerque	93.2	14.7	1.0
6. Denver	92.5	22.3	1.7
7. Miami	92.3	4.3	0.3
8. Boston	91.7	8.0	0.7
9. Raleigh-Durham, NC	90.9	3.7	0.3
10. Charlotte, NC	88.2	11.3	1.3

Source: Data from the Environmental Protection Agency.

purposes. However, the use of large numbers of wood-burning stoves and fireplaces can generate a significant air-pollution problem, called a brown cloud. Many municipalities, such as Boise, Idaho; Salt Lake City, Utah; and Denver, Colorado (and some entire states), use fines to enforce a ban on wood-burning during severe air-pollution episodes. Many other communities issue pollution alerts and request that people do not use wood-burners. Many of these communities have regulations about the number and efficiency of wood-burning stoves and fireplaces. Some

communities, such as Castle Rock, Colorado, prohibit the construction of houses with fireplaces or wood-burning stoves. Many people readily switch to gas fireplaces or high-efficiency wood-burners when they understand that their enjoyment of a rustic stove may be leading to a degraded environment, but most need to be forced to comply by official regulations and the threat of fines. Obviously, accumulations of many small sources of air pollution may cause as big a problem as one large emission source and are frequently more difficult to control.

To control sulfur dioxide, which is produced primarily by electric power generating plants, several possibilities are available. One alternative is to change from high-sulfur fuel to low-sulfur fuel. Switching from a high-sulfur coal to a low-sulfur coal reduces the amount of sulfur released into the atmosphere by 66 percent. Switching to oil, natural gas, or nuclear fuels would reduce sulfur dioxide emissions even more. However, these are not long-term solutions because low-sulfur fuels are in short supply, and nuclear power plants pose a different set of pollution problems.

A second alternative is to remove the sulfur from the fuel before the fuel is used. Chemical or physical treatment of coal before it is burned can remove nearly 40 percent of the sulfur. This is technically possible, but it increases the cost of electricity to the rate payer.

Scrubbing the gases emitted from a smokestack is a third alternative. The technology is available, but, of course, these control devices are costly to install, maintain, and operate. As with auto emissions, governments have required the installation of these devices, but when industries install them, the cost of construction and operation is passed on to the consumer. The cost of installing scrubbers on a typical power plant is about $200 million. In the United States, an estimated 20 million or more metric tons of sulfur oxides are released into the atmosphere each year. Installing scrubbers on just 50 of the largest coal-burning plants would reduce this amount by over one-third. Some utility companies question the effect of this pollutant and even deny that they are the main source of sulfur emissions.

In the past, a common solution was to build taller smokestacks. Tall stacks release their gases above the inversion layers and, therefore, add the sulfur dioxide to the upper atmosphere, where it is diluted before it comes in contact with the population downwind. This works as long as the stack is tall enough and as long as not so much pollution is added to the air that it cannot be diluted to an acceptable level. However, sulfur dioxide reacts with oxygen and dissolves in water in the atmosphere to form sulfuric acid. This acid is washed from the air when it rains or snows. It damages plants and animals and

Sources of emission

Transport and transformation

Organic acids
Hydrocarbons
Fine particulate matter
Carbon monoxide
Ammonia

SO₂

NOₓ

Natural

Stationary

Mobile

Acids deposited onto environmental surfaces

Nitric   Sulfuric

FIGURE 17.9   Acid Deposition Molecules from natural sources, power plants, and internal combustion engines react to produce the chemicals that are the source of acid deposition.

increases corrosion of building materials and metal surfaces.

ACID DEPOSITION

Acid deposition is the accumulation of potential acid-forming particles on a surface. Acids result from natural causes, such as vegetation, volcanos, and lightning; and from human activities, such as coal burning and use of the internal combustion engine. (See figure 17.9.) These combustion processes produce sulfur dioxide (SO_2) and oxides of nitrogen (NO_X). Oxidizing agents, such as ozone, hydroxyl ions, or hydrogen peroxide, along with water, are necessary to convert the sulfur dioxide or nitrogen oxides to sulfuric or nitric acid. Various reactive hydrocarbons (HC) encourage the production of oxidizing agents.

The acid-forming reactants are classified as wet or dry. Wet reactions occur in the atmosphere and come to earth as some form of precipitation: acid rain, acid snow, or acid dew. Dry deposition occurs with the settling of the precursors of the acid on a surface. An acid does not actually form until these materials mix with water. Even though the acids are formed and deposited in several different ways, all of these processes usually are referred to as **acid rain.**

Acid rain is a worldwide problem. Reports of high acid-rain damage have come from Canada, England, Germany, France, Scandinavia, and the United States. Rain is normally slightly acidic, with a pH between 5.6 and 5.7 due to atmospheric carbon dioxide that dissolves to produce carbonic acid. But acid rains sometimes have a concentration of acid a thousand times higher than normal. In 1969, New Hampshire had a rain with a pH of 2.1. In 1974, Scotland had a rain with a pH of 2.4. The average rain in much of the northeastern part of the United States and adjoining parts of Ontario has a pH between 4.0 and 4.5.

Acid rain can cause damage in several ways. Buildings and monuments are often made from materials that contain limestone (calcium carbonate, $CaCO_3$), because limestone is relatively soft and easy to work. Sulfuric acid (H_2SO_4), a major component of acid rain, converts limestone to gypsum ($CaSO_4$), which is more soluble and is eroded over many years of contact with acid rain. (See figure 17.10.) Metal surfaces can also be attacked by acid rain.

The effects of acid rain on ecosystems are often more difficult to quantify. Intense sulfur dioxide pollution around smelters is known to cause the death of many kinds of trees and other vegetation. But this is an extreme case and may not be directly comparable to less intense acid

ENVIRONMENTAL CLOSE-UP

The 1990 Clean Air Act

In 1990 the first major revision to the Clean Air Act (CAA) of 1977 was passed. While concessions were given to electrical utilities, the automobile industry, and certain urban areas with severe air-pollution problems, the final bill was considered to be strong. To date, implementation of CAA programs such as the acid rain control program and the program to control ozone-depleting substances has progressed mostly as expected. However, implementation of other CAA programs has proved to be more contentious. Based on CAA implementation, several issues related to the implementation of the 1990 amendments have emerged.

The Act was designed with a series of very tight deadlines under which states would have to respond to rules promulgated by the Environmental Protection Agency (EPA). Therefore, EPA rule-making delays became an important implementation issue. EPA delays caused states to miss deadlines for submitting their own State Implementation Plans (SIPs), causing states to delay the implementation of air-pollution control programs. By 1998, the Act had been effective at bringing nonattainment areas into compliance with national ambient air quality standards (NAAQS). More than 80 percent of the marginal nonattainment areas that were to have reached the NAAQS by 1994, were successful, and more than half of the moderate areas met the 1996 level.

States with more severe air-pollution problems have had less success implementing the requirements. Few states have been able to submit complete SIPs by the deadlines. However, most states have been able to complete SIP submissions within an 18-month "grace period."

Another important implementation issue under the CAA is the amount of flexibility the EPA has given states in allowing them to develop air-pollution control programs. In several programs, including

Control of Particulates The installation of proper pollution-control devices can significantly reduce the amount of particulate material released into the air. These photographs illustrate the difference before and after the installation of control devices.

permit programs, and automobile inspection and maintenance programs, EPA has granted more flexibility to states in complying with EPA rules. The added flexibility is intended to make the regulations less burdensome and costly for the regulated community.

There are still provisions of the CAA being debated, including the issue of cleaner burning or reformulated gasoline and the issue of stricter air-quality standards for both ground-level ozone and particulate matter. Opponents to provisions of the CAA have lobbied Congress with the argument that stricter enforcement of the CAA would have serious

---

rain. However, in many parts of the world, acid rain is suspected of causing the death of many forests and reducing the vigor and rate of growth of others. (See figure 17.11.) In Central Europe, many forests have declined significantly, resulting in the death of about 6 million hectares of trees. Northeastern North America has also seen significant tree death and reduction in vigor, particularly at higher elevations. Some areas have had 50 percent mortality of red spruce trees.

A clear link between the decline of the forests and acid rain is difficult to establish, but several hypotheses have been formulated. Molecules like sulfur dioxide and ozone are known sources of air pollution and cause direct damage to plants. The sulfur dioxide also contributes to acid formation. As soil becomes acidic, aluminum is

FIGURE 17.10 Damage Due to Acid Deposition Sulfuric acid (H_2SO_4), which is a major component of acid deposition, reacts with limestone ($CaCO_3$) to form gypsum ($CaSO_4$). Since gypsum is water-soluble, it washes away with rain. The damage to this monument is the result of such acid reacting with the stone.

economic consequences. The EPA has continued to emphasize that, under the CAA, they are not to consider costs at the stage when standards are set. The EPA's economic analysis estimates the total cost of compliance with the new standards put forth in 1997 at over $9 billion annually, $8.6 billion for particulate matter and $1.1 billion for ozone. However, other estimates by private organizations eclipse the EPA's numbers. One thing is for sure, the debate will continue into the next century.

Major provisions	Consequences
1. Reduce urban smog by 15 percent by 1996 and 3 percent per year until federal air-quality standards are met. Some cities with severe smog problems, like Los Angeles, have up to twenty years to comply.	Many changes needed, which will require improved technology and additional costs.
2. Utilities must reduce release of sulfur dioxide by one-half by the year 2000.	Utilities will need to spend $3 billion a year for low-sulfur coal and pollution-control devices.
3. Utilities must reduce release of nitrogen oxides by one-third by the year 2000, starting in 1992.	
4. Utilities can buy and sell "pollution credits." Dirty plants can purchase permits to pollute from utilities that are not using their full allotment of pollution credits.	Economic incentives will exist for utilities to clean up their emissions. In special cases, they may be able to prolong the use of a dirty plant by purchasing permits to pollute.
5. Passenger cars must emit 60 percent less nitrogen oxide and 40 percent less hydrocarbons by the year 2003.	Increased cost of fuel.
6. Auto-emission controls must last for 100,000 miles	Increased cost for new cars.
7. On-board canisters are required to capture vapors during refueling.	
8. Cleaner-burning fuels will be required in the most polluted cities (Baltimore, Chicago, Hartford, Houston, Los Angeles, Milwaukee, New York, Philadelphia, and San Diego).	Oil companies will need to develop new fuels.
9. One million special low-emission vehicles (primarily in fleets) must be present by the year 2000. Three hundred thousand private low-emission vehicles must be in California by 1999.	New automotive technology will need to be developed. Costs will be passed on to the consumer.
10. Toxic emissions must be reduced by 90 percent by the year 2000. Factories must install maximum-achievable control technology. This includes small businesses.	Many small businesses may need help to meet this standard. Additional costs will be passed on to the consumer as higher prices.
11. Production of chlorofluorocarbons (CFCs), halons, and carbon tetrachloride will be banned by the year 2000, methyl chloroform by the year 2002. Recycling of chemicals will be required.	Alternative chemicals will need to be developed. Additional costs will be passed on to the consumer for recycling equipment and higher-priced substitutes.

released from binding sites and may interfere with the plant roots' ability to absorb nutrients. A recent long-term study in New Hampshire strongly suggests that the many years of acid precipitation have reduced the amount of calcium in the soil, which is needed by plants for growth. Because there are no easy ways to replace the calcium even if acid rain were to stop, it would still take many years for the forests to return to health. Reduction in the pH of the soil may also change the kind of bacteria in the soil and reduce the availability of nutrients for plants. While none of these factors alone would necessarily result in plant death, each could add to the stresses on the plant and may allow other factors, such as insect infestations, extreme weather conditions (particularly at high elevations), or

FIGURE 17.11 **Forest Decline** Many forests at high elevations in northeastern North America have shown significant decline, and dead trees are common.

ENVIRONMENTAL CLOSE-UP

Secondhand Smoke

Smoking of tobacco has long been associated with a variety of respiratory diseases, including lung cancer, emphysema, and heart disease. Because of these strong links, tobacco products carry warning labels. Many nonsmoking people, however, are exposed to environmental tobacco smoke (secondhand smoke) because they live and work in spaces where people smoke. The U.S. Environmental Protection Agency estimates that approximately 3000 nonsmokers die each year of lung cancer as a result of breathing air that contains secondhand smoke. Young children who are exposed to secondhand smoke are much more likely to have respiratory infections.

In July 1993, the U.S. Environmental Protection Agency recommended several actions to prevent people from being exposed to secondhand indoor smoke. They include recommendations that:

people not smoke in their homes or permit others to do so;

all organizations that deal with children have policies that protect children from secondhand smoke;

every company have a policy that protects employees from secondhand smoke;

smoking areas in restaurants and bars be placed so that the smoke will have little chance of coming into contact with nonsmokers.

California has some of the most restrictive smoking laws in the United States. In 1998, California passed a no-smoking law that effectively banned all smoking in indoor public places including restaurants and bars. The law was designed for employees who did not want to inhale secondhand smoke. While there are no no-smoking police going from building to building if an employee complains, the owner of the establishment could pay a fine of up to $7,000.

Below is a chart of various compounds and their toxicity found in secondhand smoke. Note that secondhand smoke is also referred to as environmental tobacco smoke.

Compound	Type of toxicity
Vapor phase:	
Carbon monoxide	T
Carbonyl sulfide	T
Benzene	c
Formaldehyde	c
3-Vinylpyridine	SC
Hydrogen cyanide	T
Hydrazine	c
Nitrogen oxides	T
N-nitrosodimethylamine	C
N-nitrosopyrrolidine	C
Particulate phase:	
Tar	C
Nicotine	T
Phenol	TP
Catechol	CoC
o-Toluidine	C
2-Naphthylamine	C
4-Aminobiphenyl	C
Benz(a)anthracene	C
Benzo(a)pyrene	C
Quinoline	C
N'-nitrosonornicotine	C
NNK	C
N-nitrosodiethanolamine	C
Cadmium	C
Nickel	C
Polonium-210	C

Abbreviations: C, carcinogenic; CoC, cocarcinogenic; SC, suspected carcinogen; T, toxic; TP, tumor promoter. Environmental tobacco smoke (ETS) is diluted in the air before it is inhaled and thus is less concentrated than mainstream smoke (MS). However, active inhalation of MS is limited to the time it takes to smoke each cigarette, whereas exposure to ETS is constant over the period spent in the ETS-polluted environment. This fact is reflected in measurements of nicotine uptake by smokers and ETS-exposed nonsmokers.
Sources: Data from Dept. of Health and Human Services, 1989; State of California, Report on Secondhand Smoke 1997.

drought, to further weaken trees and ultimately cause their death.

The effects of acid rain on aquatic ecosystems are much more clear-cut. In several experiments, lakes were purposely converted to acid lakes and the changes in the ecosystems recorded. The experiments showed that, as lakes become more acidic, there is a progressive loss of many kinds of organisms. The food web becomes less complicated, many organisms fail to reproduce, and many others die. Most healthy lakes have a pH above 6. At a pH of 5.5, many desirable species of fish have been eliminated; at a pH of 5, only a few starving fish may be found, and none are reproducing. Lakes with a pH of 4.5 are nearly sterile.

There are several reasons for these changes. Many of the early reproductive stages of insects and fish are more sensitive to acid conditions than are the adults. In addition, the young often live in shallow water, which is most affected by a flood of acid into lakes and rivers during the spring snowmelt. The snow and its acids have accumulated over the winter, and the snowmelt releases large amounts of acid all at once. Crayfish need calcium to form their external skeleton. As the pH

FIGURE 17.12   Effects of Acid Deposition on Organisms The low pH of the water in which this fish lived caused the abnormal bone development that ultimately resulted in the death of the fish.

FIGURE 17.13   Factors That Contribute to Acid Rain Damage In an aquatic ecosystem, the following factors increase the risk of damage from acid deposition: (1) a lake is located downwind from a major source of air pollution; (2) the area around the lake is hard, insoluble bedrock covered with a layer of thin infertile soil; (3) the soil has a low buffering capacity; and (4) there is also a low watershed to lake surface area ratio.

Source: Data from U.S. Environmental Protection Agency's "Acid Rain."

of the water decreases, the crayfish are unable to form new skeletons and so they die. Reduced calcium availability also results in some fish with malformed skeletons. (See figure 17.12.) As mentioned earlier, increased acidity also results in the release of aluminum, which impairs the function of a fish's gills.

The extent to which lakes have been acidified is great. About 14,000 lakes in Canada and 11,000 in the United States have been seriously altered by becoming acidic. Many lakes in Scandinavia are similarly affected. The extent to which acid deposition affects an ecosystem depends on the nature of the bedrock in the area and the ecosystem's proximity to acid-forming pollution sources. (See figure 17.13.) Parent material derived from igneous rock is not capable of buffering the effects of acid deposition, while soils derived from sedimentary rocks such as limestone release bases that neutralize the effects of acids. Because of this, eastern Canada and the U.S. Northeast are particularly susceptible to acid rain. These areas have high amounts of granite rock and are downwind from the major air-pollution sources of North America. Scandinavian countries have a similar geology and receive pollution from industrial areas in the United Kingdom and Europe. Thousands of kilometers of streams and up to 200,000 lakes in eastern Canada and the northeastern United States are estimated to be in danger of becoming acidified because of their location and geology.

GLOBAL WARMING AND CLIMATE CHANGE

Burning coal, oil, and natural gas to heat our homes, power our cars, and illuminate our cities produces carbon dioxide and other gases as by-products. Deforestation and clearing of land for agriculture also release significant quantities of such gases. Over the last century, we have been emitting gases such as carbon dioxide and methane to the atmosphere faster than natural processes can remove them. During this time, atmospheric levels of these gases have climbed steadily and are projected to continue their steep ascent as global economies grow. (See figure 17.14.)

Records of past climate going back as far as 160,000 years indicate a close correlation between the concentration of gases in the atmosphere and global temperatures. Computer simulations of the climate indicate that global temperatures will rise as atmospheric concentrations of carbon dioxide increase.

During the 1980s, scientists, governments, and the public became concerned about the possibility that the world may be getting warmer. The United Nations Environment Programme established an Intergovernmental Panel on Climate Change (IPCC) to study the issue and make recommendations. Its First Assessment was published in 1990. In 1996 the IPCC published its Second Assessment and

Total World Emissions of Carbon Dioxide

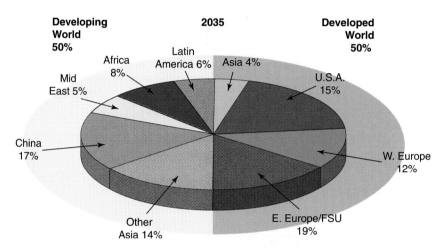

FIGURE 17.14 In 1995, the industrialized nations of the world contributed nearly three-quarters of the global emissions of carbon dioxide, with the U.S. being the largest single emitter. By 2035, developing nations will catch up and contribute half of the global emissions, with China becoming the largest single emitting country. Rapid population growth, industrialization, and increasing consumption per person in the developing world will contribute to this shift.

Source: Data from "Climate Change–State of Knowledge," October 1997, Office of Science and Technology Policy, Washington, D.C.

concluded that climate change is occurring and that it is highly probable that human activity is an important cause of the change. The IPCC has reached several important conclusions:

1. The average temperature of the earth has increased 0.3–0.6°C (1997 was the warmest year on record) and sea level has risen 10–25 cm in the last 100 years. (See figure 17.15.)

2. There is a strong correlation between the increase in temperature and the amount of greenhouse gases present in the atmosphere.

3. Human activity greatly increases the amounts of these greenhouse gases.

It is important to recognize that, although a small increase in the average temperature of the earth may seem trivial, this increase could set in motion changes that could significantly alter the climate of major regions of the world. (See figure 17.16.)

The climate changes expected from global warming are likely to have widespread effects, many of them negative, on ecological systems, human health, and socioeconomic sectors. In general, people in developing countries are more vulnerable to climate change because of limited infrastructure and capital and greater dependence on natural resources. While the specific consequences resulting from global warming will continue to be

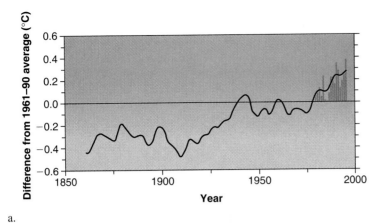

a.

b.

FIGURE 17.15 Changes in Average Global Temperature (*a*) Despite considerable variation, there has been a general trend toward increasing temperatures. 1990 was the warmest year on record. (*b*) The possible range of globally averaged surface temperature increase is shown for the period 1990 to 2100.

Source: (*a*) Data from U.K. Meteorological Office/University of East Anglia published in *Science* 12 January 1996. (*b*) Source: Data from "Global Temperature Change and Range of Globally Averaged Sea Level" in *Climate Changes—State of Knowledge* from Office of Science and Technology Policy, 1997, Office of the President, Washington, D.C.

debated, areas of concern include worsening human health effects, rising sea levels, disruption of the water cycle, changing forests and natural areas, and challenges to agriculture and the food supply.

Worsening Health Effects

Climate change will impact human health in a variety of ways. Warmer temperatures increase the risk of mortality from heat stress. For example, in July 1995, 465 deaths in Chicago were attributed to a heat wave with temperatures exceeding 32°C day and night. Today, such events occur about once every 150 years. Carbon dioxide (CO_2) concentrations of 550 ppm (parts per million) (double the pre-industrial level) could make such events six times more frequent. The potential increases in the heat index, a calculation combining temperature and humidity, illustrate the magnitude of this threat. Washington, D.C. currently has an average July heat index of 30°C, but if CO_2 levels reach 550 ppm, this could increase to 35°C, and if concentrations quadrupled to 1100 ppm, it could increase to 43°C. Climate change will also exacerbate air-quality problems, such as smog, and increase levels of airborne pollen and spores that aggravate respiratory disease, asthma, and allergic disorders. Because children and the elderly are the most vulnerable populations, they are likely to suffer disproportionately with both warmer temperatures and poorer air quality.

Diseases that thrive in warmer climates, such as malaria, dengue and yellow fevers, encephalitis, and cholera, are likely to spread due to the expansion of the ranges of mosquitoes and other disease-carrying organisms and increased rates of transmission. This could result in 50 million to 80 million additional malaria cases per year worldwide by 2100.

Rising Sea Level

Rising sea level erodes beaches and coastal wetland, inundates low-lying areas, and increases the vulnerability of coastal areas to flooding from storm surges and intense rainfall. By 2100, sea level is expected to rise by 15 to 90 cm. (See figure 17.17.) A 50 cm sea level rise will result in substantial loss of coastal land in North America, especially along the southern Atlantic and Gulf coasts, which are subsiding and are particularly vulnerable. The land area of some island nations and countries such as Bangladesh would change dramatically as flooding occurred. The oceans will continue to expand for several centuries after temperatures stabilize. Because of this, the sea level rise associated with CO_2 levels of 550 ppm (double pre-industrial levels) could eventually exceed 100 cm. A CO_2 level of 1100 ppm could produce a sea level rise of 200 cm or even more, depending on the extent to which the Greenland and Antarctic ice sheets melt.

A 50 cm sea level rise would double the global population at risk from storm surges, from roughly 45 million at present to over 90 million, and this figure does not account for any increases in coastal populations. A 100 cm rise would triple the number.

Disruption of the Water Cycle

Among the most fundamental effects of climate change are intensification and disruption of the water cycle. Droughts and floods and water quality and quantity are the areas of particular concern.

Intensification of the water cycle will produce more severe droughts in some places and floods in others. Such events are costly. Damages from the drought in the southern plains of the United States in 1996 were estimated at $4 billion, the 1993 Mississippi River flood damages at $10 billion, and the 1997 Red River flood in the northern plains of Canada and the United States about $4 billion.

Regarding the issues of water quality and quantity, the areas of greatest vulnerability are those where quality and quantity of water are already problems, such as the arid and semiarid regions of the world. Water scarcity in the Middle East and Africa is likely to be aggravated by climate change, which could increase international tension among countries that depend on water supplies originating outside their borders.

Changing Forests and Natural Areas

Climate change could dramatically alter the geographic distributions of vegetation types. The composition of one-third of the Earth's forests could undergo major changes as a result of climate changes associated with a CO_2 level of 700 ppm. Over the next 100 years, the ideal range for some North American forest species could shift by as much as 500 kilometers to the north, far faster than the forests can migrate naturally. Economically important species, such as the sugar maple, could be lost from the northeast United States by the end of the next century. (See figure 17.18.)

Such changes could have profound effects on national parks and refuges around the world leading to reductions in biological diversity and in the benefits provided by ecosystems, such as clean water and recreation. Wetlands are particularly at risk. The wetlands of the prairie pothole region, which support half the waterfowl population of North America, could diminish in area and change dramatically in character in response to climate change.

Challenges to Agriculture and the Food Supply

Climate strongly affects crop yields. A CO_2 concentration of 550 ppm is likely to increase crop yields in some areas by as much as 30 percent to 40 percent, but it will decrease yields in other places by similar amounts, even for the same crop. A warmer climate would reduce flexibility in crop distribution and increase irrigation demands. Expansions of the ranges of

a.

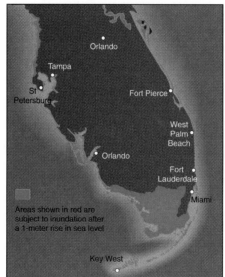

b.

FIGURE 17.16 Effects of Global Warming (*a*) Global warming would have several effects on the climate of the world. The climate changes would have important impacts on human and other living things. (*b*) Sea level rise could inundate many low-lying coastal areas in Florida, and will increase the vulnerability of all such areas to storm surges.

Source: (*b*) Elevations from USGS digital data. Prepared by the U.S. Geological Survey, 1997.

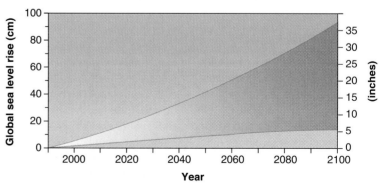

FIGURE 17.17 The possible range of globally averaged sea level rise is shown for the period 1990 to 2100.

Source: Data from "Climate Change—State of Knowledge," October 1997, Office of Science and Technology Policy, Washington, D.C

**Current and Projected
Ranges of Sugar Maple**

Present range

Overlap

Predicted range

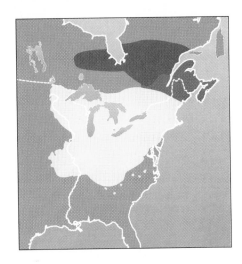

Prediction based on increased temperature

Prediction based on increased temperature and moisture reduction

FIGURE 17.18 Climatic shifts will force some species to migrate northwards or to higher elevations in order to stay in the appropriate climatic zone. The climatic zone for sugar maple, for example, could shift northwards into Canada. This would compromise the maple syrup industry and the fall foliage colors, both of which make New England famous.

Source: Data from "Climate Change—State of Knowledge," October 1997, Office of Science and Technology Policy, Washington, D.C.

TABLE 17.3 MAJOR GREENHOUSE GASES

Gas	Contribution to global warming (percentage)
Carbon dioxide	60
Chlorofluorocarbons	22
Methane	12
Nitrous oxide	6

pests could also increase vulnerability and result in greater use of pesticides. Despite these effects, total global food production is not expected to be altered substantially by climate change, but there are likely negative regional impacts. Agricultural systems in the developed countries are highly adaptable and can probably cope with the expected range of climate changes without dramatic reductions in yields. It is the poorest countries, already subject to hunger, that are the most likely to suffer significant decreases in agricultural productivity.

These predictions are based on computer models of climate. Some scientists have criticized the predictions as being inaccurate and constructed from sketchy data. However, as more accurate information is gathered and inserted into the models, the general conclusions remain the same. It is difficult for the general public

to comprehend these changes or see evidence of them since each of us experiences only our own local weather and climate. Also, there will always be short-term variations in weather patterns. The models, however, are attempting to predict the long-term trends. The consequences of a global warming would be so great that many are suggesting we alter our lifestyles regardless of whether the phenomenon is true, just to be on the safe side.

What actually causes global warming? An explanation is relatively straightforward. Several gases in the atmosphere are transparent to light but absorb infrared radiation. These gases allow sunlight to penetrate the atmosphere and be absorbed by the earth's surface. This sunlight energy is reradiated as infrared radiation (heat), which is absorbed by the gases. Because the effect is similar to what happens in a greenhouse (the glass allows light to enter

but retards the loss of heat), these gases are called greenhouse gases and the warming thought to occur from their increase is called the **greenhouse effect.** (See figure 17.19.) The most important greenhouse gases are carbon dioxide (CO_2), chlorofluorocarbons (primarily CCl_3F and CCl_2F_2), methane (CH_4), and nitrous oxide (N_2O). Table 17.3 lists the relative contribution of each of these gases to the potential for global warming.

Carbon dioxide (CO_2) is the most abundant of the greenhouse gases. It occurs as a natural consequence of respiration. However, much larger quantities are put into the atmosphere as a waste product of energy production. Coal, oil, natural gas, and biomass are all burned to provide heat and electricity for industrial processes, home heating, and cooking. These sources are increasing the amount of carbon dioxide in the atmosphere. Measurement of carbon dioxide levels at the Mauna Loa Observatory in Hawaii show that the carbon dioxide level has increased from about 315 ppm (parts per million) in 1958 to about 362 ppm in 1998. (See figure 17.20.)

A major step toward slowing global warming would be to increase the efficiency of energy utilization. This would also be of value in conserving the shrinking supplies of energy resources. It makes sense to increase energy efficiency, thus

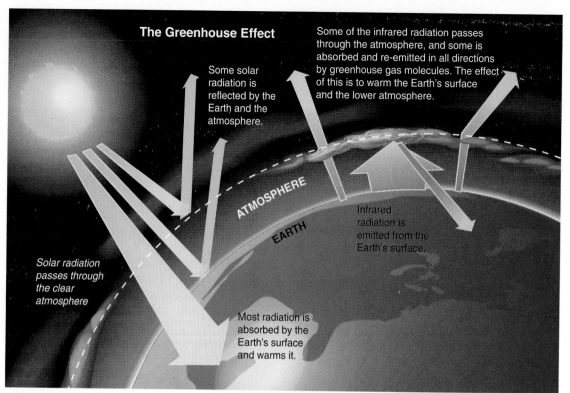

FIGURE 17.19 The greenhouse effect naturally warms the Earth's surface. Without it, Earth would be 60°F cooler than it is today—uninhabitable for life as we know it.

Source: Data from "Climate Change—State of Knowledge," October 1997, Office of Science and Technology Policy, Washington, D.C.

reducing carbon dioxide production, even if global warming is not a concern. One way to stimulate a move toward greater efficiency would be the imposition of a carbon tax. A carbon tax would increase the cost of fuels by taxing the amount of carbon put into the atmosphere by their use. This would increase the demand for fuel efficiency because the cost of fuel would rise. It would also stimulate the development of alternative fuels with a lower carbon content and generate funds for research in many aspects of fuel efficiency and alternative fuel technologies.

Another approach to the problem is to increase the amount of carbon dioxide removed from the atmosphere. If enough biomass is present, the excess carbon dioxide can be utilized by vegetation during photosynthesis, thereby reducing the impact of carbon dioxide released by fossil-fuel burning. Australia, the United States, and several other countries have announced plans to plant billions of trees to help remove carbon dioxide from the

FIGURE 17.20 **Change in Atmospheric Carbon Dioxide** Since the establishment of a carbon dioxide monitoring station at Mauna Loa Observatory in Hawaii, a steady increase in carbon dioxide levels has been observed. Since 1960 the concentration of carbon dioxide in the atmosphere has increased by nearly 14 percent.

Source: Data from Carbon Dioxide Information Analysis Center, 1997, Oak Ridge, TN.

ENVIRONMENTAL CLOSE-UP

Aesthetic Pollution

Most pollution can be measured by the amount of a particular chemical in water, air, or soil. Scientists measure these amounts and their effects on human, animal, and plant health. But some forms of pollution affect our aesthetic senses and so are more difficult to define.

Visual pollution is a sight that offends us. This type of pollution is highly subjective and is, therefore, difficult to define or control. To most people, an open garbage dump is a form of visual pollution. A dilapidated home or building may also be offensive, especially if located in an area of higher-priced homes. A heavily littered highway or street is aesthetically offensive to most people, and litter along a wilderness trail is even more unacceptable. Some sources of visual pollution are not so clear-cut, however. To many people, roadside billboards are offensive, but they can be helpful to advertisers and to travelers looking for information. This difference of opinion is typical when it comes to judging aesthetic pollution. People do not always agree on what is offensive. That makes regulation of aesthetic pollution difficult.

Taste is another subjective area. Some of the chemicals discharged into waterways have a definite taste. If chemicals do no biological injury but have an unpleasant taste, they are taste pollutants. Minute quantities of certain chemicals can affect the taste of our drinking water and our food. In fish, a concentration of 100 ppm of phenol can be tasted. Although this small amount of chemical is not harmful biologically, it does make the fish very unappetizing.

Odor pollution may originate from many sources. Various airborne chemicals have a specific odor that may be offensive. People who live near stockyards, paper mills, chemical plants, steel mills, and other industries may be offended by the odors originating from these sources. Many of these industries discharge into the water waste mate-

rials that decompose or evaporate and cause odor pollution. People who are constantly exposed to an odor are usually not as offended by it as are people who are newly exposed to it. When an odor is constantly received, the brain ceases to respond to the stimulus. In other words, the person is not aware of the odor.

As we have seen, aesthetic pollutants—sights, tastes, and odors—can be extremely difficult to define. Each of us has our own idea of what is offensive, which makes it very difficult to establish aesthetic pollution standards. This type of pollution can be controlled by educating the public so that generally unacceptable levels of aesthetic pollution are eliminated.

---

atmosphere. Many critics argue that this approach will only provide a short-term benefit since, eventually, the trees will mature and die, and their decay will release carbon dioxide into the atmosphere at some later time.

An associated concern is the destruction of vast areas of rainforest in tropical regions of the world. These ecosystems are extremely efficient at removing carbon dioxide and storing the carbon atoms in the structure of the plant. The burning of tropical rainforests to provide farm or grazing land not only adds carbon dioxide to the atmosphere, but also reduces the ability to remove carbon dioxide from the atmosphere, since the grasslands or farms created to not remove carbon dioxide as efficiently as do the original rainforests. Furthermore, the grazing lands and farms

in such regions of the world are often abandoned after a few years and do not return to their original forest condition.

Chlorofluorocarbons are entirely the result of human activity. They were widely used as refrigerant gases in refrigerators and air conditioners, as cleaning solvents, as propellants in aerosol containers, and as expanders in foam products. Although they are present in the atmosphere in minute quantities, they are extremely efficient as greenhouse gases (about 15,000 times more efficient at retarding heat loss than is carbon dioxide). Since all chlorofluorocarbons are made by people for specific purposes, their levels can be easily controlled.

Since the 1970s, when chlorofluorocarbons were linked to the depletion of the ozone layer in the upper atmosphere, their

use as propellants in aerosol cans has been banned in the United States, Canada, Norway, and Sweden, and the European Economic Community agreed to reduce use of chlorofluorocarbons in aerosol cans. However, worldwide, chlorofluorocarbons are still widely used as aerosol propellants. In foam and solvents, care can be taken to recover the chlorofluorocarbons for reuse rather than allow them to escape into the atmosphere. Alternative refrigerants are available, and care could be taken to recycle refrigerant gases rather than just venting them into the air. In 1991, DuPont announced the development of new refrigerants that would not harm the ozone layer. These have been installed in refrigerators and air conditioners in many nations, including the United States. We will need to continue to exploit all these options if we want to reduce chlorofluoro-carbon production on an international scale. Until substitutes become available, the mandatory recycling of chlorofluoro-carbons will reduce the rate at which new gases are added to the atmosphere.

In 1987, several industrialized countries, including Canada, the United States, the United Kingdom, Sweden, Norway, Netherlands, the Soviet Union, and West Germany, agreed to freeze production of chlorofluorocarbons at present levels and reduce production by 50 percent by the year 2000. This document, known as the Montreal Protocol, was ratified by the U.S. Senate in 1988. In May 1989, 82 nations signed the Helsinki Declaration, pledging to phase out the use of most chlorofluoro-carbons by the year 2000. Several companies in the United States and Japan (e.g., General Motors, AT&T, Nissan) have announced plans to either phase out chlorofluorocarbon use or reduce use significantly. In 1990 in London, international agreements were reached to further reduce the use of chlorofluorocarbons. A major barrier to these negotiations was the reluctance of the developed countries of the world to establish a fund to help less-developed countries implement technologies that would allow them to obtain refrigeration and air conditioning without the use of chlorofluorocarbons. In 1996 the United States stopped producing chloro-fluorocarbons. As a result of these international efforts and rapid changes in technology, the use of chlorofluorocarbons has dropped rapidly, and concentrations of chlorofluorocarbons in the atmosphere have stabilized and are expected to decline in the future.

Methane enters the atmosphere primarily from biological sources. Several kinds of bacteria that are particularly abundant in wetlands and rice fields release methane into the atmosphere. Methane-releasing bacteria are also found in large numbers in the guts of termites and various kinds of ruminant animals such as cattle. Some methane enters the atmosphere from fossil-fuel sources. Control of methane sources is unlikely since the primary sources involve agricultural practices that would be very difficult to change. For example, nations would have to convert rice paddies to other forms of agriculture and drastically reduce the number of animals used for meat production. Neither is likely to occur since food production in most parts of the world needs to be increased, not decreased.

Nitrous oxide, a minor component of the greenhouse gas picture, primarily enters the atmosphere from fossil fuels and fertilizers. It could be reduced by more careful use of nitrogen-containing fertilizers.

OZONE DEPLETION

In the 1970s, various sectors of the scientific community became concerned about the possible reduction in the ozone layer in the upper atmosphere surrounding the earth. **Ozone** is a molecule of three atoms of oxygen (O_3). In 1985, it was discovered that a significant thinning of the ozone layer over the Antarctic occurred during the Southern Hemisphere spring. Some regions of the ozone layer showed 95 percent depletion. Ozone depletion also has been found to be occurring farther north. Measurements in arctic regions suggest a thinning of the ozone layer there also. These findings have caused several countries to become involved in efforts to protect the ozone layer.

The presence of ozone in the outer layers of the atmosphere, approximately 12–25 kilometers from the earth's surface, shields the earth from the harmful effects of ultraviolet light radiation. Ozone (O_3) absorbs ultraviolet light and is split into an oxygen molecule and an oxygen atom:

$$O_3 \xrightarrow{\text{Ultraviolet light}} O_2 + O$$

Oxygen molecules are also split by ultraviolet light to form oxygen atoms:

$$O_2 \xrightarrow{\text{Ultraviolet light}} 2O$$

Recombination of oxygen atoms and oxygen molecules allow ozone to be formed again and to be available to absorb more ultraviolet light.

$$O_2 + O \rightarrow O_3$$

This series of reactions results in the absorption of 99 percent of the ultraviolet light energy coming from the sun and prevents it from reaching the earth's surface.

Less ozone in the upper atmosphere would result in more ultraviolet light reaching the earth's surface, causing increased skin cancers and cataracts in humans and increased mutations in all living things.

Chlorofluorocarbons are strongly implicated in the ozone reduction in the upper atmosphere. Chlorine reacts with ozone in the following way to reduce the quantity of ozone present:

$$Cl + O_3 \rightarrow ClO + O_2$$
$$ClO + O \rightarrow Cl + O_2$$

These reactions both destroy ozone and reduce the likelihood that it will be formed because atomic oxygen (O) is removed as well. It is also important to note that it can take 10 to 20 years for chlorofluorocarbon molecules to get into the stratosphere and then they can react with the ozone for up to 65 years. As a result, the World Meteorological Association predicts ozone depletion will worsen well into the next century before improvements are seen.

The use of chlorofluorocarbons in air conditioners, in refrigerators, and as propellants has resulted in the release of large amounts into the atmosphere. Because chlorofluorocarbons are implicated in both global warming and ozone depletion, considerable international attention is focused on controlling their manufacture and release. (See previous section on global warming and climate changes.)

INDOOR AIR POLLUTION

A growing body of scientific evidence indicates that the air within homes and other buildings can be more seriously polluted than outdoor air in even the largest and most industrialized cities. Many indoor air pollutants and pollutant sources are thought to have an adverse effect on human health. These pollutants include asbestos; formaldehyde, which is associated with many consumer products, including certain wood products and aerosols; airborne pesticide residues; chloroform; perchloroethylene (associated particularly with dry cleaning); paradichlorbenzene (from mothballs and air fresheners); and many disease-causing organisms. (See

TABLE 17.4 INDOOR AIR POLLUTION SUMMARY

Pollutant	Description	Sources	Effects*
Asbestos	Light, fibrous mineral; fireproof and insulative	Ceilings and floor tiles, insulation, and spackling compounds	Easily inhaled to cause lung damage Cancer
Carbon monoxide (CO)	Odorless, colorless gas	Combustion sources such as charcoal, kerosene heaters, attached garages with poor ventilation, tobacco smoke	Reduces ability of blood to carry oxygen; impairs vision and alertness Dizziness, headaches, fatigue, death by suffocation
Formaldehyde	Pungent gas; preservative and disinfectant	Foam insulation, resin in particleboards, plywood paneling, fiberboard, some carpets, drapery, and upholstery fabrics	Headaches, dizziness, nausea, lethargy, rashes, upper respiratory irritations
Lead (Pb)	Metallic element	House paints manufactured prior to 1976, old plumbing lines and solder, leaded crystal and pottery, old toys	Learning and behavior problems in children, high blood pressure, joint pain, concentration problems, reproductive problems
Biological pollutants and microorganisms	Pollen, dust mites, and pet dander Bacteria, molds, fungi, and viruses	Improperly maintained heating and cooling systems Washrooms, humidifiers, and dehumidifiers Pets and plants Wall-to-wall carpeting	Allergic diseases and skin irritations, influenza and Legionnaires' disease, acute cases of asthma
Nitrogen dioxide (NO$_2$)	Brownish gas	Gas appliances, fireplaces, wood and coal stoves	Eye and respiratory tract irritations Lowers resistance to respiratory infections Chronic bronchitis
Radon (Rn)	Naturally occurring, radioactive gas	Rocks and soils that contain the decaying radioactive elements of uranium and radium Enters through cracks in foundation and basements	Lung cancer
Tobacco smoke	Mixture of several substances including known human carcinogens	Cigarettes, pipes and cigars, secondhand smoke	Eye, nose and throat irritations, headaches and nausea, coughing and chest discomfort, respiratory infections, lung cancer
Volatile organic compounds (VOCs)	Substances produced by synthetic chemical industry and naturally; vaporize at ordinary temperatures	Some furniture, paint, adhesives, solvents, upholstery and drapery fabrics, construction materials, cleaning compounds, deodorizers, felt-tipped markers, and correction fluid, dry-cleaned clothes, pesticides, wool preservatives, tobacco smoke, etc.	Eye and respiratory irritations, kidney and liver damage in animals Long-term effects are still being studied.

*Effects from prolonged exposure or high concentrations.
Chart compiled by Earth Force, Inc. (www.earthforce.org) with data from the Environmental Protection Agency and Wisconsin Department of Natural Resources. All Rights Reserved. Reprinted by permission.

table 17.4.) Smoking is the most important air pollutant source in the United States in terms of human health. The Surgeon General estimates that 350,000 people in this country die each year from emphysema, heart attacks, strokes, lung cancer, or other diseases caused by tobacco smoking. Banning smoking probably would save more lives than would any other pollution-control measure.

A recent contributing factor to the concern about indoor air pollution is the weatherizing of buildings to reduce heat loss and save on fuel costs. In most older homes, there is a complete exchange of air every hour. This means that fresh air leaks in around doors and windows and through cracks and holes in the building. In a weatherized home, a complete air exchange may occur only once every five hours. Such a home is more energy efficient, but it also tends to trap air pollutants.

Even though we spend almost 90 percent of our time indoors, the movements to reduce indoor air pollution lag behind regulations governing outdoor air pollution. In the United States, the Environmental Protection Agency is conducting research to identify and rank the human health risks that result from exposure to individual indoor pollutants or mixtures of multiple indoor pollutants.

Radon

In 1985, the clothing worn by an engineer at the Limerick Nuclear Generating Station in Pottstown, Pennsylvania, registered a high radiation level. Initially, the generating station was believed to be the source of the radiation. However, subsequent studies indicated that radon 222 from the engineer's home was the source. Following this incident, there has been an increased interest in radon and its effects.

The source of radon is uranium-238, a naturally occurring element that makes up about three parts per million of the earth's crust. Uranium-238 goes through fourteen steps of decay before it becomes stable, nonradioactive lead 206. **Radon,** an inert radioactive gas having a half-life of 3.8 days, is one of the products formed during this process.

Since radon is an inert gas, it does not enter into any chemical reactions within the body, but it can be inhaled. Once in the lung it will undergo radioactive decay, producing other kinds of atoms called "daughters" of radon. These decay products (daughters) of radon—plutonium 218, which has a three-minute half-life; lead 214, which has a twenty-seven minute half-life; bismuth 214, which has a twenty-minute half-life; and polonium 214, which has a millisecond half-life—are solid materials that remain in the lungs and are chemically active.

Increased incidence of lung cancer is the only known health effect associated with radon decay products. It is estimated that the decay products of radon are responsible for about 15,000 lung cancer deaths annually in the United States. This is about 10 percent of lung cancer deaths.

As the radon gas is formed in the rocks, it usually diffuses up through the rocks and soil and escapes harmlessly into the atmosphere. It can also diffuse into groundwater. Radon usually enters a home through an open space in the foundation. A crack in the basement floor or the foundation, the gap around a water or sewer pipe, or a crawl space allows the radon to enter the home. It may also enter in the water supply from wells.

Only 10 percent of the homes in the United States have a potential radon problem. However, the increased publicity about radon has many people worried. In addition, the Environmental Protection Agency and the U.S. Surgeon General recommend that all Americans (other than those living in apartment buildings above the second floor) test their home for radon. If the tests indicate that the level of radon is at or above 4 picocuries/liter, EPA recommends that the homeowner take action to lower the level. This is usually not expensive and consists of blocking the places where radon is entering or venting sources of radon to the

Radon risk evaluation chart

pCi/L	WL	Estimated lung-cancer deaths due to radon exposure (out of 1,000)	Comparable exposure levels	Comparable risk
200	1.0	440–770	One thousand times average outdoor level	More than sixty times nonsmoker risk / Four-pack-a-day smoker
100	0.5	270–630	One hundred times average indoor level	Two thousand chest X rays per year
40	0.2	120–380		
20	0.1	60–120	One hundred times average outdoor level	Two-pack-a-day smoker
10	0.05	30–120	Ten times average indoor level	One-pack-a-day smoker
4	0.02	13–50	Level at which EPA suggests remedial action	Five times nonsmoker risk
2	0.01	7–30	Ten times average outdoor level	Two hundred chest X rays per year
1	0.005	3–13	Average indoor level	Nonsmoker risk of dying from lung cancer
0.2	0.001	1–3	Average outdoor level	Twenty chest X rays per year

Note: Measurement results are reported in one of two ways: (1) pCi/L (picocuries per liter) — Measurement of *radon gas*, or (2) WL (working levels) — Measurement of *radon decay products.*

Source: Data from Office of Air and Radiation Programs, U.S. Environmental Protection Agency.

outside. People who are concerned about radon should contact their state's public health department or environmental protection agency.

Source: Data from Office of Air and Radiation Programs, U.S. Environmental Protection Agency.

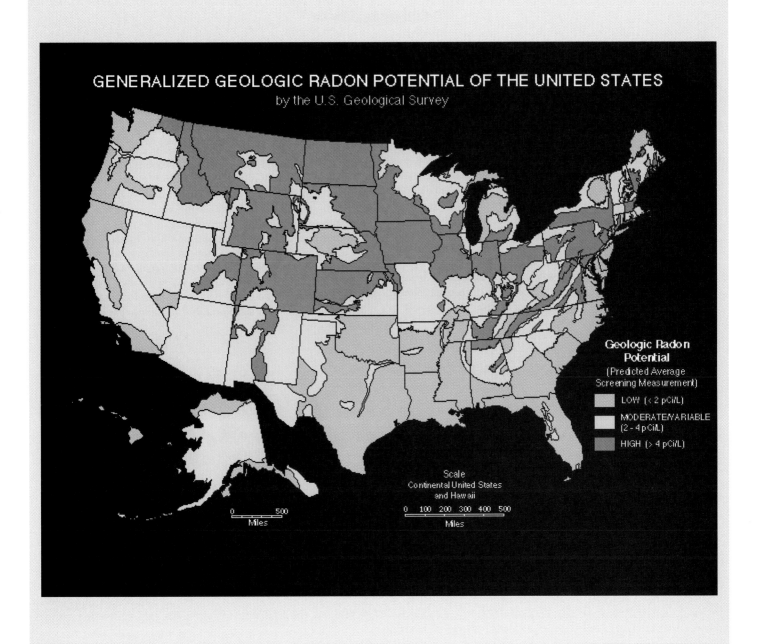

GENERALIZED GEOLOGIC RADON POTENTIAL OF THE UNITED STATES
by the U.S. Geological Survey

Geologic Radon
Potential
(Predicted Average
Screening Measurement)

LOW (< 2 pCi/L)

MODERATE/VARIABLE
(2 - 4 pCi/L)

HIGH (> 4 pCi/L)

Scale
Continental United States
and Hawaii

0 100 200 300 400 500
Miles

0   500
Miles

ENVIRONMENTAL CLOSE-UP

Noise Pollution

Noise is referred to as unwanted sound. However, noise can be more than just an unpleasant sensation. Research has shown that exposure to noise can cause physical, as well as mental, harm to the body. The loudness of the noise is measured by decibels (db). Decibel scales are logarithmic, rather than linear. Thus, the change from 40 db (a library) to 80 db (a dishwasher or garbage disposal) represents a ten-thousand-fold increase in sound loudness.

The frequency or pitch of a sound is also a factor in determining its degree of harm. High-pitched sounds are the most annoying. The most common sound pressure scale for high-pitched sounds is the A scale, whose units are written "dbA." Hearing loss begins with prolonged exposure (eight hours or more) to 80 or 90 dbA levels of sound pressure. Sound pressure becomes painful at around 140 dbA and can kill at 180 dbA. (See the table.)

In addition to hearing loss, noise pollution is linked to a variety of other ailments, ranging from nervous tension headaches to neuroses. Research has also shown that noise may cause blood vessels to constrict (which reduces the blood flow to key body parts), disturbs unborn children, and sometimes causes seizures in epileptics. The U.S. EPA has estimated that noise causes about 40 million U.S. citizens to suffer hearing damage or other mental or physical effects. Up to 64 million people are estimated to live in homes affected by aircraft, traffic, or construction noise.

The Noise Control Act of 1972 was the first major attempt made in the United States to protect the public health and welfare from detrimental noise. This act also attempted to coordinate federal research and activities in noise control, to set federal noise emission standards for commercial products, and to provide information to the public. Subsequent to the passage of the Noise Control Act, many local communities in the United States enacted their own noise ordinances. While such efforts are a step in the right direction, the United States is still controlling noise less than are many European countries. Several European countries have developed quiet construction equipment in conjunction with strongly enforced noise ordinances. The Germans and Swiss have established maximum day and night noise levels for certain areas. Regarding noise-pollution abatement, North America has much to learn from European countries.

INTENSITY OF NOISE

Source of Sound	Intensity in decibels
Jet aircraft at takeoff	145
Pain occurs	140
Hydraulic press	130
Jet airplane (160 meters overhead)	120
Unmuffled motorcycle	110
Subway train	100
Farm tractor	98
Gasoline lawn mower	96
Food blender	93
Heavy truck (15 meters away)	90
Heavy city traffic	90
Vacuum cleaner	85
Hearing loss after long exposure	85
Garbage disposal unit	80
Dishwasher	65
Window air conditioner	60
Normal speech	60

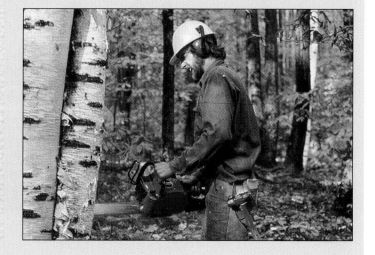

Acid rain originates as a form of air pollution, but it may damage the environment in the form of water pollution. Also, air pollution that originates in one country may result in water pollution in another country. A particular nation may have stringent environmental controls within its own boundaries but have environmental problems because of the actions of a neighboring country.

Recently, the phenomenon of acid rain and forest fires have underscored the need for international cooperation in dealing with various environmental concerns. For example, an estimated 56 percent of the acid rain falling in Sweden originates outside of that country. The main sources of the pollutants are Germany and the United Kingdom because the west coast of Sweden receives winds from these countries. When these industrialized countries release more sulfur dioxide into the atmosphere, the result is more acid rain in Sweden. Since the 1930s, lakes in western Sweden have become more acidic by a value of two pH units. Ten thousand lakes have a pH below 6.0, and 5000 lakes have a pH below 5.0. A pH below 5.5 is too acidic for many species of fish.

In the 1930s, the United Kingdom initiated a program to reduce air pollution. This program has been successful in that the air in the United Kingdom has become cleaner. But one of the country's solutions was to build taller smokestacks. As a result, more of the pollutants from the United Kingdom are transported to Sweden.

- Should the United Kingdom be permitted to disperse air pollutants in a manner that damages the Swedish environment?
- Should there be a series of international agreements to control and regulate the movement of airborne pollutants across international boundaries?
- In addition to air pollutants, are there other forms of pollutants that can naturally be transported across international boundaries?

Summary

The atmosphere has a tremendous ability to accept and disperse pollutants. Carbon monoxide, hydrocarbons, particulates, sulfur dioxide, and nitrogen compounds are the primary air pollutants. They can cause a variety of health problems. Lead and air toxics have also been identified as significant air pollutants.

Photochemical smog is a secondary pollutant formed when hydrocarbons and oxides of nitrogen are trapped by thermal inversions and react with each other in the presence of sunlight to form peroxyacetylnitrates and ozone. Elimination of photochemical smog requires changes in technology, such as more fuel-efficient automobiles, special devices to prevent the loss of hydrocarbons, and catalytic converters to more completely burn hydrocarbons.

Acid rain is caused by emissions of sulfur dioxide and oxides of nitrogen in the upper atmosphere, which form acids that are washed from the air when it rains or snows. Direct effects of acid rain on terrestrial ecosystems are difficult to prove, but changes in many forested areas are suspected of being partly the result of additional stresses caused by acid rain. Recent evidence suggests that loss of calcium from the soil may be a major problem associated with acid rain. The effect of acid rain on aquatic ecosystems is easy to quantify. As waters become more acidic, the complexity of the ecosystem decreases, and many species fail to reproduce. The control of acid rain requires the use of scrubbers, precipitators, and filters or the removal of sulfur from fuels.

Currently, many are concerned about the damaging effects of greenhouse gases: carbon dioxide, methane, and chlorofluorocarbons. These gases are likely to be causing an increase in the average temperature of the earth and consequently are leading to major changes in the climate. Human and ecological systems are already vulnerable to a range of environmental pressures, including climate extremes and variability. Global warming is likely to amplify the effects of other pressures and to disrupt our lives in numerous ways. Significant impacts on our health, the vitality of forests and other natural areas, the distribution of freshwater supplies, and the productivity of agriculture are among the probable consequences of climate change. Chlorofluorocarbons are also thought to lead to the destruction of ozone in the upper atmosphere, which results in increased amounts of ultraviolet light reaching the earth. Concern about the effects of chlorofluorocarbons has led to international efforts that have resulted in significant reductions in the amount of these substances reaching the atmosphere. Many commonly used materials release gases into closed spaces (indoor air pollution) where they cause health problems. The most important of these are associated with smoking.

Review Questions

1. List the five primary air pollutants commonly released into the atmosphere and their sources.
2. Define *secondary air pollutants,* and given an example.
3. List three health effects of air pollution.
4. Why is air pollution such a large problem in urban areas?
5. What is photochemical smog? What causes it?
6. Describe three actions that can be taken to control air pollution.
7. What causes acid rain? List three probable detrimental consequences of acid rain.
8. Why is carbon dioxide (a nontoxic normal component of the atmosphere) called a "greenhouse gas"?
9. What would the consequences be if the ozone layer surrounding the earth was destroyed?
10. How does energy conservation influence air quality?

Critical Thinking Questions

1. What could you do to limit the air pollution you create?

2. Do you agree with a ban on smoking, like in California, that includes all indoor public places, even privately owned restaurants and bars? Why?

3. Some developing countries argue that they should be exempt from limits on the production of greenhouse gases and that developed countries should bear the brunt of the changes that appear to be necessary to curb global climite change. What values, beliefs, and perspectives underly this belief? What do you think about this argument?

4. As a nation, the United States provides many subsidies to make energy cheap, figuring that development depends on cheap energy. If these subsidies were withdrawn, or taxes on energy were added, what effect would this have on your own energy consumption? Would you be willing to support high gasoline prices, in the $3-4/gallon range as in many European countries, if it would cut greenhouse gas emissions?

5. Why do you think air pollution is so much worse in developing countries than in developed countries? What should developed countries do about this, if anything?

6. What common indoor air pollutants are you exposed to? What can you do to limit this exposure?

7. What kinds of noise pollution do you encounter? How important is noise pollution to you? What can you do to reduce noise pollution?

8. Is it possible to have zero emissions of pollutants? What level of risk are you willing to live with?

Key Terms

acid deposition *339*

acid rain *339*

carbon dioxide (CO_2) *347*

carbon monoxide (CO) *334*

carcinogenic *335*

chlorofluorocarbons *349*

greenhouse effect *347*

hydrocarbons (HC) *335*

nitrous oxide *350*

oxides of nitrogen (NO and NO_2) *336*

ozone *350*

particulates *335*

photochemical smog *336*

primary air pollutants *333*

radon *352*

secondary air pollutants *333*

sulfur dioxide (SO_2) *335*

thermal inversion *337*

Related Internet Sites

Visit our website at http://www.mhhe.com/environmentalscience/ for specific information on each of the topics below.

Air pollution and radiation

Ozone depletion facts

Health effects of environmental pollution

Environmental problems

Global warming and climate change

Impact of ultraviolet radiation on the environment and human health

Pacific Northwest Pollution Prevention Resource Center

Pollution prevention and compliance

Pollution prevention technical library

Greenhouse gases and emissions

Acid rain, air quality, greenhouse gases, ozone, and smog

Gases and aerosols emitted into the atmosphere from natural and human sources

Solid Waste Management and Disposal

OBJECTIVES

After reading this chapter, you should be able to:

- Explain why solid waste has become a problem throughout the world.
- Understand that the management of municipal solid waste is directly affected by economics, changes in technology, and citizen awareness and involvement.
- Recognize that the management of municipal solid waste in the future will require an integrated approach.
- Describe the various methods of waste disposal and the problems associated with each method.
- Understand the difficulties in developing new municipal landfills.
- Define the problems associated with incineration as a method of waste disposal.
- Describe some methods of source reduction.
- Describe composting and how it fits into solid waste disposal.
- List some benefits and drawbacks of recycling.

CHAPTER OUTLINE

Introduction

The Disposable Decades

The Nature of the Problem

Methods of Waste Disposal
  Landfill

Environmental Close-Up: *Resins Used in Consumer Packaging*

  Incineration
  Source Reduction
  Recycling

Environmental Close-Up: *Container Laws*

Environmental Close-Up: *What You Can Do to Reduce Waste and Save Money*

Environmental Close-Up: *Recycling Is Big Business*

Environmental Close-Up: *Recyclables Market Basket*

Issues and Analysis: *Corporate Response to Environmental Concerns*

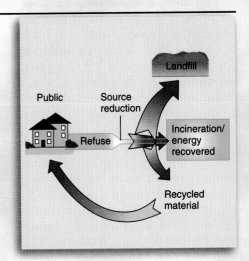

INTRODUCTION

In March 1987, a barge laden with 3200 tons of garbage set out in search of a dump. The refuse had been turned away from a landfill in Islip, New York. The barge traveled 10,000 kilometers and stopped at several foreign ports, but found no one willing to accept its noxious load. The three-month odyssey took the barge to Mexico, Belize, and the Bahamas before it returned still fully loaded to New York. (See figure 18.1.) The futile voyage made headlines, giving many North Americans their first inkling of an impending crisis.

Lack of space for dumping solid waste has become a problem for many large metropolitan areas throughout the world. Communities are concerned about the increasing costs of waste disposal, possible hazards to groundwater, and maintaining air quality. In many areas, there is not suitable land available for new landfills. Problems with solid waste have increased dramatically over the past several decades because of population increases and an attitude that convenience is a very important part of the North American lifestyle.

THE DISPOSABLE DECADES

In 1955, *Life* magazine pictured a happy family in an article entitled "Throwaway Living." A disposable lifestyle was marketed as the wave of the future and as a way to cut down on household chores. "Use it once and throw it away" became a very popular advertising slogan in the 1950s.

The solid waste problems facing us today have their roots in the economic boom

FIGURE 18.1   Wandering Garbage In 1987, a barge filled with garbage similar to this barge traveled from New York to Mexico looking for a place to dispose of its cargo. This practice of shipping unwanted garbage to other countries continues today throughout the world.

a.

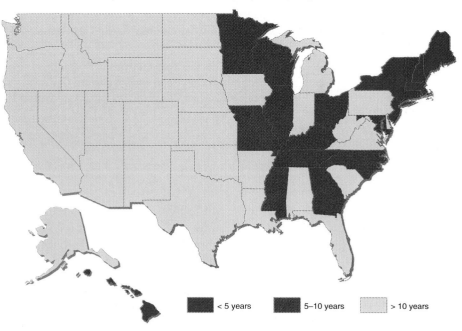

Years of Remaining Landfill Capacity

■ < 5 years   ■ 5–10 years   □ > 10 years

b.

FIGURE 18.2   Disposable Lifestyles
Landfills are the primary method of dealing with waste in the United States (*a*) but the disposable lifestyle of the past has created a shortage of space in landfills for many states in the Northeast and central portions of the United States. (*b*) The development of new landfills is expensive, and in many areas there is not available land appropriate for development.

Source: (*b*) *Directory and Atlas of Solid Waste Disposal Facilities,* 1996.

that followed World War II. Marketing experts set to work trying new tactics to get consumers to buy and toss or, as economists would say, to "stimulate consumption." In the mid-1950s, marketing consultant Victor Lebow wrote an emotional plea for "forced consumption" in the New York *Journal of Retailing:*

> Our enormously productive economy demands that we make consumption a way of life, that we convert the buying and use of goods into rituals, that we seek our spiritual satisfaction in consumption. . . . We need things consumed, burned up, worn out, replaced, and discarded at an ever-growing rate.

Consumers were quick to adapt to the lifestyle that Lebow envisioned. In fact, it was not long until a disposable, throwaway lifestyle was seen as a consumer's right. The idea was to sell "convenience" to the prosperous postwar consumers. What was initially a convenience was soon to become a "necessity"; at least, that was what the advertisements were telling people.

A good example of this way of thinking is the TV dinner, which was first marketed in 1953. The food-packaging industry was evolving into a major service, closely allied with marketing and advertising. From the first TV dinner, we progressed rapidly, so that by the late 1980s, microwave meals were available. One such meal consisted of 340 grams of edible material and six separate layers of packaging, five of them plastic.

After several decades of throwaway living, the disposable lifestyle is changing. Many cities and counties face a shortage of space in old landfills. Solid waste has become a major problem and, in a growing list of communities around the world, it has reached crisis proportions. Many states in the heavily populated Eastern United States are currently running out of places to put their garbage. (See figure 18.2.)

THE NATURE OF THE PROBLEM

As we enter the twenty-first century, the world must take steps to solve the

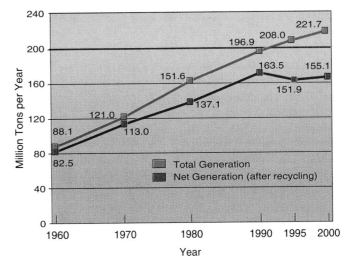

Total and Net Municipal Solid Waste Generation—1960 to 2000

Per Capita Municipal Solid Waste Generation

b.

FIGURE 18.3 Municipal Solid Waste Generation Rates The generation of municipal solid waste in the United States has increased steadily. However, because of increased recycling rates the net production rates (after recyclables have been removed) has actually fallen since 1990 (*a*) and the per capita rate has stabilized (*b*).

Source: *Characterization of MSW in the U.S.;* 1996 Update. U.S.; EPA, Washington, D.C.

ever-increasing burden of garbage—or as the professionals say, **municipal solid waste.** We are all part of the problem, but we can also be part of the solution.

The United States produces about 220 million tons of municipal solid waste each year. This equates to about 2 kilograms of trash per person per day or 0.73 tons per person per year. The amount of municipal solid waste has more than doubled since 1960 and the per capita rate has increased by nearly 70 percent in that same time, although per capita rates began to stabilize about 1990. When recycling is included the net waste produced has actually fallen since 1990. (See figure 18.3.)

Nations with high standards of living and productivity tend to have more municipal solid waste per person than less-developed countries. (See figure 18.4.) The United States and Canada, therefore, are world leaders in waste production. This high volume of solid waste and reliance on landfills for disposal have led to the problems we face today. For example, Toronto, Canada, is running out of places to put its municipal solid waste. Even with a very ambitious plan to reduce waste production by 50 percent, the metropolitan area will run out of space by 2006. The outlying districts are not willing to be the site of a new landfill for Toronto and the metropolitan area is looking for disposal sites in the United States. Many states that could serve as disposal sites are beginning to pass laws that regulate the importation of waste from other countries and Metropolitan Toronto continues to struggle with its waste problem.

Archeologists rely on the waste of past societies to tell them about the nature of the culture and lifestyle of ancient

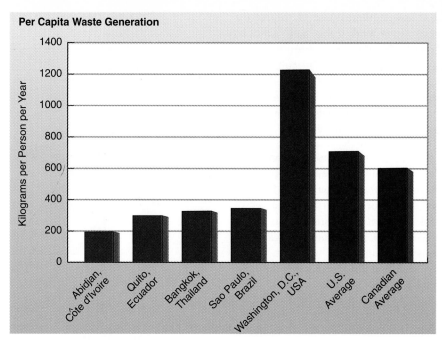

Per Capita Waste Generation

FIGURE 18.4   Waste Generation and Lifestyle The waste generation rates of people is directly related to their economic condition. People in richer countries produce more garbage than those in poorer countries.

Source: Data from World Resources 1996-97 and U.S. EPA.

Changes in Composition of Municipal Solid Waste

FIGURE 18.5   The Changing Nature of Trash While food waste was the third most important component of solid waste in 1960, it slipped to fifth as metal and plastic increased. Changes in lifestyle and packaging have led to a change in the nature of trash. Most of what is currently disposed of could be recycled.

Source: Data from the U.S. Environmental Protection Agency.

civilizations. In the same way today our municipal solid waste is a reflection of our society. Figure 18.5 shows how the composition of our trash has changed since 1960. Notice particularly the increase in the amount of paper and plastic and the effect that recycling has had on the amount of glass in the trash. In the United States the two most common items in the waste stream are paper products and yard waste, and other significant segments are wood, metal, glass, plastics, and food waste. (See figure 18.6.) An analysis of the composition of our waste will present us with possible approaches to reducing the amount of waste we generate.

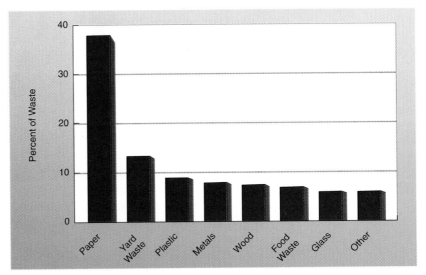

FIGURE 18.6 Composition of Trash in the United States (1995) Paper and yard waste are the most common materials disposed of, accounting for 53.5 percent of the waste stream.

Source: Data from the U.S. Environmental Protection Agency.

METHODS OF WASTE DISPOSAL

Until recently, the disposal of municipal solid waste did not attract much public attention. From prehistory through the present day, the favored means of disposal was simply to dump solid wastes outside of the city or village limits. Frequently, these dumps were in a wetlands adjacent to a river or lake. To minimize the volume of the waste, the dump was often burned. This method of waste disposal was state of the art until only recently. Unfortunately, this method is still being used in remote or sparsely populated areas in North America and the world today. (See figure 18.7.)

As better waste-disposal technologies were developed and as values changed, more emphasis was placed on the environment and quality of life. Simply dumping and burning our wastes is no longer an acceptable practice from an environmental or health perspective. While the technology of waste disposal has evolved during the past several decades, our options are still limited. Realistically, there are no ways of dealing with waste that have not been known for many thousands of years. Essentially, four techniques are used: (1) landfills, (2) incineration, (3) source reduction, and (4) recycling.

Landfills have historically been the primary method of waste disposal because it is the cheapest and most convenient, and because the threat of groundwater contamination was not initially recognized. As we

FIGURE 18.7 Burning Landfills In the past it was common practice to burn the waste in landfills to reduce the volume. Waste is still being burned in sparsely populated areas in North America and other parts of the world today.

have recognized some of the problems associated with poorly designed landfills, efforts to reduce the amount of material placed in landfills have been substantial. Although the amount of waste has increased, composting and recycling have removed significant amounts of materials from the waste stream and the amount of material entering landfills has declined.

(See figure 18.8.) However, the landfill of today is far different from a simple hole in the ground into which garbage is dumped.

Landfill

A modern **municipal landfill** is typically a depression in an impermeable clay layer that is lined with an impermeable

Resins Used in Consumer Packaging

Thermoplastics, which account for 87 percent of plastics sold, are the most recyclable form of plastics because they can be remelted and reprocessed, usually with only minor changes in their properties. Thermoplastic resins are commonly used in consumer packaging applications:

1. Polyethylene terephthalate (PET) is used extensively in rigid containers, particularly beverage bottles for carbonated beverages and medicine containers.

2. Polyethylene is the most widely used resin. High-density polyethylene (HDPE) is used for rigid containers, such as milk and water jugs, household-product containers, and motor oil bottles.

3. Polyvinyl chloride (PVC) is a tough plastic often used in construction and plumbing. It is also used in some food, shampoo, oil, and household-product containers.

4. Low-density polyethylene (LDPE) is often used in films and bags.

5. Polypropylene (PP) is used in a variety of areas, from yogurt containers to battery cases to disposable diaper linings. It is frequently interchanged for polyethylene or polystyrene.

6. Polystyrene (PS) is best known as a foam in the form of cups, trays, and food containers. In its rigid form, it is used in cutlery.

7. Other. These usually contain layers of different kinds of resins and are most commonly used for squeezable bottles (for example, ketchup).

Currently, HDPE and PET are the two most commonly recycled resins. Efforts to recycle PS are also being explored, especially within the fast-food industry.

Source: Graph data from Center for Plastics Recycling Research.

membrane. Each day's deposit of fresh garbage is covered with a layer of soil. Selection of modern landfill sites must be based on an understanding of groundwater geology, soil type, and sensitivity to local citizens' concerns. Once the site is selected, extensive construction activities are necessary to prepare it for use. New landfills have complex bottom layers to trap contaminant-laden water leaking through the buried trash. In addition, monitoring systems are necessary to detect methane gas production and groundwater contamination. In some cases, methane produced by rotting garbage is collected and used to generate electricity. The water that leaches through the site must be collected and treated. As a result, new landfills are becoming increasingly more complex and expensive. They currently cost up to $1 million per hectare to prepare. (See figure 18.9.)

Today, about 57 percent of United States and about 80 percent of Canadian municipal solid waste goes into landfills, but this method is failing to handle the volume. For example, New York City's Fresh Kills Landfill on Staten Island receives 13,000 tons of trash each day. Other trash is shipped to Pennsylvania and other states. However, the Fresh Kills Landfill is expected to close at the end of 2001 so alternative arrangements for solid waste disposal are needed.

The problem New York City faces with the closing of its Fresh Kills Landfill is similar to what many communities will be facing in the near future. The number of landfills is declining. In 1988 there were about 8000 landfills and in 1996 there were just over 3000. The number of landfills has decreased for two reasons. Many small poorly-run landfills have been closed because they were not meeting regulations. Others have closed because they reached their capacity. (See figure 18.10.)

A prolonged public debate over how to replace lost landfill capacity is developing where population density is high and available land is scarce. Siting new landfills in locations like Toronto, New York, and Los Angeles is extremely difficult because of (1) the difficulty in finding a geologically suitable site and (2) local opposition, which is commonly referred to as the NIMBY, or "not-in-my-backyard," syndrome. Resistance by the public comes from concern over groundwater contamination, odors, and truck traffic. Public officials look for alternatives to landfills to avoid problems with the public over landfill siting. Although siting a new landfill may be necessary, politicians are unwilling to take strong positions that might alienate their constituents. This is often called the NIMEY, or "not-in-my-election-year," syndrome.

Japan and many Western European countries have already moved away from landfills as the primary method of waste disposal because of land scarcities and related environmental concerns. Switzerland and Japan dispose less than 15 percent of their waste in landfills, compared to 57 percent in the United States. Instead, recycling and incineration are the primary methods. (See figure 18.11.) In addition, the energy produced by incineration can be used for electric generation or heating.

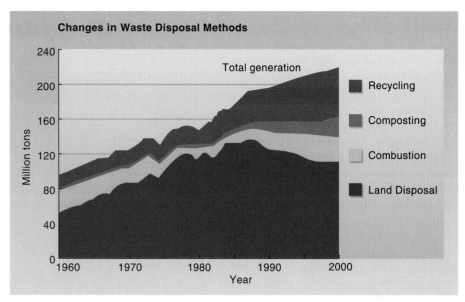

FIGURE 18.8 Changes in Waste Disposal Methods The landfill is still the primary method of waste disposal. Historically, landfills have been the cheapest means of disposal. This may not be the case in the future. Notice that recycling and composting have grown over the past decade while the amount going to landfills has declined somewhat.

Source: Data from the U.S. Environmental Protection Agency.

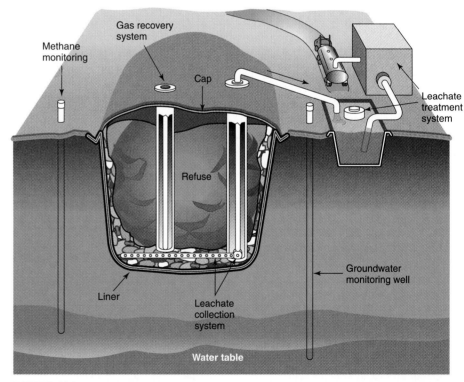

FIGURE 18.9 A Modern Well-Designed Landfill A modern sanitary landfill is far different from a simple hole in the ground filled with trash. A modern landfill is a self-contained unit that is separated from the soil by impermeable membranes and sealed when filled. Methane gas and groundwater are continuously monitored to assure that wastes are not escaping to the air or groundwater.

Source: From National Solid Waste Management Association.

Incineration

Incineration of refuse was quite common in North America and Western Europe prior to 1940. However, many incinerators were eliminated because of aesthetic concerns, such as foul odors, noxious gases, and gritty smoke, rather than for reasons of public health. Today, about 16 percent of the municipal solid waste in the United States is incinerated; Canada incinerates about 8 percent. Most incinerators are not used just to burn trash. The heat derived from the burning is converted into steam and electricity. (See figure 18.12.)

Most incineration facilities burn unprocessed municipal solid waste, which is not as efficient as some other technologies. About one-fourth of the incinerators use refuse-derived fuel—collected refuse that has been processed into pellets prior to combustion.

The newest means of incineration, a European concept, is called **mass burn.** In the mass-burn technique, municipal solid waste is fed into a furnace, where it falls onto moving grates and is burned at temperatures up to 1300°C (2400°F). The burning waste heats water, and the steam drives a turbine to generate electricity, which is sold to a utility.

Incinerators drastically reduce the amount of municipal solid waste—up to 90 percent by volume and 75 percent by weight. Primary risks of incineration, however, involve air-quality problems and the toxicity and disposal of the ash.

Though mass-burn technology works efficiently in Europe, the technology is not easily transferrable. North American municipal solid waste contains more plastic and toxic materials than European waste, thus creating air-pollution and ash-toxicity concerns. Modern incinerators have electrostatic precipitators, dual scrubbers, and fabric filters called baghouses; however, they still release small amounts of pollutants into the atmosphere, including certain metals, acid gases, and also classes of chemicals known as dioxins and furans, which have been implicated in birth defects and several kinds of cancer. The long-term risks from the emissions are still a subject of debate.

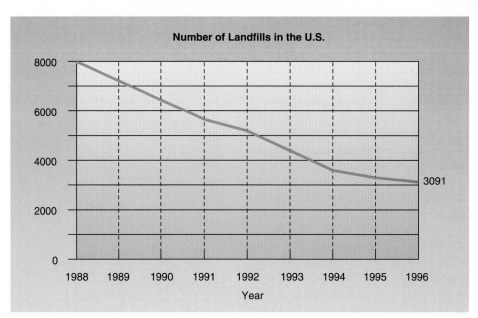

FIGURE 18.10 **Reducing the Number of Landfills** The number of landfills in the United States is declining because they are filling up or because their design and operation do not meet environmental standards.

Source: Data from the U.S. Environmental Protection Agency.

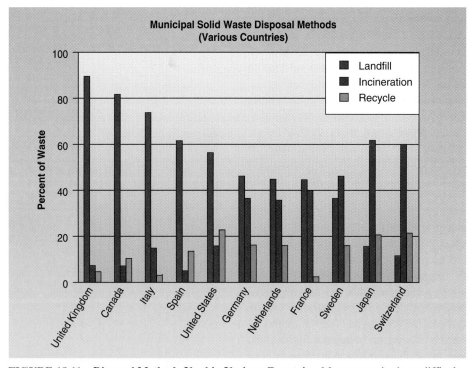

FIGURE 18.11 **Disposal Methods Used in Various Countries** Many countries have difficulty finding adequate space for landfills. Therefore they rely on other technologies such as incineration and recycling to reduce the amount of waste that must be placed in a landfill.

Source: Data from the U.S. Environmental Protection Agency.

Ash from incineration is also a major obstacle to the construction of waste-to-energy facilities. Small concentrations of heavy metals are present both in the air emissions (fly ash) and residue (bottom ash) from these facilities. Because the ash contains lead, cadmium, mercury, and arsenic in varying concentrations from such items as batteries, lighting fixtures, and pigments, this ash may need to be treated as hazardous waste. The toxic substances are more concentrated in the ash than in the original garbage and can seep into groundwater from poorly sealed landfills. Many cities have had difficulty disposing of incinerator ash, and there is still considerable debate about what is the best method of disposal.

The cost and siting of new incinerators are also major concerns facing many communities. Incinerator construction is often a municipality's single largest bond issue. Incinerator construction costs in North America in 1995 ranged from $35 million to $300 million, and the costs are not likely to decline.

Incineration is also more costly than landfills in most situations. Figure 18.13 shows the cost of landfills in comparison with solid waste incinerators. As long as landfills are available they will have a cost advantage. When cities are unable to dispose of their trash locally in a landfill and must begin to transport the trash to distant sites, incinerators become more cost effective. The U.S. Environmental Protection Agency (EPA) has not looked favorably on the construction of new waste-to-energy facilities and has encouraged recycling and source reduction as more effective ways to reduce the solid waste problem. Critics have argued that cities and towns have impeded waste reduction and recycling efforts by putting a priority on incinerators and committing resources to them. Proponents of incineration have been known to oppose source reduction. They argue that incinerators need large amounts of municipal solid waste to operate and that reducing the amount of waste generated makes incineration impractical. Many communities that have opposed incineration say that they support

a.

b.

FIGURE 18.12 A Mass Burn Incineration System (*a*) The diagram shows how a typical mass burn, waste-to-energy incinerator operates. (*b*) Incineration of municipal solid waste reduces its weight and volume significantly. However, there are concerns about air-quality problems and the toxicity and disposal of the ash.

Source: (*a*) Data from the U.S. Environmental Protection Agency.

a vigorous waste-reduction and recycling effort.

Source Reduction

The most fundamental way to reduce waste is to prevent it from ever becoming waste in the first place. Examples of

source reduction are using less material when making a product or converting from heavy packaging materials to lightweight ones. Some packagers have converted to lightweight aluminum and plastic by reducing the thickness of packages and thus the amount of packaging waste. Today, plastic milk jugs weigh

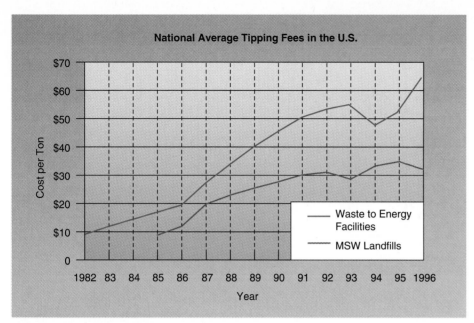

National Average Tipping Fees in the U.S.

Waste to Energy Facilities
MSW Landfills

FIGURE 18.13 Relative Costs of Landfills and Incinerators In general, the cost of disposing waste by incineration is greater than that of a landfill. However, in areas where landfills are not available and transportation costs for land disposal are high, incineration is a logical alternative.

Source: Data from the U.S. Environmental Protection Agency.

FIGURE 18.14 Source Reduction Plastic soft-drink bottles are now 35 percent lighter than in 1977. The weight of plastic milk jugs have been reduced by half since they were introduced and the weight of aluminum cans has also been reduced by 35 percent since 1965.

about half of what they weighed when they were first introduced. Since 1965, aluminum cans have been reduced in weight by 35 percent. (See figure 18.14.)

Another way companies reduce waste is by making consumer products in concentrated form. These concentrated products can then be packaged in smaller containers. This approach, however, requires consumers to favor concentrates when making purchasing decisions.

Municipal composting is another source-reduction technique that could have substantial impact, since 14 percent of material going into landfills in North America is yard waste. In order to extend the life of landfills many states have instituted bans on placing yard waste in landfills. (See figure 18.15.) Building a compost pile is a popular practice with many home gardeners because it turns waste into a useful soil additive. (See figure 18.16.) Many communities require separate collection of such yard waste as grass clippings, leaves, and even discarded Christmas trees. Such waste is either composted or shredded into wood chips. The by-products can then be used for landscaping at city parks, schools, golf courses, and cemeteries. Preserving space

in landfills by municipal composting is gaining in popularity because it is cost effective and substantially extends a landfill's useful life.

On an individual level, we can all attempt to reduce the amount of waste we generate. Every small personal commitment from each of us could have the cumulative result of a significant reduction in municipal solid waste.

Recycling

The United States recycles about 23 percent of its municipal solid waste and should approach 30 percent by 2000. Canada recycles about 10 percent. The goal for Ontario, Canada, is to reduce the amount of garbage going to the landfill sites by 50 percent by the year 2000. Recycling, along with source reduction, is a major part of the Ontario plan.

Recycling initiatives have grown rapidly in North America during the past several years. In the United States, mandatory recycling laws for all materials have passed in fifteen states. In 1990, 1000 U.S. cities had curbside recycling programs. By 1997, the number had grown to nearly 9000 cities. In Canada, Toronto, and

Mississauga, and Ontario, have comprehensive recycling programs. Recycling rates for many materials have grown significantly in recent years. (See figure 18.17.)

Benefits of Recycling

Some benefits of recycling are readily recognizable, such as conservation of resources and pollution reduction. These perceived benefits provide the primary motivation for participation in recycling programs. The following examples illustrate the resource-conservation and pollution-reducing benefits of recycling:

One Sunday edition of the *New York Times* consumes 62,000 trees. Currently, about 40 percent of all paper that enters the waste stream in North America is recycled.

The United States imports nearly all of its aluminum and recycles over 60 percent of its aluminum beverage cans.

There are no sources of tin within the United States; however, about 2 kilograms of tin can be reclaimed from each 1000 kilograms of metal food cans.

Crushed glass (cullet) reduces the energy required to manufacture new glass by 50 percent. Cullet lowers the temperature requirements of the

States with Yard Waste Bans (1997)

FIGURE 18.16 Composting Household composting of yard and other waste is a popular practice with many home gardeners. Composting reduces the amount of garbage produced and provides a material useful for gardening.

FIGURE 18.15 States with Yard Waste Bans Since yard waste is such an important segment of the solid waste stream, many states have passed laws that prohibit the deposition of yard waste in landfills. This will extend the useful life of the landfill. In order to accommodate their citizens many communities still collect yard waste but have instituted composting programs to deal with the yard waste they collect. The numbers on the map indicate the number of composting programs in each state.

Source: Data from the U.S. Environmental Protection Agency.

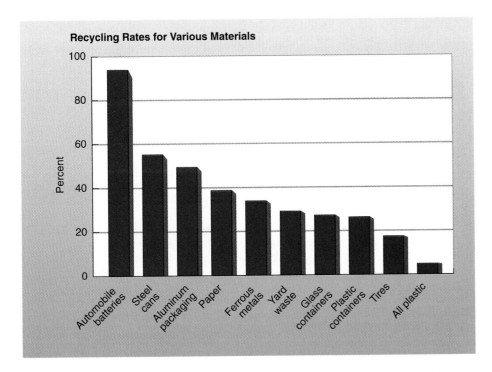

Recycling Rates for Various Materials

(Bar chart — vertical axis: Percent, 0 to 100. Categories along horizontal axis: Automobile batteries, Steel cans, Aluminum packaging, Paper, Ferrous metals, Yard waste, Glass containers, Plastic containers, Tires, All plastic.)

FIGURE 18.17 Recycling Rates for Various Materials Recycling rates for materials that have high value such as automobile batteries are extremely high. Other materials are more difficult to market. But recycling rates today are much higher than in the past as technology and markets have found uses for materials that once were considered valueless.

Source: Data from the U.S. Environmental Protection Agency, *Characterization of MSW in the U.S.*, 1996.

glassmaking process, thus conserving energy and reducing air pollution. In 1996, 36 percent of the glass in the solid waste stream was recycled.

While recycling is a viable alternative to landfilling or the incineration of municipal solid waste, recycling does present several problems.

Recycling Concerns

Problems associated with recycling tend to be either technical or economic. Technical questions are of particular concern when recycling plastics. (See figure 18.18.) While the plastics used in packaging are recyclable, the technology to do so differs from plastic to plastic. There are many different types of plastic polymers. Since each type has its own chemical makeup, different plastics cannot be recycled together. In other words, a milk container is likely to be high-density polyethylene (HDPE), while an egg container is polystyrene (PS), and a soft-drink bottle is polyethylene terephthalate (PET).

Plastic recycling is still a relatively new field. Industry is researching new technologies that promise to increase the quality of plastics recycled and that will allow mixing of different plastics. Until such technology is developed, separation of different plastics before recycling will be necessary.

Container Laws

In October 1972, Oregon became the first state to enact a "Bottle Bill." This statute required a deposit of two to five cents on all beverage containers that could be reused. It banned the sale of one-time-use beverage bottles and cans. The purpose of the legislation was to reduce the amount of litter. Beverage containers were estimated to make up about 62 percent of the state's litter. The bill succeeded in this respect: within two years after it went into effect, beverage-container litter decreased by about 49 percent.

Those opposed to the bottle bill cited a loss of jobs that resulted from the enactment of the bill. In two small can-manufacturing plants, 142 persons did lose their jobs, but hundreds of new employees were needed to handle the returnable bottles. Other arguments against bottle bills focused on the "major" inconvenience to consumers who would need to return the bottles and cans to the store and on retailers' storage problems. Neither of these problems has proven to be serious. Nine other states—Vermont, Maine, Connecticut, New York, Iowa, Rhode Island, Michigan, Delaware, and California—have enacted legislation requiring deposits on bottles, specifically beverage containers. Many states, such as New Jersey, Rhode Island, and California, are turning more toward mandating recycling laws. In 1991, Maine expanded its bottle law to all nondairy beverage containers, including all noncarbonated juice containers holding a gallon or less. Deposits are five cents except on liquor and wine bottles, which carry a fifteen-cent deposit. Excluded are containers for milk and other dairy products, cough syrup, baby formula, soap, and vinegar.

Many argue that a national bottle bill is long overdue. A national bottle bill would reduce litter, save energy and money, and create jobs. It would also help to conserve natural resources. But the lobbying efforts of the soft drink and brewing industries is very strong and the U.S. Congress currently has failed to pass a national container law.

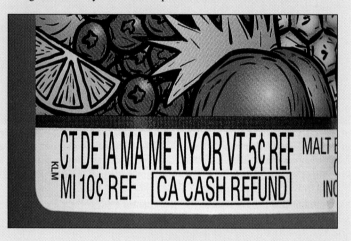

What You Can Do to Reduce Waste and Save Money

You can make a difference. While this statement is sometimes overused, it does speak the truth when it comes to your ability to lessen the stream of solid waste being generated every day. Here are a few ideas that are easy to follow, will save you money, and will help reduce waste:

- Buy things that last, keep them as long as possible, and have them repaired, if possible.
- Buy things that are reusable or recyclable, and be sure to reuse and recycle them.
- Buy beverages in refillable glass containers instead of cans or throwaway bottles.
- Use plastic or metal lunch boxes and metal or plastic garbage containers without throwaway plastic liners.
- Use rechargeable batteries.
- Skip the bag when you buy only a quart of milk, a loaf of bread, or anything you can carry with your hands.
- Use sponges and washable cloth napkins, dish towels, and handkerchiefs instead of paper ones.

- Don't use throwaway paper and plastic plates and cups, eating utensils, razors, pens, lighters, and other disposable items when reusable or refillable versions are available.
- Buy recycled goods, especially those made by primary recycling, and then recycle them.
- Recycle all newspaper, glass, and aluminum, and any other items accepted for recycling in your community.
- Reduce the amount of junk mail you get. This can be accomplished by writing to Mail Preference Service, Direct Marketing Association Inc., 1120 Avenue of the Americas, New York, NY 10036-6700, or by calling (212) 768-7277. Ask that your name not be sold to large mailing-list companies. Of the junk mail you do receive, recycle as much of the paper as possible.
- Push for mandatory trash separation and recycling programs in your community and schools.
- Choose items that have the least packaging or, better yet, no packaging ("nude products").
- Compost your yard and food wastes, and pressure local officials to set up a community composting program.

The economics of recycling are also a primary area of concern. The stepped-up commitment to recycling in many developed nations has produced a glut of certain materials on the market. Markets for collected materials fill up just like landfills. Unless the demand for recycled products keeps pace with the growing supply, recycling programs will face an uncertain future.

On the positive side, the markets for materials collected in recycling programs grew dramatically during the 1990s. The establishment of a recyclables exchange on the Chicago Board of Trade allows for consumers and producers of recyclable materials to participate in an efficient market so that it is less likely that recyclable materials will be left unclaimed. In 1995, the American Forest and Paper Association stated that it was planning to spend $10 billion by 2000 to retool and build new mills to produce recycled paper.

The long-term success of recycling programs is also tied to other economic incentives such as taxing issues and the development of and demand for products manufactured from recycled material.

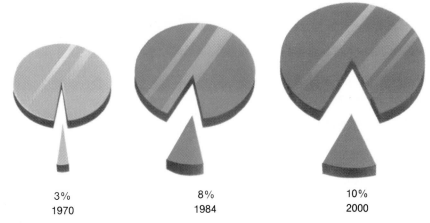

3%
1970

8%
1984

10%
2000

FIGURE 18.18 **Increasing Amounts of Plastics in Trash** Plastics are a growing component of municipal solid waste in North America. Increased recycling of plastics could reverse this trend.

Source: Data from Franklin Associates, Ltd.

Government tax policy needs to be readjusted to encourage recycling efforts. Currently in the United States it is still cheaper to transport virgin material such as fresh-cut pulp wood than to transport collected paper for recycling. Such taxing policy severely inhibits the cost-effectiveness of paper recycling. In addition, on an individual level, we can have an impact by purchasing products made from recycled materials. The demand for recycled products must grow if recycling is to succeed on a large scale. (See Environmental Close-Up: Recyclables Market Basket for photo of items made from recycled waste.)

ENVIRONMENTAL
CLOSE-UP

Recycling Is Big Business

In 1994, Weyerhaeuser Company, one of the world's largest forest products producers, began building a new paper mill in Cedar Rapids, Iowa, an area of North America not known for its abundant forests. The Cedar River Paper Company is a joint venture between Weyerhaeuser and Midwest Recycling Company and is the largest paper recycling plant in the United States. Even with a lack of local trees, the new mill will not be hurting for raw material to make paper and corrugated cardboard.

The mill is supplied with old paper, including scrap boxes from K-Mart. K-Mart signed an agreement with Weyerhaeuser in 1992 to have Weyerhaeuser purchase its cardboard. The agreement provides Weyer-haeuser with a supply of paper for its recycled paper plants and solves a waste disposal problem for K-Mart. The fact that a supply of recycled paper can determine where new paper-making factories will be built indicates just how important the recycling business is becoming. It is projected that by the turn of the century over a third of the raw material that Weyerhaeuser uses for paper making will come from recycled materials. Weyerhaeuser is the third largest paper recycler and collector in the United States. It has 37 wastepaper processing plants in both Canada and the United States. According to a company spokesperson, the recycling business is becoming as significant as the millions of hectares of forests the company owns.

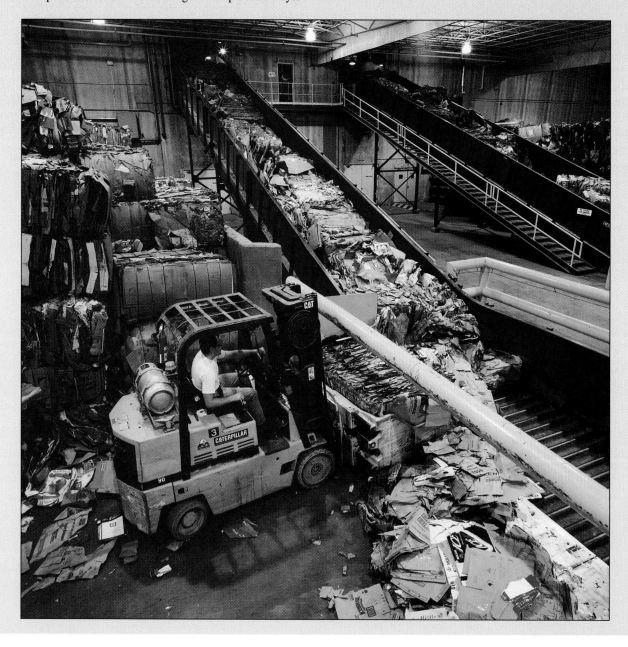

Recyclables Market Basket

Annually in the United States 17 percent of scrap used tires are recovered. Scrap tires are difficult to dispose of in landfills and waste incinerators. An estimated 2 billion to 4 billion are currently stockpiled. These stockpiles can provide convenient habitats for rodents, serve as breeding grounds for mosquitoes, and pose fire hazards. Of the scrap tires that are used, most are burned for energy. Scrap tires also are used for rubberized asphalt paving, molded rubber products, and athletic surfaces.

About 96 percent of automotive batteries are recovered each year in North America. Although these lead-acid batteries constitute a small portion of the waste stream, they contain metals that may be a concern when disposed of in landfills and incinerators. All three components of automotive batteries are recyclable: the lead, the acid, and the plastic casing.

Seventy percent of all used oil is recovered in North America. Only 10 percent of the amount generated by people who change their own motor oil is returned to collection programs. If disposed of improperly, such as being poured down sewage drains, used oil can contaminate soil, groundwater, and surface water. In some communities, used motor oil is collected at service stations, corporate or municipal collection sites, or at the curbside.

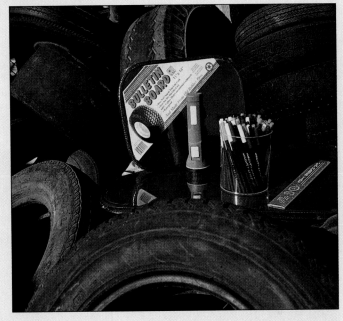

Pens and many other products can be made from recycled tires.

In 1990, McDonald's Corporation announced it would switch from polystyrene to paper for packaging its food products. In making this announcement, McDonald's stated that it was responding to consumer pressure to become more environmentally conscientious. Is using paper to wrap fast-food better for the environment than using polystyrene? The answer is not a simple yes or no.

Advocates for the switch say that polystyrene takes up space in landfills and does not decompose. They also argue that the burning of polystyrene foam in incinerators might release harmful air pollutants.

Opponents of the switch argue that polystyrene can be recycled into useful products, such as insulation board or playground equipment.

They further argue that using paper means cutting forests and that, since the paper used to wrap the food is coated with wax, it cannot be recycled.

Many grocery chains now offer several choices of carryout bags. Some encourage reuse of bags by deducting a small amount from the customer's bill. Others provide a choice of paper or plastic.

- What are the pros and cons of paper and plastic?
- Which alternative do you prefer? Why?
- Is there an alternative to both?

Which is better?

Which do you use?

Is there an alternative?

Summary

Beginning with the post-World War II era, increased consumption of consumer goods became a way of life. Products were designed to be used once and then thrown away. By the 1980s, a disposal lifestyle began to cause problems. There simply were no places to dispose of waste. Barges filled with municipal solid waste from the metropolitan areas along the eastern United States were traveling the world trying to dispose of their unwanted cargo.

Municipal solid waste is managed by landfills, incineration, waste reduction, and recycling. Landfills are the primary means of disposal; however, a contemporary landfill is significantly more complex and expensive than the simple holes in the ground of the past. The availability of suitable landfill land is also a problem in large metropolitan areas.

About 16 percent of the municipal solid waste in the United States is incinerated. While incineration does reduce the volume of municipal solid waste, the problems of ash disposal and air quality continue to be major concerns.

The most fundamental way to reduce waste is to prevent it from ever becoming waste in the first place. Using less material in packaging, producing consumer products in concentrated form, and composting yard waste are all examples of source reduction. On an individual level, we can all attempt to reduce the amount of waste we generate.

About 23 percent of the waste generated in North America is handled through recycling. Recycling initiatives have grown rapidly in North America during the past several years. As a result, the markets for some recycled materials have become very volatile. Recycling of municipal solid waste will only be successful if markets exist for the recycled materials. Another problem in recycling is the current inability to mix various plastics. The plastic industry is working on the development of a more universal plastic.

Future management of municipal solid waste will be an integrated approach involving landfills, incineration, source reduction, and recycling. The degree to which any option will be used will depend on economics, changes in technology, and citizen awareness and involvement.

Review Questions

1. How is lifestyle related to our growing municipal solid waste problem?
2. What four methods are incorporated under integrated waste management?
3. Describe some of the problems associated with modern landfills.
4. What are four concerns associated with incineration?
5. Describe examples of source reduction.
6. Describe the importance of recycling household solid wastes.
7. Name several strategies that would help to encourage the growth of recycling.

Critical Thinking Questions

1. Why do you suppose consumers were so quick to adapt the "use it once, then throw it away" lifestyle after World War II? What values, beliefs, and perceptions does this reveal?
2. How can you help solve the solid waste problem?
3. Given that you only have so much time, should you spend your time acting locally, as a recycling coordinator for example, or advocating for larger political and economic changes at the national level that would solve the waste problems, or do nothing? Why?
4. How does your school or city deal with solid waste? Can solid waste production be limited at your institution or city? How? What barriers exist that might make it difficult to limit solid waste production?
5. It is possible to have a high standard or living, as in North America and Western Europe, and not produce large amounts of solid waste? How?
6. Often environmental costs are hidden from view, and the "correct" response to an environmental problem is not readily apparent. Read the Issues and Analysis section of this chapter. Which alternative do you prefer? Are there other alternatives?
7. Incineration of solid waste is controversial. Do you support solid waste incineration in general? Would you support an incineration facility in your neighborhood?

Key Terms

incineration *363*	municipal solid waste *359*
mass burn *364*	recycling *366*
municipal landfill *361*	source reduction *365*

Related Internet Sites

Visit our website at http://www.mhhe.com/environmentalscience/ for specific information on each of the topics below.

Information on composting

Recycling and waste management

Solid waste management and litter prevention

Canadian waste management

Canadian waste exchanges

Solid waste management

Solid and hazardous waste

Recycling, treatment, and cleanup of solid waste

Sustainable waste management

Solid waste policy and worldwide waste issues

Composting resources

Global Recycling Network

Legislative and regulatory issues and waste management solutions

Waste reduction and prevention and coalition projects

Information on everything from composting to zero emissions

chapter 19

Regulating Hazardous Materials

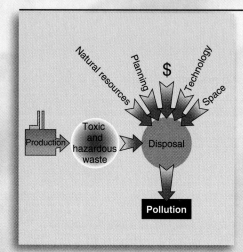

OBJECTIVES

After reading this chapter, you should be able to:

- Distinguish between hazardous substances and hazardous wastes.
- Distinguish between hazardous and toxic materials.
- Explain the complexity in regulating hazardous materials.
- Describe the four characteristics by which hazardous materials are identified.
- Describe the environmental problems of hazardous and toxic materials.
- Understand the difference between persistent and nonpersistent pollutants.
- Describe the health risks associated with hazardous wastes.
- Explain the problems associated with hazardous-waste dump sites and how such sites developed.
- Describe how hazardous wastes are managed, and list five technologies used in their disposal.
- Describe the importance of source reduction with regard to hazardous wastes.

HAZARDOUS AND TOXIC MATERIALS IN OUR ENVIRONMENT

Our modern technological society makes use of a large number of substances that are hazardous and toxic. The benefits gained from using these materials must be weighed against the risks associated with their use.

- Pesticides thought to degrade in soils turn up in rural drinking-water wells.
- Underground plumes of toxic chemicals emanating from abandoned waste sites contaminate city water supplies.
- A gas leak at a chemical production plant in Bhopal, India, killed more than 2000 people.
- Pesticides spilled into the Rhine River from a warehouse near Basel, Switzerland, destroyed a half million fish, disrupted water supplies, and caused considerable ecological damage.
- The collapse of an oil storage tank in Pennsylvania spilled over 2 million liters of oil into the Monongahela River and threatened the water supply of millions of residents.

Toxic and hazardous products and by-products are becoming a major issue of our time. At sites around the world, accidental or purposeful releases of hazardous and toxic chemicals are contaminating the land, air, and water. The potential health effects of these chemicals range from minor, short-term discomforts such as headaches and nausea to serious health problems such as cancers and birth defects (that may not manifest themselves for years) to major accidents that cause immediate injury or death. Today, names like Love Canal, New York, and Times Beach, Missouri, in the United States; Lekkerkerk in the Netherlands; Vac, Hungary; and Minamata Bay, Japan, are synonymous with the problems associated with the release of hazardous and toxic wastes into the environment.

Increasingly, governments and international agencies are attempting to control the growing problem of hazardous substances in our environment. Controlling the release of these substances is difficult

FIGURE 19.1 The Life Cycle of Toxic Substances Controlling the problems of hazardous substances is complicated because of the many steps involved in a substance's life cycle.

since there are so many places in their cycles of use at which they may be released. (See figure 19.1.)

HAZARDOUS AND TOXIC SUBSTANCES— SOME DEFINITIONS

To begin, it is important to clarify various uses of the words *hazardous* and *toxic* as well as distinguish between things that are wastes and those that are not. **Hazardous substances** are those that can cause harm to humans or the environment. The U.S. Environmental Protection Agency (EPA) defines hazardous materials as having one or more of the following characteristics:

1. **Ignitability**—Describes materials that pose a fire hazard during routine management. Fires not only present immediate dangers of heat and smoke but also can spread harmful particles over wide areas. Common examples are gasoline, paint thinner, and alcohol.

2. **Corrosiveness**—Describes materials requiring special containers because of their ability to corrode standard materials, or requiring segregation from other materials because of their ability to dissolve toxic contaminants. Common examples are strong acids and bases.

3. **Reactivity** (or explosiveness)—Describes materials that, during routine management, tend to react spontaneously, to react vigorously with air or water, to be unstable to shock or heat, to generate toxic gases, or to explode. Common examples are gunpowder, which will burn or explode; the metal sodium, which reacts violently with water; and nitroglycerine, which explodes under a variety of conditions.

4. **Toxicity**—Describes materials that, when improperly managed, may release toxicants (poisons) in sufficient quantities to pose a substantial hazard to human health or the environment.

Almost everything that is hazardous is toxic in high enough quantities. For example, tiny amounts of carbon dioxide in the air are not toxic, but high levels are.

Some hazardous materials fall into several of these categories. Gasoline, for example, is ignitable, can explode, and is toxic. It is even corrosive to certain kinds of materials. Other hazardous materials meet only one of the criteria. Polychlorinated biphenyls (PCBs) are toxic but will not burn, explode, or corrode other materials. While the terms *toxic* and *hazardous* are often used interchangeably, there is a difference. **Toxic** commonly refers to a narrow group of substances that are poisonous and cause death or serious injury to humans and other organisms by interfering with normal body physiology. **Hazardous,** the broader term, refers to all dangerous materials, including toxic ones, that present an immediate or long-term human health risk or environmental risk.

Another important distinction that needs to be made is the difference between hazardous substances and hazardous wastes. Although the health and safety considerations regarding hazardous substances and hazardous wastes are similar, the legal and regulatory implications are quite different. Hazardous substances are materials that are used in business and industry for the production of goods and services. Typically, hazardous substances are consumed or modified in industrial processes. **Hazardous wastes** are by-products of industrial, business, or household activities for which there is no immediate use. These materials must be disposed of in an appropriate manner, and there are stringent regulations pertaining to their production, storage, and disposal.

DEFINING HAZARDOUS WASTE

The definition of hazardous waste varies from one country to another. One of the most widely used definitions, however, is contained in the U.S. **Resource Conservation and Recovery Act** of 1976 (RCRA). The RCRA considers wastes toxic and/or hazardous if they:

cause or significantly contribute to an increase in mortality or an increase in serious irreversible, or incapacitating reversible, illness; or pose a substantial present or potential hazard to human health or the environment when improperly treated, stored, transported, disposed of, or otherwise managed.

This definition gives one an appreciation for the complexity of hazardous-waste regulation.

Working from the RCRA definition, the EPA compiled a list of hazardous wastes. Listing is the most common method for defining hazardous waste in European countries and in some state laws. The EPA has also proposed that a hazardous waste be identified by testing it to determine if it possesses any one of the four characteristics discussed earlier: ignitability, reactivity, corrosiveness, and toxicity. If it does, it is subject to regulation under RCRA.

ISSUES INVOLVED IN SETTING REGULATIONS

Whether a hazardous substance is a raw material, an ingredient in a product, or a waste, there are problems associated with determining regulations that pertain to it. In the industrialized countries of Europe and North America, chemical and petrochemical industries produce nearly 70 percent of all hazardous wastes; in developing countries, the figure is 50 to 66 percent. These industries produce many useful materials that are converted into the everyday products we use. Most toxic and hazardous wastes come from chemical and related industries that produce plastics, soaps, synthetic rubber, fertilizers, medicines, paints, pesticides, herbicides, and cosmetics. (See figure 19.2.)

Identification of Hazardous and Toxic Materials

In attempting to regulate the use of toxic and hazardous substances and the generation of toxic and hazardous wastes, most countries simply draw up a list of specific

substances for which there is sufficient scientific evidence linking them to adverse human health or environmental effects. However, since many potentially harmful chemical compounds have yet to be tested adequately, most lists include only the known offenders. Historically, we have often identified toxic materials only after their effects have shown up in humans or other animals. Asbestos was identified as a cause of lung cancer in humans who were exposed on the job, and DDT was identified as toxic to birds when robins began to die and eagles and other fish-eating carnivores failed to reproduce. Once these substances are identified as toxic, their use is regulated. Countries contemplating regulation of hazardous and toxic materials and wastes must consider not only how toxic each one is but also how flammable, corrosive, and explosive it is, and whether it will produce mutations or cause cancer.

Setting Exposure Limits

Even after a material is identified as hazardous or toxic, there are problems in determining appropriate exposure limits. Nearly all substances are toxic in sufficiently high doses. The question is, When does a chemical cross over from safe to toxic? There is no easy way to establish acceptable levels. For any new compounds that are to be brought on the market, extensive toxicology studies must be done to establish their ability to do harm. Usually these are tests on animals. (See Environmental Close-Up: Exposure to Toxins.) Typically, the regulatory agency will determine the level of exposure at which none of the test animals is affected (**threshold level**) and then set the human exposure level lower to allow for a safety margin. This safety margin is important because it is known that threshold levels vary significantly among species, as well as among members of the same species. Even when concentrations are set, they may vary considerably from country to country. For example, in the Netherlands, 50 milligrams of cyanide per kilogram of waste is considered hazardous; in neighboring Belgium, the toxicity standard is fixed at 250 milligrams per kilogram.

The products we use...	The potentially hazardous waste they generate...
Plastics	Organic chlorine compounds
Pesticides	Organic chlorine compounds, organic phosphate compounds
Medicines	Organic solvents and residues, heavy metals (mercury and zinc, for example)
Paints	Heavy metals, pigments, solvents, organic residues
Oil, gasoline, and other petroleum products	Oils, phenols, and other organic compounds, heavy metals, ammonia salts, acids, caustics
Metals	Heavy metals, fluorides, cyanides, acid and alkaline cleaners, solvents, pigments, abrasives, plating salts, oils, phenols
Leather	Heavy metals, organic solvents
Textiles	Heavy metals, dyes, organic chlorine compounds, solvents

FIGURE 19.2 Common Materials Can Produce Hazardous Wastes Many commonly used materials can release toxic wastes if not properly disposed of.

Acute and Chronic Toxicity

Regulatory agencies must look at both the effects of one massive dose of a substance (**acute toxicity**) and the effects of exposure to small doses over long periods (**chronic toxicity**). Acute toxicity is readily apparent because organisms respond to the toxin shortly after being exposed. Chronic toxicity is much more difficult to determine because the effects may not be seen for years. Furthermore, an acute exposure may make an organism ill but not kill it, while chronic exposure to a toxic material may cause death. A good example of this effect is alcohol toxicity. Consuming extremely high amounts of alcohol can result in death (acute toxicity and death). Consuming moderate amounts may result in illness (acute toxicity and full recovery). Consuming moderate amounts over a number of years may result in liver damage and death (chronic toxicity and death).

Another example of chronic toxicity involves lead. Lead has been used in paints, in gasoline, and in pottery glazes for many years, but researchers discovered that it has harmful effects. The chronic effects on the nervous system are most noticeable in children, particularly when children eat paint chips.

Synergism

Another problem in regulating hazardous materials is assessing the effects of mixtures of chemicals. Most toxicological studies focus on a single compound, even though industry workers may be exposed to a variety of chemicals, and in waste dumps, the compounds are usually found in mixtures. Although the materials may be relatively harmless as separate compounds, once mixed, they may become highly toxic and cause more serious problems than do individual pollutants. This is referred to as **synergism.** For example, all uranium miners are exposed to radioactive gases, but those who smoke tobacco and thus are exposed to the toxins in tobacco smoke have unusually high incidences of lung cancer. Apparently, the radioactive gases found in uranium mines interact synergistically with the carcinogens found in tobacco smoke.

Persistent and Nonpersistent Pollutants

The regulation of hazardous and toxic materials is also influenced by the degree of persistence of the pollutant. **Persistent pollutants** are those that remain in the environment for many years in an unchanged condition. Most of the persistent pollutants are human-made materials. An estimated 30,000 synthetic chemicals are used in the United States. They are mixed in an endless variety of combinations to produce all types of products used in every aspect of daily life. They are part of our food, transportation, clothing, building materials, home appliances, medicine, recreational equipment, and many other items. Our way of life is heavily dependent upon synthetic materials.

An example of a persistent pollutant is DDT. It was used as an effective pesticide worldwide and is still used in some countries because it is so inexpensive. However, once released into the environment, it accumulates in the food chain and causes death when its concentration is high enough. (See chapter 14 for a discussion of DDT as a pesticide.)

Another widely used group of synthetic compounds of environmental concern are polychlorinated biphenyls (PCBs). PCBs are highly stable compounds that resist changes from heat, acids, bases, and oxidation. These characteristics make PCBs desirable for industrial use but also make them persistent pollutants when released into the environment. At one time these materials were commonly used in transformers and electrical capacitors. Other uses included inks, plastics, tapes, paints, glues, waxes, and polishes. PCBs are harmful to fish and other aquatic forms of life because they interfere with reproduction. In humans, PCBs produce liver ailments and skin lesions. In high concentration, they can damage the nervous system, and they are suspected carcinogens. In 1970, PCB production was limited to those cases where satisfactory substitutes were not available.

In addition to synthetic compounds, our society uses heavy metals for many purposes. Mercury, beryllium, arsenic, lead, and cadmium are examples of heavy metals that are toxic. When released into the environment, they enter the food chain and become concentrated. In humans, these metals can produce kidney and liver disorders, weaken the bone structure, damage the central nervous system, cause blindness, and lead to death. Because these materials are persistent, they can accumulate in the environment even though only small amounts might be released each year. When industries use these materials in a concentrated form, it presents a hazard not found naturally.

A **nonpersistent pollutant** does not remain in the environment for very long. Most nonpersistent pollutants are biodegradable. Others decompose as a result of inorganic chemical reactions. Still others quickly disperse to concentrations that are too low to cause harm. A biodegradable material is chemically changed by living organisms and often serves as a source of food and energy for decomposer organisms, such as bacteria and fungi. Phenol and many other kinds of toxic organic materials can be destroyed by decomposer organisms.

Other toxic materials such as many of the insecticides are destroyed by sunlight or reaction with oxygen or water in the atmosphere. These include the "soft biocides." For example, organophosphates are a type of pesticide that usually decomposes within several weeks. As a result, organophosphates do not accumulate in food chains because they are pollutants for only a short period of time.

Other toxic and hazardous materials such as carbon monoxide, ammonia, or hydrocarbons can be dispersed harmlessly into the atmosphere (as long as their concentration is not too great) where they eventually react with oxygen.

Because persistent materials can continue to do harm for a long time (chronic toxicity), they are particularly important to regulate. Nonpersistent materials need to be kept below threshold levels to protect the public from acute toxicity. They are not likely to present a danger of chronic toxicity since they either disperse or decompose. Therefore, regulations must be aimed at eliminating acute toxicity.

ENVIRONMENTAL PROBLEMS CAUSED BY HAZARDOUS WASTES

Hazardous wastes contaminate the environment in several ways. Many hazardous materials are released directly to the environment. Many molecules that evaporate readily are vented directly to the atmosphere or escape from faulty piping and valves. These materials are often not even thought of as being hazardous waste. Once hazardous wastes are produced, they must be stored. Improper storage or even poor bookkeeping may inadvertently result in a release. Uncontrolled or improper incineration of hazardous wastes, whether on land or at sea, can contaminate the atmosphere and the surrounding environment. The discharge of hazardous substances into the sea or into lakes and rivers often kills fish and other aquatic life. Further, disposal on land in dumps that are later abandoned, or in improperly controlled landfills, can pollute both the soil and the groundwater as materials leach below the site.

Because most hazardous wastes are disposed of on or in land, the most serious environmental effect is contaminated groundwater. In the United States alone, an

Exposure to Toxins

We are all exposed to materials that are potentially harmful. The question is, at what levels is such exposure harmful or toxic? One measure of toxicity is LD_{50}, the dosage of a substance that will kill (lethal dose) 50 percent of a test population. Toxicity is measured in units of poisonous substance per kilogram of body weight. For example, the deadly chemical that causes botulism, a form of food poisoning, has an LD_{50} in adult human males of 0.0014 milligrams per kilogram. This means that if each of 100 human adult males weighing 100 kilograms consumed a dose of only 0.14 milligrams—about the equivalent of a few grains of table salt—approximately 50 of them will die.

Lethal doses are not the only danger from toxic substances. During the past decade, concern has been growing over minimum harmful dosages, or threshold dosages, of poisons, as well as their sublethal effects.

The length of exposure further complicates the determination of toxicity values. Acute exposure refers to a single exposure lasting from a few seconds to a few days. Chronic exposure refers to continuous or repeated exposure for several days, months, or even years. Acute exposure usually is the result of a sudden accident, such as the tragedy at Bhopal,

India, mentioned at the beginning of the chapter. Acute exposures often make disaster headlines in the press, but chronic exposure to sublethal quantities of toxic materials presents a much greater hazard to public health. For example, millions of urban residents are continually exposed to low levels of a wide variety of pollutants. Many deaths attributed to heart failure or such diseases as emphysema may actually be brought on by a lifetime of exposure to sublethal amounts of pollutants in the air.

---

estimated 100,000 active industrial landfill sites may be possible sources of groundwater contamination, along with 200 special facilities for disposal of both liquid and solid hazardous wastes, and some 180,000 surface impoundments (ponds) for all types of waste. (See chapter 15.) Nearly 2 percent of North America's underground aquifers could be contaminated with such chemicals as chlorinated solvents, pesticides, trace metals, and PCBs. Once groundwater is polluted with hazardous wastes, the cost of reversing the damage is prohibitive. In fact, if an aquifer is contaminated with organic chemicals, restoring the water to its original state is seldom physically or economically feasible.

HEALTH RISKS ASSOCIATED WITH HAZARDOUS WASTES

Because most hazardous wastes are chemical wastes, controlling chemicals and their waste products is a major issue in most developed countries. Every year,

roughly 1000 new chemicals join the nearly 70,000 in daily use. Many of these hazardous chemicals are toxic, but they pose little threat to human health unless they are used or disposed of improperly. For example, many insecticides are extremely toxic to humans. However, if they are stored, used, and disposed of properly they do not constitute a human health hazard. Unfortunately, at the center of the hazardous-waste problem is the fact that the products and by-products of industry are often handled and disposed of improperly. Table 19.1 is a list of 15 top toxic materials as identified by the Agency for Toxic Substances and Disease Registry.

Establishing the medical consequences of exposure to toxic wastes is extremely complicated. The problem of linking a particular chemical or other hazardous waste to specific injuries or diseases is further compounded by the lack of toxicity data on most hazardous substances.

Although assessing environmental contamination from toxic wastes and determining health effects is extremely difficult, what little is known is cause for concern. Most older hazardous-waste dump sites, for

example, contain dangerous and toxic chemicals along with heavy-metal residues and other hazardous substances.

HAZARDOUS-WASTE DUMPS—A LEGACY OF ABUSE

In the United States prior to the passage of the Resource Conservation and Recovery Act (RCRA) in 1976, hazardous waste was essentially unregulated. Similar conditions existed throughout most of the industrial nations of the world. The solution to hazardous-waste disposal was simply to bury or dump the wastes without any concern for potential environmental or health risks. Such uncontrolled sites included open dumps, landfills, bulk storage containers, and surface impoundments. These sites were typically located convenient to the industry and were often in environmentally sensitive areas, such as floodplains or wetlands. Rain and melting snow soaked through the sites, carrying chemicals that contaminated underground waters. When

Toxic Chemical Releases

In 1987, as the result of a new EPA requirement, industries in the United States had to report toxic chemicals released into the environment. Any industrial plant that released 23,000 kilograms or more of toxic pollutants was required to file a report. Industrial plants that released under 23,000 kilograms were not required to file; thus, the data are incomplete. In 1998, 22,400 reports were filed, covering 332 toxic chemicals.

About 2 billion kilograms of toxic chemicals were reported released into the environment by industry in 1988, compared with nearly 2.3 billion in 1987. Another 800 million kilograms—not counted in the overall figure—were sent to municipal-waste treatment centers or private treatment and storage facilities.

According to the 1998 data, among the chemicals routinely emitted from industrial sources were 77 carcinogens. The most widely released cancer-causing chemical—52 million kilograms—was dichloromethane, a chemical often used as an industrial solvent and paint stripper. Industry released a total of 128 million kilograms of carcinogens, including such chemicals as arsenic, benzene, and vinyl chloride. Chemical and allied industries accounted for about 60 percent of the total releases.

Toxic Substances Commonly Released into the Environment

Name of Substance	Number of states that listed it in the top five
Methanol	37
Toluene	37
Ammonia	34
Xylene	22
Hydrochloric Acid	17

Toxic Releases to the Environment (U.S. 1998)
Environmental Compartment Affected

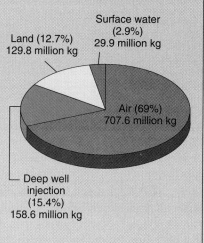

Land (12.7%) 129.8 million kg

Surface water (2.9%) 29.9 million kg

Air (69%) 707.6 million kg

Deep well injection (15.4%) 158.6 million kg

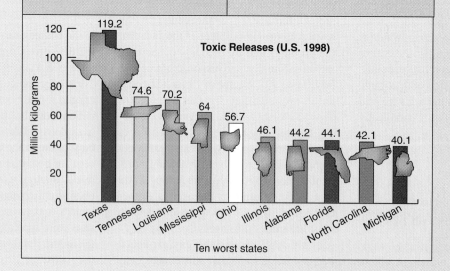

Toxic Releases (U.S. 1998)

Ten worst states

State	Million kilograms
Texas	119.2
Tennessee	74.6
Louisiana	70.2
Mississippi	64
Ohio	56.7
Illinois	46.1
Alabama	44.2
Florida	44.1
North Carolina	42.1
Michigan	40.1

Lead and Mercury Poisoning

Lead and mercury are naturally present in the environment and probably have been a source of pollution for centuries. For example, the lead drinking and eating vessels used by the wealthy Romans may have caused many of their deaths. In 1865, when Alice in Wonderland was published, one of the characters was the Mad Hatter. At that period in history, mercury was widely used in the treatment of beaver skins for making hats. As a result of exposure to mercury, hat makers often suffered from a variety of mental problems; hence, the phrase "mad as a hatter."

In 1953, a number of physical and mental disorders in the Minamata Bay region of Japan was diagnosed as being caused by mercury: 52 people developed symptoms of mercury poisoning, 17 died, and 23 became permanently disabled. In 1970, an outbreak of mercury poisoning in North America was traced to mercury in the meat of swordfish and tuna. In both incidents, the toxic material was not metallic mercury but a mercurous compound, methylmercury. Metallic mercury is converted to methylmercury by bacteria in the water. Methylmercury enters the food chain and may become concentrated as the result of biological amplification. Sufficient amounts in humans can cause brain damage, kidney damage, or birth defects. Today, regulations reduce the release of mercury into the environment and set allowable levels in foods. The problem still persists, however, because it is impossible to eliminate the large amounts of mercury already present in the environment, and it is difficult to prevent the release of mercury in all cases. For example, burning coal releases 3200 metric tons of mercury into the earth's atmosphere each year, and mercury is still "lost" when it is used for various industrial purposes.

Like mercury, lead is a heavy metal and has been a pollutant for centuries. Studies of the Greenland Ice Cap indicate a 1500 percent increase in the lead content today as compared to 800 B.C. These studies reveal that the first large increase occurred during the Industrial Revolution and the second great increase occurred after the invention of the automobile. Oil companies added lead to gasoline to improve performance, and burning gasoline is a major source of lead pollution.

There has been a reduction in airborne lead as a result of North America and Europe reducing lead content in gasoline. Another source of lead pollution is older paints. Prior to 1940, indoor and outdoor paints often contained lead.

There is still disagreement over what levels of lead and mercury can cause human health problems. Although ingested lead from paint can cause death or disability, such a strong correlation cannot be made for atmospheric lead.

There are concerns about fish contamination and public health as a result of methylmercury.

these groundwaters reached streams and lakes, they were contaminated as well. When the sites became full or were abandoned, they were frequently left uncovered, thus increasing the likelihood of water pollution from leaching or flooding, and increasing the chances of people having direct contact with the wastes. At some sites, specifically the uncovered ones, the air was also contaminated as toxic vapors rose from evaporating liquid wastes or from uncontrolled chemical reactions. (See figure 19.3.) In North America alone, the number of abandoned or uncontrolled sites is over 25,000, and the list grows yearly. The costs involved in cleaning up the sites are high. Nearly all industrialized countries are faced with costly, even massive, cleanup bills.

As is to be expected from the amount of toxic wastes generated each year, the United States has the highest number of hazardous-waste dumps needing immediate attention. Europeans are also paying a heavy price for their negligence. Every country in Europe (except Sweden and Norway) is plagued by an abundance of toxic-waste sites—both old and new—needing urgent attention. Holland is a good example. Authorities estimate that up to 8 million metric tons of hazardous chemical wastes may be buried in Holland. Estimates for cleaning up those wastes run as high as $7 billion.

In the republics of the former Soviet Union and Eastern Europe, many hazardous waste sites have been identified recently, and there is no money to pay for cleanup.

In the United States, the federal government has become the principal participant in the cleanup of hazardous-waste sites. The program that deals with the cleanup has popularly become known as **Superfund.** Superfund was established when Congress responded to public pressure to clean up hazardous-waste dumps and protect the public against the dangers of such wastes. The **Comprehensive Environmental Response, Compensation, and Liability Act (CERCLA)**

TABLE 19.1 TOP FIFTEEN HAZARDOUS SUBSTANCES 1998

Substance	Source	Toxic effects
Lead	Lead-based paint Lead additives in gasoline	Neurological damage. Affects brain development in children. Large doses affect brain and kidneys in adults and children.
Arsenic	From elevated levels in soil or water	Multiple organ systems affected. Heart and blood vessel abnormalities, liver and kidney damage, impaired nervous system function.
Metallic mercury	Air or water at contaminated sites	Permanent damage to brain, kidneys, developing fetus.
Vinyl chloride	Plastics manufacturing Air or water at contaminated sites	Acute effects: dizziness, headache, unconsciousness, death. Chronic effects: liver, lung, and circulatory damage.
Benzene	Industrial exposure Glues, cleaning products, gasoline	Acute effects: drowsiness, headache, death at high levels. Chronic effects: damages blood forming tissues and immune system; also carcinogenic.
Polychlorinated biphenyls (PCBs)	Eating contaminated fish Industrial exposure	Probable carcinogens. Acne and skin lesions.
Cadmium	Released during combustion Living near a smelter or power plant Picked up in food	Probable carcinogen, kidney damage, lung damage, high blood pressure.
Benzo[a]pyrene	Product of combustion of gasoline or other fuels In smoke and soot	Probable carcinogen, possible birth defects.
Chloroform	Contaminated air and water Many kinds of industrial settings	Affects central nervous system, liver, and kidneys; probable carcinogen.
Benzo[b]fluoranthene	Product of combustion of gasoline and other fuels Inhaled in smoke	Probable carcinogen.
DDT	From food with low levels of contamination Still used as pesticide in parts of world	Probable carcinogen; possible long-term effect on liver; possible reproductive problems.
Aroclor 1260 (a mixture of PCBs)	From food and air	Probable carcinogens. Acne and skin lesions.
Trichloroethylene	Used as a degreaser, evaporates into air	Dizziness, numbness, unconsciousness, death.
Aroclor 1254 (a mixture of PCBs)	From food and air	Probable carcinogens. Acne and skin lesions.
Chromium (+6)	From food, water, and air Originates from combustion source	Ulcers of the skin, irritation of nose and gastrointestinal tract, also affects kidney and liver.

Source: Data from Agency for Toxic Substances and Disease Registry.

(Superfund) was enacted in 1980. CERCLA had several key objectives:

1. To develop a comprehensive program to set priorities for cleaning up the worst existing hazardous-waste sites.

2. To make responsible parties pay for those cleanups whenever possible.

3. To set up a $1.6 billion Hazardous Waste Trust Fund—popularly known as Superfund—to support the identification and cleanup of abandoned hazardous-waste sites.

4. To advance scientific and technological capabilities in all aspects of hazardous-waste management, treatment, and disposal.

A **National Priority List** of hazardous-waste dump sites requiring urgent attention was drawn up for Superfund action. The size of the list varies, however, depending on which governmental agency is making the list. The U.S. Office of Technology Assessment (OTA) estimates that 10,000 sites may eventually be placed on the National Priority List requiring Superfund cleanup. The OTA believes that cleaning up these hazardous dumps may take 50 years and cost up to $100 billion because of the complex mixture of contaminants in most sites. (See table 19.2.)

TABLE 19.2 COMMON CONTAMINANTS FOUND AT SUPERFUND SITES (IN ORDER OF OCCURRENCE)

Chemical
Lead
Cadmium
Toluene
Mercury
Benzene
Trichloroethylene
Ethylbenzene
Benzo[a]anthracene
Bromodichloromethane
Polychlorinated biphenyls
Toxaphene

Source: Data from USEPA, Office of Emergency and Remedial Resource, 1998.

The Government Accounting Office (GAO) believes that the National Priority List could reach more than 4000 sites with cleanup costs of around $40 billion. The EPA has the shortest list of sites (2500) that it says should be placed on the National Priority List at a cost of nearly $30 billion. Table 19.3 lists Superfund sites considered priority one.

By the late 1990s, the Superfund program was still controversial. Millions of dollars have been spent by both the federal government and industry, but most of the money has involved litigation. The money has paid lawyers but has not paid for cleanup. One of the primary reasons for this is the way CERCLA was written. It provided that anyone who contributed to a specific hazardous-waste site could be required to pay for the cleanup of the entire site regardless of the degree to which it contributed to the problem. Since many industries that contributed to the problem had gone out of business or could not be identified, those who could be identified were asked to pay for the cleanup. Most businesses found it cost-effective to hire lawyers to fight their inclusion in a cleanup effort rather than to pay for the cleanup. Consequently, cleanup has been slow.

In 1997, the U.S. Congress attempted to reauthorize CERCLA. Unable to reach an agreement, CERCLA has been authorized on a year-to-year basis referred to as a "continuing resolution."

MANAGING HAZARDOUS WASTES

In the past, the management of hazardous waste was always added on to the end of the industrial process. The effluents from pipes or smokestacks were treated to reduce their toxicity or concentration. For 1998, the EPA reported that the United States produced about 225 million metric tons of hazardous waste. This was 52 million metric tons less than in 1991. In recent years it has become obvious that a better way to deal with the problem of hazardous waste is to not produce it in the first place. To this end, the EPA and regulatory agencies in other countries have emphasized pollution prevention and waste minimization. Strong regulatory control requires that

FIGURE 19.3 Toxic Chemical Storage This is a site where toxic wastes were improperly stored. Local governments often assume the problems and costs of correcting bankrupt companies' errors.

industries report the hazardous wastes they produce and that the wastes be stored, transported, and disposed of properly.

The EPA now fosters a **pollution-prevention hierarchy** that emphasizes reducing the amount of hazardous waste produced. This involves the following strategy:

First—reduce the amount of pollution at the source.

Second—recycle wastes wherever possible.

Third—treat wastes to reduce their hazard or volume.

Fourth—dispose of wastes on land or incinerate them as a last resort. (See figure 19.4.)

Pollution Prevention

Pollution prevention encourages changes in the operations of business and industry that prevent hazardous wastes from being produced in the first place. Many of these actions are simple to perform and cost little. Primary among them are activities that result in fewer accidental spills, leaks from pipes and valves, loss from broken containers, and similar mishaps. These reductions often can be achieved through better housekeeping and awareness training for employees at little cost. Many industries actually save money because they need to buy less raw material because less is being lost.

Waste Minimization

Waste minimization involves changes that industries could make in the way they manufacture products that would reduce the waste produced. For example, it may be possible to change a process so that a solvent that is a hazardous material is replaced with water, which is not a hazardous

TABLE 19.3 PROPOSED NATIONAL LIST OF SUPERFUND CLEANUP SITES IN THE UNITED STATES (PRIORITY ONE)

State/City/County	Site name	State/City/County	Site name
Alabama		**Minnesota**	
Limestone and Morgan	Triana, Tennessee River	Brainerd Baxter	Burlington Northern
Arkansas		Fridley	FMC
Jacksonville	Vertac, Inc.	New Brighton/Arden	New Brighton
California		St. Louis Park	Reilly Tar
Glen Avon Heights	Stringfellow	**Montana**	
Delaware		Anaconda	Anaconda-Anaconda
New Castle	Army Creek	Silver Bow/Deer Lodge	Silver Bow Creek
New Castle County	Tybouts Corner	**New Hampshire**	
Florida		Epping	Kes-Epping
Jacksonville	Pickettville Road Landfill	Nashua	Sylvester, Nashua
Plant City	Schuylkill Metals	Somersworth	Somersworth Landfill
Indiana		**New Jersey**	
Gary	Midco I	Bridgeport	Bridgeport Rent & Oil
Iowa		Fairfield	Caldwell Trucking
Charles City	Labounty Site	Freehold	Lone Pine Landfill
Kansas		Gloucester Township	Gems Landfill
Cherokee County	Tar Creek, Cherokee County	Mantua	Helen Kramer Landfill
Maine		Marlboro Township	Burnt Fly Bog
Gray	McKin Company	Old Bridge Township	CPS/Madison Industries
Massachusetts		Pittman	Lipari Landfill
Acton	W.R. Grace	Pleasantville	Price Landfill
Ashland	Nyanza Chemical	**New York**	
East Woburn	Wells G&H	Oswego	Pollution Abatement Services
Holbrook	Baird & McGuire	Oyster Bay	Old Bethpage Landfill
Woburn	Industri-Plex	Wellsville	Sinclair Refinery
Michigan		**Ohio**	
Swartz Creek	Berlin & Farro	Arcanum	Arcanum Iron & Metal
Utica	Liquid Disposal Inc.	**Oklahoma**	
		Ottawa County	Tar Creek
		Pennsylvania	
		Bruin Boro	Bruin Lagoon
		Grove City	Osborne
		McAdoo	McAdoo
		Rhode Island	
		Coventry	Picillo Coventry
		South Dakota	
		Whitewood	Whitewood Creek
		Texas	
		Crosby	French, Ltd.
		Crosby	Sikes Disposal Pits
		Houston	Crystal Chemical
		La Marque	Motco

Source: Data from U.S. Environmental Protection Agency, "Proposed National Priorities List: As Provided for in Section 105(8)(b) of CERCLA," December 2, 1982, Washington, D.C.

material. This is an example of source reduction: any change or strategy that reduces the amount of waste produced.

Another strategy is to use the waste produced in a process in another aspect of the process, thus reducing the amount of waste produced. For example, water used to clean equipment might be included as a part of the product rather than being discarded as a contaminated waste.

Another technique that can be used to reduce the amount of waste produced is to clean solvents used in processes. Using a still to purify solvents results in a lower total volume of hazardous waste being produced because the same solvent can be used over and over again.

The simple process of allowing water to evaporate from waste can reduce the total amount of waste produced. Obviously,

the hazardous components of the waste are concentrated by this process.

Recycling of Waste

Often it is possible to use a waste for another purpose and thus eliminate it as a waste. Many kinds of solvents can be burned as a fuel in other kinds of operations. For example, waste oils can be used

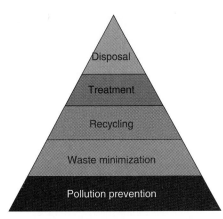

FIGURE 19.4 Pollution-Prevention Hierarchy The simplest way to deal with hazardous wastes is to not produce them in the first place. The pollution-prevention hierarchy stresses reductions in the amount of hazardous waste produced by employing several different strategies.

as fuels for power plants, and other kinds of solvents can be burned as fuel in cement kilns. Care needs to be taken that the contaminants in the oils or solvents are not released into the environment during the burning process, but the burning of these wastes destroys them and serves a useful purpose at the same time.

Similarly, many kinds of acids and bases are produced as a result of industrial activity. Often these can be used by other industries that have a need for them.

Ash or other solid wastes can often be incorporated into concrete or other building materials and therefore do not require disposal. Thus the total amount of waste is reduced.

Treatment of Wastes

Wastes can often be treated in such a way that their amount is reduced or their hazardous nature is modified. Dangerous acids and bases can be reacted with one another to produce materials that are not hazardous.

Hazardous wastes that are biodegradable can be subjected to the actions of microorganisms that destroy the hazardous chemicals. Many kinds of organic molecules can be handled in this way.

Incineration (thermal treatment) can be used to treat a variety of kinds of wastes, although many people are suspicious of this technique because they feel that toxic materials may be escaping from the smokestacks. Wastes are heated in a flame-powered incinerator. Under controlled conditions, incineration can destroy 99.999 percent of organic wastes, and hazardous waste incinerators must burn 99.9999 percent of certain hazardous materials. Incineration accounts for the disposal of only about 2 percent of the hazardous wastes in North America. In Europe, the amount of hazardous wastes destroyed by incineration is higher, but still amounts to less than 50 percent. The relatively high costs of incineration (compared with landfills) and concerns for the safety of surrounding areas in case of accidents have kept incineration from becoming a major method of treatment or disposal.

Air stripping is sometimes used to remove volatile chemicals from water. Volatile chemicals, which have a tendency to vaporize easily, can be forced out of liquid when air passes through it. Steam stripping works on the same principle, except that it uses heated air to raise the temperature of the liquid and force out volatile chemicals that ordinary air would not. The volatile compounds can be captured and reused or disposed of.

Carbon absorption tanks contain specifically activated particles of carbon to treat hazardous chemicals in gaseous and liquid waste. The carbon chemically combines with the waste or catches hazardous particles just as a fine wire mesh catches grains of sand. Contaminated carbon must then be disposed of or cleaned and reused.

Precipitation involves adding special materials to a liquid waste. These bind to hazardous chemicals and cause them to precipitate out of the liquid and form large particles called floc. Floc that settles can be separated as sludge; floc that remains suspended can be filtered and the concentrated waste can be sent to a hazardous-waste landfill.

Land Disposal

When all other options have been exhausted, any remaining hazardous wastes are typically disposed of on land. (See table 19.4.) For over 80 percent of their hazardous wastes, North America, Europe, and Japan still rely principally on six methods of disposal:

1. Deep-well injection into porous geological formations or salt caverns.

2. Discharge of treated and untreated liquids into municipal sewers, rivers, and streams.

3. Placement of liquid wastes or sludges in surface pits, ponds, or lagoons.

4. Storage of solid wastes in specially lined dumps covered by soil.

5. Storage of liquid and solid wastes in underground caverns and abandoned salt mines.

6. Sending wastes to sanitary landfills not designated for toxic or hazardous wastes.

There are techniques that reduce the chance that hazardous materials will escape from these locations and become a problem for the public. Immobilizing a waste puts it into a solid form that is easier to handle and less likely to enter the surrounding environment. Waste immobilization is useful for dealing with wastes, such as certain metals, that cannot be destroyed. Once a waste has been immobilized, it can be disposed of in a hazardous-waste landfill. Two popular methods of immobilizing waste are fixation and solidification. Engineers and scientists mix materials such as fly ash or cement with hazardous wastes. This either "fixes" hazardous particles, in the sense of immobilizing them or making them chemically inert, or "solidifies" them into a solid mass. Solidified waste is sometimes made into solid blocks that can be stored more easily than can a liquid.

HAZARDOUS-WASTE MANAGEMENT CHOICES

Today, the two most common methods for disposing of hazardous wastes are land disposal and incineration. The choice between these two methods involves both economic decisions and acceptance by the public. In North America there is abundant land available for land disposal, making it the most economical and most widely used method. In Europe and Japan where land is in short supply and is expensive, incineration is more economical and is a major method for dealing with hazardous waste. Because of concerns about the emissions from incinerators, significant amounts of hazardous

TABLE 19.4 HAZARDOUS-WASTE MANAGEMENT METHODS, NORTH AMERICA

Management method	Share of total waste managed (percent)
Land disposal[1]	67
Discharge to sewers, rivers, streams	22
Distillation for recovery of solvents	4
Burning in industrial boilers	4
Chemical treatment by oxidation	1
Land treatment of biodegradable waste	1
Incineration	1
Recovery of metals through ion exchange	less than 1
Total	**100**

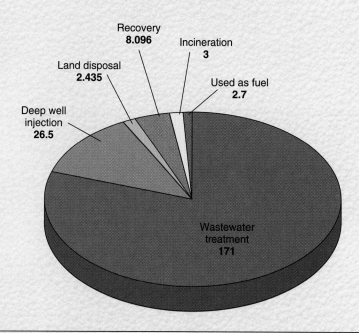

Hazardous Waste Treatment Methods U.S. 1998 (Million tons of waste)

Recovery 8.096
Incineration 3
Land disposal 2.435
Used as fuel 2.7
Deep well injection 26.5
Wastewater treatment 171

[1]Includes injection wells (25 percent of total), surface impoundments (19 percent), hazardous-waste landfills (13 percent), and sanitary landfills (10 percent).

Source: Data from U.S. Environmental Protection Agency, Office of Solid Waste, *Treatment, Storage, and Disposal Facts, 1997*. U.S. Government Printing Office, 1998, Washington, D.C.

waste are incinerated at sea on specially designed ships.

Laws and regulations dealing with hazardous-waste disposal are driving industrial behavior toward pollution prevention and waste minimization. In addition, because the costs of safe disposal are mounting, waste-handling firms—both private and public—are looking for better and cheaper ways to treat and dispose of hazardous wastes. Strong, enforceable laws have eliminated the economic incentives to pollute. Unfortunately, as strong laws have been enacted, some small companies that were unable or, more likely, unwilling to properly dispose of their wastes have turned to illegal nighttime dumping.

The environmental costs of not managing hazardous wastes, as witnessed in virtually every industrialized country, are astronomical. And because major generators of hazardous wastes remain liable for past mistakes, economic and regulatory incentives for complying with hazardous-waste regulations should continue to encourage responsible management.

Love Canal, near Niagara Falls, New York, was constructed as a waterway in the nineteenth century. It was subsequently abandoned and remained unused for many years. In the 1930s, it became an industrial dump. The Hooker Chemical Company purchased the area in 1947 and used it as a burial site for 20,000 metric tons of chemicals.

Hooker, later to become part of Occidental Petroleum, then sold the property to the local government for one dollar. A housing development and an elementary school were constructed on the site. Soon after the houses were constructed, people began to complain about chemicals seeping into their basements. In 1978, 80 different chemicals were found in this seepage. Approximately one dozen probable carcinogens were identified among these chemicals.

As a result of these findings, $27 million in government funds were appropriated in 1978 to purchase homes and permanently relocate 237 families. The funds also provided for the construction of a series of ditches to contain the chemicals and a clay cap to prevent the fumes from entering the atmosphere. But the problems continued.

The remaining 710 families in the Love Canal area were not satisfied with the government's approach to the problem. They cited the fact that women in the area had a 50 percent higher rate of miscarriages. Of 17 reported pregnancies in the area during 1979, two children were born normal, nine had defects, two were stillborn, and four ended in miscarriage. In addition to the abnormalities in birth, there are other biological problems in the Love Canal area.

Neurologists determined that the speed of the nerve impulses in 37 residents who were examined were slower than normal. They stated that chemical exposure could have caused this damage. In 1980, the EPA released the findings of a study that found that 11 out of 36 residents tested in the Love Canal area had broken chromosomes, which are linked to cancer and birth defects. As a result, the federal government released $5 million to temporarily relocate Love Canal residents to motels or other quarters.

By 1990, a $150 million cleanup effort had sealed off the leaky dump, demolished 238 homes nearest the chemical graveyard, and scoured toxins from neighborhood storm sewers and streams (the two major sources of danger to area homes).

In early 1991, some families began to move back into the area. The Love Canal Area Revitalization Agency is planning to sell 236 homes in Love Canal (renamed Black Creek Village). One of the incentives is that the houses can be purchased at a low cost. The families are confident that the houses are safe, but many environmental groups oppose them. In March of 1998 the last of 2300 families received compensation for medical claims. Individual compensation ranged from $400,000 to as little as $83.

- Who is responsible for providing treatment for the physical and mental problems experienced by the residents in this community?

- There is a question no one ever seems to ask. Why were permits ever awarded to construct a thousand-unit housing development on top of a site known to contain 20,000 metric tons of toxic wastes?

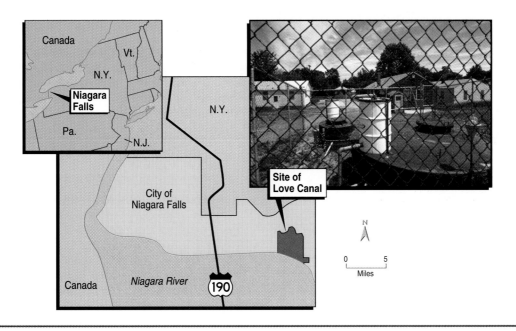

Hazardous Wastes and Toxic Materials in China

Large areas of land are currently utilized in China for the uncontrolled disposal of industrial wastes. Approximately 600 million tons of industrial waste, of which 50 to 70 percent are hazardous, is generated in China annually. Estimates indicate that a total of 5.9 billion tons of industrial waste occupying 540 million cubic meters have been improperly stored or discarded in recent years. The majority of the waste is simply piled on unprotected areas, which causes leaching to surface and groundwater bodies. As a result, environmental accidents are prevalent. For instance, a chromium residue disposal site in Jinzhou caused groundwater pollution in a 12.5-square kilometer area; as a result, water from 1800 wells in nine villages is no longer potable.

The production and use of chemicals is developing rapidly in China. More than 30,000 classes of chemicals are now produced, of which many are toxic. During production, transportation, storage and use, many releases and spills occur. For example, in 1995, the toxic chemical storage in Shenzhen exploded, causing significant damage to life, property, and the environment.

Management of hazardous and toxic materials in China is still in its early stages, therefore treatment and disposal technologies are primitive and equipment is poor. None of China's current hazardous-waste disposal sites meet environmental standards. In part this is due to a lack of funds and management capabilities. A demonstration project, however, is currently underway in China that hopefully will assist them in developing a management system for hazardous and toxic materials.

The "Law of Pollution Prevention and Control of Solid Wastes" has been made a priority item of legislation by the National People's Congress. Research on hazardous-waste management and disposal have been classified as key items in national scientific and technological development plans. Experiments with solid-waste declaration and registration, and waste exchanges are being performed, and standard protocols for chemical testing, toxicity evaluation of synthetic chemicals, laboratory analysis, and risk assessment have begun to be formulated. The long- and short-term objectives of the new project include the following:

- Formulate and strengthen China's hazardous- and toxic-materials control laws and regulations, and establish criteria for sound environmental management of wastes.
- Establish technical support system for hazardous-waste management.
- Design hazardous-waste treatment and disposal demonstration projects.

Formulate laws and regulations, criteria, and policies for toxic-chemicals management.

Construct a hazardous-waste incineration plant with an annual capacity of 3000 tons in Beijing.

This project will introduce hazardous- and toxic-materials control laws and regulations, antipollution criteria, and mitigation measures. The capabilities for the management and control of hazardous materials will be developed and regulatory enforcement will be initiated. The establishment of declaration, registration, licensing, and wastes exchange for hazardous-waste procedures, will hopefully result in a reduction in the volume of waste generated in China while providing an incentive for recovery and reuse.

The hazardous-waste disposal demonstration project will provide a model, and increased capabilities and experience can be utilized by China in establishing regional central-disposal facilities for hazardous wastes. Dissemination of the experience gained in the demonstration project should produce a significant improvement in waste-disposal management strategies resulting in greater awareness, regulation, and environmental protection.

Summary

Public awareness of the problems of hazardous substances and hazardous wastes is relatively recent. The industrialized countries of Europe and North America began major regulation of hazardous materials only during the past 20 years, and most developing countries exercise little or no control over such substances. As a result, many countries are living with serious problems from prior uncontrolled dumping practices, while current systems for management of hazardous and toxic waste remain incomplete and incapable of even identifying all hazardous waste.

A number of fundamental problems are involved in hazardous-waste management. First, there is no agreement as to what constitutes a hazardous waste. Moreover, little is known about the amounts of hazardous wastes generated throughout the world. The issue is further complicated by our limited understanding of the health effects of most hazardous wastes and the fact that large numbers of potentially hazardous chemicals are being developed faster than their health risks can be evaluated.

Hazardous-waste management must move beyond burying and burning. Industries need to be encouraged to generate less hazardous waste in their manufacturing processes. Although toxic wastes cannot be entirely eliminated, technologies are available for minimizing, recycling, and treating wastes. It is possible to enjoy the benefits of modern technology while avoiding the consequences of a poisoned environment. The final outcome rests with governmental and agency policymakers, as well as with an educated public.

Review Questions

1. Explain the problems associated with hazardous-waste dump sites and how such sites developed.
2. Distinguish between acute and chronic toxicity.
3. Give two reasons why regulating hazardous wastes is difficult.
4. In what ways do hazardous wastes contaminate the environment?
5. Describe how hazardous wastes contaminate groundwater.
6. Why is there often a problem in linking a particular chemical or hazardous waste to a particular human health problem?
7. Describe what is meant by the U.S. National Priority List.
8. Describe five technologies for managing hazardous wastes.
9. What is meant by pollution prevention and waste minimization?
10. Describe the pollution-prevention hierarchy.
11. What are RCRA and CERCLA? Why is each important for managing hazardous wastes?

Critical Thinking Questions

1. Scientists at the EPA have to make decisions about thresholds to identify which materials are toxic materials. What thresholds would you establish for various toxic materials? What is your reasoning for establishing the limits you do?
2. Go to the EPA's web site (www.epa.gov/enviro/html/ef_overview.html) and identify the major releasers of toxic materials in your area. Were there any surprises? Are there other releasers of toxic materials that might not be required to list their releases?
3. According to the textbook, in North America alone there are over 25,000 abandoned and uncontrolled hazardous waste dumps. Many were abandoned before the RCRA law of 1976 was passed. Who should be responsible for cleaning up these dumps?
4. Look at the textbook's discussion called "Hazardous Waste—A Legacy of Abuse." Do the authors present the information from a particular point of view? What other points of view might there be to this issue? What information do you think these other viewpoints would provide?
5. Many economically deprived areas, Native American reservations and developing countries that need an influx of cash, have agreed, over significant local opposition, to cite hazardous waste facilities in their areas. What do you think about this practice? Should "outsiders" have a say in what happens within these sovereign territories?
6. After reading about the problem with hazardous wastes and toxic materials in China, should the United States, or any other country, have the right to intervene if another country is creating significant environmental damage? Why?

Key Terms

acute toxicity *377*
chronic toxicity *377*
Comprehensive Environmental Response, Compensation, and Liability Act (CERCLA) *381*
corrosiveness *375*
hazardous *376*
hazardous substances *375*
hazardous wastes *376*
ignitability *375*
incineration *385*
LD$_{50}$ *379*
National Priority List *382*
nonpersistent pollutant *378*
pollution-prevention hierarchy *383*
persistent pollutant *378*
reactivity *375*
Resource Conservation and Recovery Act (RCRA) *376*
Superfund *381*
synergism *378*
threshold level *376*
toxic *376*
toxicity *375*
waste minimization *383*

Related Internet Sites

Visit our website at http://www.mhhe.com/environmentalscience/ for specific information on each of the topics below.

Hazardous Materials magazine

Nuclear waste, land use, and military toxics

Waste sites, health concerns, and hazardous substances

Environmental organizations involved with toxic waste cleanup

Toxicology information

Environmental toxins and toxic substances, from pesticides and heavy metals to fragrances

Safe use, disposal, and recycling of household hazardous waste

Toxins—compliance, enforcement, and storage and disposal of wastes

Federal legislation dealing with hazardous waste

Superfund site information and current programs

20
chapter

Environmental Policy and Decision Making

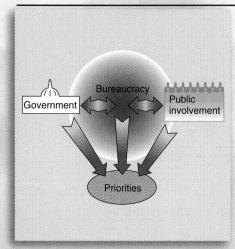

OBJECTIVES

After reading this chapter, you should be able to:

- Explain how the executive, judicial, and legislative branches of the U.S. government interact in forming policy.
- Understand how environmental laws are enforced in the United States.
- Describe the forces that led to changes in environmental policy in the United States during the past three decades.
- Understand the history of the major U.S. environmental legislation.
- Understand why there are concerns about environmental regulations by some individuals in the United States.
- Understand what is meant by "green" politics.
- Describe the reasons why environmentalism is a growing factor in international relations.
- Understand the factors that could result in "eco-conflicts."
- Understand why it is not possible to separate politics and the environment.
- Explain how citizen pressure can influence governmental environmental policies.

CHAPTER OUTLINE

New Challenges for a New Century

Learning from the Past

Thinking about the Future

Defining the Future

The Development of Environmental Policy in the United States

Environmental Policy and Regulation

Environmental Close-Up: *The National Environmental Policy Act*

The Greening of Geopolitics

Environmental Close-Up: *Changing the Nature of Environmental Regulation— The Safe Drinking Water Act*

Environmental Close-Up: *Environmental Backlash—The Wise Use Movement*

Global Perspective: *Eco-Terrorism*

International Environmental Policy

Global Perspective: *Environmental Policy and the European Union*

It All Comes Back to You

Global Perspective: *Overview of an International Organization— The International Whaling Commission (IWC)*

Global Perspective: *Eco-Labels*

NEW CHALLENGES FOR A NEW CENTURY

We live in remarkable times. This is an era of rapid and often bewildering alterations in the forces and conditions that shape human life. This is evident both in the altered nature of geopolitics in the post-Cold War era and in the growing understanding of the relationship between human beings and the natural world. This relationship varies, however, between the developing and developed countries of the world. One of the major challenges of the foreseeable future will be centered around worldwide environmental impact as developing nations evolve economically.

The end of the Cold War has been accompanied by the swift advance of democracy in places where it was previously unknown and an even more rapid spread of market-based economies. The authority of central governments is eroding, and power has begun to shift to local governments and private institutions. In some countries, freedom and opportunity are flourishing, while in others these changes have unleashed the violence of old conflicts and new ambitions.

Internationally, trade, investment, information, and even people flow across borders largely outside of governmental control. Domestically, deregulation and the shift of responsibilities from federal to state and local governments are changing the relationships among levels of government and between government and the private sector.

Communications technology has enhanced people's ability to receive information and influence events that affect them. This has sparked explosive growth in the number of organizations, associations, and networks formed by citizens, businesses, and communities seeking a greater voice for their interests. As a result, society outside of government—civil society—is demanding a greater role in governmental decisions, while at the same time impatiently seeking solutions outside government's power to decide.

But technological innovation is changing much more than communication. It is changing the ways in which we live, work, produce, and consume. Knowledge has

FIGURE 20.1 Struggling to survive in places that can no longer sustain them, growing populations overfish, overharvest, and overgraze.

become the economy's most important and dynamic resource. It has rapidly improved efficiency as those who create and sell goods and services substitute information and innovation for raw materials. During the past 20 years, the amount of energy and natural resources the U.S. economy uses to produce each constant dollar of output has steadily declined, as have many forms of pollution. When U.S. laws first required industry to control pollution, the response was to install cleanup equipment. The shift to a knowledge-driven economy has emphasized the positive connection among efficiency, profits, and environmental protection and helped launch a trend in profitable pollution prevention. More and more people today now understand that pollution is waste, waste is inefficient, and inefficiency is expensive.

Even as their access to information and to means of communication has increased, citizens of the more developed nations are becoming cynical about, and frustrated with, traditional political arrangements that no longer seem responsive to their needs. The confidence of many citizens in the large institutions that affect their lives—such as business; government; the media; and environmental, labor, and civic organizations—is eroding. Individual citizens have lost faith in their ability to influence events and have surrendered to apathy, or worse, to anger.

Since the end of World War II, the world's economic output has increased substantially, allowing widespread improvements in health, education, and opportunity, but also creating growing disparities between rich and poor. Even in the highly developed nations the gap between rich and poor is widening.

Tomorrow's world will be shaped by the aspirations of a much larger global population. The number of people living on Earth has doubled in the last 50 years. Growing populations demand more food, goods, services, and space. Where there is scarcity, population increase aggravates it. Where there is conflict, rising demand for land and natural resources exacerbates it. Struggling to survive in places that can no longer sustain them, growing populations overfish, overharvest, and overgraze. (See figure 20.1.)

As we move into the new century it is important that we recognize that economic, environmental, and social goals are integrally linked and that we develop policies that reflect that interrelationship. Thinking narrowly about jobs, energy, transportation, housing, or ecosystems—as if they were not connected—creates new problems even as it attempts to solve old ones. Asking the wrong questions is a sure way to get misleading answers that result in short-term remedies for symptoms, instead of cures for long-term problems.

All of this will require new modes of decision making ranging from the local to the international level. While trend is not always destiny, the trend that has been evolving over the past several years has been towards more collaborative forms of decision making. Perhaps such collaborative structures will involve more people and a broader range of interests in shaping and making public policy. This will hopefully improve decisions, mitigate conflict, and begin to counteract the corrosive trends of cynicism and civic disengagement that seems to be growing.

More collaborative approaches to making decisions can be arduous and

FIGURE 20.2 Scenes such as these were very common in North America only a short time ago. Fortunately, for the most part, such photos are today only historic in nature. Positive change is possible. As was mentioned in the text, trend is not destiny.

time-consuming, and all of the players must change their customary roles. For government, this means using its power to convene and facilitate, shifting gradually from prescribing behavior to supporting responsibility by setting goals, creating incentives, monitoring performance, and providing information.

For their part, businesses need to build the practice and skills of dialogue with communities and citizens, participating in community decision making and opening their own values, strategies, and performance to their community and the society.

Advocates, too, must accept the burdens and constraints of rational dialogue built on trust, and communities must create open and inclusive debate about their future.

Does all of this sound too idealistic? Perhaps it is; however, without a vision for the future where would we be? As was stated previously, trend is not destiny. In other words we are capable of change regardless of the status quo. This is perhaps nowhere more important than in the world of environmental decision making. (See figure 20.2.)

LEARNING FROM THE PAST

For the past quarter century, the basic pattern of environmental protection in economically developed nations has been to react to specific crisis. Institutions have been established, laws passed, and regulations written in response to problems that already were posing substantial ecological and public health risks and costs, or that already were causing deep-seated public concern.

The United States is no exception. The U.S. Environmental Protection Agency (EPA) has focused its attention almost exclusively on present and past problems. The political will to establish the agency grew out of a series of highly publicized, serious environmental problems, like the fire on the Cuyahoga River in Ohio, smog in Los Angeles, and the near extinction of the bald eagle. During the 1970s and 1980s Congress enacted a series of laws intended to solve these problems, and the EPA, which was created in 1970, was given the responsibility for enforcing most environmental laws.

Despite success in correcting a number of existing environmental problems, there has been a continuing pattern of not responding to environmental problems until they pose immediate and unambiguous risks. Such policies, however, will not adequately protect the environment in the future. People are recognizing that the agencies and organizations whose activities affect the environment must begin to anticipate future environmental problems, and then take steps to avoid them. One of the most important lessons learned during the past quarter century of environmental history is that the failure to think about the future environmental consequences of prospective social, economic, and technological changes may impose substantial and avoidable economic and environmental costs on future generations.

THINKING ABOUT THE FUTURE

Thinking about the future is more important today than ever before, because the accelerating rate of change is shrinking the distance between the present and the

FIGURE 20.3 The plight of these fishermen, resulting from the collapse of the cod fishery in the northeast United States and Canada and the salmon fishery in the northwest United States and Canada, is an environmental debt inherited from past generations of abuse and misuse.

future. Technological capabilities that seemed beyond the horizon just a few years ago are now outdated. Scientific developments and the flow of information are accelerating. For example, who would have envisioned cellular phones, voice-activated computers, or a widely used Internet only 10 years ago? Similarly, the environmental effects of changes in global economic activity are being felt more rapidly by both nations and individuals. Examples include the collapse of the cod fishing industry in the Maritimes of Canada and the northeast United States, the threatened salmon fishing industry in the Pacific northwest of the United States and British Columbia, Canada, the severe air pollution problems in Mexico, and the shortage of safe drinking water supplies in Russia.

Initiating thought and analysis well in advance of anticipated change can shorten the time needed to respond to such change or allow us to avoid the problem entirely. Because some damage is irreversible, response time is critical.

Thinking about the future also is valuable because the cost of avoiding a problem is often far less than the cost of solving it later. The U.S. experience with hazardous-waste disposal provides a compelling example. Some private companies and federal facilities undoubtedly saved money in the short term by disposing of hazardous wastes inadequately, but those savings were dwarfed by the cost of cleaning up hazardous-waste sites years later. In that case, foresight could have saved private industry, insurance companies, and the federal government (i.e., taxpayers) billions of dollars, while reducing exposure to pollutants and public anxieties in the affected communities.

Thinking about the future has another value, one that goes beyond the immediate costs and benefits of environmental protection. Environmental foresight can preserve the environment for future generations. When one generation's behavior necessitates environmental remediation in the future, an environmental debt is bequeathed to future generations just as

surely as unbalanced government budgets bequeath a burden of financial debt. (See figure 20.3.)

By anticipating environmental problems, and by taking steps now to prevent them, the present generation can minimize the environmental and financial debts that its children will incur.

Today, we face new classes of environmental problems that are more diffuse than those of the past and thus demand different approaches. Since the first Earth Day in 1970, the vast majority of the significant "point" sources of air and water pollution—large industrial facilities and municipal sewage systems—that once spewed untreated wastes into the air, rivers, and lakes have been controlled. The most important remaining sources of pollution are diffuse and widespread: sediment, pesticides, and fertilizers that run off farmland; oil and toxic heavy metals that wash off city streets and highways; and air pollutants from automobiles, outdoor grills, and woodstoves. Pollution from these sources cannot always be controlled with sewage treatment plants or the same regulatory techniques used to check emissions from large industries. To make matters more complicated, we now have environmental problems on a global scale—biodiversity loss, ozone depletion, and climate change. These problems will require cooperative international responses. We are also recognizing that controlling pollutants alone, no matter how successful, will not achieve an environmentally sustainable economy, since many global concerns are related to the size of the human population and the unequal distribution of resources.

DEFINING THE FUTURE

We are progressing from an environmental paradigm based on cleanup and control to one including assessment, anticipation, and avoidance. Expenditures to develop technologies that prevent environmental harm are beginning to pay off. Agricultural practices are becoming less wasteful and more sustainable, manufacturing processes are becoming more efficient in the use of resources, and consumer products are being designed with the environment in mind. The infrastructures that supply energy, transportation services, and water supplies are becoming more resource efficient and environmentally benign. Remediation efforts are cleaning up a large portion of existing hazardous-waste sites. Our ability to respond to emerging problems is being aided by more advanced monitoring systems and data analysis tools that continually assess the state of the local, regional, and global environment. Finally, we are developing effective ways of restoring or recreating severely damaged ecosystems to preserve the long-term health and productivity of our natural resource base.

This trend will continue if we are willing to develop strong environmental policies, develop new strategies, and closely coordinate the actions of the public and private sectors toward a set of shared, long-term goals. Such a strategy should include three aims. The first is to articulate a vision of the future and the role environmental technology will play in shaping that future. This vision must be built on an understanding of the strengths

and weaknesses of past policies and actions. The second is to define the roles of the many individuals and organizations who are needed to implement the goals. The third is to chart a course by offering suggestions for strategic goals for all the partners in this endeavor.

In the long run, environmental quality is not determined solely by the actions of government, regulated industries, or non-government organizations. It is largely a function of the decisions and behavior of individuals, families, businesses, and communities everywhere. Consequently, the extent of environmental awareness and the strength of environmental institutions will be two critical factors driving changes in environmental quality in the future.

A concerned, educated public, acting through responsive local, national, and international institutions, will serve as effective agents for avoiding future environmental problems, no matter what they are. Environmental institutions, strengthened by informed public support, will play a critical role in devising and implementing effective national and international responses to emerging issues.

THE DEVELOPMENT OF ENVIRONMENTAL POLICY IN THE UNITED STATES

Public **policy** is the general principal by which government branches—the **legislative, executive,** and **judicial**—are guided in their management of public affairs. The legislature (Congress), is directed to declare and shape national policy by passing legislation, which is the same as enacting law. The executive (President), is directed to enforce the law while the judiciary (Courts), interprets the law when a dispute arises. (See figure 20.4.)

When Congress considers certain conduct to be against public policy and against the public good, it passes legislation in the form of acts or statutes. Congress specifically regulates, controls, or prohibits activity in conflict with public policy and attempts to encourage desirable behavior.

Through legislation, Congress regulates behavior, selects agencies to implement new programs, and sets general procedural guidelines. When Congress passes environmental legislation it also declares and shapes the national environmental policy, thus fulfilling its policy-making function. (See figure 20.5.)

Over 80 years ago President Teddy Roosevelt declared that nothing short of defending your country in wartime "compares in leaving the land even better land for our descendants than it is for us." The environmental issues that Roosevelt strongly believed in, however, did not become major political issues until the early 1970s.

While the publication of Rachel Carson's *Silent Spring* in 1962 is considered to be the beginning of the modern environmental movement, the first Earth Day on April 22, 1970 was perhaps the single event that put the movement into high gear. In 1970, as a result of mounting public concern over environmental deterioration—cities clouded by smog, rivers on fire, waterways choked by raw sewage—many nations, including the United States, began to address the most obvious, most acute environmental problems.

Public opinion polls indicate that a permanent change in national priorities followed Earth Day 1970. When polled in May 1971, 25 percent of the U.S. public declared protecting the environment to be an important goal—a 2500 percent increase over 1969.

During the 1970s many important pieces of environmental legislation were enacted in the United States. (See figure 20.5.) Many of the identified environmental problems were so immediate, so obvious, that it was relatively easy to see what had to be done and to summon the political will to do it. (See table 20.1.)

Just as it was beginning to gain momentum, however, the environmental movement began to decline. When the energy crisis threatened to stall the North American economy in the early 1970s, environmental concerns quickly faded. By 1974, President Gerald Ford had proposed an acceleration of his administration's leasing program for offshore gas and oil drilling. A turnaround in environmental policy was even more pronounced during

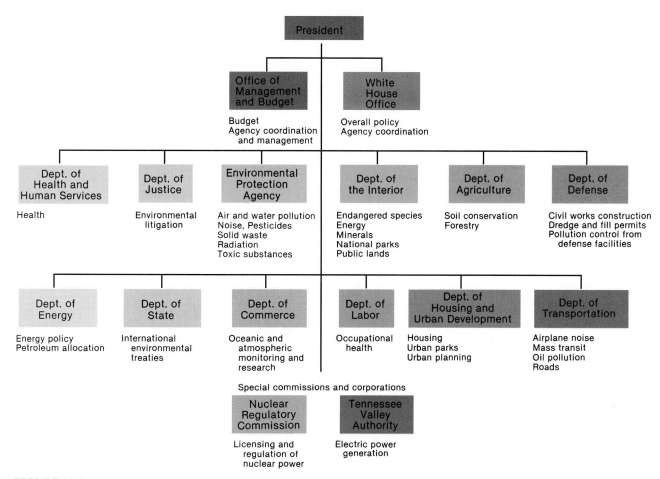

FIGURE 20.4 **Major Agencies of the Executive Branch** Major agencies of the executive branch are shown with their environmental responsibility.

the 1980s during the Reagan administration. Former Vice President Walter Mondale was fond of noting that President Reagan "would rather take a polluter to lunch than to court." During the mid-1980s, the environment was not a priority in the Reagan administration.

Despite the wavering of political will toward environmental concerns in the period from 1970 to 1990, there were some very tangible accomplishments. Among the most visible and quantifiable is the expansion of protected areas. During this period, federal parklands in the United States—excluding Alaska—increased 800,000 hectares, to 10.5 million. In Alaska, 18.3 million additional hectares were protected, bringing the state's total to over 232 million hectares. Also, the extent of the waterways included in the National Wild and Scenic Rivers System increased by more than 12 times, to some 15,000 kilometers.

By the late 1980s, however, a new environmental awareness and concern began to surface as a major political issue. This was in part due to a number of highly visible environmental problems that appeared nightly on the evening news. Images of toxic waste, including hospital waste such as used syringes washing up on the nation's beaches and the pristine waters of Alaska covered in oil from the *Exxon Valdez* spill, made an impact on the public. Once again, the public reacted by organizing and putting pressure on the political system, and, as in 1970, the politicians began to respond. For the first time in the history of the United States, the environment became a key issue in a presidential campaign. In 1988, the environmental records of the two major candidates were hotly debated. Environmentalism was evolving as a major public issue. By the 1992 U.S. Presidential election the environment was established as a

major campaign issue, a trend that continues today.

In many respects, the environmental movement of the 1970s and 1980s came of age in the 1990s. The linking of politics and science, and emotionalism and logic, together in a new environmental movement, represented a significant integration of human thinking. It has been said that politics have always forged science. Prioritization of issues and political will determine where money will be spent. By the mid-1990s, it appeared that the United States political will to address environmental concerns was on the rise.

A 1996 survey revealed that most Americans would prefer a federal government more actively involved in environmental protection. Ninety-two percent desired stronger active participation by leaders in government and business in environmental concerns. Forty-one percent believed the general public should bear the primary responsibility for a clean environment, followed by industry at 34 percent and government at 22 percent. It should be noted, however, that these polls, like most polls, often reflect attitudes and not necessarily actions. For example, saying that you support stronger environmental laws does not always translate into individual actions such as purchasing environmentally "friendly" products, active recycling, or support of higher taxes for more governmental regulation of the environment. Saying and doing are not always the same.

While the public was supportive of environmental protection the 1990s also witnessed a decline in membership of the major environmental organizations. By 1998 membership declines in many of the large environmental organizations was forcing the layoff of staff and the closure of offices. A major part of the problem with the larger environmental groups may be that they had actually grown into larger bureaucracies and in the process lost the trust of many grassroots environmentally concerned individuals. For example, the National Wildlife Federation has an annual budget of around $100 million. This includes profits on merchandise of nearly $25 million and nearly $12 million spent on magazines. By any definition, the National Wildlife Federation is a large bureaucracy.

While many of the larger, established environmental organizations are still declining somewhat, many newer, smaller, local and grassroots organizations are being created or are expanding, often in response to environmental threats in their own communities. Estimates are that some 7000 local environmental organizations are active in the United States. On college campuses, interest in the environment is growing. Environmental studies programs are expanding at both the undergraduate and graduate level. Students are not just concerned with local recycling programs or nuclear power plants but are focusing on the broader issues to be faced in the new century. The primary issue is the achievement of long-term sustainability and the fundamental changes in society that this will entail.

FIGURE 20.5 Passage of a Law This figure illustrates the path of a bill from organization to becoming a law. As we can see, the process is not a quick one.

ENVIRONMENTAL POLICY AND REGULATION

Environmental laws are not a recent phenomenon. As early as 1306, London adopted an ordinance limiting the burning of coal because of the degradation of local air quality. Such laws became more common as industrialization created many sources of air and water pollution throughout the world.

Environmental law in the United States is governed by administrative law. Administrative law is a relatively new concept, having been developed only during this century. In 1946 Congress passed the Federal Administrative Procedure Act (APA). This act designated general procedures to be used by federal agencies when they exercised their rulemaking, adjudicatory, and enforcement powers. This is a rapidly expanding area of law and defines how governmental organizations such as agencies, boards, and commissions develop and implement the regulatory programs they are legislatively authorized to create. Some of the many U.S. federal agencies that impact environmental issues include the Environmental Protection Agency (EPA), the Council on Environmental Quality (CEQ), the National Forest Service and the Bureau of Land Management.

Administrative law applies to government agencies and to those that are affected by agency actions. In the United

TABLE 20.1 MAJOR U.S. ENVIRONMENTAL AND RESOURCE CONSERVATION LEGISLATION

Wildlife conservation
Anadromous Fish Conservation Act of 1965
Fur Seal Act of 1966
National Wildlife Refuge System Act of 1966, 1976, 1978
Species Conservation Act of 1966, 1969
Marine Mammal Protection Act of 1972
Marine Protection, Research, and Sanctuaries Act of 1972
Endangered Species Act of 1973, 1982, 1985, 1988, 1995
Fishery Conservation and Management Act of 1976, 1978, 1982, 1996
Whale Conservation and Protection Study Act of 1976
Fish and Wildlife Improvement Act of 1978
Fish and Wildlife Conservation Act of 1980 (Nongame Act)

Land use and conservation
Taylor Grazing Act of 1934
Wilderness Act of 1964
Multiple Use Sustained Yield Act of 1968
Wild and Scenic Rivers Act of 1968
National Trails System Act of 1968
National Coastal Zone Management Act of 1972, 1980
Forest Reserves Management Act of 1974, 1976
Forest and Rangeland Renewable Resources Act of 1974, 1978
Federal Land Policy and Management Act of 1976
National Forest Management Act of 1976
Soil and Water Conservation Act of 1977
Surface Mining Control and Reclamation Act of 1977
Antarctic Conservation Act of 1978
Endangered American Wilderness Act of 1978
Alaskan National Interests Lands Conservation Act of 1980
Coastal Barrier Resources Act of 1982
Food Security Act of 1985
Coastal Development Act of 1990

General
National Environmental Policy Act of 1969 (NEPA)
International Environmental Protection Act of 1983
Environmental Education Act of 1990

Energy
National Energy Act of 1978, 1980

Water quality
Water Quality Act of 1965
Water Resources Planning Act of 1965
Federal Water Pollution Control Acts of 1965, 1972
Ocean Dumping Act of 1972
Safe Drinking Water Act of 1974, 1984, 1996
Clean Water Act of 1977, 1987
Great Lakes Critical Programs Act of 1990
Oil Spill Prevention and Liability Act of 1990

Air quality
Clean Air Act of 1963, 1965, 1970, 1977, 1990

Noise control
Noise Control Act of 1965
Quiet Communities Act of 1978

Resources and solid-waste management
Solid Waste Disposal Act of 1965
Resources Recovery Act of 1970
Resource Conservation and Recovery Act of 1976
Waste Reduction Act of 1990

Toxic substances
Toxic Substances Control Act of 1976
Resource Conservation and Recovery Act of 1976
Comprehensive Environmental Response, Compensation, and Liability (Superfund) Act of 1980, 1986, 1990
Nuclear Waste Policy Act of 1982

Pesticides
Federal Insecticide, Fungicide, and Rodenticide Control Act of 1972, 1988

States, many federal environmental programs are administered by the states under the authority of federal and related state laws. States often differ from both the federal government and often each other in the way they interpret, implement, and enforce federal laws. In addition, each state has its own administrative guidelines that govern and define how state agencies act. All actions of federal agencies must comply with the 1946 Administrative Procedure Act.

Beginning in 1970 Congress passed a series of environmental statutes including the National Environmental Policy Act (NEPA), the Resource Conservation and Recovery Act (RCRA), the Comprehensive Environmental Response, Compensation, and Liability Act (CERCLA) and the Clean Air Act (CAA). All of these acts are broadly worded to identify existing problems that Congress believes can be corrected to protect human health, welfare, and the environment.

Protecting human health, welfare, and the environment is the national policy that Congress has chosen to encourage. For example, under NEPA, the national policy is to "promote efforts which will prevent or eliminate damage to the environment and biosphere and stimulate the health and welfare of man." Similarly, under the Clean Air Act, the policy is to "protect and enhance the quality of the Nation's air resources so as to promote the public health and welfare and the productive capacity of its population."

Each of the above statutes declares national policy on environmental issues and addresses distinct problems. These statutes also authorize the use of some or all of the administrative functions discussed above, such as rulemaking, adjudication, administrative, civil and criminal enforcement, citizen suits, and judicial review. (See figure 20.6.)

Congress established the Environmental Protection Agency in 1970 as the primary agency to implement the statutes. Administrative functions empower EPA, the states, and private citizens to take the responsibility to enforce the various authorized programs. These administrative functions not only shape environmental law, but also control the daily operations of both the regulated industry and agencies authorized to protect the environment.

To date, much environmental law has reflected the perception that environmental problems are localized in time, space, and media (i.e., air, water, soil). For example, many hazardous-waste sites in the United States have been "cleaned" by simply shipping the contaminated dirt someplace else, which not only does not solve the problem, but creates the danger of

The National Environmental Policy Act

The National Environmental Policy Act of 1969 was enacted in 1969 and signed into law by President Nixon on New Year's Day, 1970. It is a short, general statute designed to institutionalize within the federal government a concern for the "quality of the environment." NEPA helps encourage environmental awareness among all federal agencies, not just those that prior to NEPA had to consider environmental factors in their planning and decision making. Until 1970, most federal agencies acted within their delegated authority without considering the environmental impacts of their actions. However, in the 1960s, Congress seriously began to study pollution problems. Because Congress has found that the federal government is both a major cause of environmental degradation and a major source of regulatory activity, all actions of the federal government now fall under NEPA.

NEPA forces federal agencies to consider the environmental consequences of their actions before implementing a proposal or recommendation. NEPA has two purposes: first, to advise the president on the state of the nation's environment, and second, to create an advisory council called the Council on Environmental Quality (CEQ). The CEQ outlines NEPA compliance guidelines. The CEQ also provides the president with consistent expert advice on national environmental policies and problems.

Between 1970 and 1977, the CEQ served only as an advisory council to the president. However, in 1977, through an executive order, President Carter granted the CEQ authority to issue binding regulations. These regulations set out details for matters that are broadly addressed by NEPA. The CEQ applies to all federal agencies except Congress, the judiciary, and the president.

NEPA has been interpreted narrowly by the federal courts. As a result, many states have passed much stronger state environmental protection acts (SEPAs) as well. Today, NEPA analysis is undertaken as part of almost every recommendation or proposal for federal action. This not only includes actions by agencies of the federal government, but also states, local municipalities, and private corporations. NEPA is Congress's mission statement that mandates the means by which the federal government, through the guidance of the CEQ, will achieve its national environmental policy.

EPA Enforcement Options

1. Warning Letter
- Describes alleged violation
- Action that must be taken to remedy
  - Deadline for compliance or enforcement action
  - Allow for an opportunity to discuss the situation

2. Administrative Order
- Requires certain activity be ceased, or
- Requires compliance by a certain date, or
- Requires certain tests be performed
- Provides Rights
  - Respond by citing defenses or objections
  - Right to confer with EPA

3. Permit Action
- Revoke or modify existing permit
- Seek to add conditions to permits that are being negotiated

4. Civil Enforcement
- Injunctive
- Monetary relief varies by statute per violation, per day

5. Criminal Proceedings
- Fines
- Incarceration

FIGURE 20.6 **Enforcement Options of the U.S. Environmental Protection Agency** The enforcement options of the U.S. Environmental Protection Agency range from a warning letter to a jail sentence.

Source: U.S. Environmental Protection Agency, Office of Enforcement, Washington, D.C.

Criticism of Environmental Regulation

- **Unrealistic nature of regulations**

- **Failure to use market incentives**

- **Inadequate assignment of responsibility for pollution**

- **Disproportionate harm to small businesses**

- **Technology-based rather than performance-based standards**

- **"Business vs Environment" mentality**

FIGURE 20.7   Criticisms of Environmental Regulation While some of the above criticisms are being addressed others still need to be studied and discussed.

Source: U.S. Environmental Protection Agency, Office of Enforcement, Washington, D.C.

incidents during removal and transportation. Environmental regulation has focused on specific phenomena and adopted the so-called "command and control" approach, in which restrictive and highly specific legislation and regulation are implemented by centralized authorities and used to achieve narrowly defined ends.

Such regulations generally have very rigid standards, often mandate the use of specific emission-control technologies, and generally define compliance in terms of "end-of-pipe" requirements. (See figure 20.7.) Examples in the United States include the Clean Water Act (which applied only to surface waters), the Clean Air Act (urban air quality), and the Comprehensive Environmental Response, Compensation, and Liability Act (Superfund), which applied to specific landfill sites.

If properly implemented, command and control methods can be effective in addressing specific environmental problems. For example, rivers such as the Potomac and Hudson in the United States are much cleaner as a result of the Clean Water Act. (See figure 20.8.) Moreover, where applied against particular substances, such as the ban on tetraethyl lead in gasoline in the United States, the command and control approach has clearly worked well.

THE GREENING OF GEOPOLITICS

Environmental or "green" politics has emerged from minority status and become a political movement in many nations.

Issues such as transboundary water supply and pollution, acid precipitation, and global climate change, have served to bolster the emergence of green politics. Even in Russia and Eastern Europe, the public has demanded more environmental protection, and leaders seem to be listening.

Concern about the environment is not limited to developed nations. A 1989 treaty signed in Switzerland limits what poorer nations call toxic terrorism—use of their lands by richer countries as dumping grounds for industrial waste. In 1990, more than 100 developing nations called for a "productive dialogue with the developed world" on "protection of the environment." As was covered in chapter 2, the Earth Summit in Rio brought together nearly 180 governments to address world environmental concerns and the 1997 conference on

FIGURE 20.8 The environmental quality of rivers has dramatically improved over the past 20 years as a direct result of the Clean Water Act. The upper photo shows a visibly impaired river in New England in the 1970s. The bottom photo shows an environmentally healthy Hudson River in the late 1990s.

FIGURE 20.9 If China's current "modernization" continues, the boom will be fueled by coal to the possible detriment of the planet as a whole.

global warming in Kyoto brought together some 120 nations.

Environmental concern is also a growing factor in international relations. Many world leaders see the concern for environment, health, and natural resources as entering the policy mainstream. A sense of urgency and common cause about the environment is leading to cooperation in some areas. Ecological degradation in any nation is now understood almost inevitably to impinge on the quality of life in others. Drought in Africa and deforestation in Haiti have resulted in large numbers of refugees, whose migrations generate tensions both within and between nations. From the Nile to the Rio Grande, conflicts flare over

water rights. The growing megacities of the developing world are areas of potential civil unrest. Sheer numbers of people overwhelm social services and natural resources. The government of the Maldives has pleaded with the industrialized nations to reduce their production of green house gases, fearing that the polar ice caps may melt and inundate the island nation.

Economic progress in developing nations could also bring the possibility for environmental peril and international tension. China, which accounts for 21 percent of the world's population, has the world's third largest recoverable coal reserves. If China's current "modernization" campaign succeeds, the boom will be fueled

by coal, to the possible detriment of the planet as a whole. (See figure 20.9.)

Some experts estimate that the developing world, which today produces one-fourth of all greenhouse gas emissions, could be responsible for nearly two-thirds by the middle of the next century. Developing nations have repeatedly indicated that they are not prepared to slow down their own already weak economic growth to help compensate for decades of environmental problems caused largely by the industrialized world.

Some developing countries may resist environmental action because they see a chance to improve their bargaining leverage with foreign aid donors and international bankers. Where before the poor nations never had a strategic advantage, they now may have an ecological edge. Ecologically, there could be more parity than there ever was economically or militarily.

A former U.S. ambassador to the United Nations stated that, just as the Cold War between the East and West seems to be winding down, "eco-conflicts" between the industrialized North and the developing South may pose a comparable challenge to world peace. National security may no longer be about fighting forces and

Changing the Nature of Environmental Regulation—The Safe Drinking Water Act

According to the American Water Works Association (AWWA) the 1996 amendments to the Safe Drinking Water Act (SDWA) amount to a radical rewrite of the 1986 amendments. The 1986 amendments were written with little substantive input from the regulated community. The 1996 amendments, however, were developed with significant contributions from water suppliers and state and local officials. The new amendments embody a partnership approach that includes major new infusions of federal funds to help water utilities, especially the thousands of smaller systems, comply with the law.

What this all means is that the amended SDWA is more focused on what it was initially intended to achieve. For example, rather than require the EPA to set standards for 25 new contaminants every three years regardless of need or actual threat to public health (in cases, it has been argued, this requirement impeded public health protection by diverting resources from the most critical needs), the new law focuses the EPA's efforts on regulating contaminants known to pose health risks and requires cost-benefit analyses and risk assessments to take place before new standards are set. The act also allows the EPA to adopt interim regulations for contaminants on an emergency basis if they pose urgent health threats. The act provided millions of dollars in new funding for crucial research; links source water to drinking water (and sets out a state-administered system designed to protect drinking water from becoming contaminated in the first place); authorizes a new system of federal grants to states which will enable water utilities to borrow funds to upgrade their systems; and requires water systems to provide customers with annual updates delineating the sources and quality of the water they provide. It has been stated that the 1996 amendments are good for the public and drinking water industry alike. The law has more flexibility, more state responsibility, more cooperative approaches, and is achieved through partnerships.

All too often we take clean, safe drinking water for granted.

weaponry alone. It also relates increasingly to watersheds, croplands, forests, climate, and other factors rarely considered by military experts and political leaders, but that, when taken together, deserve to be viewed as equally crucial to a nation's security as are military factors. It is interesting to note that the North Atlantic Treaty Organization (NATO) has developed an office for Scientific and Environmental Affairs, and the U.S. Department of Defense has created an Office of Environmental Security. Environmental offices within traditional military organizations would have been viewed as very unusual until only recently.

The increased attention to the environment as a foreign policy and national security issue is only the beginning of what will be necessary to avert problems in the future. The most formidable obstacle may be the entrenched economic and political interests of the world's most advanced nations. If the United States, for example, asks others not to cut their forests, then it will have to be more judicious about cutting its own. If North Americans wish to stem the supply of hardwood from a fragile jungle or furs from endangered species, then they will have to stem demand for fancy furniture and fur coats. If they wish to preserve wilderness from the intrusions of the oil industry, then they will have to find alternative sources of energy and use all fuels more efficiently.

What may be needed is self-discipline on the part of the world's haves and increased assistance to the have-nots. In the world today, a billion people live in a degree of poverty that forces them to deplete the environment without regard to its future. Their governments often are too crippled by international debt to afford the short-term costs of environmental safeguards.

William Ruckelshaus, a former administrator of the U.S. Environmental Protection Agency, believes a historical watershed may be at hand. If the industrialized and developing countries did everything they should, he says, the resulting change would represent a "modification of

Environmental Backlash—The Wise Use Movement

During the late 1980s there began to develop a backlash or anti-environmental attitude amongst sectors of the U.S. populace. A loose-knit organization of several hundred of these groups became known as the Wise Use Movement. The majority of funding for this coalition comes from various interests including the timber, oil and coal industries, real estate developers, and ranchers.

In his 1988 book, *The Wise Use Agenda,* Ron Arnold, an early leader of the movement, laid out some of its goals including:

Elimination of the National Park Service—to be replaced with privately operated parks.

Removal of government restrictions on development in wetlands.

The opening of all national parks, wilderness areas and wildlife refuges to off-road vehicles, commercial development, mining and drilling for oil.

Cutting of all the remaining old-growth forests in national forests and replacing them with tree plantations.

Recognize private property rights to water, grazing permits, and mining claims on public lands.

In addition to the above, the Wise Use Movement has been in the forefront of advocating private property rights. They argue that regulations protecting environmentally sensitive areas on private property are unconstitutional "takings." They cite the Fifth Amendment to the U.S. Constitution, which states in part: "nor shall private property be taken for public use, without just compensation." That clause is the basis for the concept of eminent domain, which allows government entities to take land for public projects by paying property owners the land's fair market value.

Wise Use pamphlets argue that extinction is a natural process and that some species were not meant to survive. It has been argued that the

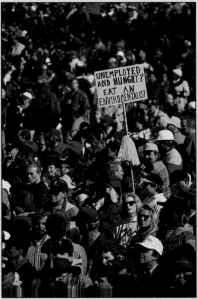

movement's signature public relations tactic is to frame complex environmental and economic issues in simple, scapegoating terms that benefit its corporate backers. In the Pacific Northwest, for example, the movement has continually dwelled on a supposed "battle" for survival between spotted owls and the families of the workers involved in harvesting the old growth timber. "Jobs versus owls" is a good sound bite but it is not the issue. The real issue is much more complicated.

As would be expected, the environmental community has come out strongly against the Wise Use Movement. Environmentalists have renamed it the "earth- and people-abuse movement" and stated that the basic principles of Wise Users are: "If it grows, cut it; if it's swampy, fill it; if it moves, kill it."

What are your thoughts on the Wise Use Movement? Can you identify local issues or supporters of their philosophy? Would you call the Wise Use Movement a "threat to" or a "protector of" future generations? Why?

society comparable in scale to the agricultural revolution of the Neolithic age and to the Industrial Revolution of the past two centuries."

Let's put all of this into another perspective that may ring closer to home. By even conservative estimates the population of the U.S. is expected to double in the next 50 years. It could be argued then that with twice as many people projected for 2050, it will be necessary to double the total U.S. infrastructure in the next 50 years. Potentially then this translates to the following: twice as many cars, trucks, planes, airports, parking lots, streets, and freeways; twice as

many houses and apartment buildings; twice as many landfills, wastewater treatment plants, hazardous waste treatment facilities and chemicals (pesticides and herbicides) for agriculture; in short, twice as much of everything. Under this scenario it will be necessary to take over and develop in the next 50 years an amount of farmland and scenic countryside equal to the total already developed in the past 200 years. (See figure 20.10.) Megacities such as New York and Los Angeles will double in size and many new megacities will develop. In short, this vision is not very appealing. What happens to wilderness areas, remote and quiet places, habitat for

songbirds, waterfowl and other wild creatures? What happens to our quality of life?

Another way of looking at the future would be along the lines of what was previously mentioned about the analogy to the agricultural and industrial revolutions. This translates into a future of profound change. A future in which virtually everything we do will change. This future of profound change will also be one of phenomenal opportunity and excitement. It is a future in which we will farm, build, and transport in entirely new ways. This vision of the future could well become known as the environmental revolution.

Eco-Terrorism

The effects of war on the natural environment are frequently catastrophic. By its nature, war is destructive, and the environment is an innocent victim. The 1990–91 war in the Persian Gulf, however, witnessed a new role for the environment in war: its degradation was used as a weapon.

The phrase "eco-terrorism" was applied to Iraqi leader Saddam Hussein's use of oil as a weapon in the war. Iraq dumped an estimated 1.1 billion liters of crude oil into the Persian Gulf from Kuwait's Sea Island terminal before the United States made bombing raids on the pumps feeding the facility. The oil spill that resulted from the dumping was the world's largest—over 30 times the size of the 1989 *Exxon Valdez* disaster in Alaska.

Saddam Hussein may have engineered the spill to block allied plans for an amphibious invasion of Kuwait, but he was also probably trying to shut down seaside desalination plants that provide much of the freshwater for Saudi Arabia's Eastern Province. Whatever the military or political effect, the environmental effect was much greater.

The Persian Gulf is a sensitive ecosystem. The water is very salty, and temperatures vary widely from one season to another. As a result, indigenous animals and plants are finely attuned to specialized conditions.

The gulf is also rather isolated, with only one narrow outlet—the Strait of Hormuz, just 55 kilometers across. The Gulf takes up to five years to flush out.

The Persian Gulf waters, shores, and islands are dotted with coral reefs, mangrove swamps, and beds of sea grass, alive with many unique species of birds, fish, and marine mammals. Many of the region's species, such as the bottlenose dolphin, dugong, green turtle, and caspian tern were already classified as threatened.

Another act of eco-terrorism that took place during the Iraq-Kuwait war was the intentional burning of hundreds of oil wells in Kuwait. The air pollution resulting from the burning oil covered a large area. It took several months to extinguish the fires following the war. Using environmental damage as a weapon of war is a frightening concept. The environmental consequences of such acts of eco-terrorism will not be fully understood for years. Though the war in the Persian Gulf lasted less than one year, the environmental effects on some species could be irreversible.

Sources: *International Organizations: A Dictionary and Directory* by Giuseppe Schiavone (St. James Press, 1983).

INTERNATIONAL ENVIRONMENTAL POLICY

"If there must be a war, let it be against environment contamination, nuclear contamination, chemical contamination, against the bankruptcy of soil and water systems; against the driving of people away from the lands as environmental refugees. If there must be war, let it be against those who assault people and other forms of life by profiteering at the expense of nature's capacity to support life. If there must be war, let the weapons be your healing hands, the hands of the world's youth in defense of the environment."

Mustafa Tolba
Former Secretary General
United Nations Environment Programme

 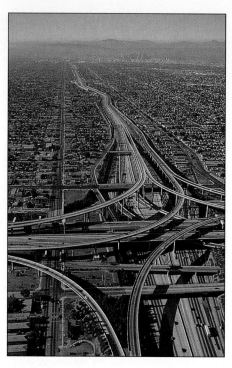

FIGURE 20.10 Options and Tradeoffs for the New Century Unless we become more creative in areas such as transportation and land use, it will be necessary to develop in the United States alone, in the next 50 years, an amount of farmland and scenic countryside as was developed over the past 200 years.

International coordination and political resolve are necessary if the goals of preserving and protecting the global environment are to be realized. A major step in that direction was the 1972 United Nations Conference in Stockholm, Sweden. This was the first international conference specifically dealing with global environmental concerns. The UN Environment Programme, a separate department of the United Nations that deals with environmental issues, developed out of that conference. United Nations Conference on the Law of the Sea produced a comprehensive convention that addressed many of the issues concerning jurisdiction over ocean waters and use of ocean resources. This treaty is viewed by many as a model for international environmental protection. Positive results are already evident from application of the agreements on pollution control, marine mammal protection, navigation safety, and other aspects of the marine environment. The issue of deep ocean mining, primarily of manganese nodules, has to date kept many industrial nations, such as Germany and the United States, from ratifying the treaty.

There have been several other successful international conventions and treaties. The Antarctic Treaty of 1961 reserves the Antarctic continent for peaceful scientific research and bans all military activities in the region. The 1979 Convention on Long-Range Transboundary Air Pollution was the first multilateral agreement on air pollution and the first environmental accord involving all the nations of Eastern and Western Europe and North America. The 1987 Montreal Protocol on Substances that Deplete the Stratospheric Ozone Layer addresses the ozone protective shield problem.

There is no international legislature with authority to pass laws; nor are there international agencies with power to regulate resources on a global scale. An international court at the Hague in the Netherlands has no power to enforce its decisions. Nations can simply ignore the court if they wish. However, a growing network of multilateral environmental organizations have developed a greater sense of their roles and a greater incentive to work together. These include not only the United Nations Environment Programme but also the Environment Committee of the Organization for Economic Cooperation and Development (OECD) and the Senior Advisors on Environmental Problems of the Economic Commission for Europe. Such institutions perform unique functions that cannot be carried out by governments acting alone or bilaterally.

Is the goal of an environmentally healthy world realistic? There is growing optimism that the community of nations is slowly maturing with regard to our common environment. We have all suffered a loss on innocence about "earth management." Laissez-faire may be good economics, but it can be a prescription for disaster in ecology.

This environmental "coming of age" is reflected in the broadening of intellectual perspective. Governments used to be preoccupied with domestic environmental affairs. Now, they are beginning to broaden their scope to confront problems that cross international borders, such as transboundary air and water pollution, and threats of a planetary nature, such as stratospheric ozone depletion and climatic warming. It is becoming increasingly evident that only decisive mutual action can secure the kind of world we seek.

GLOBAL PERSPECTIVE

Environmental Policy and the European Union

"The environment knows no frontiers" was the slogan of the 1970s, when the European Community—better known today as the European Union—began to develop its first environmental legislation.

The early laws focused on the testing and labeling of dangerous chemicals, testing drinking water, and controlling air pollutants such as SO_2, NO_x, and particulates from power plants and automobiles. Many of the directives from the 1970s and 1980s were linked to Europe's desire to improve the living and working conditions of its citizens.

In 1987, the Single European Act gave this growing body of environmental legislation a formal legal basis, and set three objectives: protection of the environment, human health, and prudent and rational use of natural resources.

The treaty reflected what many governments had already understood: that countries are part of an interconnected and interdependent world of people who are bound together by the air they breathe, the food they eat, the products they use, the wastes they throw away, and the energy they consume.

Similarly, a factory in one European nation may import supplies and raw material from several other neighboring nations; it may consume energy produced from imported gas, produce wastes that affect the air and water quality across the border or downstream, and export products whose wastes become the risk and responsibility of governments and peoples several hundred or many thousand kilometers away.

The 1992 Maastricht Treaty formally established the concept of sustainable development in European Union law. Then, in 1997, the Amsterdam Treaty made sustainable development one of the overriding objectives of the European Union. The Treaty considerably strengthened the commitment to the principle that the European Union's future development must be based on the principle of sustainable development and a high level of protection of the environment. The environment must be integrated into the definition and implementation of all of the Union's other economic and social policies, including trade, industry, energy, agriculture, transport, and tourism.

Under certain conditions, the member states may maintain or introduce environmental standards and requirements which are stricter than those of the Union. The Commission verifies the compatibility of these stricter national programs and the potential effects on other member states.

IT ALL COMES BACK TO YOU

No set of policies, no system of incentives, no amount of information can substitute for individual responsibility when it comes to ensuring that our grandchildren will enjoy a quality of life that comes from a quality environment. Information can provide a basis for action. Vision and ideas can influence perceptions and inspire change. New ways to make decisions can empower those who seek a role in shaping the future. However, all of this will be meaningless unless individuals acting as citizens, consumers, investors, managers, workers, and professionals decide that it is important to them to make choices on the basis of a broader, longer view of their self-interest; to get involved in turning those choices into action; and, most importantly, to be held accountable for their actions.

The combination of political will, technological innovation, and a very large investment of resources and human ingenuity in pursuit of environmental goals has produced enormous benefits over the past two decades. This is an achievement to celebrate, but in a world that steadily uses more materials to make more goods for more people, we must recognize that we will have to achieve more in the future for the sake of the future. We must move towards a world in which zero waste will become the ideal for society that zero defects has become for manufacturing. Finally we must recognize that the pursuit of one set of goals affects others and that we must pursue policies that integrate economic, environmental, and social goals.

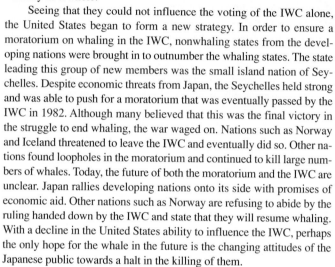

Overview of an International Organization—The International Whaling Commission (IWC)

The first records of whaling can be traced back hundreds of years to the twelfth century in the Bay of Biscay of Spain. During whaling's heydays of the 1930s, 85 percent of the world's catch of whales was coming from the Antarctic, with 95 percent of these catches dominated by Norway and the United Kingdom.

With the possible exhaustion of whale stock, the problem of whales as a common property resource came to the forefront. As a common property resource no single user has a right to the resource, nor can they prevent others from sharing in its exploitation. These doctrines of freedom of the seas and common property resources date back to Roman times and were constantly used by states as an argument to justify the overfishing of whale stocks in previous centuries. Seeing a need to keep statistical data on the populations of different whale species, Norway, at the request of the International Council for the Exploration of the Seas (ICES), set up the Bureau of International Whaling Statistics in 1930 to deal with whaling problems. This eventually led to the drafting and signing of the Convention for the Regulation of Whaling in 1931. This convention "applied to all waters . . . but was only applicable to baleen whales," and provided exemptions for aboriginal subsistence whaling. This convention had little effect on the amount of whaling because important nations such as Germany and Japan failed to adhere to the rules stated. A new draft of the convention was brought into effect in 1937 and included such countries as Norway, the United Kingdom, and Germany. In 1946 the International Whaling Conference (IWC) was convened by the United States to address the post-World War II whaling issues.

The measures to be taken by this commission were to completely protect certain whale species, specify certain whale sanctuaries, set limits on the number of whales taken, decide when open and closed hunting seasons would begin, and to specify the size and age of those whales that could be taken. It is important to point out that the majority of the nations who were original members of the IWC were whaling states and had no real interest in protecting the whale in order to preserve biologic diversity. The formation and actions of the IWC in its early years were purely economically motivated.

The ineffectiveness of the IWC to control the slaughter of whale populations continued into the 1970s until environmentally oriented nongovernmental organizations began to champion the struggle of the whale. Soon millions of people in the world were chanting "save the whale" and began demanding that the IWC pass a moratorium to end whaling. But a powerful veto coalition made up of Japan, Norway, the Soviet Union, Iceland, Chile, and Peru were able to control the vote in the IWC and defeat any measures that were harmful to the whaling industry as a whole.

Seeing that they could not influence the voting of the IWC alone, the United States began to form a new strategy. In order to ensure a moratorium on whaling in the IWC, nonwhaling states from the developing nations were brought in to outnumber the whaling states. The state leading this group of new members was the small island nation of Seychelles. Despite economic threats from Japan, the Seychelles held strong and was able to push for a moratorium that was eventually passed by the IWC in 1982. Although many believed that this was the final victory in the struggle to end whaling, the war waged on. Nations such as Norway and Iceland threatened to leave the IWC and eventually did so. Other nations found loopholes in the moratorium and continued to kill large numbers of whales. Today, the future of both the moratorium and the IWC are unclear. Japan rallies developing nations onto its side with promises of economic aid. Other nations such as Norway are refusing to abide by the ruling handed down by the IWC and state that they will resume whaling. With a decline in the United States ability to influence the IWC, perhaps the only hope for the whale in the future is the changing attitudes of the Japanese public towards a halt in the killing of them.

* What are your thoughts on the taking of whales?
* Do individual nations have a right to take whales?
* How would you define "common property resource"?

Eco-Labels

Green consumerism is the concept of rational consumption of our scarce resources for the benefit of the environment and future generations. The old saying "the world is enough for everyone's need but not for everyone's greed" calls for a change in our behavior and lifestyle in favor of a sustainable future. Eco-labels have been introduced in a number of countries to help consumers choose products with a proven environmental edge, determined by the product's choice of raw materials, production process, product life cycle, and associated disposal problems. Eco-labels provide evidence that products have met the safety, quality, and environmental protection requirements of the authority that issues the label.

The first eco-label was introduced in Germany in 1978. This distinctive Blue Angel label was followed by other environmental certification, including the White Swan of Northern European countries, Environmental Choice of Canada, Green Label of Singapore, and Environmental Label of China. By 1995, over a dozen countries had adopted this system of providing consumers with information enabling them to become responsible green consumers. The eco-label of the European Union is significant because it is the world's first regional scheme to apply the same minimum standards across national markets.

1992
European Union - 'Eco-label'

1992
Singapore - 'Green Label'

1993
China - 'Environmental Label'

1978
Germany - 'Blue Angel'

1988
Canada - 'Environmental Choice'

1989
Nordic Council - 'White Swan'

Source: From ECCO, Bulletin of the Environmental Campaign Committee, Issue 49, October 1995.

Summary

Politics and the environment cannot be separated. In the United States, the government is structured into three separate branches, each of which impacts environmental policy.

The increase in environmental regulations in the U.S. over the past 30 years has caused concerns in some sectors of society.

The late 1980s and early 1990s witnessed a new international concern about the environment, both in the developed and developing nations of the world. Environmentalism is also seen as a growing factor in international relations. This concern is leading to international cooperation where only tension had existed before.

While there exists no world political body that can enforce international environmental protection, the list of multilateral environmental organizations is growing.

It remains too early to tell what the ultimate outcome will be, but progress is being made in protecting our common resources for future generations. Several international conventions and treaties have been successful. In the final analysis, however, each of us has to adjust our lifestyle to clean up our own small part of the world.

Review Questions

1. What are the major responsibilities of each of the three branches of the U.S. government?
2. What are some of the enforcement options in U.S. environmental policy?
3. What role does administrative law play in U.S. environmental policy?
4. What are some of the criticisms of U.S. environmental policy?
5. How has public opinion in the United States changed concerning the protection of the environment in the past 10 years?
6. Why is environmentalism a growing factor in international relations?
7. Give some examples of some international environmental conventions and treaties.

Critical Thinking Questions

1. Does Chapter 20 have an overall point of view? If you were going to present the problems of environmental policy making and enforcement to others, what framework would you use?
2. The authors of this text say that "we are progressing from an environmental paradigm based on cleanup and control to one including assessment, anticipation, and avoidance." Do you agree with this assessment? Are there environmental problems that are harder to be proactive about than others?
3. Does a command and control approach to environmental problems, an approach that emphasizes regulation and remediation, make sense with global environmental problems such as global climite change, habitat destruction, and ozone depletion?
4. What values, perspectives, and beliefs does the Wise Use Movement exhibit in their response to environmental legislation? How are they similar, and/or different than your own?

5. How is it best, as a global society with many political demarcations, to preserve the resources that are held in common? What special problems does this kind of preservation entail?
6. Do you agree with William Ruckelshaus that current environmental problems require a change on the part of industrialized and developing countries that would be "a modification in society comparable in scale to the agricultural revolution . . . and Industrial Revolution?" What kinds of changes might that mean in your life? Would these be positive, or negative, changes?
7. New treaties regarding free trade might enable some nations to argue that other nations' environmental legislation is too restrictive, thereby imposing a barrier to trade that is subject to sanction. What special problems and possibilities might the new global economy provide environmental preservation? What do you think about that?

Key Terms

executive branch *394*

legislative branch *394*

judicial branch *394*

policy *394*

Related Internet Sites

Visit our website at http://www.mhhe.com/environmentalscience/ for specific information on each of the topics below.

Environmental policy and sustainability

U.S. environmental laws and policies

Public finance policies for land and resources

Legislation, laws, and codes pertaining to the environment

Policies and frameworks for sustainable development

Organizations active in trade and sustainable development

Environment and development policymakers

Sustainable environment and economy

Sustainable living on a personal level

Businesses, governmental agencies, and laws and regulations regarding the environment

Environmental regulations, education, and career resources

Sustainable development activities in Canada and worldwide

APPENDIX ONE

CRITICAL THINKING

We live in an age of information. Computers, e-mail, the Internet, CD-ROMs, instant news, and fax machines bring us information more quickly than ever before. A simple search of the Internet will provide huge amounts of information. Some of the information has been subjected to scrutiny and is quite valid, some is well-informed opinion, some is naive misinformation, and some is even designed to mislead. How do we critically evaluate the information we get?

Critical thinking involves a set of skills that help us to evaluate information, arguments, and opinions in a systematic and thoughtful way. Critical thinking also can help us better understand our own opinions as well as the points of view of others. It can help us evaluate the quality of evidence, recognize bias, characterize the assumptions behind arguments, identify the implications of decisions, and prevent us from jumping to conclusions.

CHARACTERISTICS OF CRITICAL THINKING

Critical thinking involves a set of skills that allow us to sort information in a meaningful way and discard invalid or useless information while recognizing that which is valuable. Some key components of critical thinking are:

Recognize the importance of context. All information is based on certain assumptions. It is important to recognize what those assumptions are. Critical thinking involves looking closely at an argument or opinion by identifying the historical, social, political, economic, and scientific context in which the argument is being made. It is also important to understand the kinds of bias contained in the argument and the level of knowledge the presenter has.

Consider alternative views. A critical thinker must be able to understand and evaluate different points of view. Often these points of view may be quite varied. It is important to keep an open mind and to look at all the information objectively and try to see the value in alternative points of view. Often people miss obvious solutions to problems because they focus on a certain avenue of thinking and unconsciously dismiss valid alternative solutions.

Expect and accept mistakes. Good critical thinking is exploratory and speculative, tempered by honesty and a recognition that we may be wrong. It takes courage to develop an argument, engage in debate with others, and admit that your thinking contains errors or illogical components. By the same token, be willing to point out what you perceive to be shortcomings in the arguments of others. It is always best to do this with good grace and good humor.

Have clear goals. When analyzing an argument or information, keep your goals clearly in mind. It is often easy to get sidetracked. A clear goal will allow you to quickly sort information into that which is pertinent and that which may be interesting but not germane to the particular issue you are exploring.

Evaluate the validity of evidence. Information comes in many forms and has differing degrees of validity. When evaluating information it is important to understand that not all the information from a source may be of equal quality. Often content about a topic is a mix of solid information interspersed with less certain speculations or assumptions. Apply a strong critical attitude to each separate piece of information. Often what appears to be a minor, insignificant error or misunderstanding can cause an entire argument to unravel.

Critical thinking requires practice. As with most skills, you become better if you practice. At the end of each chapter in the text there are a series of questions that allow you to practice critical thinking skills. Some of these questions are straightforward and simply ask you to recall information from the chapter. Others ask you to apply the information from the chapter to other similar contexts. Still others ask you to develop arguments that require you to superimpose the knowledge you have gained from the chapter on quite different social, economic, or political contexts from your own.

Practice, Practice, Practice.

APPENDIX TWO

METRIC UNIT CONVERSION TABLES

THE METRIC SYSTEM

Standard metric units   Abbreviations

Standard unit of mass	Gram	g
Standard unit of length	Meter	m
Standard unit of volume	Liter	l

Common prefixes   Examples

Kilo	1,000	A kilogram is 1,000 grams.
Centi	0.01	A centimeter is 0.01 meter.
Milli	0.001	A milliliter is 0.001 liter.
Micro (μ)	One-millionth	A micrometer is 0.000001 (one-millionth) of a meter.
Nano (n)	One-billionth	A nanogram is 10^{-9} (one-billionth) of a gram.
Pico (p)	One-trillionth	A picogram is 10^{-12} (one-trillionth) of a gram.

UNITS OF LENGTH

Unit	Abbreviation	Equivalent
Meter	m	Approximately 39 in
Centimeter	cm	10^{-2} m
Millimeter	mm	10^{-3} m
Micrometer	μm	10^{-6} m
Nanometer	nm	10^{-9} m
Angstrom	Å	10^{-10} m

Length conversions

1 in = 2.5 cm	1 mm = 0.039 in
1 ft = 30 cm	1 cm = 0.39 in
1 yd = 0.9 m	1 m = 39 in
1 mi = 1.6 km	1 m = 1.094 yd
	1 km = 0.6 mi

To convert	Multiply by	To obtain
Inches	2.54	Centimeters
Feet	30	Centimeters
Centimeters	0.39	Inches
Millimeters	0.039	Inches

UNITS OF VOLUME

Unit	Abbreviation	Equivalent
Liter	1	Approximately 1.06 qt
Milliliter	ml	10^{-3} (1 ml = 1 cm^3 = 1 cc)
Microliter	μl	10^{-6} 1

Volume conversions

1 tsp = 5 ml	1 ml = 0.03 fl oz
1 tbsp = 15 ml	1 l = 2.1 pt
1 fl oz = 30 ml	1 l = 1.06 qt
1 cup = 0.24 l	1 l = 0.26 gal
1 pt = 0.47 l	
1 qt = 0.95 l	
1 gal = 3.8 l	

To convert	Multiply by	To obtain
Fluid ounces	30	Milliliters
Quarts	0.95	Liters
Milliliters	0.03	Fluid ounces
Liters	1.06	Quarts

UNITS OF WEIGHT

Unit	Abbreviation	Equivalent
Kilogram	kg	10^3 g (approximately 2.2 lb)
Gram	g	Approximately 0.035 oz
Milligram	mg	10^{-3} g
Microgram	μg	10^{-6} g
Nanogram	ng	10^{-9} g
Picogram	pg	10^{-12} g

Weight conversions

1 oz = 28.3 g	1 g = 0.035 oz
1 lb = 453.6 g	1 kg = 2.2 lb
1 lb = 0.45 kg	

To convert	Multiply by	To obtain
Ounces	28.3	Grams
Pounds	453.6	Grams
Pounds	0.45	Kilograms
Grams	0.035	Ounces
Kilograms	2.2	Pounds

Temperature conversions

$$°C = \frac{(°F - 32) \times 5}{9}$$

$$°F = \frac{°C \times 9}{5} + 32$$

Some equivalents

0°C = 32°F
37°C = 98.6°F
100°C = 212°F

APPENDIX THREE

THE PERIODIC TABLE OF THE ELEMENTS

Traditionally, elements are represented in a shorthand form by letters. For example, the formula for water, H_2O, shows that a molecule of water consists of two atoms of hydrogen and one atom of oxygen. These chemical symbols for each of the atoms can be found on any periodic table of the elements. Using the periodic table, we can determine the number and position of the various parts of atoms. Notice that

atoms number 3, 11, 19, and so on are in column one. The atoms in this column act in a similar way since they all have one electron in their outermost layer. In the next column, Be, Mg, Ca, and so on act alike because these metals all have two electrons in their outermost electron layer. Similarly, atoms number 9, 17, 35, and so on all have seven electrons in their outer layer. Knowing how fluorine, chlorine,

and bromine act, you can probably predict how iodine will act under similar conditions. At the far right in the last column, argon, neon, and so on all act alike. They all have eight electrons in their outer electron layer. Atoms with eight electrons in their outer electron layer seldom form bonds with other atoms.

Periodic Table of the Elements

APPENDIX FOUR

WHAT *YOU* CAN DO TO MAKE THE WORLD A BETTER PLACE IN WHICH TO LIVE

It is easy to complain; in fact, we all tend to do so almost on a daily basis. Complaining, however, never corrects a problem or helps to resolve a dispute. There are actions you can take that *will* help improve the environment. The following list offers a few suggestions, and others will probably come to mind once you have thought about it. Try to expand the list with others in your class, and give your suggestions to your instructor. Have your instructor send your suggestions to the authors at Delta College, University Center, Michigan 48710 or Huxley College of Environmental Studies, Western Washington University, Bellingham, Washington 98226. We will expand the list in the next edition of the textbook. Remember, you do not have to start big to help!

1. **Continue your education.** Learning should not stop when you leave school. Become informed about issues; then you can begin to bring about change.

2. **Do not feel responsible for every problem in the world.** You cannot do everything. Concentrate on issues that you feel strongly about and that you can do something about. Focus your energy.

3. **Think about the consequences of your profession and your lifestyle.** If they are damaging to other people or to the environment, adjust your behavior accordingly. Try to persuade friends, family, and coworkers to do the same. Make environmental awareness a family affair.

4. **Work with others.** Attend meetings of your local government and ask officials about their plans to prevent pollution. Often, officials are very responsive to visits of this kind. Being part of a group of people with similar interests gives you support and increases your effectiveness. If you cannot find an appropriate group, start one of your own.

5. **Become active in your community.** Organize a community conference to discuss positive approaches to pollution prevention. Invite public officials, industry and labor representatives, other interested groups, and individual citizens. Get all the facts, and then try to get the appropriate action programs initiated.

6. **Learn about the ecology of your bioregion.** Develop a sense of place that puts you in contact with your local physical environment. Learn about the unique environmental features of your area. What are the most urgent environmental problems?

7. **Vote.** You cannot improve your world by not voting. If you do not like the choices available, work to get individuals on the ballot who represent your interests.

8. **Think globally and act locally.** You need to be aware of global conditions, but you should also work to improve your own particular place.

9. **Do not be discouraged.** It is important to face facts honestly and to be realistic about the state of the world, but it does not help to wallow in despair. Do not dwell on negatives. Do what you can to improve the world, and take pleasure and pride in the small victories and elements of success.

10. **Try to leave things better than you found them.** Pick up a piece of litter on a beach or on your street, plant a tree, recycle your papers . . . the list goes on. Don't wait for the next person to begin—you can make a difference!

APPENDIX FIVE

HOW TO WRITE TO YOUR ELECTED OFFICIALS

1. Address your letter properly.

   a. Your representative:

   The Honorable_____
   House Office Building
   Washington, DC 20515
   Dear Representative_____,

   b. Your senators:

   The Honorable_____
   Senate Office Building
   Washington, DC 20510
   Dear Senator_____,

   c. The president:

   The President
   The White House
   1600 Pennsylvania Ave. NW
   Washington, DC 20500
   Dear Mr. President,

2. Tell who you are and why you are interested in this subject. Be sure to give your return address.

3. Always be courteous and reasonable. You can disagree with a particular position, but be respectful in doing so. You will gain little by being hostile, or abrasive.

4. Be brief. Keep letters to one page or less. Cover only one subject, and come to the point quickly. Trying to cover several issues confuses the subject and dilutes your impact.

5. Write in your own words. It is more important to be authentic than polished. Don't use form letters or stock phrases provided by others. Try to show how the issue affects the legislator's own district and constituents.

6. If you are writing about a specific bill, identify it by number (for instance, H.R. 321 or S. 123). You can get a free copy of any bill or committee report by writing to the House Document Room, U.S. House of Representatives, Washington, DC 20515 or the Senate Document Room, U.S. Senate, Washington, DC 20510. Copies of bills are also available online from *The Congressional Record.*

7. Ask your legislator to vote a specific way, support a specific amendment, or take a specific action. Otherwise you will get a form response that says: "Thank you for your concern.

Of course I support clean air, pure water . . . "

8. If you have expert knowledge or specific relevant experience, share it. But do not try to intimidate, threaten, or dazzle your representative. Legislators see through artifice and posturing; they are professionals in this field!

9. If possible, include some reference to the legislator's past action on this or related issues. Show that you are aware of his or her past record and are following the issue closely.

10. Follow up with a short note of thanks after a vote on an issue that you support. Show your appreciation by making campaign contributions or working for candidates who support issues important to you.

11. Try to meet your senators and representatives when they come home to campaign, or visit their office in Washington if you are able. If they know who you are personally, you will have more influence when you call or write.

GLOSSARY

A

abiotic factors   Nonliving factors that influence the life and activities of an organism.

abyssal ecosystem   The collection of organisms and the conditions that exist in the deep portions of the ocean.

acid   Any substance that, when dissolved in water, releases hydrogen ions.

acid deposition   The accumulation of potential acid-forming particles on a surface.

acid mine drainage   A kind of pollution, associated with coal mines, in which bacteria convert the sulfur in coal into compounds that form sulfuric acid.

acid rain (acid precipitation)   The deposition of wet acidic solutions or dry acidic particles from air.

activated sludge sewage treatment   Method of treating sewage in which some of the sludge is returned to aeration tanks, where it is mixed with incoming wastewater to encourage degradation of the wastes in the sewage.

activation energy   The initial energy input required to start a reaction.

active solar system   A system that traps sunlight energy as heat energy and uses mechanical means to move it to another location.

acute toxicity   A serious effect, such as a burn, illness, or death, that occurs shortly after exposure to a hazardous substance.

age distribution   The comparative percentages of different age groups within a population.

agricultural products   Any output from farming; milk, grain, meat, etc.

agricultural runoff   Surface water that carries soil particles, nutrients, such as phosphate, nitrates, and other agricultural chemicals as it runs off agricultural land to lakes and streams.

air stripping   The process of pumping air through water to remove volatile materials dissolved in the water.

alpha radiation   A type of radiation consisting of a particle with two neutrons and two protons.

alpine tundra   The biome that exists above the tree line in mountainous regions.

anthropocentric   Human-centered, it is the first of three primary theories of moral responsibility to the environment.

aquiclude   An impermeable layer in an aquifer.

aquifer   A layer of earth material that can transmit water sufficient for water supply purposes.

aquitard   A partially permeable layer in an aquifer.

artesian aquifer   The result of a pressurized aquifer intersecting the surface or being penetrated by a pipe or conduit, from which water gushes without being pumped.

atom   The basic subunit of elements, composed of protons, neutrons, and electrons.

auxin   A plant hormone that stimulates growth.

B

base   Any substance that, when dissolved in water, removes hydrogen ions from solution; forms a salt when combined with an acid.

benthic   Describes organisms that live on the bottom of marine and freshwater ecosystems.

benthic ecosystems   A type of marine or freshwater ecosystem consisting of organisms that live on the bottom.

beta radiation   A type of radiation consisting of electrons released from the nuclei of many fissionable atoms.

bioaccumulation   The buildup of a material in the body of an organism.

biocentric   Life-centered, the second theory of moral responsibility states that all forms of life have an inherent right to exist.

biochemical oxygen demand (BOD)   The amount of oxygen required by microbes to degrade organic molecules in aquatic ecosystems.

biocide   A kind of chemical that kills many different types of living things.

biodegradable   Able to be broken down by natural biological processes.

biogenetic law   The generally accepted concept that living things can only be produced by other living things. Life cannot be created from non-life.

biological amplification   The increases in the amount of a material in the bodies of organisms at successively higher trophic levels.

biomass   Any accumulation of organic material produced by living things.

biome   A kind of plant and animal community that covers large geographic areas. Climate is a major determiner of the biome found in a particular area.

biotic factors   Living portions of the environment.

biotic potential   The inherent reproductive capacity.

birthrate   The number of individuals born per thousand individuals in the population per year.

black lung disease   A respiratory condition resulting from the accumulation of large amounts of fine coal dust particles in miners' lungs.

boiling-water reactor (BWR)   A type of light water reactor in which steam is formed directly in the reactor, which is used to generate electricity.

boreal forest   A broad band of mixed coniferous and deciduous trees that stretches across northern North America (and also Europe and Asia); its northernmost edge is integrated with the arctic tundra.

C

carbamate   A class of soft pesticides that work by interfering with normal nerve impulses.

carbon absorption   The use of carbon particles to treat chemicals by having the chemicals attach to the carbon particles.

carbon cycle   The cyclic flow of carbon from the atmosphere to living organisms and back to the atmospheric reservoir.

carbon dioxide (CO_2)   A normal component of the earth's atmosphere that in elevated concentrations may interfere with the earth's heat budget.

carbon monoxide (CO)   A primary air pollutant produced when organic materials, such as gasoline, coal, wood, and trash, are incompletely burned.

carcinogen   A substance that causes cancer.

carcinogenic   The ability of a substance to cause cancer.

carnivores   Animals that eat other animals.

carrying capacity   The optimum number of individuals of a species that can be supported in an area over an extended period of time.

catalyst   A substance that alters the rate of a reaction but is not itself changed.

chemical bond The physical attraction between atoms that results from the interaction of their electrons.

chemical weathering Processes that involve the chemical alteration of rock in such a manner that it is more likely to fragment or to be dissolved.

chlorinated hydrocarbon A class of pesticide consisting of carbon, hydrogen, and chlorine, which are very stable.

chlorofluorocarbons (CFC) Used as a refrigerant, a propellant in aerosol containers, and an expander in foam products, they are linked to the depletion of the ozone layer.

chronic toxicity A serious effect, such as an illness or death, that occurs after prolonged exposure to small doses of a toxic substance.

clear-cutting A forest harvesting method in which all the trees in a large area are cut and removed.

climax community Last stage of succession; a relatively stable, long-lasting, complex, and interrelated community of plants, animals, fungi, and bacteria.

combustion The process of releasing chemical bond energy from fuel.

commensalism The relationship between organisms in which one organism benefits while the other is not affected.

community Interacting groups of different species.

competition An interaction between two organisms in which both require the same limited resource, which results in harm to both.

composting A waste disposal system whereby organic matter is allowed to decay to a usable product.

compound A kind of matter composed of two or more different kinds of atoms bonded together.

Comprehensive Environmental Response, Compensation, and Liability Act (CERCLA) The 1980 U.S. law that addressed the issue of cleanup of hazardous-waste sites.

confined aquifer An aquifer that is bounded on the top and bottom by impermeable confining layers.

conservation To use in the best possible way so that the greatest long-term benefit is realized by society.

conservation ethic An environmental ethic that stresses a balance between total development and absolute preservation.

consumers Organisms that rely on other organisms for food.

contour farming A method of tilling and planting at right angles to the slope, which reduces soil erosion by runoff.

controlled experiment An experiment in which two groups are compared. One, the control, is used as a basis of comparison and the other, the experimental, has one factor different from the control.

coral reef ecosystem A tropical, shallow-water, marine ecosystem dominated by coral organisms that produce external skeletons.

corporation A business structure that has a particular legal status.

corrosiveness Ability of a chemical to degrade standard materials.

cost-benefit analysis A method used to determine the feasibility of pursuing a particular project by balancing estimated costs against expected benefits.

cover A term used to refer to any set of physical features that conceals or protects animals from the elements or their enemies.

crust The outer, solid surface of the Earth.

D

death phase The portion of the population growth curve that shows the population declining.

death rate The number of deaths per thousand individuals in the population per year.

decommissioning Decontaminating and disassembling a nuclear power plant and safely disposing of the radioactive materials.

decomposers Small organisms, like bacteria and fungi, that cause the decay of dead organic matter and recycle nutrients.

demand Amount of a product that consumers are willing and able to buy at various prices.

demographic transition The hypothesis that economies proceed through a series of stages, resulting in stable populations and high economic development.

demography The study of human populations, their characteristics, and changes.

denitrifying bacteria Bacteria that convert nitrogen compounds into nitrogen gas.

density-dependent limiting factors Those limiting factors that become more severe as the size of the population increases.

density-independent limiting factors Those limiting factors that are not affected by population size.

desert A biome that receives less than 25 centimeters of precipitation per year.

desertification The conversion of arid and semiarid lands into deserts by inappropriate farming practices or overgrazing.

detritus Organic material that results from fecal waste material or the decomposition of plants and animals.

development ethic Philosophy that states that the human race should be the master of nature and that the earth and its resources exist for human benefit and pleasure.

dispersal Migration of organisms from a concentrated population into areas with lower population densities.

domestic water Water used for domestic activities, such as drinking, air conditioning, bathing, washing clothes, washing dishes, flushing toilets, and watering lawns and gardens.

E

ecology A branch of science that deals with the interrelationship between organisms and their environment.

ecocentrism The third approach to environmental responsibility maintains that the environment deserves direct moral consideration and not one merely derived from human interests.

economic costs Those monetary costs that are necessary to exploit a natural resource.

economic growth The perceived increase in monetary growth within a society.

ecosystem A group of interacting species combined with the physical environment.

ectoparasite A parasite that is adapted to live on the outside of its host.

electron The lightweight, negatively charged particle that moves around the nucleus of an atom.

element A form of matter consisting of a specific kind of atom.

emergent plants Aquatic vegetation that is rooted on the bottom but has leaves that float on the surface or protrude above the water.

emigration Movement out of an area that was once one's place of residence.

endangered species Those species that are present in such small numbers that they are in immediate jeopardy of becoming extinct.

endoparasite A parasite that is adapted to live within a host.

energy The ability to do work.

energy cost The amount of energy required to exploit a resource.

environment Everything that affects an organism during its lifetime.

environmental costs Damage done to the environment as a resource is exploited.

environmental justice Fair application of laws designed to protect the health of human beings and ecosystems; that no groups suffer unequal environmental harm.

Environmental Protection Agency (EPA) U.S. government organization responsible for the establishment and enforcement of regulations concerning the environment.

environmental resistance The combination of all environmental influences that tend to keep populations stable.

environmental science An interdisciplinary area of study that includes both applied and theoretical aspects of human impact on the world.

enzyme Protein molecules that speed up the rate of specific chemical reactions.

erosion The processes that loosen and move particles from one place to another.

estuaries Marine ecosystems that consist of shallow, partially enclosed areas where freshwater enters the ocean.

ethics A discipline that seeks to define what is fundamentally right and wrong.

euphotic zone The upper layer in the ocean where the sun's rays penetrate.

eutrophication The enrichment of water (either natural or cultural) with nutrients.

eutrophic lake A usually shallow, warm-water lake that is nutrient rich.

evolution A change in the structure, behavior, or physiology of a population of organisms as a result of some organisms with favorable characteristics having greater reproductive success than those organisms with less favorable characteristics.

executive branch The office of the President of the United States.

experiment An artificial situation designed to test the validity of a hypothesis.

exponential growth phase The period during population growth when the population increases at an ever-increasing rate.

external costs Expenses, monetary or otherwise, borne by someone other than the individuals or groups who use a resource.

extinction The death of a species; the elimination of all the individuals of a particular kind.

F

first law of thermodynamics A statement about energy that says that under normal physical conditions, energy is neither created nor destroyed.

fissionable The property of the nucleus of some atoms that allows them to split into smaller particles.

fixation A form of waste immobilization in which materials, such as fly ash or cement, are mixed with hazardous waste.

floodplain Lowland area on either side of a river.

floodplain zoning ordinances Designations that restrict future building in floodplains.

food chain The series of organisms involved in the passage of energy from one trophic level to the next.

food web Intersecting and overlapping food chains.

fossil fuels The organic remains of plants, animals, and microorganisms that lived millions of years ago that are preserved as natural gas, oil, and coal.

free-living nitrogen-fixing bacteria Bacteria that live in the soil and can convert nitrogen gas (N_2) in the atmosphere into forms that plants can use.

freshwater ecosystem Aquatic ecosystems that have low amounts of dissolved salts.

friability The ability of a soil to crumble.

fungicide A pesticide designed to kill or control fungi.

G

gamma radiation A type of electromagnetic radiation that comes from disintegrating atomic nuclei.

gas-cooled reactor (GAR) A type of reactor that uses graphite as a moderator and carbon dioxide or helium as a coolant.

geothermal energy The heat energy from the earth's molten core.

grasslands Areas receiving between 25 and 75 centimeters of precipitation per year. Grasses are the dominant vegetation, and trees are rare.

greenhouse effect The property of carbon dioxide (CO_2) that allows light energy to pass through the atmosphere but prevents heat from leaving; similar to the action of glass in a greenhouse.

gross national product (GNP) An index that measures the total goods and services generated annually within a country.

groundwater Water that infiltrates the soil and is stored in the spaces between particles in the earth.

groundwater mining Removal of water from an aquifer faster than it is replaced.

H

habitat An identifiable region in which a particular kind of organism lives.

habitat management The process of changing the natural community to encourage the increase in populations of certain desirable species.

hard pesticide A pesticide that persists for long periods of time; a persistent pesticide.

hazardous All dangerous materials, including toxic ones, that present an immediate or longterm human health risk or environmental risk.

hazardous substances Substances that can cause harm to humans or the environment.

hazardous-waste dump A disposal site for hazardous waste in a dump, landfill, or surface impoundment without any concern for potential environmental or health risks.

hazardous wastes Substances that could endanger life if released into the environment.

heavy-water reactor (HWR) A type of reactor that uses the hydrogen isotope deuterium in the molecular structure of the coolant water.

herbicide A pesticide designed to kill or control plants.

herbivores Primary consumers; animals that eat plants.

horizon A horizontal layer in the soil. The top layer (A horizon) has organic matter. The lower layer (B horizon) receives nutrients by leaching. The C horizon is partially weathered parent material.

host The organism a parasite uses for its source of food.

humus Soil organic matter.

hydrocarbons (HC) Group of organic compounds consisting of carbon and hydrogen atoms that are evaporated from fuel supplies or are remnants of the fuel that did not burn completely, and that act as a primary air pollutant.

hydrologic cycle Constant movement of water from surface water to air and back to surface water.

hydroxyl ion A negatively charged particle consisting of a hydrogen and an oxygen atom, commonly released from materials that are bases.

hypothesis A logical statement that explains an event or answers a question that can be tested.

I

ignitability Characteristic of materials that results in their ability to combust.

immigration Movement into an area in which one has not previously resided.

incineration Method of disposing of solid waste by burning.

industrial ecology A concept that stresses cycling resources rather than extracting and eventually discarding them.

Industrial Revolution A period of history during which machinery replaced human labor.

industrial uses Uses of water for cooling and for dissipating and transporting waste materials.

insecticide A pesticide designed to kill or control insects.

in-stream uses Use of a stream's water flow for such purposes as hydroelectric power, recreation, and navigation.

integrated pest management A method of pest management in which many aspects of the pest's biology are exploited to control its numbers.

interspecific competition Competition between members of different species for a limited resource.

intraspecific competition Competition among members of the same species for a limited resource.

ion An atom or group of atoms that has an electric charge because it has either gained or lost electrons.

irrigation Adding water to an agricultural field to allow certain crops to grow where the lack of water would normally prevent their cultivation.

isotope Atoms of the same element that have different numbers of neutrons.

J

judicial branch That portion of the U.S. government that includes the court system.

K

kinetic energy Energy of moving objects.

kinetic molecular theory The widely accepted theory that all matter is made of small particles that are in constant movement.

K-strategists Large organisms that have relatively long lives, produce few offspring, provide care for their offspring, and typically have populations that stabilize at the carrying capacity.

L

lag phase The initial stage of population growth during which growth occurs very slowly.

land The surface of the earth not covered by water.

landfill A method of disposing of solid wastes that involves burying the wastes in specially constructed sites.

land-use planning The construction of an orderly list of priorities for the use of available land.

law A fundamental rule of nature that is central to the understanding of many related aspects of nature and is considered true by nearly everyone although it is still possible that it could be disproved. Example; law of gravity.

LD$_{50}$ A measure of toxicity; the dosage of a substance that will kill (lethal dose) 50 percent of a test population.

leaching The movement of minerals from the A horizon to the B horizon by the downward movement of soil water.

legislative branch That portion of the U.S. government that is responsible for the development of laws.

light-water reactor A reactor that uses ordinary water as a coolant.

limiting factor The one primary condition of the environment that determines the success of an organism.

limnetic zone Region that does not have rooted vegetation in a freshwater ecosystem.

liquefied natural gas Natural gas that has been converted to a liquid by cooling to $-162°C$.

liquid metal fast-breeder reactor (LMFBR) Nuclear fission reactor using liquid sodium as the moderator and heat transfer medium; produces radioactive plutonium-235, which can be used as a nuclear fuel.

littoral zone Region with rooted vegetation in a freshwater ecosystem.

loam A soil type with good drainage and good texture that is ideal for growing crops.

M

macronutrient A nutrient, such as nitrogen, phosphorus, and potassium, that is required by plants in relatively large amounts.

mangrove swamp ecosystems Marine shoreline ecosystems dominated by trees that can tolerate high salt concentrations.

mantle The layer of the Earth between the crust and the core.

marine ecosystems Aquatic ecosystems that have high salt content.

marsh Area of grasses and reeds that is either permanently flooded or flooded for a major part of the year.

mass burn A method of incineration of solid waste in which material is fed into a furnace on movable metal grates.

matter Substance with measurable mass and volume.

mechanical weathering Physical forces that reduce the size of rock particles without changing the chemical nature of the rock.

megalopolis A large, regional urban center.

micronutrient A nutrient needed in extremely small amounts for proper plant growth; examples are boron, zinc, and magnesium.

migratory birds Birds that fly considerable distances between their summer breeding areas and their wintering areas.

mixture A kind of matter consisting of two or more kinds of matter intermingled with no specific ratio of the kinds of matter.

moderator Material that absorbs the energy from neutrons released by fission.

molecule Two or more atoms combined to form a stable unit.

monoculture A system of agriculture in which large tracts of land are planted with the same crop.

morals Predominant feeling of a culture about ethical issues.

mortality The number of deaths per year.

multiple land use Land uses that do not have to be exclusionary, so that two or more uses of land may occur at the same time.

municipal landfill An area used for the containment of solid wastes.

municipal solid waste All the waste produced by the residents of a community.

mutualism The association between organisms in which both benefit.

N

natality The number of individuals added to the population through reproduction.

National Priority List A listing of hazardous-waste dump sites requiring urgent attention as identified by Superfund legislation.

natural resources Those structures and processes that can be used by humans for their own purposes, but cannot be created by them.

natural selection A process that determines which individuals within a species will reproduce more effectively.

nature centers Teaching institutions that provide a variety of methods for people to learn about and appreciate the natural world.

negligible risk A point at which there is no significant health or environmental risk.

neutralization Reacting acids with bases to produce relatively safe end products.

neutron Neutrally charged particle located in the nucleus of an atom.

niche The total role an organism plays in a habitat.

nitrogen cycle The series of stages in the flow of nitrogen in ecosystems.

nitrogen-fixing bacteria Bacteria that are able to convert the nitrogen gas (N_2) in the atmosphere into forms that plants can use.

nitrous oxide N_2O, one of the oxides of nitrogen.

nonpersistent pesticide A pesticide that degrades in a short period of time.

nonpersistent pollutants Those pollutants that do not remain in the environment for long periods.

nonpoint source Diffuse pollutants, such as agricultural runoff, road salt, and acid rain, that are not from a single, confined source.

nonrenewable energy sources Those energy sources that are not replaced by natural processes within a reasonable length of time.

nonrenewable resources Those resources that are not replaced by natural processes, or those whose rate of replacement is so slow as to be noneffective.

nontarget organism An organism whose elimination is not the purpose of pesticide application.

northern coniferous forest See boreal forest.

nuclear breeder reactor Nuclear fission reactor designed to produce radioactive fuel from nonradioactive uranium and at the same time release energy to use in the generation of electricity.

nuclear chain reaction A continuous process in which a splitting nucleus releases neutrons that strike and split the nuclei of other atoms, releasing nuclear energy.

nuclear fission The decomposition of an atom's nucleus with the release of particles and energy.

nuclear fusion The union of smaller nuclei to form a heavier nucleus accompanied by the release of energy.

nuclear reactor A device that permits a controlled nuclear fission chain reaction.

nucleus The central region of an atom that contains protons and neutrons.

O

observation Ability to detect events by the senses or machines that extend the senses.

oligotrophic lakes Deep, cold, nutrient-poor lakes that are low in productivity.

omnivores Organisms that eat both plants and animals.

organophosphate A class of soft pesticides that work by interfering with normal nerve impulses.

outdoor recreation Recreation that uses the natural out-of-doors for leisure-time activities.

overburden The layer of soil and rock that covers deposits of desirable minerals.

oxides of nitrogen (NO, N$_2$O, and NO$_2$) Primary air pollutants consisting of a variety of different compounds containing nitrogen and oxygen.

ozone (O$_3$) A molecule consisting of three atoms of oxygen, which absorb much of the sun's ultraviolet energy before it reaches the earth's surface.

P

parasite An organism adapted to survival by using another living organism (host) for nourishment.

parasitism A relationship between organisms in which one, known as the parasite, lives in or on the host and derives benefit from the relationship while the host is harmed.

parent material Material that is weathered to become the mineral part of the soil.

particulates Small pieces of solid materials, such as smoke particles from fires, bits of asbestos from brake linings and insulation, dust particles, or ash from industrial plants, that are dispersed into the atmosphere.

passive solar system A design that allows for the entrapment and transfer of heat from the sun to a building without the use of moving parts or machinery.

patchwork clear-cutting A forest harvest method in which patches of trees are clear-cut among patches of timber that are left untouched.

peat The first stage in the conversion of organic material into coal.

pelagic Those organisms that swim in open water.

pelagic ecosystem A portion of a marine or freshwater ecosystem that occurs in open water away from the shore.

periphyton Attached organisms in freshwater streams and rivers, including algae, animals, and fungi.

permafrost Permanently frozen ground.

permeability A measure of how rapidly water can flow through a material.

persistent pesticide A pesticide that remains unchanged for a long period of time; a hard pesticide.

persistent pollutant A pollutant that remains in the environment for many years in an unchanged condition.

pest An unwanted plant or animal that interferes with human activity.

pesticide A chemical used to eliminate pests; a general term used to describe a variety of different kinds of pest killers, such as insecticides, fungicides, rodenticides, and herbicides.

pH The negative logarithm of the hydrogen ion concentration; a measure of the number of hydrogen ions present.

pheromone A chemical produced by one animal that changes the behavior of another.

photochemical smog A yellowish-brown haze that is the result of the interaction of hydrocarbons, oxides of nitrogen, and sunlight.

photosynthesis The process by which plants manufacture food. Light energy is used to convert carbon dioxide and water to sugar and oxygen.

photovoltaic cell A means of directly converting light energy into electricity.

phytoplankton Free-floating, microscopic, chlorophyll-containing organisms.

pioneer community The early stages of succession that begin the soil-building process.

plate tectonics The concept that the outer surface of the Earth consists of large plates that are slowly moving over the surface of a plastic layer.

plutonium-239 (Pu-239) A radioactive isotope produced in a breeder reactor and used as a nuclear fuel.

point source Pollution that can be traced to a single source.

policy Planned course of action on a question or a topic.

pollution Waste material that people produce in such large quantities that it interferes with their health or well-being.

pollution costs The private or public expenditures undertaken to avoid pollution damage once pollution has occurred and the increased health costs and loss of the use of public resources because of pollution.

pollution prevention To prevent either entirely or partially the pollution that would otherwise result from some production or consumption activity.

pollution-prevention hierarchy Regulatory controls that emphasize reducing the amount of hazardous waste produced.

population density A measure of how close organisms are to one another, generally expressed as the number of organisms per unit area.

porosity A measure of the size and number of spaces in an aquifer.

postwar baby boom A large increase in the birthrate immediately following World War I.

potable waters Unpolluted freshwater supplies.

potential energy The energy of position.

prairies Grasslands.

precipitation Removal of materials by mixing with chemicals that cause the materials to settle out of the mixture.

predator An organism that kills and eats another organism.

preservation To keep from harm or damage; to maintain in its original condition.

preservation ethic Philosophy that considers nature to be so special that it should remain intact.

pressurized-water reactor (PWR) A type of light-water reactor in which the water in the reactor is kept at high pressure and steam is formed in a secondary loop.

prey An organism that is killed and eaten by a predator.

primary consumer An organism that eats plants (producers) directly.

primary air pollutants Types of unmodified materials that, when released into the environment in sufficient quantities, are considered hazardous.

primary sewage treatment Process that removes larger particles by settling or filtering raw sewage through large screens.

primary succession Succession that begins with bare mineral surfaces or water.

probability A mathematical statement about how likely it is that something will happen.

producer An organism that can manufacture food from inorganic compounds and light energy.

profitability The extent to which economic benefits exceed the economic costs of doing business.

proton The positively charged particle located in the nucleus of an atom.

public resources Those parts of the environment that are owned by everyone.

R

radiation Energy that travels through space in the form of waves or particles.

radioactive Describes unstable nuclei that release particles and energy as they disintegrate.

radioactive half-life The time it takes for half of the radioactive material to spontaneously decompose.

radon Radioactive gas emitted from certain kinds of rock; can accumulate in very tightly sealed buildings.

range of tolerance The ability organisms have to succeed under a variety of environmental conditions. The breadth of this tolerance is an important ecological characteristic of a population.

reactivity The property of materials that indicates the degree to which a material is likely to react vigorously to water or air, or to become unstable or explode.

recycling The process of reclaiming a resource and reusing it for another or the same structure or purpose.

reforestation The process of replanting areas after the original trees are removed.

rem A measure of the biological damage to tissue caused by certain amounts of radiation.

renewable energy sources Those energy sources that can be regenerated by natural processes.

renewable resources Those resources that can be formed or regenerated by natural processes.

repeatability An important criterion in the scientific method, it requires that independent investigators be able to repeat an experiment and get the same results.

replacement fertility The number of children per woman needed to replace just the parents.

reserves The known deposits from which materials can be extracted profitably with existing technology under present economic conditions.

Resource Conservation and Recovery Act (RCRA) The 1976 U.S. law that specifically addressed the issue of hazardous waste.

resource exploitation The use of natural resources by society.

resources Naturally occurring substances that can be utilized by people but may not be economic.

respiration The process that organisms use to release chemical bond energy from food.

ribbon sprawl Development along transportation routes that usually consists of commercial and industrial building.

risk assessment The use of facts and assumptions to estimate the probability of harm to human health or the environment that may result from exposures to specific pollutants, toxic agents, or management decisions.

risk management Decision-making process that uses input such as risk assessment, technological feasibility, economic impacts, public concerns, and legal requirements.

rodenticide A pesticide designed to kill rodents.

r-strategist Typically, a small organism that has a short life span, produces a large number of offspring, and does not reach a carrying capacity.

runoff The surface water that enters a river system.

S

salinization An increase in the amount of salt in soil due to the evaporation of irrigation water.

savanna Tropical biome having seasonal rainfall of 50 to 150 centimeters per year. The dominant plants are grasses, with some scattered fire- and drought-resistant trees.

science A method for gathering and organizing information that involves observation, hypothesis formation, and experimentation.

scientific method A way of gathering and evaluating information. It involves observation, hypothesis formation, hypothesis testing, critical evaluation of results, and the publishing of findings.

secondary consumers Organisms that eat animals that have eaten plants.

secondary air pollutants Pollutants produced by the interaction of primary air pollutants in the presence of an appropriate energy source.

secondary recovery Techniques used to obtain the maximum amount of oil or natural gas from a well.

secondary sewage treatment Process that involves holding the wastewater until the organic material has been degraded by bacteria and other microorganisms.

secondary succession Succession that begins with the destruction or disturbance of an existing ecosystem.

second law of thermodynamics A statement about energy conversion that says that, whenever energy is converted from one form to another, some of the useful energy is lost.

selective harvesting A forest harvesting method in which individual high-value trees are removed from the forest, leaving the majority of the forest undisturbed.

septic tank Underground holding tank into which sewage is pumped and where biological degradation of organic material takes place; used in places where sewers are not available.

seral stage A stage in the successional process.

sere A stage in succession.

sewage sludge A mixture of organic material, organisms, and water in which the organisms consume the organic matter.

sex ratio Comparison between the number of males and females in a population.

soft pesticide A nonpersistent pesticide that breaks down into harmless products in a few hours or days.

soil A mixture of mineral material, organic matter, air, water, and living organisms; capable of supporting plant growth.

soil profile The series of layers (horizons) seen as one digs down into the soil.

soil structure Refers to the way that soil particles clump together. Sand has little structure because the particles do not stick to one another.

soil texture Refers to the size of the particles that make up the soil. Sandy soil has large particles, and clay soil has small particles.

solidification The conversion of liquid wastes to a solid form to allow for more safe storage or transport.

solid waste Unusable or unwanted solid products that result from human activity.

source reduction Reducing the amount of solid waste generated by using less, or converting from heavy packaging materials to lightweight ones.

speciation The process of developing a new species.

species A group of organisms that can interbreed and produce offspring capable of reproduction.

stable equilibrium phase The phase in a population growth curve in which the death rate and birthrate become equal.

standard of living The necessities and luxuries essential to a level of existence that is customary within a society.

steam stripping The use of heated air to drive volatile compounds from liquids.

steppe A grassland.

storm-water runoff Storm-water that runs off of streets and buildings and is often added directly to the sewer system and sent to the municipal wastewater treatment facility.

strip farming The planting of crops in strips that alternate with other crops. The primary purpose is to reduce erosion.

submerged plants Aquatic vegetation that is rooted on the bottom and has leaves that stay submerged below the surface of the water.

subsidy A gift given to private enterprise by government when the enterprise is in temporary economic difficulty but is viewed as being important to the public.

succession Regular and predictable changes in the structure of a community, ultimately leading to a climax community.

successional stage A stage in succession.

sulfur dioxide (SO₂) A compound containing sulfur and oxygen produced when sulfur-containing fossil fuels are burned. When released into the atmosphere, it acts as a primary pollutant.

Superfund The common name given to the U.S. 1980 Comprehensive Environmental Response, Compensation, and Liability Act, which was designed to address hazardous-waste sites.

supply Amount of a good or service available to be purchased.

supply/demand curve The relationship between available supply of a commodity or service and its price.

surface impoundment Pond created to hold liquid materials. Some may hold only water while others may be used to contain polluted water or liquid contaminants.

surface mining (strip mining) A type of mining in which the overburden is removed to procure the underlying deposit.

sustainable development Using renewable resources in harmony with ecological systems to produce a rise in real income per person and an improved standard of living for everyone.

swamp Area of trees that is either permanently flooded or flooded for a major part of the year.

symbiosis A close, long-lasting physical relationship between members of two different species.

symbiotic nitrogen-fixing bacteria Bacteria that grow within a plant's root system and that can convert nitrogen gas (N₂) from the atmosphere to nitrogen compounds that the plant can use.

synergism The interaction of materials or energy that increases the potential for harm.

T

taiga Biome having short, cool summers and long winters with abundant snowfall. The trees are adapted to winter conditions.

target organism The organism a pesticide is designed to eliminate.

technological advances Increasing use of machines to replace human labor.

temperate deciduous forest Biome that has a winter-summer change of seasons and that receives 100 centimeters or more of relatively evenly distributed precipitation throughout the year.

terrace A level area constructed on steep slopes to allow agriculture without extensive erosion.

tertiary sewage treatment Process that involves a variety of different techniques designed to remove dissolved pollutants left after primary and secondary treatments.

theory A unifying principle that binds together large areas of scientific knowledge.

theory of evolution The widely accepted concept that populations of living things can change genetically over time and that this change can lead to a population that is very well adapted to its environment.

thermal inversion The condition in which warm air in a valley is sandwiched between two layers of cold air and acts like a lid on the valley.

thermal pollution Waste heat that industries release into the environment.

thermal treatment A form of hazardous-waste destruction involving heating waste.

threatened species Those species that could become extinct if a critical factor in their environment were changed.

threshold level The minimum amount of something required to cause measurable effects.

total fertility rate The number of children born per woman per lifetime.

toxic A narrow group of substances that are poisonous and cause death or serious injury to humans and other organisms by interfering with normal body physiology.

toxicity A property of materials which, in sufficient quantity, can cause danger to human health or the environment.

toxic waste Substances that are poisonous and cause death or serious injury to humans and animals when released into the environment.

tract development The construction of similar residential units over large areas.

transpiration Transportation of water to leaves and its evaporation from the surfaces of plants.

transuranic waste Nuclear wastes of the U.S. weapons program that consist primarily of isotopes of plutonium.

trickling filter system A secondary sewage treatment technique that allows polluted water to flow over surfaces harboring microorganisms.

trophic level A stage in the energy flow through ecosystems.

tropical rainforest A biome with warm, relatively constant temperatures where there is no frost. These areas receive more than 200 centimeters of rain per year in rains that fall nearly every day.

tundra A biome that lacks trees and has permanently frozen soil.

U

unconfined aquifer An aquifer that usually occurs near the land's surface and may be called a water table aquifer.

underground mining A type of mining in which the deposited material is removed without disturbing the overburden.

underground storage tank Tank located below ground level for the storage of materials, such as oil, gasoline, or other chemicals.

uranium-235 (U-235) A naturally occurring radioactive isotope of uranium used as fuel in nuclear reactors.

urban sprawl Unplanned urban growth.

W

waste destruction Destruction of a portion of hazardous waste with harmful residues still left behind.

waste immobilization Putting hazardous wastes into a solid form that is easier to handle and less likely to enter the surrounding environment.

waste minimization A process that involves changes that industries could make in the way they manufacture products that would reduce the waste produced.

waste separation Either separating one hazardous waste from another, or separating hazardous waste from nonhazardous material that it has contaminated.

water diversion The physical process of transferring water from one area to another.

water table The top of the layer of water in an aquifer.

waterways Low areas that water normally flows through.

weathering The physical and chemical breakdown of materials; involved in the breakdown of parent material in soil formation.

weed An unwanted plant.

wetlands Areas that include swamps, tidal marshes, coastal wetlands, and estuaries.

wilderness Designation of land use for the exclusive protection of the area's natural wildlife; thus, no human development is allowed.

windbreak The planting of trees or strips of grasses at right angles to the prevailing wind to reduce erosion of soil by wind.

Z

zero population growth The stabilized growth stage of human population during which births equal deaths and equilibrium is reached.

zoning Type of land-use regulation in which land is designated for specific potential uses, such as agricultural, commercial, residential, recreational, and industrial.

zooplankton Weakly swimming microscopic animals.

CREDITS

Photographs

Part Opener:

© McGraw-Hill Higher Education/Barry Barker, photographer.

Front Matter

Pages xxii-1: Map © 1998 Rand McNally, R.L. #98-S-173.

Chapter 1

Figure 1.3: © Kaku Kurita/Gamma Liaison; p. 8(bottom), p. 9: © Leonard Rue Enterprises/Animals, Animals/Earth Scenes; 1.5(left): © Steve McCutcheon/Visuals Unlimited , (middle): © Frank Hanna/Visuals Unlimited, (right): © John Shaw/Tom Stack & Associates; 1.6(left, middle): © Red Diamond Photography, (right): © Link/Visuals Unlimited; 1.7(top left): © Bayard H. Brattstrom/Visuals Unlimited, (top right): © Gary Ladd/Photo Researchers, Inc., (bottom left): © David Halpern/Photo Researchers, Inc., (bottom right): © J. Eastcott & Y. Momatiuk/The Image Works; 1.8(top left): © Bob Pool/Tom Stack & Associates, (top right): © Joe McDonald/Tom Stack & Associates, (bottom): © William J. Weber/Visuals Unlimited; 1.9(left): © Richard Choy/Peter Arnold, Inc., (middle): © Matt Bradley/Tom Stack & Associates, (right): © A. Louis Goldman/Photo Researchers, Inc.; 1.10 (left): © Malcolm S. Kirk/Peter Arnold, Inc., (top right): © Brian Parker/Tom Stack & Associates, (bottom right): © Martin G. Miller/Visuals Unlimited.

Chapter 2

Figure 2.1: © Earth Imaging/Tony Stone Images; 2.2(top left): © Terry Donnelly/Tom Stack & Associates, (bottom left): © Joanne Lotter/Tom Stack & Associates, (top right): © Bob Pool/Tom Stack & Associates, (bottom right): © Chase Swift/Tom Stack & Associates; p. 21(top left): Granger Collection, (top right, middle left): Corbis-Bettmann Archive, (middle right, bottom): AP/Wide World Photos; p. 24: courtesy General Motors; 2.3(top left): © Toni Michels, (top right): © Stephen Frisch/Stock Boston, (bottom): © McGraw-Hill Higher Education/James Shaffer, photographer; 2.4a: © Greg Vaughn/Tom Stack & Associates; 2.4b: © Natalie Fobes/Tony Stone Images; 2.5(top left): © Nigel J.H. Smith/Animals Animals/Earth Scenes, (top right): © E.R. Degginger/Color-Pic/Animals Animals/Earth Scenes, (middle left): © L.L.T. Rhodes/Animals Animals/Earth Scenes, (middle right): © Bruce Davidson/Animals Animals/Earth Scenes, (bottom left): © George Gainsburgh Bernard Photo Productions/Animals Animals/Earth Scenes, (bottom right): © E.R.

Degginger/Color-Pic/Animals Animals/Earth Scenes; p. 30(top): © Anna Zuckerman/Tom Stack & Associates.

Chapter 3

Page 38(top, bottom): © McGraw-Hill Higher Education/Bob Coyle, photographer; 3.8: © Carl Purcell/Photo Researchers, Inc.

Chapter 4

Figure 4.2(top): © H. Berthoule-Scott/JSC/Photo Researchers, Inc., (right): © Gary Meszaros/Visuals Unlimited, (bottom): © S. Maslowski/VU; 4.3: © Joanne Lotter/Tom Stack & Associates; 4.5: © Doug Sokell/VU; 4.6: © Michael Tweedie/Photo Researchers, Inc.; 4.7: © Stephen Krasemann/Photo Researchers, Inc.; 4.8: © Fritz Polking/Peter Arnold, Inc.; 4.9(top): © SPL/Photo Researchers, Inc., (bottom): © Manfred Kage/Peter Arnold, Inc.; 4.10: © Noble Proctor/Photo Researchers, Inc.; 4.11: © J. Burgess/SPL/Photo Researchers, Inc.; p. 60(left): © Sinclair Stammers/SPL/Photo Researchers, Inc., (middle): © Charles Sykes/Visuals Unlimited; (right): © K. Maslowski/VU; 4.16(top left): © Alexander Lowry/Photo Researchers, Inc., (top middle): © Maresa Pryor/Lightwave Photography, Inc./Animals Animals/Earth Scenes, (top right): © Paul Souders/Tony Stone Images, (middle left): © Fred Hossler/Visuals Unlimited, (middle right): © Knolan Benfield/Visuals Unlimited, (bottom left): © David Phillips/Visuals Unlimited, (bottom right): © R. Kessel & G. Shih/VU; p. 64: Marc A. Blovin, National Biological Survey/U.S. Dept.of Interior Fish & Wildlife Service.

Chapter 5

Figure 5.1: © William E. Ferguson; 5.4: © Larry Mellichamp/Visuals Unlimited; 5.9b: © Leonard Lee Rue, Jr./Photo Researchers, Inc.; 5.10b: © William E. Ferguson/Stephanie Ferguson, photographer; 5.11b.: © Jacques Jangoux/Peter Arnold, Inc.; 5.12b: © Gregory Dimijian/Photo Researchers, Inc.; p. 77(left, middle): © Harold Hungerford/University of Southern Illinois-Carbondale, (right): © Carl Bollwinkel/University of Northern Iowa; p. 78: © Michael J. Balick/Peter Arnold, Inc.; p. 79: © Jerry Franklin; 5.13b: © William E. Ferguson; 5.14b: © Walt Anderson/VU; p. 81: © Peter K. Ziminski/Visuals Unlimited; 5.15b: © John Shaw/Tom Stack & Associates; 5.19: © Tammy Peluso/Tom Stack & Associates; 5.20: © William E. Ferguson.

Chapter 6

Figure 6.4(left): © Betty Derig/Photo Researchers, Inc., (middle): © John Colwell/Grant Heilman Photography, (right): © Kees Van Den

Berg/Photo Researchers, Inc.; p. 98 (elephant): © Jeff Foott/Tom Stack & Associates; p. 101(left): © David M. Dennis/Tom Stack & Associates, (top right): © John Cunningham/Visuals Unlimited, (upper mid-right): © David M. Dennis/Tom Stack & Associates, (lower mid-right): © Maslowski/Visuals Unlimited, (bottom right): © William S. Omerod/Visuals Unlimited.

Chapter 7

Page 115(top): © Sue Cunningham/Tony Stone Images; p. 119: © Jeff Greenberg/Photo Researchers, Inc.

Chapter 8

Figure 8.1: © Irven Devore/Anthro-Photo; 8.2: © Nimatellah/Art Resource, NY; 8.3b: © Field Museum of Natural history, Chicago, Neg. #75400c; 8.5(left): © Joseph Nettis/Photo Researchers, Inc., (right): © Van Bucher/Photo Researchers, Inc.; 8.6: © Steve McCurry/Magnum Photos; 8.7: © Margot Granitsas/The Image Works; p. 132(left, right): General Motors Corporation.

Chapter 9

Figure 9.8: © Michaud/Photo Researchers, Inc.; 9.10a: © Matt Meadows/Peter Arnold. Inc.; 9.10b: © David J. Cross/Peter Arnold, Inc.; 9.11a: © Sam Pierson/Photo Researchers, Inc.; 9.13: © McGraw-Hill Higher Education/Bob Coyle, photographer; 9.14(background): © Natalie Fobes/Tony Stone Images; 9.15: © John Running/Stock, Boston; 9.16: © Michelango Durazzo/Magnum Photos; 9.17: © Paolo Koch/Photo Researchers, Inc.; 9.18: © Stephen Krasemann/Photo Researchers, Inc.; 9.20(bottom): © Tom McHugh/Photo Researchers, Inc.; 9.21: © Tom McHugh/Photo Researchers, Inc.; 9.22: © Hank Morgan/Photo Researchers, Inc.; 9.24: © Susan Meiselas/Magnum Photos; 9.25: © Hank Morgan/Photo Researchers, Inc.; 9.26 (left): © McGraw-Hill Higher Education/Bob Coyle, photographer, (middle): © Toni Michaels, (right): Courtesy Andersen Windows, Inc.; p. 162(right): © B. & C. Alexander/Photo Researchers, Inc.

Chapter 10

Figure 10.1(top): U.S. Air Force, (middle): © Tom McHugh/Photo Researchers, Inc., (bottom): © SIU/VU; 10.12: TASS/Sovfoto/Eastfoto; 10.13: © Alain Morvan/Gamma Liaison; 10.14: © Alex Tsiaras/Photo Researchers, Inc.; 10.16: © Y. Arthus-Bertrand/Peter Arnold, Inc.

Chapter 11

Figure 11.1(from top left down): © Ulrike Welsh/Photo Researchers, Inc., © Toni Michaels, © J.C. Allen & Son, © Bob Daemmrich/The Image Works, (from top right down) © Joe

Sohm/The Image Works, © Mark Antman/The Image Works, © C.J. Collins/Photo Researchers, Inc., © Toni Michaels; 11.2: © Grant Heilman Photography; 11.3: © Byron Augustin/Tom Stack & Associates; 11.4: © John Cancalosi/Peter Arnold, Inc.; 11.5: © Jack Swenson/Tom Stack & Associates; p. 194(top): © Kevin Schafer/Tom Stack & Associates; 11.6: © Julia Sims/Peter Arnold, Inc.; 11.7(bottom left): © Eagy Landau/Photo Researchers, Inc., (bottom right): © Les Christman/Visuals Unlimited; 11.8(bottom): © J. Eastcott & Y. Momatiuk/The Image Works; p. 198: © Francois Gohier/Photo Researchers, Inc.; 11.12(top left): © Tom Stack/Tom Stack & Associates, (bottom left): © Patrice/Tom Stack & Associates, (top right): © Jack Fields/Photo Researchers, Inc., (bottom right): Courtesy Dept. of Fisheries, Olympia, WA; 11.16: © Dennis Paulson/VU; 11.18(left): © Gary Milburn/Tom Stack & Associates, (top right): © Richard H. Smith/Photo Researchers, Inc., (bottom right): © Tim Davis/Photo Researchers, Inc., (bottom): © Ken Highfill/Photo Researchers, Inc.; p. 207(left): © Brian Milne/Animals Animals/Earth Scenes, (right): © Robert Winslow/Tom Stack & Associates; p. 210: © W. Perry Conway/Tom Stack & Associates; 11.21(top left): © Tom McHugh/Photo Researchers, Inc., (top right): © Steven C. Kaufman/Peter Arnold, Inc., (middle left): © Edward Ross, (middle right): © Miguel Castro/Photo Researchers, Inc., (bottom left): © Stan Wayman/Photo Researchers, Inc., (bottom right): © Jeff Lepore/Photo Researchers, Inc.

Chapter 12

Figure 12.2(top right): © Red Diamond Photos/Bob Coyle, photographer, (bottom right): © Breck Kent/Earth Scenes; 12.3: © Bob Daemmrich/The Image Works; 12.5: © W.T. Sullivan III/SPL/Photo Researchers, Inc.; 12.4a: © Mike Okoniewski/The Image Works; 12.4b: © Mark Phillips/Photo Researchers, Inc.; 12.4c: © Jim Pickerell/Gamma Liaison; 12.6(background): © James Shaffer; 12.7: © McGraw-Hill Higher Education/Bob Coyle, photographer; 12.9(left): © Thomas Hollyman/Photo Researchers, Inc., (right): © Bruce Berg/Visuals Unlimited; 12.10(top): © Toni Michaels, (bottom): © Vanessa Vick/Photo Researchers, Inc.; 12.11: USDA; 12.12(top): © Sharon Gerig/Tom Stack & Associates, (bottom) © Robert Winslow/Tom Stack & Associates; p. 228: © Eric Hartman/Magnum Photos.

Chapter 13

Figure 13.5: © Spencer Swanger/Tom Stack & Associates; 13.11: © John D. Cunningham/Visuals Unlimited; 13.12: © Larry Miller/Photo Researchers, Inc.; 13.13 (bottom): © P. Newman/Visuals Unlimited; 13.14: © Grant Heilman Photography; 13.15: © Steve McCurry/Magnum Photos; 13.16a: © Mark Boulton/Photo Researchers, Inc.; 13.16b, 13.18: © Grant Heilman Photography; 13.19: © Joe Munroe/Photo Researchers, Inc.; 13.20a: © Bruno Zehnder/Peter

Arnold, Inc.; 13.20b: USDA Soil Conservation Service; 13.21a: © Grant Heilman Photography; 13.21b: © Larry Lefever/Grant Heilman Photography; 13.22a: © Grant Heilman Photography; 13.22b: © Link/VU; 13.23: © John Griffin/The Image Works; 13.24: © Francis de Richemond/The Image Works; p. 248: © Earl Roberge/Photo Researchers, Inc.

Chapter 14

Figure 14.1: © Jacques Jangoux/Peter Arnold, Inc.; 14.2: © Walter H. Hodge/Peter Arnold, Inc.; 14.3: © Earl Roberge/Photo Researchers, Inc.; 14.7: © Larry Lefever/Grant Heilman Photography; 14.8: © Toni Michaels; 14.9: courtesy J.E. Nugent, NW Michigan Horticulture Research Foundation; p. 265(top): © S. Wittwer/Visuals Unlimited, (middle): © Erica Lansner/Tony Stone Images, (bottom): © Dave Bartruff/Corbis Images; 14.13: © Jeremy Burgess/SPL/Photo Researchers, Inc.; 14.15 a-d: © Dr. Bernd Blossey/Dept. of Natural Resources, Cornell University.

Chapter 15

Figure 15.8: © Pat & Tom Leeson/Photo Researchers, Inc.; 15.10a: © Grant Leduc/Stock Boston; 15.10b: © Thomas Kitchin/Tom Stack & Associates; 15.10c: © John Colwell/Grant Heilman Photography; 15.11: © Theodore Clutter/Photo Researchers, Inc.; 15.12: © Francois Gohier/Photo Researchers, Inc.; p. 286(bottom): © William Campbell/Peter Arnold, Inc.; p. 288(top): © Steve Elmore/Tom Stack & Associates; p. 293(top right): © Tom & Pat Leeson/Photo Researchers, Inc., (bottom right): © Wendell Mentzen/Index Stock Photography; 15.16a: © Ray Pfortner/Peter Arnold, Inc.;15.16b: © Bob Daemmrich/Stock Boston; 15.16c: © Ray Pfortner/Peter Arnold, Inc.; p. 296(left): TASS/Sovfoto/Eastfoto; 15.17: © Mark Gibson; 15.18: © M. Timothy O'Keefe/Tom Stack & Associates; 15.20: © Nada Pecnik/VU; p. 300: Brad Smith; p. 303(top left): © Toni Michaels, (middle left): © Gene Marshall/Tom Stack & Associates, (bottom): © William E. Ferguson, (top right): © Ken W. Davis/Tom Stack & Associates.

Chapter 16

Figure 16.4: (counter-clockwise from top) © William Ferguson, © Mark Antman/The Image Works, © Thomas Kitchin/Tom Stack & Associates, © McGraw-Hill Higher Education/Bob Coyle, photographer, © McGraw-Hill Higher Education/Bob Coyle, photographer, © Scott Blackman/Tom Stack & Associates, © Jon Feingersh/Tom Stack & Associates; 16.7(top left): © Cedric Max Dunham/Photo Researchers, (top middle): © Chris Caswell/Photo Researchers, Inc., (top right): © Jan Halaska/The Image Works, (bottom left): © Didier Givois/Vandystadt/Photo Researchers. Inc., (bottom middle): © Clyde H. Smith/Peter Arnold, Inc., (bottom right): © George E. Jones III/Photo Researchers, Inc.; 16.8(right): AP/Wide World Photos; 16.9b: © Runk/Schoenberger/Grant

Heilman Photography; 16.9c: © Victoria McCormick/Animals Animals/Earth Scenes; p. 328(left): © C. Allan Morgan/Peter Arnold, Inc., (right): © John R. MacGregor/Peter Arnold.

Chapter 17

Figure 17.4: © John Cunningham/Visuals Unlimited; 17.5: © Bard Wrisley/Gamma Liaison; 17.6(top, bottom): © Peggy/Yoram Kahana/Peter Arnold, Inc.; p. 340(top left): © Thomas Kitchin/Tom Stack & Associates, (top right): © Toni Michaels; 17.10: © Don & Pat Valenti/Tom Stack & Associates; 17.11: © John Shaw/Tom Stack & Associates; p. 342: © Billy E. Barnes/Stock Boston; 17.12: Canapress Photo Service; p. 349: © Robert W. Ginn/Unicorn Stock Photos; p. 354: © Bruce M. Wellman/Stock Boston.

Chapter 18

Figure 18.1: © R. Maiman/Sygma; 18.2a: © Ray Pfortner/Peter Arnold, Inc.; 18.7: © Rapho Agence/Photo Researchers, Inc.; p. 362(all): © McGraw-Hill Higher Education/Bob Coyle, photographer; 18.12b: © David M. Dennis/Tom Stack & Associates; 18.16: © Marie Mills/Unicorn Stock Photos; p. 368: © McGraw-Hill Higher Education/Bob Coyle, photographer; p. 370: David Putnam, courtesy Weyerhaeuser; p. 371: © Pat Toth-Smith; p. 372(all): © McGraw-Hill Higher Education/Bob Coyle, photographer.

Chapter 19

Page 377(from top): © McGraw-Hill Higher Education/Bob Coyle, photographer, © John Colwell/Grant Heilman Photography, © Larry Kolvoord/The Image Works, © McGraw-Hill Higher Education/Bob Coyle, photographer, © McGraw-Hill Higher Education/James Shaffer, photographer, © Benelux Press BV/Photo Researchers, Inc., © McGraw-Hill Higher Education/James Shaffer, photographer; p. 379: © Kerry T. Givens/Tom Stack & Associates; p. 381: © Lowell Georgia/Photo Researchers, Inc.; 19.3: © Gary Milburn/Tom Stack & Associates; p. 384: © William Campbell/Peter Arnold, Inc.; p. 387(right): © Ken Sherman/Bruce Coleman/PNI.

Chapter 20

Figure 20.1(left): © Michael Dwyer/Stock Boston, (right): © Ashvin Mehta/Dinodia Picture Agency/The Image Works; 20.2(left): © Red Diamond Photos/Bob Coyle, photographer, (right): © Laurence Lowry/Stock Boston; 20.3: © John Neubauer/Rainbow/PNI; 20.9: © Forrest Anderson/Gamma Liaison; 20.8(top): New York State Department of Environmental Conservation, (bottom): © Paul David Mozell/Stock Boston; p. 401(top): © C. Osbourne/Photo Researchers, Inc., (bottom): © Mary Kate Denny/PhotoEdit; p. 402: © Chip Vinai/Gamma Liaison; p. 403(right): © Peter Turnley/Black Star; 20.10(left): © Red Diamond Photos/Bob Coyle, photographer, (right): © A. Ramey/Stock Boston; p. 406: © Greenpeace/Morgan.

INDEX

Oil use–*Cont.*
supply predictions and, 138
20th century, 135
Okita, Saburo, 319
Oligotrophic lakes, 88
Omnivores, 57
Ore processing, 190
Organic agriculture
description of, 266
soil and water protection using, 267–268
Organic Foods Production Act of 1990, 266
Organic matter, 40
Organisms
effects of acid deposition on, 343
habitat of, 50
interactions in, 53–56, 60
niche of, 50–51
soil formation and, 233
Organization for Economic Cooperation and Development (OECD), 135, 404
Organization of the Petroleum Exporting Countries (OPEC)
oil supply and, 138, 143
overview of, 134
Organophosphates, 257–258
Osprey, 262
Outdoor recreation, 226
Overburden, 145
Overgrazing, effects of, 195
Oxides of nitrogen, 336
Oxygen, in streams, 49
Ozone
depletion of, 350
description of, 350
Mexico City and, 334
photochemical smog and, 336, 338

P

Pacific Gas and Electric Company (PG&E), 154
Packaging
reduction in thickness of, 365–366
resins use in, 362
Paint, 337
Painted Desert, 233, 234
Pakistan, 239
Parasites
description of, 54
humans as, 56
insecticide use and, 264, 265
types of, 54–55
Parasitism, 54–55
Parathion, 258
Parent material, 233–234, 238
Paris Green, 257
Particulates, 335, 338
Passenger pigeons, 208
Passive solar system, 156

Patchwork clear-cutting, 193
PCBs, 271
Pelagic, 83
Pelagic ecosystems, 83
Pelagic marine ecosystems, 83
Pelicans, 262
Peoples Republic of China. *See* China
Peppered moths, 52
Periphyton, 88
Permafrost, 82
Permeability, 277
Persistent pesticides, 257, 261
Persistent pollutants, 378
Persistent trillium, 211
Pest, 257
Pesticides
as air pollutants, 291, 337
description of, 256
fungicides and rodenticides as, 260–261
herbicides as, 258–260
insecticides as, 256–258
in integrated pest management, 270
legislation on, 397
natural plant products as, 257
nonpersistent, 257–258
persistent, 257, 261
regulation of, 258
Pesticide use
bioaccumulation and bioamplification and, 261–263
changes in use of, 264–266
in former Soviet Union, 266
human health issues and, 264
nontarget organisms and, 263–264
pollution from, 291, 337, 375
resistance and, 263
Pests, 256
Petrochemicals, 150
Pfiesteria piscicida, 202
pH
acid rain and, 339, 342–343
description of, 40
Pheromone, 268
Phosphates, in detergents, 284
Phosphorus cycle
description of, 62, 65
human impact on, 63
illustration of, 63
Photochemical smog
control of, 338
description of, 336
events producing, 336–337
Photosynthesis
carbon cycle and, 60
carbon dioxide pollution and, 348–349
description of, 41, 125
Photovoltaic cells, 157–158
Phytoplankton, 83
PICOS program (General Motors), 24
Pine martin, 193

Pioneer community, 69
Plant growth, 62
Plants
emergent, 85
as parasites, 55
submerged, 85
Plastics, recycling of, 362, 367, 369
Plate tectonics, 231–232
Plimsoll line, 327
Plutonium-239 (Pu-239), 168, 169
Point source, 283
Policy, 394. *See also* Environmental policy
Politics. *See also* Environmental policy
endangered species and, 212
"greening" of, 399–404
human population growth and, 113
risk management and, 312
Pollutants, 378. *See also* Air pollutants
Pollution. *See also* Air pollution; Thermal pollution; Water pollution
aesthetic, 349
as consequence of energy conversion, 44
costs of, 323–324
description of, 187
economic solutions to, 324
groundwater, 290–291
historical basis of, 187
marine oil, 148, 290
noise, 354
prevention of, 322–323
thermal, 176
Pollution-prevention hierarchy, 383
Ponds, 85, 87–88
Population, 113
Population characteristics
carrying capacity as, 95–96
density as, 93–94, 107
growth curve, 94–95
influences on growth rates and, 94
natality and mortality as, 92
reproductive strategies and, 96–99
sex ratio and age distribution as, 92–93
Population growth
in Canada, 119
in China, 112
factors influencing, 94
governmental policy and, 112
human, 99–100, 103, 106 (*See also* Human population growth)
of invading species, 101
in Mexico, 119
Porosity, 276–277
Positive crankcase ventilation valve (PCV), 338
Postwar baby boom, 114

Potable water, 275
Potential energy, 42, 43
Poverty, 13–14, 118
Prairie dogs, 50
Prairies, 75. *See also* Grasslands
Precipitation
influence on vegetation of, 71, 73
terrestrial ecosystem distribution and, 82
Predation, 53–54
Predators
description of, 53–54
extinction of, 207
humans as, 56
insecticide use and, 264, 265
management of, 204
Preservation ethic, 20
Pressurized-water reactors (PWR), 167–168
Prey, 53–54
Primary air pollutants
carbon monoxide as, 334–335
description of, 333
hydrocarbons as, 335
oxides of nitrogen as, 336
particulates as, 335
sulfur dioxide as, 335–336
Primary consumers, 57
Primary sewage treatment, 292, 295
Primary succession, 69–70
Probability, 309
Producers, 56
Products
concentrated, 366
eco-labels on, 407
"green" advertising of, 321
life cycle of, 316
Profitability, 22
Propane, 133
Property tax, 221
Propoxur, 258
Protons, 39, 165
Protozoa, 235
Public land, 226–227
Purple loosestrife (*Lythrum salicaria*), 269

Q

Quail, 202, 204

R

Radiation, 165, 174–175
Radioactive, 165
Radioactive half-life, 166
Radioactive waste disposal. *See also* Hazardous waste management
high-level radioactive, 177–179
low-level, 180
Radon, 337, 351–352
Rainforests. *See also* Tropical rainforests
carbon dioxide and, 349
destruction of, 78, 208